Energy, Environment, and Sustainability

Series Editor

Avinash Kumar Agarwal, Department of Mechanical Engineering, Indian Institute of Technology Kanpur, Kanpur, Uttar Pradesh, India

This books series publishes cutting edge monographs and professional books focused on all aspects of energy and environmental sustainability, especially as it relates to energy concerns. The Series is published in partnership with the International Society for Energy, Environment, and Sustainability. The books in these series are edited or authored by top researchers and professional across the globe. The series aims at publishing state-of-the-art research and development in areas including, but not limited to:

- Renewable energy
- Alternative fuels
- Engines and locomotives
- Combustion and Propulsion
- Fossil Fuels
- Carbon capture
- Control and automation for energy
- Environmental Pollution
- Waste management
- Transportation sustainability

More information about this series at http://www.springer.com/series/15901

Tarun Gupta · Swatantra Pratap Singh ·
Prashant Rajput · Avinash Kumar Agarwal
Editors

Measurement, Analysis and Remediation of Environmental Pollutants

Springer

Editors
Tarun Gupta
Department of Civil Engineering
Indian Institute of Technology Kanpur
Kanpur, Uttar Pradesh, India

Prashant Rajput
Department of Civil Engineering
Indian Institute of Technology Kanpur
Kanpur, Uttar Pradesh, India

Swatantra Pratap Singh
Centre for Environmental Science
and Engineering
Indian Institute of Technology Bombay
Mumbai, Maharashtra, India

Avinash Kumar Agarwal
Department of Mechanical Engineering
Indian Institute of Technology Kanpur
Kanpur, Uttar Pradesh, India

ISSN 2522-8366 ISSN 2522-8374 (electronic)
Energy, Environment, and Sustainability
ISBN 978-981-15-0539-3 ISBN 978-981-15-0540-9 (eBook)
https://doi.org/10.1007/978-981-15-0540-9

This Springer imprint is published by the registered company Springer Nature Singapore Pte Ltd.
The registered company address is: 152 Beach Road, #21-01/04 Gateway East, Singapore 189721, Singapore

Preface

Energy demand has been rising remarkably due to increasing population and urbanization. Global economy and society are significantly dependent on the energy availability because it touches every facet of human life and activities. Transportation and power generation are two major examples. Without the transportation by millions of personalized and mass transport vehicles and availability of 24 × 7 power, human civilization would not have reached contemporary living standards.

The International Society for Energy, Environment and Sustainability (ISEES) was founded at Indian Institute of Technology Kanpur (IIT Kanpur), India, in January 2014, with an aim to spread knowledge/awareness and catalyze research activities in the fields of energy, environment, sustainability and combustion. The society's goal is to contribute to the development of clean, affordable and secure energy resources and a sustainable environment for the society, spread knowledge in the above-mentioned areas and create awareness about the environmental challenges, which the world is facing today. The unique way adopted by the society was to break the conventional silos of specializations (engineering, science, environment, agriculture, biotechnology, materials, fuels, etc.) to tackle the problems related to energy, environment and sustainability in a holistic manner. This is quite evident by the participation of experts from all fields to resolve these issues. ISEES is involved in various activities such as conducting workshops, seminars and conferences. in the domains of its interests. The society also recognizes the outstanding works done by the young scientists and engineers for their contributions in these fields by conferring them awards under various categories.

Third International Conference on "Sustainable Energy and Environmental Challenges" (III-SEEC) was organized under the auspices of ISEES from December 18 to 21, 2018, at Indian Institute of Technology Roorkee. This conference provided a platform for discussions between eminent scientists and engineers from various countries including India, USA, Norway, Finland, Sweden, Malaysia, Austria, Hong Kong, Bangladesh and Australia. In this conference, eminent speakers from all over the world presented their views related to different aspects of energy, combustion, emissions and alternative energy resource for sustainable development and cleaner environment. The conference presented five high-voltage plenary talks from globally renowned experts on topical themes, namely "The Evolution of Laser Ignition Over more than Four Decades" by Prof. Ernst Wintner, Technical University of Vienna, Austria; "Transition to Low Carbon Energy Mix for India," Dr. Bharat Bhargava, ONGC Energy Center; "Energy Future of India," By Dr. Vijay Kumar Saraswat, Hon. Member (S&T), NITI Aayog, Government of India; "Air Quality Monitoring and Assessment in India" by Dr. Gurfan Beig, Safar; and "Managing Large Technical Institutions and Assessment Criterion for Talent Recruitment and Retention" by Prof. Ajit Chaturvedi, Director, IIT Roorkee.

The conference included 24 technical sessions on topics related to energy and environmental sustainability including 5 plenary talks, 27 keynote talks and 15 invited talks from prominent scientists, in addition to 84 contributed talks and 50 poster presentations by students and researchers. The technical sessions in the conference included advances in IC engines, solar energy, environmental biotechnology, combustion, environmental sustainability, coal and biomass combustion/gasification, air and water pollution, biomass to fuels/chemicals, combustion/gas turbines/fluid flow/sprays, energy and environmental sustainability, atomization and sprays, sustainable transportation and environmental issues, new concepts in energy conservation, waste to wealth. One of the highlights of the conference was the rapid fire poster sessions in (i) engine/fuels/emissions, (ii) renewable and sustainable energy and (iii) biotechnology, where 50 students participated with great enthusiasm and won many prizes in a fiercely competitive environment. Two hundred-plus participants and speakers attended this four-day conference, which also hosted Dr. Vijay Kumar Saraswat, Hon. Member (S&T), NITI Aayog, Government of India, as the chief guest for the book release ceremony, where 14 ISEES books published by Springer, Singapore, under a special dedicated series "Energy, Environment and Sustainability" were released. This was second time in a row that such significant and high-quality outcome has been achieved by any society in India. The conference concluded with a panel discussion on "Challenges, Opportunities and Directions for National Energy Security," where the panelists were Prof. Ernst Wintner, Technical University of Vienna; Prof. Vinod Garg; Central University of Punjab, Bhatinda; Prof. Avinash Kumar Agarwal, IIT Kanpur; and Dr. Michael Sauer, University of Natural Resources and Life Sciences (BOKU), Austria. The panel discussion was moderated by Prof. Ashok Pandey, Chairman, ISEES. This conference laid out the road map for technology development, opportunities and challenges in energy, environment and sustainability domain. All these topics are very relevant to the country and the world in the present context. We acknowledge the support received from various funding agencies and organizations for the successful conduct of the Third ISEES Conference, III-SEEC, where these books germinated. We would, therefore, like to acknowledge NIT Srinagar, Uttarakhand (TEQIP) (special thanks to Prof. S. Soni, Director, NIT, UK); SERB, Government of India (special thanks to Dr. Rajeev Sharma, Secretary); UP Bioenergy Development Board, Lucknow (special thanks to Sh. P. S. Ojha); CSIR; and our publishing partner Springer (special thanks to Swati Meherishi).

We would like to express their sincere gratitude to a large number of authors from all over the world for submitting their high-quality work in a timely manner and revising it appropriately at a short notice. We would like to express our special thanks to various academicians who reviewed different chapters of this monograph and provided their valuable suggestions to improve the manuscripts.

Individual access to the clean water, food and air becomes one of the most critical challenges around the world owing to the presence of various toxic pollutants in various media. Although many pollution abatement efforts have been taken up and burgeoning expenditures made, no comprehensive solution has been implemented for the masses. A comprehensive discussion on the latest developments in the detection and analysis of contaminants that have enabled the scientists to understand the evolution of these pollutants in real time and that led to more accurate source apportionment of these pollutants has also been attempted in this work.

Kanpur, India — Tarun Gupta
Mumbai, India — Swatantra Pratap Singh
Kanpur, India — Prashant Rajput
Kanpur, India — Avinash Kumar Agarwal

Contents

Editors and Contributors

About the Editors

Dr. Tarun Gupta is currently Associate Dean of Research and Development and N C Nigam Chair Professor in the Department of Civil Engineering, Indian Institute of Technology (IIT) Kanpur, India. He holds a Doctor of Science (2004) in environmental health from Harvard University, USA, and Master of Technology (2000) in environmental science and engineering from IIT Bombay. He has published more than 125 articles in ISI indexed journals, 4 books, 8 chapters, and filed 8 Indian patents. A submicron aerosol sampler designed, developed and evaluated at the IIT Kanpur has since been commercialized by Envirotech, Delhi, and another technology transferred to BARC, Mumbai. He has developed several low-flow-rate and high-flow-rate impaction-based samplers and a non-selective membrane-based diffusion denuder. He has won INAE Entrepreneur and Innovator (2018), Membership of INYAS (2016), PK Kelkar Research Fellowship (2015), NASI Scopus Young Scientist (2015), INSA Young Scientist (2011), INAE Young Engineer (2009) and IEI Young Engineer (2008) Awards.

Dr. Swatantra Pratap Singh is Assistant Professor at IIT Bombay in the Centre for Environmental Science and Engineering. Before this, he was Postdoctoral Scholar in the Zuckerberg Institute of Water Research at Ben-Gurion University, Israel. He is Environmental Engineer with training in pollution control and has received his Ph.D. and M.Tech from the Indian Institute of Technology Kanpur, India. His research focuses on the low-cost membrane-based treatment units for water and wastewater treatment, and the generation of catalytic membranes with high flux and better selectivity and the use nanomaterials to address environmental challenges. He has published 14 articles in reputed journals and 2 chapters and holds 3 patents.

Dr. Prashant Rajput is CSIR-Senior Research Associate in the Department of Civil Engineering at IIT Kanpur. He received his M.Sc. in chemistry from University of Allahabad and his PhD from Physical Research Laboratory, Ahmedabad, India. Post to PhD., he has also worked in the Department of Civil and Environmental Engineering at University of Surrey, Guildford, UK; Lovely Professional University, Jalandhar, Punjab, India; and Physical Research Laboratory, Ahmedabad, Gujarat, India. His research interests are in atmospheric and aerosol chemistry, air pollution, source apportionment, risk assessment and particle measurement. He has published 30 journal articles and 4 chapters.

Prof. Avinash Kumar Agarwal joined IIT Kanpur in 2001. He worked at the Engine Research Center, University of Wisconsin at Madison, USA, as Postdoctoral Fellow (1999–2001). His interests are IC engines, combustion, alternative and conventional fuels, lubricating oil tribology, optical diagnostics, laser ignition, HCCI, emissions and particulate control, and large bore engines. He has published 270+ peer-reviewed international journal and conference papers, 35 edited books and 63 chapters. He is Associate Editor of ASME Journal of Energy Resources Technology and has edited the Handbook of Combustion, Wiley-VCH, Germany. He is Fellow of SAE, ASME, NASI, Royal Society of Chemistry, ISEES and INAE. He has been the recipient of several prestigious awards such as Clarivate Analystics India Citation Award-2017 in engineering and technology; NASI- Reliance Industries Platinum Jubilee Award-2012; INAE Silver Jubilee Young Engineer Award-2012; Dr. C. V. Raman Young Teachers Award-2011; SAE Ralph R. Teetor Educational Award-2008; INSA Young Scientist Award-2007; UICT Young Scientist Award-2007; and INAE Young Engineer Award-2005. He received Prestigious Shanti Swarup Bhatnagar Award-2016 in engineering sciences.

Contributors

Avinash Kumar Agarwal Department of Civil Engineering, Indian Institute of Technology, Kanpur, UP, India

Tirthankar Banerjee Institute of Environment and Sustainable Development, Banaras Hindu University, Varanasi, India

Abhishek Chakraborty ESE, IIT Bombay, Mumbai, India

Saurabh Chaubey School of Environmental Sciences, Jawaharlal Nehru University, New Delhi, India

Sneha Gautam Department of Civil Engineering, Karunya Institute of Technology and Sciences, Coimbatore, India

Tarun Gupta Department of Civil Engineering, Indian Institute of Technology, Kanpur, UP, India

Subrata Hait Department of Civil and Environmental Engineering, Indian Institute of Technology, Patna, Bihar, India

Charmi Humbal Department of Environmental Science and Engineering, Marwadi University, Rajkot, Gujarat, India

Manish Jangid School of Environmental Sciences, Jawaharlal Nehru University, New Delhi, India

Hema Joshi Indian Institute of Technology, Kanpur, Uttar Pradesh, India

Neha Kamal Department of Environmental Science and Engineering, Indian Institute of Technology (Indian School of Mines), Dhanbad, India

Suneel Kumar Joshi National Institute of Hydrology, Roorkee, India

Uthradevi Kannan Department of Civil and Environmental Engineering, Indian Institute of Technology Tirupati (IITT), Tirupati, AP, India

Ansaf V. Karim Environmental Science and Engineering Department, Indian Institute of Technology Bombay, Mumbai, India

A. K. Keshari Department of Civil Engineering, Indian Institute of Technology Delhi, New Delhi, India

Tahmeena Khan Department of Chemistry, Integral University, Lucknow, UP, India

V. Khan Department of Civil Engineering, Indian Institute of Technology Kanpur, Kanpur, India

S. Krishna Prashanth Department of Civil and Environmental Engineering, Indian Institute of Technology Tirupati (IITT), Tirupati, AP, India

Sukanya Krishnan Environmental Science and Engineering Department, Indian Institute of Technology Bombay, Mumbai, India

Manish Kumar Institute of Environment and Sustainable Development, Banaras Hindu University, Varanasi, India

Vipin Kumar Department of Environmental Science and Engineering, Indian Institute of Technology (Indian School of Mines), Dhanbad, India

Alfred J. Lawrence Department of Chemistry, Isabella Thoburn College, Lucknow, UP, India

Animesh Maitra Institute of Radio Physics and Electronics, University of Calcutta, Kolkata, India

Shihabudheen M. Maliyekkal Department of Civil and Environmental Engineering, Indian Institute of Technology Tirupati (IITT), Tirupati, AP, India

Alaa Mhawish Institute of Environment and Sustainable Development, Banaras Hindu University, Varanasi, India

Amit Kumar Mishra School of Environmental Sciences, Jawaharlal Nehru University, New Delhi, India

Swati Mishra School of Water Resources, Indian Institute of Technology Kharagpur, Kharagpur, India

M. S. V. Naga Jyothi Department of Civil and Environmental Engineering, Indian Institute of Technology Tirupati, Tirupati, AP, India

Manish Naja Aryabhatta Research Institute of Observational Sciences, Nainital, Uttarakhand, India

Dipteek Parmar Department of Civil Engineering, Harcourt Butler Technical University, Kanpur, UP, India

Harish C. Phuleria IDP in Climate Studies, Centre for Environmental Science and Engineering, IIT Bombay, Mumbai, India

Pragati Rai Department of Civil Engineering, Indian Institute of Technology, Kanpur, UP, India

S. Rajesh Department of Civil Engineering, Indian Institute of Technology Kanpur, Kanpur, India

Mahendrapal Singh Rajput Department of Microbiology, Marwadi University, Rajkot, Gujarat, India

Prashant Rajput Department of Civil Engineering, Indian Institute of Technology, Kanpur, UP, India

B. J. Ramaiah Department of Civil and Environmental Engineering, Indian Institute of Technology Tirupati, Tirupati, AP, India

Karthik Rathinam BASF SE, RAA/OS, Ludwigshafen, Germany

S. Roy Department of Civil Engineering, Indian Institute of Technology Kanpur, Kanpur, India

Pallavi Saxena Department of Environmental Sciences, Hindu College, University of Delhi, Delhi, India

Amritanshu Shriwastav Environmental Science and Engineering Department, Indian Institute of Technology Bombay, Mumbai, India

Shweta Department of Chemistry, Hindu College, University of Delhi, Delhi, India

Lakhwinder Singh Department of Chemistry, Hindu College, University of Delhi, Delhi, India

Minu Singh Pesticide Toxicology Laboratory, Regulatory Toxicology Group, CSIR—Indian Institute of Toxicology Research (CSIR-IITR), Lucknow, Uttar Pradesh, India

Nandita Singh Institute of Environment and Sustainable Development, Banaras Hindu University, Varanasi, India

Sheelendra Pratap Singh Pesticide Toxicology Laboratory and Analytical Chemistry Laboratory, Regulatory Toxicology Group, CSIR—Indian Institute of Toxicology Research (CSIR-IITR), Lucknow, Uttar Pradesh, India

Shiv Singh Light Weight Metallic Materials, Council of Scientific and Industrial Research-Advanced Materials and Processes Research Institute, Bhopal, Madhya Pradesh, India;
Nanomaterial Toxicology Group CSIR-Indian Institute of Toxicology Research (CSIR-IITR), Lucknow, Uttar Pradesh, India

Swatantra Pratap Singh Department of Environmental Science and Engineering, Indian Institute of Technology Bombay, Mumbai, India

Alok Sinha Department of Environmental Science and Engineering, Indian Institute of Technology (Indian School of Mines), Dhanbad, India

Saurabh Sonwani School of Environmental Sciences, Jawaharlal Nehru University, New Delhi, India

Anju Srivastava Department of Chemistry, Hindu College, University of Delhi, Delhi, India

Anshuman Srivastava Pesticide Toxicology Laboratory, Regulatory Toxicology Group, CSIR—Indian Institute of Toxicology Research (CSIR-IITR), Lucknow, Uttar Pradesh, India

Shamitaksha Talukdar National Atmospheric Research Laboratory, Gadanki, India

Manoj Kumar Tiwari School of Water Resources, Indian Institute of Technology Kharagpur, Kharagpur, India

Amber Trivedi Department of Civil and Environmental Engineering, Indian Institute of Technology, Patna, Bihar, India

Shivangi Verma Department of Chemistry, Hindu College, University of Delhi, Delhi, India

K. S. Vinjamuri DST-Mahamana Centre of Excellence in Climate Change Research, Banaras Hindu University, Varanasi, India

Suman Yadav IDP in Climate Studies, IIT Bombay, Mumbai, India

Dina Zaman School of Water Resources, Indian Institute of Technology Kharagpur, Kharagpur, India

Chapter 1
Introduction of Measurement, Analysis and Remediation of Environmental Pollutants

Tarun Gupta, Swatantra Pratap Singh, Prashant Rajput and Avinash Kumar Agarwal

Abstract This book deals with contaminated **W**ater, **A**ir, **S**oil media and their **H**ealth **E**ffects on population and **R**emedial measures to improve this situation. This book presents a synthesis with recent findings on air pollution, water pollution, soil pollution and human health. Each chapter, as applicable, discusses on abundance of pollutants in different media and highlights the plausible and effective tool to be implemented in order to control for it. There are twenty one chapters in this book. Initially, we discuss about major challenges in atmospheric research and then in second half part we focus on water and soil pollutants and their control as well as synthesis on human health perspectives. We have attempted to incorporate ongoing research from conventional to advance measurement techniques in all the aforementioned areas of scientific and engineering research. Our efforts have led to compilation of recent data set and its practical implications from many research aspects.

Keywords Air · Water · Soil · Health · Exposure

1.1 Introduction

Chapter 2 summarizes indoor air quality and emphasizes on hazardous effects of particulate matter (PM) comprised of fine and ultrafine particles. Exposure to indoor air pollutants causes poor health in many developing countries. Nano particles found in indoor air have been linked with cardiovascular diseases and premature deaths in India. Transition metals, one of the main components of small particles, are considered to be hazardous because of their potential role to produce reactive oxygen species (ROS) in respiratory system. This study has been conducted at a typical urban location at Lucknow in central part of Indo-Gangetic Plain (IGP). Three microenvironments namely (1) well planned, (2) densely populated and (3) roadside have been assessed for a period of 2 long years (2012–2014). Source identification and

T. Gupta (✉) · S. P. Singh · P. Rajput · A. K. Agarwal
Department of Civil Engineering, Indian Institute of Technology, Kanpur, UP 208016, India
e-mail: tarun@iitk.ac.in

T. Gupta et al. (eds.), *Measurement, Analysis and Remediation of Environmental Pollutants*, Energy, Environment, and Sustainability,
https://doi.org/10.1007/978-981-15-0540-9_1

quantification has been carried out by principal component analysis (PCA). Enrichment factors for different species were also estimated to better assess the sources. The issue needs more focus to understand the mechanisms of damage and to propose measures to control the anthropogenic emissions and decrease the personal exposure.

Chapter 3 deals with the aerosol data set over Himalayan region in India. The ground-based measurements of aerosols along with satellite retrieval are shown to augment aerosol characteristics over Himalayan region. The role of boundary layer, ambient meteorology, and air-mass transport is discussed in detail. In addendum to this, the aerosol variability over the foothills of the Himalayas is also studied and the role of updraft/downdraft is elaborated. The role of absorbing aerosols in influencing the atmospheric radiation budget over the central Himalaya region is also discussed.

Chapter 4 deals with atmospheric measurements study during Diwali festival. Air pollutants like nitrogen oxides (NO_x), particulate matter ($PM_{2.5}$ and PM_{10}) and ozone (O_3) were studied on pre-Diwali, Diwali and post-Diwali periods for two consecutive years i.e. October 2016 and October 2017 at capital city of India, Delhi. The high background condition due to local emissions and air-mass transport of biomass burning emissions from upwind source-region poses a challenge in quantifying short-term low-level emission source. The ambient concentrations are compared on a 1-to-1 scale with National Ambient Air Quality Standards (NAAQS).

Chapter 5 presents measurements and control of organic aerosols. It gives a wide spectrum of information on conventional and advanced measurement techniques which are currently being used for organic aerosols measurement and characterization. This chapter highlights urgent actions to be taken in mitigating emissions from anthropogenic sources like vehicular exhaust, industrial emission, reduction in biomass burning activities etc. It appeals to act for reduction of emission by public awareness and stringent emission regulation policy coupled with technological innovation. It highlights role of control technologies such as diesel particulate filter (DPF) equipped with diesel oxidation catalyst (DOC) and selective catalytic reduction (SCR) system to be utilized to reduce vehicular emissions.

Chapter 6 discusses about chemical characteristics and source of coarse fraction atmospheric aerosols from an urban location at Kanpur in central Indo-Gangetic Plain. The mass concentrations of $PM_{10\text{-}2.5}$, black carbon (BC), water soluble inorganic carbon (WSIC), water soluble organic carbon (WSOC), water-soluble ions and several metals have been studied. The data set was analyzed using positive matrix factorization (PMF) to identify possible sources and estimate the contribution of these sources to the coarse PM mass concentration. The identified sources and their contributions for the $PM_{10\text{-}2.5}$ are found as paved road dust (53%), vehicular emission (7%), coal combustion and brick kilns (2.5%), construction activities & incineration (0.5%), crustal dust (32%) and biomass burning and oil combustion (5%).

Chapter 7 documents an analysis of the aerosol over Kolkata, a densely-populated metropolitan area in the eastern part of India, located near the land-ocean boundary of the Bay of Bengal (BoB). Besides near continuous active natural sources of aerosols, anthropogenic aerosols constantly contribute to the total aerosol loading over this region. This chapter shows how the spectral dependence pattern of aerosol optical parameters determines the dominant type of aerosols over the region. Furthermore, the synthesis exhibits the impact of the dominant type of aerosols on the boundary layer meteorology utilizing ground based observations. The role of atmospheric black carbon (BC) aerosols and its impact are also focused. This region experiences high BC concentrations predominantly from incomplete combustion of fossil fuel. This chapter discusses about inter play of aerosols impact with regional meteorology and thermodynamic processes.

Chapter 8 discusses aerosol monitoring through satellite retrieval and highlights large-scale implications of remote sensing from the study conducted over Indo-Gangetic Plain. A synthesis in detail has been shown on annual, seasonal and diurnal scenarios wherein varying types and amount of aerosols burden influence the atmospheric budget and related processes. The study highlights satellite aerosol remote sensing as a powerful tool to characterize the optical and microphysical properties of aerosols. The importance of active and passive remote sensing techniques has also been discussed.

Chapter 9 summarizes with recent findings on atmospheric brown clouds (ABCs). It discusses about role of ABCs on human health, air quality and regional climate. The Indo-Gangetic Plains (IGP) is also one of the ABCs hotspots in the world. The study highlights need on understanding of spatio-temporal characteristics of ABCs over the region to decipher its socio-economic impact on human population residing within the IGP. This study gives an overview on the frequency of ABCs occurrences and associated optical and microphysical properties using data from seven ground-based remote sensors situated across the IGP. A well-defined algorithm based on extinction and absorbing properties of particles has been used to characterize extreme pollution days (ABCs days) for study sites. Spectral dependency of aerosol optical depth (AOD) and absorption optical depth (AAOD) have been utilized to characterize aerosol types during ABCs days over IGP.

Chapter 10 gives an overview of fungal spores as one of the aeroallergens adversely affecting human health and environment. The study reports fungal spore concentration levels probably first time from an urban city situated in the western part of India. Five locations (i.e., residential area, poultry farm, slaughter house, industrial area and dump site) selected to cover probably all major environment of an urban city. The variability of fungal concentrations has been presented here systematically. In addition, meteorological parameters (i.e., wind speed, temperature, and relative humidity) were assessed to understand their role on fungal spore distribution.

Chapter 11 discusses about monitoring and control of biological pollutants in the water. The water pollution and associated diseases could result into loss of ~2.1 million human lives every year. Basically, the outbreak of water-related microbial infections such as diarrhoea, typhoid, and cholera are the primary cause of the loss of lives. Towards this, a comprehensive analysis of the source, occurrence, fate, and control of biological contaminants in drinking water is discussed. This chapter elucidates the growing significance to address the issue of microbial contamination in drinking water and its associated health implications from the past to the present, recent developments in the technologies for the detection, analysis and the remediation of pathogens in the water. This chapter also gives a brief overview of the research efforts required on detection of pathogens.

We have included twelfth chapter about selenium (Se) as an antioxidant and anti-carcinogenic agent and an essential micronutrient for animals and human. However, even a little high dose of the Se will be toxic in nature. This chapter describes the possible sources and pathway of Se release in the environment along with the possible analytical techniques for its measurement.

The printed circuit board (PCB) is one of the vital components of electronic waste (e-waste). PCB is full of metallic content and makes them an ideal candidate for recycling metals to conserve the environment. In Chap. 13, bioleaching process is discussed for the leaching of metals form PCB with pure and mix culture of *Aspergillus* species.

Pesticides are a supplementary input to control the pest and increase productivity. The contamination of these pesticides in the food matrix is the biggest problem. The accurate measurement of these pesticides in different food matrixes also important to evaluate the toxicity. Chapter 14 summarizes the modern analytical technique for the pesticide and metabolite sample preparation and analysis via chromatography and mass spectrometry techniques.

The pharmaceutical compounds and their metabolites are one of the classes of emerging environmental pollutants. Chapter 15 deals with the available option for the biotransformation and biodegradation of the anticonvulsant drugs in the environment.

Particulate matter (PM) can cause many adverse health effects; this PM generates reactive oxygen species (ROS), also known as oxidative potential (OP), which are one of the most relevant indicators of PM toxicity. Chapter 16 summarizes the current understanding of OP for PM. The effect of particle size, chemical composition, and its effect on human health.

Chapter 17 describes the background problem of waste load allocation modeling (WLA), followed by the existing challenges and possible solutions. The rivers in the developing countries facing heavy waste load because of uncontrolled urbanization and industrialization. The pollution abatement efforts have been made without WLA.

Chapter 18 talks about how organic fraction within the solid waste leads to generation of methane. It is the second largest anthropogenic methane source and needs a much accurate estimation and control in the near future. This chapter provides the details of various factors which affect the methane generate rates such as waste temperature, waste composition and density, pH within the landfill. Additionally, numerous techniques for mitigation have been discussed in the chapter.

In the last 3–4 decades, the production and consumption of pharmaceuticals and personal care products (PPCPs) have grown threateningly. The ninteenth chapter of this book discusses the available low cost adsorption based technologies for the remediation of PPCPs. PPCPs are a new class of emerging contaminants, and have a serious risk to the ecosystem.

Chapter 20 is on the measurement, analysis, and remediation of Bisphenol-A from environmental matrices. Bisphenol-A (BPA), an endocrine disrupter and an important emerging contaminant, widely used as a raw material for the epoxy and polycarbonate preparation. These epoxy and polycarbonates are used in our daily products such as canned food, water containers and beverages due to its excellent heat resistance and elasticity property.

Chromium (Cr) in the water has several adverse health effects to human and other species because of its toxicity. The adsorption process is one of the best technique to effective removal of Cr form the aqueous phase. Chapter 21 summarizes the potential of eco-friendly bioadsobents for the effective treatment of Cr along with the synthesis and removal mechanism.

Chapter 2
Quantification of Airborne Particulate and Associated Toxic Heavy Metals in Urban Indoor Environment and Allied Health Effects

Alfred J. Lawrence and Tahmeena Khan

Abstract The present chapter is an attempt to summarize the importance of indoor air quality, which can be considered as IAQ and to emphasize on the hazardous effects of particulate matter (PM) consisting of fine and ultrafine particles. Indoor air pollution is a leading cause of poor health outcome in India. Nano particles in indoor air have been linked to growing cardiovascular diseases and premature deaths in India. Metals are associated with particulate matter. Heavy metals are produced by non-exhaust discharges, fuel additives and by extraction processes. They are considered to be hazardous mainly because of their potential to produce reactive oxygen species in respiratory system. The purpose of this chapter is explained by a case study undertaken in Lucknow city where particulate matter ($PM_{2.5}$, PM_{10}) and associated heavy metals viz. Fe, Zn, Pb, Cr, Ni, Cu and Mn were analyzed in three microenvironments namely (1) well planned, (2) densely populated and (3) roadside, over a period of two years (2012–2014). Identification of the main sources of the heavy metals was done through principal component analysis. Calculation of enrichment factors was also done for heavy metals to know their source of origin. Human health is generally affected by the accumulation of pollutants in the body. Even though the hazardous effects of heavy metals are known, still there is limited knowledge on association of a disease with inhalation exposure, particularly in indoor environment. The issue needs more focus to understand the causes, harm and to recommend actions to check the emissions and lower the ill effects.

Keywords Indoor air · Metals · Particulate matter · Urban emissions

A. J. Lawrence (✉)
Department of Chemistry, Isabella Thoburn College, Lucknow, UP 226007, India
e-mail: alfred_lawrence@yahoo.com

T. Khan
Department of Chemistry, Integral University, Lucknow, UP 226026, India

T. Gupta et al. (eds.), *Measurement, Analysis and Remediation of Environmental Pollutants*, Energy, Environment, and Sustainability,
https://doi.org/10.1007/978-981-15-0540-9_2

2.1 Introduction—Indoor Air Quality and Its Importance

India, as a developing country is witnessing poor outdoor as well as indoor air quality. Indoor air pollution (Census of India 2011) contributes to approximately 2.6% of the global burden of disease and 1.6 million pre-mature deaths annually (Smith and Mehta 2003; Balakrishnan et al. 2011). World Health Organization (WHO) has designated the problem of indoor air pollution (IAP) in top five perilous global environmental issues in progressing nations (World Health Organization 2002). The report on Global Burden of Disease (GBD) has reported 2 million early deaths yearly in India owing to IAP (Global Burden of Disease 2016). Though, indoor usage of biomass burning is accountable for less than one-fourth of the mass of particulate matter suspended in ambient air, suggesting importance of other sources which are responsible for particulate emission. Hence country specific solutions are needed to tackle the emission (Chafe et al. 2014). Indoor air quality (IAQ) is dependent on the level of ambient air pollution to an appreciable extent, owing to infiltration and penetration. Household sources also significantly contribute to outdoor environment, hence making them inter-dependent. Thus the two phenomena cannot be accurately studied in isolation or considered as separate entities (Balakrishnan et al. 2014). Air pollution exposure whether occurred outdoor or indoor has a common impact on health. The risk of exposure is equal for both rural and urban populations, though the sources contributing to the pollution may be different in both environments. However due to partial significance to urban centres, the overall understanding of the nature and distribution of pollutants is somewhat lacking (Garaga et al. 2018). Another important aspect of consideration is exposure to local sources like use of biomass, garbage burning, emerging small scale industries which contribute significantly to large spatial gradients that cannot be accounted for, while monitoring at the specific or central sites (Pant et al. 2016). For instance, metropolitan cities usually have high ambient levels of particulate contamination, whereas in rural settings, higher values of the same can be obtained in households, due to biomass exposure, hence exposure-response relations will be unlike, for different settings and should be tackled separately in urban and rural populations. Other than criteria pollutants many hazardous pollutants like toxic heavy metals, poly aromatic hydrocarbons (PAHs) can be associated with particulate matter, which are usually prevalent in urban industrial areas. They are released along with other particulate species and lead to severe health outcomes (Guo et al. 2017). A comprehensive air pollution assessment study would include all factors of significance including emission inventory, exposure pathway, exposure to population, toxicity factors and health outcomes (Gordona et al. 2018). The objectives of the present investigations were to carry a distribution of heavy metals including Fe, Pb, Mn, Cu, Ni, Cr and Zn in the indoor environment of urban Lucknow homes, in two size ranges of particulate viz. $PM_{2.5}$ and PM_{10} and to elucidate the relationship between the levels of metals in these sizes. The results have practical importance in identifying sources and processes that control levels of fine particulate matter.

2.1.1 Monitoring of Indoor Air Pollution

Slight modifications may be made in the tools and instruments used for ambient pollution monitoring, making them adaptable for indoor pollution monitoring. Personalized monitors are constantly used for the determination of individual exposure to air pollutants. The indoor environment has a limited air volume and comparatively lesser rate of air exchange. Hence while drawing an indoor sample, it is important to adjust the sampling flow at a slower rate so that the indoor air movement and air exchange rates are not affected. Instruments can measure pollutants directly, or they may be absorbed for further laboratory analysis (https://www.ncbi.nlm.nih.gov/books/NBK234059/; U.S. Environmental Protection Agency 1980). Air sampling techniques have been categorized as follows:

Continuous Monitoring
Continuous monitoring is needed to measure both temporal and spatial concentrations variation which is essential for indoor air pollution modeling. The continuous sampling also gives an insight to occupant's exposure and the levels of variability. The key advantage of continuous monitoring is the determination of short-term varying concentrations. The average concentration of pollutant measured over a period of time may depend on the factors such as infiltration, ventilation and source generation. Continuous monitors do require regular calibrations. They are also costly and need high maintenance. Safety hazards are also associated with them and they work at a set power level. They can also produce noise and heat, thereby creating hindrance in daily measurements (Moschandreas et al. 1978). Continuous monitoring is not suitable for large-scale surveys, owing to time and cost limitations.

Integrated Sampling
In this method an average data is obtained over a specific period of time. Integrated sampling is used when the mean concentration is either necessary or acceptable. The time period can vary from minutes to weeks or months. Analysis can be done at the sampling site, which may be indoor or outdoor. The results are presented as average concentration for the sampling period. A variety of pollutants are monitored by integrated sampling. Particulate matter can be sampled on a filter paper which can be subjected to gravimetric or chemical analysis. Size-specific samplers are also used for indoor sampling which can collect particulate matter of less than 2.5 μm diameter (Loo et al. 1979) or particles smaller than 3.5 μm in size. The particle samplers range from handy units which are manually operated to completely automated devices that are programmed to work automatically without manual handling. Gaseous pollutants can be measured by both passive (diffusion-controlled) and active (powered bulk airflow) samplers. Integration sampling is cost effective and requires less manpower. Pollutants in very low concentrations can also be measured by adopting a suitable sampling/measurement technique. On the contrary, there is loss of short-term temporal information. Frequent sampling must be done if temporal variability is to be studied (Schuette 1967).

2.2 Criteria Pollutants and Their Effects

Environmental Protection Agency (EPA) has established national ambient air quality standards (NAAQS) for six pollutants known as criteria pollutants viz. carbon monoxide, lead, ground-level ozone, particulate matter, nitrogen dioxide and sulfur dioxide. They can be emitted in air by various sources. The standard limits are set to check their emission and to protect public health. The main sources and effects associated with the six criteria pollutants are given in Table 2.1.

2.2.1 Particulate Contamination—A Death Trap

Particulate matter (PM) can be derived from an array of heterogeneous sources differing in chemical composition and physical state and is differentiated or classified by it's size (Ingrid et al. 2014) as given below:

- PM_{10}- less than 10 μm (10 microns) in diameter
- $PM_{2.5}$- less than 2.5 μm in diameter,
- $PM_{1.0}$- less than 1 μm in diameter.

Particles ranging between 2.5 and 10 μm in diameter are known as coarse particles. Particles less than 2.5 μm in diameter are fine particles including ultrafine (less than 0.1 μm) ($PM_{0.1}$).

Primary particulate matter: is released into the atmosphere by a number of anthropogenic and natural sources.

Secondary particulate matter: is formed by physical and chemical reactions via other pollutants.

Re-suspended particulate matter: after its deposition, particulate matter can be re-suspended into the air through the movement of wind or through vehicular exhaust (Tolis et al. 2014).

Particles with large size can be removed by mechanical methods, like cyclone and gravity collectors etc. whereas electro-filters are used to remove finer particles. It has been found that the productivity of an electro-filter relies upon the resistivity of the particles, which depends upon their chemistry (Zhang et al. 2018; Qin and Wang 2006) and can also serve as a carrier of harmful elements such as As, Be, Cd, Pb, Mn and Ni (Hu et al. 2012).

2.2.2 Generation of Particulate Matter

Particulate matter is generated from the following sources:

1. **Mechanical sources**: Through metallurgical operations like powdering, crushing, cracking etc.

Table 2.1 Emission and effects of criteria pollutants

Criteria pollutant	Emitting source	Health effect	Environmental effect
Sulfur dioxide (SO_2)	Volcanic emissions, fuel combustion, metal smelting, industries (http://cfpub.epa.gov/ncea/isa/recordisplay.cfm?deid=1988)	Breathing difficulties, Respiratory Problems including wheezing, chest tightness, or shortness of breath, heart and lung disorders, visual impairment (http://cfpub.epa.gov/ncea/isa/recordisplay.cfm?deid=1988)	Acid rain
Nitrogen dioxide (NO_2)	Fossil Fuel burning, biomass, high temperature combustion (ENVIS Centre on Control of Pollution 2016) processes power plants, non-road engines and equipment	Respiratory symptoms, Pulmonary disorders	Precursor of ozone formation in troposphere (ENVIS Centre on Control of Pollution 2016)
Particulate Matter (PM)	Vehicular exhaust, Industrial commercial and residential (ENVIS Centre on Control of Pollution 2016) combustion, construction activities	Respiratory problems, exacerbation of allergic symptoms (ENVIS Centre on Control of Pollution 2016), liver fibrosis, lung/liver cancer, heart stroke, premature mortality	Reduced visibility
Carbon monoxide (CO)	Gasoline-fueled vehicles, Combustion of carbonaceus fuels, emission from IC engines (http://www.epa.gov/air/caa/)	Reduced oxygen carrying capacity, anorexia, cardiovascular problems, myocardial ischemia (http://www.epa.gov/air/caa/)	Climate change and green- house effect

(continued)

Table 2.1 (continued)

Criteria pollutant	Emitting source	Health effect	Environmental effect
Ozone (O_3)	Hydrocarbons and NO_X react with sunlight to produce O_3, pollutants emitted by industrial facilities, electric utilities, and vehicles, substances generating ozone can also be emitted by anthropogenic sources, especially plants (http://cfpub.epa.gov/ncea/isa/recordisplay.cfm?deid=1499)	Respiratory problems, asthma, bronchitis etc. inflammation of lung and symptoms such as cough, wheezing, chest pain and shortness of breath etc. (Saadeh and Klaunig 2015)	O_3 in upper troposphere produces greenhouse effect, hazardous effects on plants in the form of interference in photosynthesis (ENVIS Centre on Control of Pollution 2016)
Lead (Pb)	Metallurgical operations, vehicular exhausts, lead-acid batteries, industrial wastes etc. (ENVIS Centre on Control of Pollution 2016)	Central nervous system disorders in children, anemia, toxic for soft tissues and bones (http://cfpub.epa.gov/ncea/isa/recordisplay.cfm?deid=1588)	Disturbed soil functioning

2. **Chemical or thermal sources**: These sources generate particulate matter through chemical or high-temperature evaporation reactions followed by the condensation.
3. **Biological sources**: Include pollen grains, fungi and bacteria.

Generation of particulates human activities: Generation of particulate matter takes place usually through the following activities:

1. Ploughing
2. Construction activities
3. Mining
4. Combustion of solid fuels, liquid fuels and biomass
5. Vehicular exhaust

Particulate emission from the natural sources is usually associated with volcanic eruptions, forest fires and soil erosion etc. (Pio et al. 2008; Oucher et al. 2015). According to WHO air quality guidelines, a reduction in yearly average $PM_{2.5}$ concentrations up to 10 $\mu g/m^3$ could lead to reduced air pollution related deaths by 15% in developing countries (WHO 2018). The low and middle income group countries have faced a lot of exposure to particulate matter usually originating from the household combustion of inferior fuel in traditional inefficient stoves, thereby increasing the risk for respiratory and cardiovascular diseases including acute lower respiratory infections, chronic obstructive pulmonary disease (COPD) and lung cancer (WHO 2018). An estimation done by WHO's International Agency for Research on Cancer (IARC) in 2013 revealed that particulate matter is the vital factor for increased vulnerability for cancer, especially pertaining to lungs. A relationship has been established between ambient pollution and risk of urinary tract/bladder cancer (WHO 2018). The particulate matter has bad effect on human health as well as on environment in the form of reduced visibility which results due to absorption and scattering of sunlight. In India, if the particulate level remains at its current status, per-capita mortality rate would increase by 21% by the year 2030. The average particulate emission should be reduced by 20–30% in the coming fifteen years to reduce the mortality rates. A large number of epidemiologic studies have demonstrated a correlation between enhanced ambient particulate levels and hospital admissions for respiratory diseases (Fusco et al. 2001; Spix et al. 1998; Atkinson et al. 2001; Migliaretti et al. 2005; Sun et al. 2006) (Table 2.2).

By following the air quality guidelines a gradual shift can be made in the concentrations at present. Hence it should be the major objective to follow the guidelines

Table 2.2 WHO 2005 guidelines for ambient particulate matter (Ambient (outdoor) air quality and health WHO, 2nd May 2018)

Particulate matter	Annual mean ($\mu g/m^3$)	24-hour mean ($\mu g/m^3$)
Fine particulate matter ($PM_{2.5}$)	10	25
Coarse particulate matter (PM_{10})	20	50

compulsorily. The risk exposure to the particulate matter is same for the people residing in urban and rural areas, although people in growing cities are at higher risk than those in developed ones.

2.2.3 *Indoor Particulate Concentration and Health Impact*

Not only that the outdoor particulate matter can infiltrate indoors, but it can also be generated by indoor sources as well. There is a growing interest to study about the specific impacts of indoor PM on human health.

Health Effects of Inhalable Particles
Respiratory and cardiovascular systems are affected largely due to the exposure to particulate matter. Studies have suggested association between health problems and size of the inhalable particles. Particles having small size viz. less than 10 μm in diameter have a tendency to penetrate deep into the lungs even reaching to the bloodstream. Those with heart or lung diseases are more vulnerable to get affected from PM exposure. Children and older adults are the most susceptible groups to be affected. Researches have linked PM exposure to a variety of health impacts, including:

1. Eye, nose and throat irritation.
2. Aggravation of coronary and respiratory disease symptoms.
3. Premature death in people with heart or lung disease.

2.2.4 *Fine and Ultrafine Particles*

Particulate matter is mostly sub-classified in two main classes groups: coarse and fine particles. Generally particulate matter is defined as either $PM_{2.5}$ (size less than 2.5 μm) or PM_{10} (size less than 10 μm). Though limits have been defined to control $PM_{2.5}$ and PM_{10}, there are no such limits or standards for ultrafine particles, as PM_1 (less than 0.1 μm or 100 nm to 1 μm) and $PM_{0.1}$. The defense mechanism of human respiratory system does not work against ultrafine particles (UFPs). WHO has not defined threshold limit for ultrafine particles like PM_1, which is seventy times finer than a human hair. Fine and ultrafine particulate matter (PM_1) is found in the form of a complex mixture in ambient atmospheric aerosol (Mauderly and Chow 2008).The particles ranging in size below <0.1 μm are known as the ultrafine particles (Hinds 1999). Because of their large surface area and high number density, they have high pulmonary deposition efficiency. Ultrafine particles are important in atmospheric chemistry research, particularly because of the associated health factors. They also carry substantial quantities of adsorbed or condensed toxic air pollutants including organic compounds and transition metals (Oberdörster et al. 2002). Vehicular load is identified as the key reason of ultrafine particulate emission (Shi et al. 2001; Keogh et al. 2009) contributing to around 60% particle number concentration (PNC) and 90% along the roadside in cities with high pollution level. People residing near main roads have a high exposure rate to ultrafine particulate matter and have shown escalation in the incidences of severe health issues, because of heavy traffic flow (Kumar et al. 2014; Pey et al. 2009; Pérez et al. 2010). Owing to their fine size these ultrafine particles can establish contact with the body through skin. Studies have linked skin diseases with prolonged exposure to ultrafine particles among the industry labors. PM_1 is efficient to penetrate through the cardiovascular system, prompting heart diseases. Cardiovascular mortality has also been associated with the inhalation of ultrafine particles. PM_1 has also been associated with premature births and abnormal foetal development. There has been limited research in developing countries like India to assess the health impact and severity of ultrafine particles in environment (Jayanthi et al. 2012; Verma et al. 2014; Bhardawaj et al. 2016). India's present vehicle count is around forty million, with an increase of 5% per annum, thus exposing its population to hazards of UFPs (Kumar et al. 2012).

2.2.5 Mechanism of Formation of Ultrafine Particles

They may be released in the environment by at these processes:

(a) **Direct emission**—Through vehicular or industrial sources (Kumar et al. 2012).
(b) **Combustion**—Combustion may also emit hot supersaturated vapors, which undergo nucleation and condensation while getting cooled to normal temperatures.
(c) **Chemical reactions**—Chemical reactions in the atmosphere may lead to the formation of UFPs that are more "localized," with their concentrations diminishing with distance to their sources.

Sources of Ultrafine Particles

Vehicular exhaust—The main source of atmospheric ultrafine particles is automobile exhaust (Pennington and Johnston 2012). In conjugation with roadside atmosphere, UFPs may result in the formation of condensable organic compounds (COCs) (Minoura et al. 2009). Increasing pollution episodes and closeness to high-traffic roads significantly contribute to the concentration of UFPs (Gramotnev and Ristovski 2004; Oberdörster et al. 2005). Particles generated from diesel driven vehicular exhaust are between 20 and 130 nm in size (Morawska et al. 1998) and 20–60 nm for gasoline engines (Ristovski et al. 1998). Highest concentrations are usually reported during morning when people are leaving for work and children are going to school (Imhof et al. 2005; Jain et al. 2011).

Photochemical reactions—UFPs can also result by photochemical reactions at ambient temperature. Low-volatility species are formed due to these reactions which lead to the formation of UFPs by nucleation (Kulmala et al. 2004) which may occur

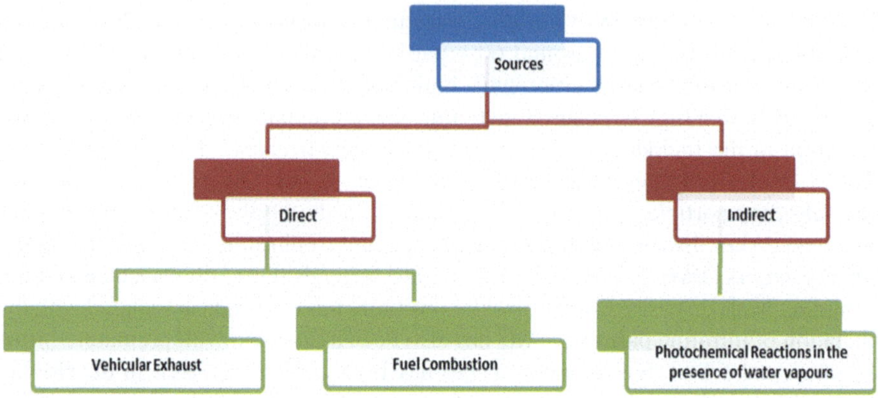

Fig. 2.1 Sources of ultrafine particles

on ions and possibly is a multicomponent process. Evidences suggest that sulphuric acid vapors help in nucleation. Binary water-sulphuric acid mechanism has been proposed for nucleation (Kulmala and Laaksonen 1990). Ternary water-sulphuric acid-ammonia nucleation has also been suggested (Yu and Turco 2000). The small sized particles can condense through low-volatility organic species condensation which is produced through photochemical oxidation of low volatility organic species (Kulmala et al. 2004).

Biofuel usage—Biofuels have been used as an alternative to overcome pollution from conventional fuels (Bhardawaj et al. 2016). Though biofuels produce reduced particle mass but consist of higher PNC including ultrafine particles (Bhardawaj et al. 2013). Biofuels with low calorific values are required in higher volumes with an amplified fuel flow rate. Hence biofuels must be introspected and appropriate mitigation strategies should be evolved, which diminish UFP production from them and minimize health risks (Lapuerta et al. 2008) (Figs. 2.1 and 2.2).

2.2.6 Health Effects of Ultrafine Particulates

Ultrafine particulate matter can enter the human body through inhalation, dermal, oral, olfactory or through digestive routes. They deposit in the respiratory tract and travel to other organs. They penetrate into lungs causing acute and chronic illnesses including asthma, COPD and childhood leukemia etc. (Ezz et al. 2015; Qiao et al. 2015; Ajmani et al. 2016; Peters et al. 1997). A negative correlation has been established between ultrafine particles and peak expiratory flow (Penttinen et al. 2001), whereas a positive correlation between cardiovascular mortality and fine and ultrafine particles has been established in two pollutant models (Wichmann et al. 2000; Oberdörster et al. 2002). The biological response of UFPs is dependent on its miniscule size and high surface area where damaging organic chemicals can be adsorbed

Fig. 2.2 Release of particles in environment

(Block and Calderón-Garcidueñas 2009) which are further moved to the pulmonary system and are finally mixed into blood stream which can affect heart and central nervous system through inflammation (Genc et al. 2012).

2.2.7 *Indoor Particulate Sources*

Indoor particulate matter can be originated both from outdoor as well as indoor sources. Common indoor particulate matter originates from cooking and combustion activities like, use of unvented heaters or stoves, fireplaces and use of inferior quality fuel for various household purposes. Cigarette smoking is another factor that contributes to the indoor particulate matter. Biological sources may also contribute significantly in the production of particulate matter (Pekkanen et al. 1997; Adams et al. 2015). In India, the studies have mostly concentrated on indoor air pollution due to combustion of biomass fuels in inefficient cooking stoves in rural and semi urban areas. In these microenvironments indoor air quality has been linked with their socio-economic status which makes it difficult for the rural population to afford environment friendly fuel (Awasthi et al. 1996; Balakrishnan et al. 2002; Bruce et al. 2000; Kulshreshtha et al. 2008; Singh and Jamal 2012). Exposure to household generated smoke is more severe in developing countries as large proportions of households are still dependent on biofuels for cooking and other household purposes. Poor ventilation in such households exacerbates the situation producing large volumes of smoke indoors (Bruce et al. 2000). Women are more prone to the exposure of particulate contamination because they are mostly involved in cooking (Behera et al. 1988) and young children who tend to stay indoors (Albalak et al. 1999). Indoor particulate quantification is important for the assessment of the total human exposure to pollutants (Varghese et al. 2005). Many studies have assessed exposure of particulate matter generated from cooking in many Indian areas (Kulkarni and Patil 1999; Saksena et al. 2002) including the potential health risks in different microenvironments. Health effects of particulate matter depend upon several factors including physiological, physical, chemical and biological (Dreher 2000). Though previous studies have established a strong relationship between the biofuel usage and production of particulate matter, especially in rural households, still there is no concrete evidence which reveals the generation and size distribution of the particulate matter generated in urban areas of India where energy resources are different. Less data is available which highlights the problem of indoor air pollution in urban areas as the studies have been more focused on the ambient pollution. To predict indoor air quality, we have to take an account of a variety of indoor sources, other than biomass fuel burning (Goyal et al. 2012). Some of the chief elements affecting the IAQ are: (i) concentrations of ambient pollutants infiltrating indoors (ii) building constituents such as asbestos, cement etc. (iii) indoor characteristics such as ventilation, (iv) occupancy inside a dwelling (v) equipments used indoors (vi) rituals, habits and practices of the occupants, and (vii) the financial standing of the dwellers (Kumar 2011).

2.3 Particulate Matter and Associated Heavy Metals

Heavy metals are mostly associated with particulate matter and get absorbed in human body via inhalation. These heavy metals have severe physiological effects. Strong evidences have suggested that heavy metal adsorption is crucial to the adverse effects associated with particulate matter (Bollati et al. 2010). Heavy metals are metallic elements with high density, specific gravity and atomic weights (Duffus 2002). They enter the environment mainly through deposition of atmospheric particulates; disposal of metal containing sewage, traffic emission, re-suspension of road dust and through mining operations (Shrivastav 2001). Re-suspension of roadside dust is a chief source of heavy metals (Schroeder et al. 1987). Mn, Cu, Zn, Cd, Cr, Fe, Ni, K, Ca, V, Ba, As, Se and Sr are some of the metals which have been commonly studied. Ultrafine particulate emitted from diesel exhaust is considered to be very toxic due to its high metal content which causes oxidative stress leading to endothelial dysfunction and increased risk of heart dysfunction (Miller et al. 2012). $PM_{2.5}$ is considered more dangerous because of its longer retention time and deeper penetration in lungs (WHO 2006). A person is more susceptible to the exposure of particulate matter indoors because people spend 90% of their time indoors (Massey et al. 2013). Influence of the inhaled aerosols depend on the chemical species, its concentration, extent of exposure in terms of time and its deposition within the respiratory tract (Salma et al. 2002). Fine particles have a higher tendency to absorb toxic metals than coarse particles (Massey et al. 2013). Outdoor particulate matter can reach indoors and affect the indoor concentration potentially. Ventilation has also been found to effect the indoor PM concentration, which is also dependent on the source characteristics and physical nature of pollutants (Leung 2015). Heavy metals exert toxic effects (Erisman et al. 1994) hence their assessment is of enormous importance for toxicological, environmental and occupational studies concentrating on health outcomes (Pandey et al. 1998).

2.3.1 Hazardous Effects of Heavy Metal Borne Particulates

A number of adverse health consequences have been linked with heavy metal borne particulate contamination. Incidences of myocardial infarction have been linked with particulate contamination. Ultrafine particles induce oxidative stress in the endothelium, and reach other structures through the nervous system. Thrombosis, heart rate fluctuations and blood pressure changes have been related with particulate matter (Anderson et al. 2012) and lately irregularities in glucose metabolism have been reported. Low birth-weight, infertility, genotoxicity and cancer (Loomis et al. 2014) are also associated with PM exposure. Some of the health effects associated with heavy metals carried by inhaled particles have been summarized in Table 2.3.

2.3.2 Mechanism of Action of Heavy Metals

The exact mechanism through which heavy metals exert hazardous effects is not known, still it has been found that metals cause oxidative stress, inflammation and changes in the immune response (Wiseman and Zereini 2009). Cr, Cd and Ni have been found to alter the DNA's repairing ability which is an outcome of oxidative stress. Alterations in the microstructure of the non-ciliated bronchilar cells have also been reported after Pb inhalation (Fortoul et al. 1999). Cardiovascular effects of the metals have been found to occur through inflammation and the initiation of hypercoagulability and endothelial dysfunction (Martinelli et al. 2013). High concentration of Ni, Cu and As has been linked with high level of C-receptor protein, interleukin-6 and vascular endothelial growth factor showing low capacity of endothelial repair (Niu et al. 2013). Cd carcinogenicity has been associated with the generation of ROS owing to the inactivation of detoxifying enzymes like catalase, glutathione peroxidase etc. Cu has been identified as the rate-limiting nutrient for tumors. The presence of Cu in tumors affects cytochrome c oxidase activity (Ishida et al. 2013). Mercury compounds have been found to inhibit mitotic spindle and alter DNA repair process. Hg has also been seen to deplete the immune system and could decrease immune tumor

Table 2.3 Health symptoms of common metals

S. No.	Metal	Source	Health Symptoms (short and long term)
1	Pb	Anti-knocking agent in gasoline	1. Changes in bronchiolar cells (Soto-Jimenez and Flegal 2011) 2. Increased cancer risk 3. Alterations in DNA synthesis, mutations (Silbergeld 2003) 4. Cell proliferation and oxidative stress (Garcia-Leston et al. 2010)
2	Ni	Refineries, industries, cigarette smoke	1. Cancer risk 2. Chromosome aberration in rats (Chorvatovicova and Kovacikova 1992) 3. Cardiovascular diseases
3	Cd	Cigarette smoke	1. Genotoxicity (Fortoul et al. 2015) 2. Disruption of cell adhesion causing tumour progression (Koedrith and Seo 2011) 3. Liver damage through ROS generation (Fowler 2009) 4. Elongation of estrous cycle in rats 5. Affects the synthesis of placental hormones like progesterone (Caserta et al. 2013) 6. Oxidative stress
4	Fe	Steel manufacturing industries, cement mills	1. Alterations in immune functions (Nairz et al. 2014) 2. Overload results in insulin resistance, hyperglycemia, elevated risk of Diabetes Mellitus (Fernandez-Real et al. 2002; Ferrannini 2000) 3. Overload results in hypertension and increased cardiovascular risk (Jomova and Valko 2011)
5	Cr	Metallurgical processes- in the manufacturing of steel, spray painting, impurities in fuel additives, non-exhaust emission, cigarette smoke	1. Genotoxic 2. Alteration in DNA's repair capacity 3. Carcinogenic 4. Cause hypoglycemia (Mertz 1993) 5. Immunotoxic-causes inhibition of immunoglobin secretion (Mitra et al. 2012)

(continued)

Table 2.3 (continued)

S. No.	Metal	Source	Health Symptoms (short and long term)
6	Hg	Forest fires, fossil fuels, hydroelectric, mining, pulp, and paper industries. Incineration of municipal and medical waste and emissions from coal-using power plants also contribute to mercury	1. Methyl mercury is a known carcinogen 2. Produces oxidative stress 3. Genotoxic 4. Lethargy, irritability, depression (Magos and Clarkson 2006) 5. Hyperglycemia (Chang et al. 2011) 6. Reproductive effects including abortion, stillbirth and menstrual disorders and reduced fertility (De Rosis et al. 1985)
7	Cu	Copper extraction, metal and electrical manufacturing, agricultural and domestic use of pesticides and fungicides, leather processing, and automotive brake pads	1. Formation of ROS (Theophanides and Anastassopoulou 2002) 2. Neurodegenerative diseases such as Parkinson (Modgil et al. 2014) 3. Wilson's disease- accumulation of copper in liver (Pietrangelo 1996)
8.	Mn	Gasoline additive methylcyclopentadienyl manganese tricarbonyl (MMT)	1. Neurotoxicity 2. Decrease in intellectual ability 3. Alteration of psychomotor development (Lin et al. 2012) 4. Tetragonic effect 5. Growth restrictor, embryonic death and bone alterations (Gerber et al. 2002)
9.	Zn	Industrial activities	1. Neurotoxin (Sorsa 2011) 2. Impaired glucose metabolism (Praveena et al. 2013) 3. Teratogen (Basu et al. 2014)
10	V	Burning of fossil fuel including petroleum, oil, tar and asphaltite etc.	1. Respiratory effects (Cakmak et al. 2014) 2. Carcinogenicity 3. DNA disruption (Galanis et al. 2009). 4. Initiation of hypoxia (Assem and Levy 2009) 5. Behavioral and cognitive disorders (Barceloux 1999)

response (Koedrith and Seo 2011). Cr(VI) induced carcinogenicity has been associated with free radical production and establishing the cancer phenotype (Hartwig 2013). Mn has been observed to interfere with DNA polymerases, mitochondrial function and activation of some cytokinases and MAPK cell signaling cascades in its cancer causing ability (Valko et al. 2005). The genotoxicity of heavy metals has been linked with chromosomal aberration and DNA damage (Sorsa 2011). Neurotoxicity of the metals has been credited to molecular level metal dishomeostatis and mitochondrial dysfunction. Metals like Mn have been found to cross the blood-brain barrier inducing tremors and production of other schizophrenic symptoms (Braidy et al. 2014). Metals have also been found to effect retina where phototransduction process takes place. In addition a gradual reduction in PL and enhancement of the oxidative stress biomarker-4-hydroxinonenal has been observed on exposure to V (Avila et al. 2013). Metals also exert hazardous effects on the reproductive system. Cd has been linked with the increase in the duration of estrous cycle in rats (Cervantes-Yépez et al. 2018; Baranski and Sitarek 1987). Cadmium accumulates in the placenta and effects the expression of metallothionein, which retains Cd, thereby preventing it from reaching the foetus. But overexpression of metallothionein stops Zn transportation to the placenta and decreases the placental permeability. (Caserta et al. 2013). High levels of iron during the human embryonic period may be teratogenic. Researches in mice have shown that an escalation of iron produces histological variations in the encephalon, as well as spine and ribs abnormalities (Weinberg 2010) (Fig. 2.3).

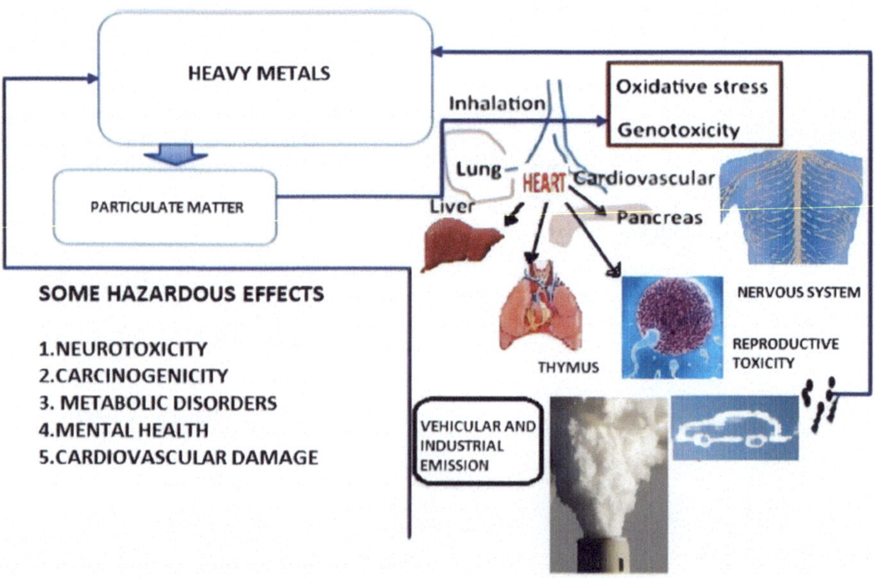

Fig. 2.3 Hazardous effects of heavy metals

2.4 Estimation and Analysis of Heavy Metals Associated with Particulate Matter—Parameters Employed

2.4.1 *Source Apportionment*

Source Apportionment (SA) is employed to identify sources of pollution and the quantification of their impact to pollution levels in ambient as well as indoor environment. The source apportionment can be done using different methods: emission inventories, source-oriented models and receptor-oriented models and is usually performed to assess the contributing sources to pollutant level at representative monitoring site. Among the variety of models being used to achieve source apportionment, receptor-oriented models are widely used as they have the benefit of providing information derived from online measurements. Receptor based models usually work to identify source contribution estimation at local and regional level all over the world. The outline for using receptor models to solve air quality problems consists of: formulating a conceptual model, finding potential sources, characterizing source emissions, sampling and analysis of ambient particulate matter samples, determination of source types using multivariate receptor models; quantifying source contributions with the chemical mass balance; assessing profile changes and reconciling receptor modeling results with source models, emissions inventories and eventually analysis of receptor data (Watson et al. 2002). The basic principle of receptor modeling is that mass conservation between the emission source and the study site can be anticipated, and a mass balance analysis can be done to recognize and apportion sources of pollutants. RMs recognize sources through the following mass balance equation

$$x_{ij} = \sum_{k=1}^{p} g_{ik} f_{kj} + e_{ij} \tag{2.1}$$

Where x_{ij} is the concentration of the j-th species in the i-th sample, g_{ik} is the contribution of k-th source to the i-th sample, f_{kj} is the concentration of the j-th species in the k-th source, and e_{ij} is the residual term. RMs are generally used to allocate PM concentration as categorized according to its chemical composition and heavy metals etc. From the analysis of particulate matter, it is enabled to assess the sources responsible for the total PM_{10} and $PM_{2.5}$ concentration accompanying several uncertainties (Karagulian et al. 2015).

2.4.2 *Enrichment Factor Analysis*

The enrichment factors (EF) of the elements are calculated to identify the source of the elements in the atmosphere. Other than that, the EF values can also define whether the elements have an additional contamination from the anthropogenic activities or not. Aluminium (Al) and iron (Fe) are usually selected as reference elements for the crust based on previous studies (Enamorado-Báez et al. 2015). The equation for calculation of EF was proposed by Rahn (1976). The analysis is used to compare the concentrations of atmospheric aerosol elements, when they are present in higher concentrations. The EF can be calculated by the following formula

$$\mathrm{EF} = \frac{[\mathrm{C}]\mathrm{sample}/[\mathrm{Cref}]\mathrm{sample}}{[\mathrm{C}]\mathrm{crust}/[\mathrm{Cref}]\mathrm{crust}} \tag{2.2}$$

where

$[\mathrm{C}]_{\mathrm{sample}}$ = Concentration of C (analyte element)
$[\mathrm{C}_{\mathrm{ref}}]_{\mathrm{sample}}$ = Concentration of the normalization reference element in the sample
$[\mathrm{C}]_{\mathrm{crust}}$ = Concentration in the continental rust
$[\mathrm{C}_{\mathrm{ref}}]_{\mathrm{crust}}$ = Concentration in the continental crust for the reference element

Five categories have been defined on the basis of EFs: $\mathrm{EF} < 2$, minimal enrichment; $2 \leq \mathrm{EF} < 5$, moderate enrichment; $5 \leq \mathrm{EF} < 20$, significant enrichment; $20 \leq \mathrm{EF} < 40$, very high enrichment; and $\mathrm{EF} \geq 40$, extremely high enrichment.

2.4.3 Chemical Mass Balance Modeling

The chemical mass balance (CMB) model is a popular receptor model for source apportionment. The model works on using solutions to linear equations attributing each receptor chemical concentration as a linear sum of products of source profile loads and source contributions. (Hidy and Venkataraman 1996). The CMB receptor method works on extracting information about a source's contribution on the basis of the profile variability of elements measured (Henry et al. 1984).

2.4.4 Principal Component Analysis (PCA)

Principal component Analysis is done to reduce the dimensionality of multivariate data without losing information. PCA analysis can be used to identify more clearly the various sources of particulate matter. Results can be plotted as different variables and interpretation can be done to obtain positive and negative correlations among them. The data set are subjected to Varimax rotation and Kaiser normalization for source apportionment. PCA data can be split up in specific groups which are known as factors and combine on the basis of similar features corresponding to similar originating sources and modes of transportation. The number of factors explains the highest maximum variance of the data (Vaio et al. 2018; Yang et al. 2016). To lessen the number of variables, Varimax rotation method is used to produce one rotated component matrix.

2.4.5 Instrumentation and Principle of Analysis of Heavy Metals in Particulate Matter

2.4.5.1 Flame-Atomic Absorption Spectroscopy

Atomic Absorption Spectrometry (AAS) is used for quantitative estimation of elements. The technique is based on measuring the absorbed radiation by the species under consideration. The atoms absorb ultraviolet or visible light and get promoted to higher energy levels. A detector is employed to measure the wavelength of light transmitted by the sample and makes a comparison with the originally passed light through the sample. The concentration is assessed by plotting a calibration curve, using standards of known concentration. The working principle of AAS depends on Beer-Lambert's law (Sammut et al. 2010). For precision, a light beam from a lamp whose cathode is made of the element being determined is passed through the flame. A photon multiplier detects the quantity of light intensity owing to absorption by the analyte, which is directly related to the concentration of the element in the sample (Welz and Sperling 1999).

2.5 Assessment of Heavy Metals Associated with Particulate Matter in Urban Environment—A Case Study in Lucknow Region

Lucknow is the capital of Uttar Pradesh, India. The city was earlier known as Awadh and is recognized for its rich cultural heritage, monuments and chicken work. The city is growing at a fast pace among all the nonmetropolitan cities. The city is a centre for industrial and real estate development depicting an economic boon. The growing urbanization has put the city under scanner as the level of air pollution has also shot up in the last few years (Verma et al. 2015). Lucknow City has a population of 2.82 millions as per 2011 census and emerging as one of the fastest growing cities. Deforestation is another major issue in the district which is leading to ecological imbalance. The actual forest cover in the district is about 115 km^2, which is less than 5% of the total area of the district (2500 km^2) as estimated in 2015. With the increase in vehicular load, roads need to be widened leading to felling of trees. The main causes of air pollution in Lucknow have been identified due to:

1. Vehicular exhaust and progressive increment in vehicular load.
2. Fast urbanization and loss of vegetation.
3. Nearby Industries.
4. Negligence of government and public.
5. Biomass combustion.
6. Open dumping and burning of the garbage (Saluja 2017).

2.5.1 Weather and Meteorological Factors

Lucknow has subtropical climate. Cool dry winter prevails from December to February and summers prevail from March to June. The maximum temperature in summer and winter is about 45 °C and 3 °C respectively. Annual rainfall is about 100 cm. Upon release in the ambient environment, a pollutant interacts with other pollutants and the concentration is influenced by micrometeorological factors which may affect its complexity making it more hazardous in nature. So it is necessary to identify the pollutants, their sources, transformation as well as fate of each pollutant and subsequently their impact on environment including living beings (Assessment of ambient Air Quality of Lucknow City 2017). Climate data retrieved from Lucknow Airport by Meteorological Department of India is shown in Table 2.4.

Foggy conditions prevail from late December to January. Summers are quite hot. The seasonal variability pattern of the pollutants shows a dynamic variation with gaseous pollutants being in high concentrations in winters and low concentration in monsoon. The difference in concentration level may be attributed to the wind speed. In winter season the comparatively highest concentrations may be attributed to low wind speed that is <5 km/h and high humidity. The stable atmospheric conditions hinder the vertical and horizontal instability for appropriate mixing of pollutants making them more stagnant in atmosphere. As a result, atmospheric dispersion is not adequate and pollutants accumulate near the ground breathing zone. The lack of precipitation during winters also decreases the capacity of wet deposition and related

Table 2.4 Meteorological data from Lucknow, India (http://www.imd.gov.in/section/climate/extreme/)

Month	Temperature (°C)			Rainfall (mm)
	Avg. high	Daily mean	Avg. low	
January	19	11.3	3.5	21
February	23.3	14.8	6.3	26
March	31.1	21.8	12.5	16
April	36.8	27.9	19	15
May	40	32	24	23.4
June	38.4	32.6	26.7	122.9
July	33.9	29.8	26	276.2
August	33.2	29	25.6	278.9
September	33.1	28.6	24.1	175.9
October	31	23.4	17.7	24.7
November	27.7	18.8	10	7.4
December	20.5	12.2	4.7	12.6
Yearly	30.7	21.7	16.7	1000

cleansing mechanisms. Washout effect is another factor playing an important role in monsoon which accounts for the low concentration of SO_2 and NO_2.

2.5.2 *Air Pollution Status of Lucknow*

During the recent years air quality in Lucknow has witnessed a down fall. Studies have correlated air pollution status with environmental, human and plant health impacts. The metallic concentration associated with particulate matter has been found to fluctuate with meteorological conditions in the region. A survey based study done in early 2000 showed that particulate concentration exceeded the NAAQS limits. Particulate fraction PM_{10} analyzed for heavy metals showed the presence of heavy metals in all the seasons (Pradhan et al. 2004). The study revealed significant variation in the metal concentrations at different locations. The high PM_{10} concentration was attributed to diesel driven vehicles. In another study in Trans-Gomti area, during peak traffic hours (11.00 a.m.–1.00 p.m.) the recorded PM_{10} level was 499 $\mu g/m^3$. The highest PM_{10} level (990 $\mu g/m^3$) was observed at the most crowded crossing with most traffic density (6723 vehicle/h) and least (150 $\mu g/m^3$) at low traffic density crossing with least vehicular load (52 vehicle/h). This investigation revealed that the Trans-Gomti area is mainly polluted with PM_{10} and advised to ban diesel driven vehicles in the city areas to cut the PM_{10} level within the permissible limit (Verma et al. 2003). Vehicular exhaust is one of the major reasons of particulate contamination, which affects the air quality and the health. Uttar Pradesh State Road Transport Corporation (UPSRTC) introduced bus services under the banner "Lucknow Mahanagar Parivahan Sewa" on different routes of Lucknow city. The Green Peace Report in 2016 has highlighted that in 20 cities of Uttar Pradesh, from where the data was collected by Pollution Control Board, the PM_{10} concentration was higher than the annual average of 60 $\mu g/m^3$ as suggested under NAAQS. PM_{10} concentrations in Gaziabad, Barielly, Allahabad, Kanpur, Agra, Lucknow, Varanasi and Sonebhadra were respectively 258, 240, 250, 201, 186, 169, 145 and 132 $\mu g/m^3$ for year 2015. A comprehensive observation of the data suggests that the PM_{10} level has been dangerously high throughout the year for from October 2015 to September 2016, with October to February being the most polluted months when the PM_{10} reached to 400 $\mu g/m^3$ (http://ueppcb.uk.gov.in/files/Ambient_Air_Quality_2015__(2).pdf). Some of the recent episodes of air pollution in Lucknow are as follows:

1. In December, 2015 AQI of the city reached to 489. The AQI between 401 and 500 has been categorized as severe for the exposed population (http://timesofindia.indiatimes.com/city).
2. Lucknow recorded a "very poor" air quality condition on Diwali night, 2018 with $PM_{2.5}$—(International Business Times).
3. With an average $PM_{2.5}$ concentrations of 138$\mu g/m^3$, Lucknow was placed at 7th position out of ten most polluted cities of the world, according to WHO's latest report in 2018.

4. Lucknow climbed up to the position of second most polluted city in the country on November 28, 2018 with the AQI ‘very poor (Times of India). As per the data released by the Central Pollution Control Board, the AQI of the city was 363.

2.5.3 *Objectives of the Study*

The study was conducted in fifteen urban houses of Lucknow city from 2012 to 2014 to assess PM_{10} and $PM_{2.5}$ concentration to analyze seven heavy metals viz. Fe, Zn, Cu, Pb, Mn, Ni and Cr associated with PM_{10} and $PM_{2.5}$. Principal component analysis and enrichment of elements/metals from outdoor sources were also performed to identify the sources of metals in particulate matter as a part of source apportionment.

Selected Microenvironments
Fifteen urban houses were selected on the basis of house characteristics, traffic rush and population distribution and categorized under three microenvironments as:

1. Unplanned or densely populated
2. Well planned
3. Roadside

Houses in densely populated microenvironment were constructed according to old architectural designs. They were closely built in congested narrow lanes. Parking space was very limited and mostly people parked their vehicles outside the house. Frequent power cuts were observed in these localities and people usually used diesel driven generators as power back up. Outdoor characteristics were marked by crowded market places. There were frequent waste dumping and incineration activities and piles of garbage sighted. In contrast to the architectural pattern of the houses in densely populated microenvironment, those in well planned microenvironment were relatively newly constructed. The houses mostly had parking spaces within the house premises. Small parks and abundant greenery marked the surroundings. Electricity supply was also regular and most of the households relied on invertors for power back up. Roadside microenvironment was marked by heavy traffic rush as the houses were situated along the highways experiencing heavy vehicular load throughout the day and reaching to the peak between 9:00–11:00 a.m. and 4:00–7:00 p.m. Mainly natural ventilation was observed in the houses and they were constructed according to modern architectural patter. For outdoor sampling, the instrument was placed either on roof top or on the verandah as per convenience. House characteristics are summarized in Table 2.5.

Table 2.5 House characteristics in different microenvironments

Microenvironment	Sampling site	Site description
Well planned	LDA	Proper new houses with ventilation, Negligible traffic flow
	Gomtinagar	Developed area with proper ventilation and greenery
	Mahanagar	Moderate flow of traffic
	Indiranagar	Less population, new houses, parks and low vehicle traffic
	Jankipuram	New Houses, with low vehicle traffic and shops
Densely populated	Chowk	Dense population, Congested area with least greenery. Old type of houses, lots of small shops
	Rajajipuram	Dense population, bad roads, many small schools, low vehicular traffic
	Muftiganj	Congested area with no proper roads, Approximately 20 year old congested houses
	Triveni Nagar	Busy area with old types of commercial shops and houses
	Aminabad	Very Busy market area (shops are present inside houses).
Roadside	I.T. Chowraha	Heavy traffic flow during day time
	Bakshi ka Talab	Heavy diesel traffic flow 24 h
	Balaganj	Bus and truck flow mainly in daytime
	Alambagh	Heavy traffic flow 24 h. Mainly due to buses
	Mushi Pulia	Traffic rush from Buses and trucks mainly in daytime

2.5.4 Sampling and Analytical Method

The study was conducted for a span of two years viz. March 2012 to Feb 2014. Particulate matter was measured through 24 h continuous monitoring in the selected houses representing three distinct microenvironments. The purpose of continuous monitoring was to cover major household activities and their influence on the indoor particulate variation as well as the effect of outdoor activities. The instrument was set up in the living area where people spend maximum of their time. PM_{10} and $PM_{2.5}$ samples for chemical analysis were collected with fine particulate dust sampler (APM 550, Envirotech). The air inlet has a circular symmetrical hood designed to keep out rain, insects and very large particles. Inlet section leads to an Impact stage with trap particles larger than 10 microns. For the sampling of $PM_{2.5}$ a well shaped (WINS) Impactor can be attached to the down tube. It is designed to trap medium size (between 2.5 and 10 micron) particles. To avoid sampling error due to bouncing of small size particulate from impaction surface, a 37 mm diameter GF/A (Glass Fibre) paper immersed in silicon oil is used. APM 550 used oil-less rotary pump to produce the suction pressure and critical flow control or office (as recommended by USEPA) for maintaining a constant air-flow rate of 1 m^3/hr or 16.7 L/min. The particles were collected on 47 mm diameter, 2 μm pore size polytetrafluoroethylene Whatmann filters. Glass fiber filters are cheap and have been used in many studies previously for determination of metals in coarse particle. Sampling was done at a height of 1 m from ground level for 24 h with frequency of once a week at all sites, that is 4 times a month, 16 times a season. The air inlet hood of collection sampler was kept at a height of 1–1.5 m above the ground level to simulate the human breathing zone. The instrument was set at least 1 m away from any significant pollution sources.

Analysis of Heavy Metals
Gravimetric Analysis

The filters were weighed thrice before and after sampling using balance (Citizen, Model No. ISO 900:2000) with a sensitivity limit of ±0.2 mg. Before weighing, the samples were equilibrated in desiccator at 20–30 °C and relative humidity of 30–40% in a humidity controlled room for 24 h. Field blank filters were collected to reduce gravimetric bias due to filter handling while sampling and afterward. Filters were handled only with tweezers coated with Teflon to reduce the possibility of contamination. After weighing, samples were kept in fridge till further analysis. It was assumed that particulate matter accumulated on filter papers was evenly distributed over the entire area.

Chemical Analysis

The exposed filter was digested in 6–8 mL of analytical grade (AR) (Merck) HNO_3 on hot plate at 40–60 °C for 90 min. The solution was diluted up to 50 mL with distilled de-ionized water and preserved in polypropylene bottles till analysis. Analysis of the metals was done on Atomic Absorption Spectrophotometer (AAS) (Perkin Elmer, AAnalyst 100) (Kulshrestha et al. 2014).

Quality Assurance

The particulate sampler was calibrated in the beginning and at the end of sampling. Filter in the wins impactor was usually changed after 72 h of sampling. The minimum detection limit of the heavy metals is given in Table 2.6.

Table 2.6 Detection limit of the heavy metals

Metal	Detection limit (mg/L)
Pb	0.10
Zn	0.018
Ni	0.063
Fe	0.060
Cr	0.055
Mn	0.052
Cu	0.077
Cd	0.60

2.6 Results and Discussion

Average concentrations of particulate matter are given in Tables 2.7, 2.8 and 2.9 in summer, rainy and winter seasons. The results revealed that average yearly indoor concentration of PM_{10} and $PM_{2.5}$ were 79 $\mu g/m^3$ and 59 $\mu g/m^3$ respectively in well planned houses, whereas houses in densely populated areas had an average yearly indoor concentrations of 90 $\mu g/m^3$, and 60 $\mu g/m^3$ for PM_{10} and $PM_{2.5}$ respectively. Roadside houses had an average yearly concentration of 100 $\mu g/m^3$ and 66.4 $\mu g/m^3$ respectively for PM_{10} and $PM_{2.5}$. Roadside houses were found to have maximum particulate contamination followed by densely populated and well planned, Indoor particulate concentration was found to be higher than the WHO limits. Overall analysis showed that roadside microenvironment was most polluted, having the poorest air quality.

2.6.1 Analysis of Heavy Metal Concentration in Particulate Samples

Fe, Zn, Pb, Cr, Ni, Cu and Mn were analyzed in PM_{10} and $PM_{2.5}$ samples with the help of AAS in all the three microenvironments. Fe was found to have highest concentration in both particulates $PM_{2.5}$. The levels of assessed metals in all the three microenvironments were found to be in the following order:

Well planned microenvironment—Fe > Zn > Pb > Cr > Ni > Cu > Mn
Densely populated microenvironment—Fe > Cr > Zn > Pb > Ni > Cu > Mn
Roadside microenvironment—Fe > Pb > Zn > Cr > Ni > Cu > Mn
Overall trend of heavy metals—Fe > Zn > Pb > Cr > Ni > Cu > Mn

The average concentration of heavy metals is given in Tables 2.10, 2.11 and 2.12. The sum of the seven analyzed metals ΣMetals was calculated and the trend exceeded up to five times at the roadside houses than the threshold limits. Excess of Pb in the roadside houses can be attributed to mixing of residual Pb in soil and its re-suspension in the air due to heavy vehicles. It was also observed that the concentration of Ni associated with $PM_{2.5}$ was much higher in the months of November and December. Low temperature and relative humidity causes dilution of pollutants in these seasons. Higher concentration of Ni is a major concerning issue as people can easily get exposed to it by inhalation or dermal contact. Atmospheric concentrations of metals did not differ substantially between each microenvironment. In densely populated microenvironment Cr was in higher concentration because of coal and kerosene burning in winter months. Nickel finds its way into the ambient air as a result of the combustion of coal, diesel oil and fuel oil, the incineration of waste and sewage, and miscellaneous sources. Tobacco smoke in the form of gaseous nickel carbonyl, stainless steel in kitchen, inexpensive jewelry utensils also contributes to lower level of nickel (Kulshrestha et al. 2014). Although Zn and Cu are essential elements, their

Table 2.7 Particulate concentration in summer season (μg/m^3)

Microenvironment	Pollutant	March			April			May			June		
		Indoor	Outdoor	I/O	Indoor	Outdoor	I/O	Indoor	Outdoor	I/O	Indoor	Outdoor	I/O
Densely populated	PM_{10}	82 ± 3.2	102 ± 7.4	0.8	74 ± 0.06	144 ± 7.7	0.40	90 ± 6.8	115 ± 7.1	0.74	84 ± 4.9	178 ± 5.1	0.47
Well planned		62 ± 4	94 ± 5	0.63	67 ± 9	107 ± 21	0.63	87 ± 13	121 ± 14	0.76	78 ± 18	103 ± 12	0.84
Roadside		82 ± 3.2	121 ± 7.8	0.91	104 ± 0.06	184 ± 7.1	0.57	127 ± 6.8	191 ± 7.1	0.66	124 ± 4.9	186 ± 5.8	0.7
Densely populated	$PM_{2.5}$	42 ± 1.6	89 ± 1.1	0.61	34 ± 0.04	79 ± 0.2	0.43	53 ± 1.7	92 ± 0.03	0.65	55 ± 0.8	82 ± 0.6	0.67
Well planned		62 ± 9	84 ± 26	0.66	64 ± 30	156 ± 14	0.41	57 ± 23	109 ± 35	0.52	51 ± 8	75 ± 34	0.85
Roadside		72 ± 9	98 ± 26	1.0	74 ± 30	86 ± 14	0.86	69 ± 21	119 ± 35	0.64	71 ± 9	89 ± 34	0.77

Table 2.8 Particulate concentration in rainy season (μg/m^3)

Microenvironment	Pollutant	July			August			September			October		
		Indoor	Outdoor	I/O	Indoor	Outdoor	I/O	Indoor	Outdoor	I/O	Indoor	Outdoor	I/O
Densely populated	PM_{10}	133 ± 42	163 ± 28	0.54	77 ± 28	112 ± 36	0.69	68 ± 42	124 ± 12	0.55	62 ± 21	96 ± 17	0.65
Well planned		82 ± 42	95 ± 90	0.91	67 ± 8	82 ± 3	0.81	59 ± 19	101 ± 20	0.58	61 ± 43	97 ± 63	0.63
Roadside		89 ± 32	91 ± 90	0.95	78 ± 12	108 ± 40	0.72	92 ± 11	117 ± 26	0.79	71 ± 52	109 ± 74	0.65
Densely populated	$PM_{2.5}$	103 ± 1.7	113 ± 62	0.51	56 ± 3.5	61 ± 0.09	0.92	45 ± 3.2	83 ± 1.1	0.54	54 ± 2.1	75 ± 0.8	0.72
Well planned		57 ± 13	45 ± 67	1.22	39 ± 1.3	62 ± 1.8	0.63	54 ± 1.7	82 ± 2.1	0.65	56 ± 2.7	74 ± 3.7	0.76
Roadside		47 ± 7.3	86 ± 7.5	0.55	64 ± 3.8	81 ± 4.6	0.79	44 ± 2.8	89 ± 5.7	0.49	79 ± 7.9	88 ± 7.1	0.9

Table 2.9 Particulate concentration in winter season (μg/m^3)

Microenvironment	Pollutant	November			December			January			February		
		Indoor	Outdoor	I/O	Indoor	Outdoor	I/O	Indoor	Outdoor	I/O	Indoor	Outdoor	I/O
Densely populated	PM_{10}	118 ± 17	165 ± 23	0.72	102 ± 40	153 ± 17	0.67	109 ± 30	163 ± 13	0.67	81 ± 9	104 ± 84	0.78
Well planned		103 ± 51	312 ± 46	0.33	98 ± 27	247 ± 38	0.4	89 ± 13	156 ± 12	0.57	96 ± 36	129 ± 61	0.74
Roadside		108 ± 13	232 ± 54	0.46	107 ± 43	215 ± 28	0.50	102 ± 26	205 ± 17	0.5	114 ± 40	212 ± 36	0.54
Densely populated	$PM_{2.5}$	64 ± 0.4	94 ± 0.07	0.68	70 ± 1.2	82 ± 0.04	0.85	72 ± 0.03	83 ± 0.5	0.87	71 ± 2.1	87 ± 0.9	0.82
Well planned		66 ± 5.65	105 ± 7.1	0.63	69 ± 6.9	97 ± 1.1	0.71	53 ± 7.1	64 ± 2.6	0.83	80 ± 6.4	95 ± 2.1	0.84
Roadside		75 ± 11.2	81 ± 13.4	0.93	69 ± 12.1	81 ± 17.7	0.85	74 ± 18.7	94 ± 14.1	0.79	54 ± 9.4	89 ± 8.4	0.61

Table 2.10 Average mass concentration (μg/m^3) of metals in PM_{10} and $PM_{2.5}$ at well planned microenvironment

Months	Fe		Pb		Mn		Cu		Ni		Cr		Zn	
	PM_{10}	$PM_{2.5}$	PM_{10}	$PM_{2.5}$	PM_{10}	$PM_{2.5}$	PM_{10}	$PM_{2.5}$	PM_{10}	$PM_{2.5}$	PM_{10}	$PM_{2.5}$	PM_{10}	$PM_{2.5}$
January	8.33 ± 2.18	4.20 ± 1.79	0.02 ± 0.13	0.98 ± 0.16	0.01 ± 0.02	0.05 ± 0.03	0.10 ± 0.02	0.50 ± 0.22	0.02 ± 0.04	0.24 ± 0.29	0.03 ± 0.12	0.48 ± 0.63	0.21 ± 0.19	1.69 ± 0.21x
February	9.36 ± 3.52	5.73 ± 2.72	0.01 ± 0.17	0.73 ± 0.21	0.21 ± 0.24	0.06 ± 0.05	0.01 ± 0.03	0.23 ± 0.17	0.32 ± 0.26	0.36 ± 0.28	0.97 ± 0.54	0.72 ± 0.50	0.05 ± 0.08	2.45 ± 0.98
March	6.79 ± 1.51	3.68 ± 1.23	0.04 ± 0.02	0.80 ± 0.23	0.16 ± 0.23	0.07 ± 0.05	0.03 ± 0.03	0.16 ± 0.14	0.11 ± 0.15	0.38 ± 0.24	0.06 ± 0.17	0.50 ± 0.03	0.96 ± 0.54	1.13 ± 0.78
April	1.32 ± 0.21	4.11 ± 1.09	0.03 ± 0.11	0.62 ± 0.22	0.32 ± 0.24	0.06 ± 0.04	0.03 ± 0.02	0.24 ± 0.13	0.21 ± 0.24	0.43 ± 0.22	0.72 ± 0.53	0.62 ± 0.13	0.04 ± 0.03	1.39 ± 0.11
May	0.18 ± 0.03	1.33 ± 2.01	0.03 ± 0.04	0.82 ± 0.24	0.13 ± 0.05	0.06 ± 0.08	0.02 ± 0.02	0.12 ± 0.14	0.15 ± 0.16	0.34 ± 0.22	0.09 ± 0.11	0.12 ± 0.03	0.02 ± 0.52	2.45 ± 0.68
June	0.39 ± 0.06	1.32 ± .23	0.06 ± 0.14	0.79 ± 0.24	0.09 ± 0.34	0.03 ± 0.04	0.02 ± 0.05	0.25 ± 0.13	0.36 ± 0.28	0.31 ± 0.24	0.21 ± 0.23	0.18 ± 0.32	0.02 ± 0.03	1.43 ± 0.58
July	0.13 ± 0.17	1.45 ± 1.74	0.02 ± 0.01	0.93 ± 0.09	0.09 ± 0.22	0.03 ± 0.02	0.04 ± 0.02	0.12 ± 0.14	0.13 ± 0.12	0.35 ± 0.25	0.13 ± 0.13	0.42 ± 0.43	0.04 ± 0.52	1.39 ± 0.15

(continued)

Table 2.10 (continued)

Months	Fe		Pb		Mn		Cu		Ni		Cr		Zn	
	PM_{10}	$PM_{2.5}$	PM_{10}	$PM_{2.5}$	PM_{10}	$PM_{2.5}$	PM_{10}	$PM_{2.5}$	PM_{10}	$PM_{2.5}$	PM_{10}	$PM_{2.5}$	PM_{10}	$PM_{2.5}$
August	0.29 ± 0.08	3.82 ± 2.12	0.04 ± 0.15	0.35 ± 0.35	0.14 ± 0.23	0.02 ± 0.04	0.04 ± 0.02	0.22 ± 0.16	0.15 ± 0.26	0.43 ± 0.38	0.08 ± 0.24	0.35 ± 0.35	0.05 ± 0.08	2.35 ± 0.28
September	0.21 ± 0.03	3.48 ± 1.63	0.02 ± 0.05	0.82 ± 0.13	0.05 ± 0.45	0.02 ± 0.04	0.04 ± 0.01	0.14 ± 0.15	0.42 ± 0.14	0.28 ± 0.22	0.02 ± 0.17	0.25 ± 0.57	0.16 ± 0.34	1.33 ± 0.08
October	1.90 ± 0.45	1.78 ± 0.60	0.35 ± 0.48	0.75 ± 0.20	0.07 ± 0.02	0.07 ± 0.03	0.02 ± 0.08	0.23 ± 0.17	0.45 ± 0.33	0.43 ± 1.24	0.01 ± 0.01	0.09 ± 0.07	0.19 ± 0.03	2.10 ± 0.95
November	1.19 ± 0.78	1.05 ± 1.41	0.26 ± 0.06	0.91 ± 0.56	0.12 ± 0.15	0.04 ± 0.03	0.01 ± 0.10	0.39 ± 0.50	0.06 ± 0.14	0.09 ± 0.07	0.03 ± 0.02	1.16 ± 1.53	0.21 ± 0.22	0.90 ± 0.83
December	2.72 ± 0.21	2.76 ± 0.71	0.03 ± 0.07	0.89 ± 0.77	0.09 ± 0.07	0.08 ± 0.06	0.05 ± 0.07	0.06 ± 0.02	0.06 ± 0.21	0.68 ± 0.81	0.08 ± 0.14	1.35 ± 1.54	0.65 ± 0.22	0.42 ± 0.54

Table 2.11 Average mass concentration (μg/m^3) of metals in PM_{10} and $PM_{2.5}$ at densely populated microenvironment

Months	Fe		Pb		Mn		Cu		Ni		Cr		Zn	
	PM_{10}	$PM_{2.5}$	PM_{10}	$PM_{2.5}$	PM_{10}	$PM_{2.5}$	PM_{10}	$PM_{2.5}$	PM_{10}	$PM_{2.5}$	PM_{10}	$PM_{2.5}$	PM_{10}	$PM_{2.5}$
January	3.18 ± 2.531	3.19 ± 0.81	0.11 ± 0.06	0.51 ± 0.16	0.12 ± 0.26	0.08 ± 0.07	0.08 ± 0.09	0.14 ± 0.29	0.02 ± 0.05	0.01 ± 0.02	0.71 ± 0.81	1.83 ± 0.71	1.64 ± 1.53	0.59 ± 0.03
February	4.45 ± 2.61	4.65 ± 1.24	0.18 ± 0.07	0.89 ± 0.24	0.01 ± 0.06	0.02 ± 0.05	0.08 ± 0.06	0.50 ± 0.06	0.33 ± 0.17	0.05 ± 0.09	1.61 ± 0.98	1.53 ± 0.34	0.10 ± 0.05	0.66 ± 0.17
March	2.54 ± 1.73	1.63 ± 0.23	0.08 ± 0.09	0.76 ± 0.29	0.10 ± 0.09	0.20 ± 0.45	0.06 ± 0.09	0.48 ± 0.05	0.01 ± 0.02	0.21 ± 0.14	0.56 ± 0.33	0.54 ± 0.67	1.32 ± 1.48	1.74 ± 1.25
April	1.32 ± 0.11	2.1 ± 1.9	0.23 ± 0.11	0.42 ± 0.22	0.22 ± 0.14	0.02 ± 0.01	0.03 ± 0.72	0.41 ± 0.33	0.22 ± 0.28	0.33 ± 0.32	0.73 ± 0.17	0.45 ± 0.12	0.01 ± 0.03	1.31 ± 0.12
May	0.18 ± 0.08	1.43 ± 2.01	0.33 ± 0.54	0.62 ± 0.24	0.34 ± 0.35	0.01 ± 0.03	0.12 ± 0.12	0.22 ± 0.34	0.12 ± 0.19	0.44 ± 0.62	0.02 ± 0.28	0.21 ± 0.03	0.03 ± 0.51	2.1 ± 0.08
June	0.39 ± 0.01	1.22 ± .22	0.56 ± 0.64	0.79 ± 0.24	0.08 ± 0.36	0.04 ± 0.08	0.32 ± 0.005	0.05 ± 0.43	0.39 ± 0.09	0.25 ± 0.25	0.28 ± 0.24	0.21 ± 0.32	0.01 ± 0.01	0.33 ± 0.28
July	0.13 ± 0.27	1.15 ± 1.74	0.08 ± 0.01	0.63 ± 0.09	0.03 ± 0.21	0.02 ± 0.02	0.06 ± 0.01	0.12 ± 0.44	0.10 ± 0.12	0.33 ± 0.26	0.05 ± 0.13	0.34 ± 0.17	0.04 ± 0.56	1.27 ± 0.12

(continued)

Table 2.11 (continued)

Months	Fe		Pb		Mn		Cu		Ni		Cr		Zn	
	PM_{10}	$PM_{2.5}$	PM_{10}	$PM_{2.5}$	PM_{10}	$PM_{2.5}$	PM_{10}	$PM_{2.5}$	PM_{10}	$PM_{2.5}$	PM_{10}	$PM_{2.5}$	PM_{10}	$PM_{2.5}$
August	0.29 ± 0.08	3.32 ± 2.12	0.08 ± 0.1	0.25 ± 0.35	0.11 ± 0.22	0.02 ± 0.04	0.01 ± 0.01	0.32 ± 0.16	0.25 ± 0.26	0.06 ± 0.38	0.05 ± 0.22	0.68 ± 0.41	0.05 ± 0.46	2.13 ± 0.18
September	0.21 ± 0.03	3.18 ± 1.33	0.01 ± 0.04	0.42 ± 0.13	0.02 ± 0.41	0.01 ± 0.03	0.04 ± 0.03	0.34 ± 0.15	0.32 ± 0.12	0.06 ± 0.11	0.02 ± 0.09	0.16 ± 0.78	0.14 ± 0.24	1.13 ± 0.009
October	1.95 ± 1.66	0.39 ± 0.86	0.11 ± 0.05	0.88 ± 0.79	0.04 ± 0.05	0.03 ± 0.03	0.06 ± 0.02	0.04 ± 0.28	0.13 ± 0.16	0.33 ± 0.56	0.85 ± 0.54	2.59 ± 1.42	0.59 ± 0.72	1.69 ± 0.63
November	2.77 ± 0.91	1.84 ± 0.77	0.15 ± 0.06	0.31 ± 0.05	0.07 ± 0.04	0.05 ± 0.09	0.02 ± 0.03	0.05 ± 0.60	0.09 ± 0.06	0.72 ± 0.24	0.53 ± 0.47	2.07 ± 0.87	0.08 ± 0.15	0.34 ± 0.12
December	2.73 ± 1.55	2.24 ± 0.67	0.21 ± 0.55	0.32 ± 0.08	0.10 ± 0.39	0.03 ± 0.21	0.03 ± 0.12	0.66 ± 0.16	0.13 ± 0.18	0.43 ± 0.52	0.74 ± 0.68	1.01 ± 0.92	0.11 ± 0.01	0.73 ± 0.63

Table 2.12 Average mass concentration (μg/m^3) of heavy metals in PM_{10} and $PM_{2.5}$ at roadside microenvironment

Months	Fe		Pb		Mn		Cu		Ni		Cr		Zn	
	PM_{10}	$PM_{2.5}$	PM_{10}	$PM_{2.5}$	PM_{10}	$PM_{2.5}$	PM_{10}	$PM_{2.5}$	PM_{10}	$PM_{2.5}$	PM_{10}	$PM_{2.5}$	PM_{10}	$PM_{2.5}$
January	2.67 ± 0.21	2.24 ± 0.28	2.04 ± 1.03	1.11 ± 1.23	0.12 ± 0.23	0.05 ± 0.3	0.10 ± 0.16	0.48 ± 0.68	0.03 ± 0.01	0.65 ± 0.56	0.11 ± 0.06	1.13 ± 0.97	0.74 ± 0.53	1.58 ± 0.63
February	1.76 ± 0.32	0.87 ± 0.51	1.24 ± 1.56	1.07 ± 0.25	0.08 ± 0.05	0.05 ± 0.23	0.07 ± 0.04	0.39 ± 0.03	0.02 ± 0.02	0.56 ± 0.36	0.08 ± 0.04	0.67 ± 0.07	0.51 ± 0.69	0.75 ± 0.01
March	1.21 ± 0.23	0.87 ± 0.23	1.36 ± 0.33	1.22 ± 1.13	0.11 ± 0.09	0.02 ± 0.02	0.09 ± 0.16	0.09 ± 0.16	0.06 ± 0.09	0.59 ± 0.23	0.11 ± 0.06	0.49 ± 0.36	0.08 ± 0.6	0.44 ± 0.32
April	1.32 ± 0.11	0.52 ± 0.9	1.24 ± 0.19	1.32 ± 0.16	0.31 ± 0.10	0.03 ± 0.12	0.13 ± 0.32	0.44 ± 0.33	0.23 ± 0.45	0.32 ± 0.25	0.22 ± 0.13	0.42 ± 0.53	0.43 ± 0.03	1.29 ± 0.21
May	0.18 ± 0.08	0.18 ± 0.14	1.01 ± 0.26	1.09 ± 0.16	0.13 ± 0.02	0.04 ± 0.12	0.45 ± 0.03	0.22 ± 0.44	0.05 ± 0.16	0.09 ± 0.25	0.09 ± 0.81	0.32 ± 0.43	0.12 ± 0.12	1.31 ± 0.68
June	0.39 ± 0.01	0.82 ± 0.16	0.91 ± 0.31	0.86 ± 0.21	0.13 ± 0.09	0.08 ± 0.21	0.24 ± 0.24	0.23 ± 0.24	0.34 ± 0.23	0.04 ± 0.24	0.11 ± 0.03	0.28 ± 0.12	0.32 ± 0.03	2.03 ± 0.13
July	0.13 ± 0.27	0.1 ± 0.06	1.32 ± 1.4	0.91 ± 0.93	0.03 ± 0.11	0.03 ± 0.32	0.24 ± 0.24	0.24 ± 0.56	0.09 ± 0.12	0.35 ± 0.25	0.03 ± 0.12	0.12 ± 0.33	0.03 ± 0.51	1.41 ± 0.11

(continued)

Table 2.12 (continued)

Months	Fe		Pb		Mn		Cu		Ni		Cr		Zn	
	PM_{10}	$PM_{2.5}$	PM_{10}	$PM_{2.5}$	PM_{10}	$PM_{2.5}$	PM_{10}	$PM_{2.5}$	PM_{10}	$PM_{2.5}$	PM_{10}	$PM_{2.5}$	PM_{10}	$PM_{2.5}$
August	0.29 ± 0.08	0.42 ± 0.19	0.98 ± 1.23	0.83 ± 0.19	0.05 ± 0.02	0.38 ± 0.13	0.12 ± 0.32	0.22 ± 0.15	0.06 ± 0.24	0.43 ± 0.38	0.05 ± 0.21	0.55 ± 0.55	0.03 ± 0.01	2.32 ± 0.21
September	0.21 ± 0.03	0.14 ± 0.32	0.97 ± 0.94	0.95 ± 0.11	0.06 ± 0.12	0.08 ± 0.06	0.24 ± 0.41	0.15 ± 0.12	0.04 ± 0.12	0.08 ± 0.42	0.01 ± 0.17	0.25 ± 0.97	0.36 ± 0.44	1.05 ± 0.18
October	0.14 ± 0.02	0.12 ± 0.07	1.45 ± 0.67	0.67 ± 0.34	0.01 ± 0.04	0.09 ± 0.05	0.08 ± 0.03	0.74 ± 0.64	0.05 ± 0.04	0.08 ± 0.05	0.10 ± 0.01	0.31 ± 0.08	0.07 ± 0.13	0.73 ± 0.40
November	1.95 ± 0.51	1.87 ± 0.91	1.98 ± 0.36	0.68 ± 1.24	0.09 ± 0.31	0.04 ± 0.23	0.07 ± 0.18	0.24 ± 0.09	0.02 ± 0.08	0.11 ± 0.25	0.08 ± 0.04	0.55 ± 0.04	0.53 ± 0.32	0.61 ± 0.07
December	7.42 ± 1.6	2.67 ± 0.3	2.18 ± 0.76	1.87 ± 1.3	0.26 ± 0.13	0.12 ± 0.37	0.02 ± 0.05	0.34 ± 0.40	0.06 ± 0.04	0.61 ± 0.41	0.08 ± 0.33	0.87 ± 0.85	1.24 ± 0.64	2.27 ± 0.77

excessive concentration in food and feed plants are of great concern because of their toxicity to humans and animals (Kulshrestha et al. 2014).

2.6.2 Correlation Among Heavy Metals

Univariate Pearson correlations were determined using SPSS 11 (Tables 2.13, 2.14 and 2.15) to find association among heavy metals so as to identify the common sources of these metals. The positive correlation have been shown by * sign.

Correlation of Metals in PM_{10}

In well planned microenvironment, Fe, Zn and Mn showed good correlation with each other probably originating from common sources including the outdoor dust. Some other known outdoor sources of Zn include lubricating oils and tyres of automobiles while re-suspended road dust is a major source of Fe and Mn. Pb, Cu, Ni and Cr depicted correlation with each other, which may be due to common indoor sources like paint and household dust. In densely populated microenvironment Ni, Cr and Fe showed good correlation. Common sources of Ni and Cr in these houses may be the combustion of polluting fuels such as kerosene and diesel various proposes (Joshi et al. 2017).

Correlation of Metals in $PM_{2.5}$

Pb, Ni, Cr and Mn showed positive correlation in roadside microenvironment which can be attributed to vehicular emission along with re-suspension of soil dust. Fe, Cu, Ni, Cr and Pb showed positive correlation owing to construction work, as Fe is used for pillar framing and the roofing. Cement and paint are known sources of Pb. In another study on similar pattern (Mahish et al. 2015) it was found that concentration of Pb and other trace elements was several times higher in PM_{10} sample near a cement plant. Cr showed a good correlation with Ni (0.95) attributed to the combustion of coal and other inferior fuels. Zn showed correlation with Mn and Pb in PM_{10} mainly due to vehicular exhaust and its infiltration. Pb, Mn, Ni and Cr showed good correlation with $PM_{2.5}$ whereas Fe, Mn and Zn showed good correlation with PM_{10} at roadside microenvironment. House dust originated from outdoor or indoor sources was identified as a potent source of particulate concentration. Activities like wear and tear of tyres, burning of oil, abrasion of mechanical parts and oil lubricants etc. also contribute to the release of metals. Pb, Cu, Ni and Cr showed good inter-correlation indicating their mutual originating sources.

Table 2.13 Inter correlation between metals at well planned microenvironment

Metals	Fe	Pb	Mn	Cu	Ni	Cr	Zn	PM_{10}	$PM_{2.5}$
Fe	1								
Pb	−0.68	1							
Mn	0.99*	−0.69	1						
Cu	−0.85	0.80*	−0.79	1					
Ni	−0.98	0.80*	−0.96	0.92*	1				
Cr	−0.5	0.97*	−0.5	0.71*	0.66*	1			
Zn	0.97*	−0.68	0.99*	−0.72	−0.93	−0.49	1		
PM_{10}	0.74*	−0.46	0.80*	−0.45	−0.69	−0.32	0.82*	1	
$PM_{2.5}$	0.59*	0.77*	0.66*	0.18	0.86*	0.75*	0.09	0.69*	1

The sign * represents values which are positive.

Table 2.14 Inter correlation between metals at densely populated microenvironment

Metals	Fe	Pb	Mn	Cu	Ni	Cr	Zn	PM_{10}	$PM_{2.5}$
Fe	1								
Pb	−0.46	1							
Mn	0.3	−0.32	1						
Cu	0.43	0.54*	−0.59	1					
Ni	−0.11	0.77*	−0.01	−0.54	1				
Cr	0.64*	0.53*	0.71*	−0.29	0.24	1			
Zn	0.03	−0.16	0.13	0.17	−0.36	−0.4	1		
PM_{10}	0.51*	−0.22	0.52*	0.83*	−0.49	−0.1	0.54*	1	
$PM_{2.5}$	−0.29	0.07	−0.22	0.29	−0.25	0.16	0.12	0.09	1

Table 2.15 Inter correlation between metals at roadside microenvironment

Metals	Fe	Pb	Mn	Cu	Ni	Cr	Zn	PM_{10}	$PM_{2.5}$
Fe	1								
Pb	0.51*	1							
Mn	0.59*	−0.32	1						
Cu	0.74*	−0.26	−0.2	1					
Ni	0.79*	0.67*	−0.88	0.24	1				
Cr	0.94*	0.51*	−0.82	0.54*	0.95*	1			
Zn	−0.23	−0.79	0.69*	0.45	−0.76	−0.51	1		
PM_{10}	0.57*	0.62*	0.23	0.22	0.52*	0.59*	−0.05	1	
$PM_{2.5}$	−0.12	0.59*	0.51*	0.33	−0.24	−0.24	0.83*	0.56*	1

2.6.3 *Principal Component Analysis for Heavy Metals Associated with PM_{10} and $PM_{2.5}$ and Identification of Their Sources*

PCA was used to identify the key sources of the heavy metals associated with the particulate concentrations. Varimax rotated factor analysis was done to identify three chief sources of the heavy metals associated with the particulate matter. Only those factors were included in the matrix whose Eigen value was > than one ranging from 3.11 to 1.18 in the three microenvironments. Mass distribution of the metals was considered for Factor Analysis.

Factor Analysis in Well Planned Microenvironment

In PM_{10} samples Cu and Cr with 34% variance represented Factor 1 and road dust was identified as their main source. Fe, Mn, Ni and Cr represented Factor 2 with 31% variance which was linked to solid waste dumping and burning (Kulshrestha et al. 2014). Zn and Pb with 27% variance represented Factor 3 and household dust was their main source. Household paint was identified as a chief source of Pb (Wang et al. 2000). These three factors attributed to 92% of the sources responsible for heavy metal contamination as represented in Fig. 2.4a. In $PM_{2.5}$ samples Cu represented Factor 1 with 19% variance and road dust was attributed as its major source. Fe, Mn and Cr with 36% variance represented Factor 2 due to solid waste incineration (Zhang et al. 2008). Pb and Ni represented Factor 3 with 32% variance and household dust and some indoor activities were attributed as their major sources. The results obtained from the PCA analysis were well justified by inter-correlation of metals (Fig. 2.4b).

Factor Analysis in Densely Populated Microenvironment

Ni and Cr with 42% variance represented Factor 1 owing to fuel combustion (Massey et al. 2013). Fe and Cu with 29% variance represented Factor 2. Construction activities and road dust were identified as their major sources. Pb and Zn with 25% variance

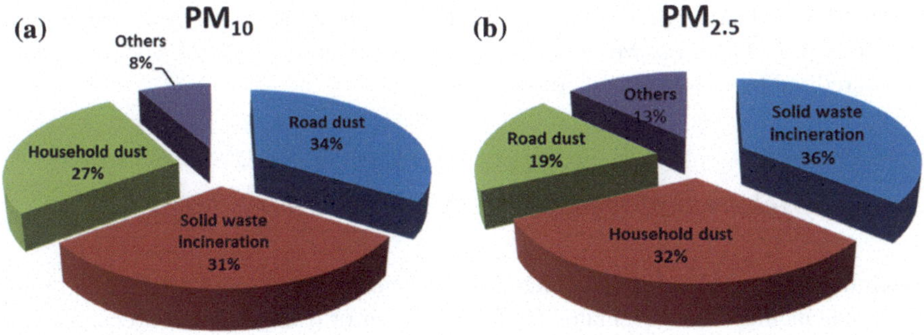

Fig. 2.4 **a–b** Pie diagram showing source distribution of particle constituent at well planned microenvironment in **a** PM_{10}, **b** $PM_{2.5}$

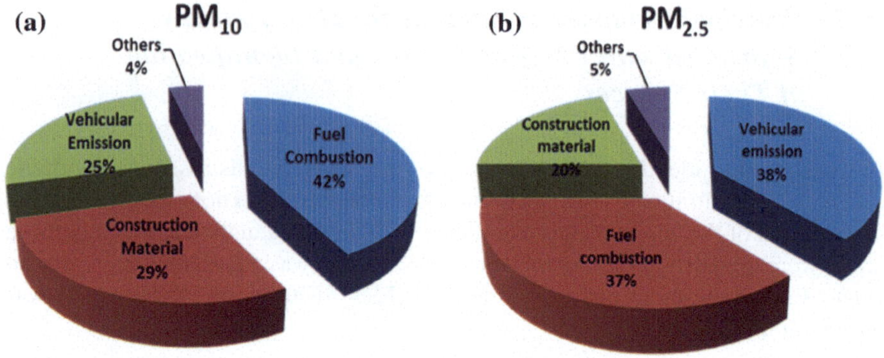

Fig. 2.5 **a–b** Pie diagram showing source distribution of particle constituent at densely populated microenvironment in **a** PM_{10}, **b** $PM_{2.5}$

represented Factor 3 attributing to vehicular exhaust and its infiltration (Park and Dam 2010). These three factors contributed to 96% of source contribution of metals in PM_{10} (Fig. 2.5a). In $PM_{2.5}$ samples three main factors explained 80% of the source contribution of heavy metals. Pb with 38% variance represented Factor 1 and was attributed to vehicular exhaust. Fe, Ni and Cr with 37% variance represented Factor 2 owing to combustion. Fe, Cu and Cr with 5% variance represented Factor3 being associated with construction activities (Fig. 2.5b).

Factor Analysis in Roadside Microenvironment
Two main factors analyzed in this microenvironment contributed to 93% of heavy metal concentration in PM_{10} samples (Fig. 2.6a). Fe and Zn with 58% variance represented Factor 1 attributed to infiltration of outdoor dust. Pb, Mn, Cu, Ni, and Cr represented Factor 2 with 35% variance owing to household dust. Cr is also found to be emitted from domestic heating (Homa et al. 2016) while Mn and Pb contamination can be sourced from smoking and cooking. Zn is usually emitted from abrasion of tyres while resuspended road dust is a potential source of Fe. In $PM_{2.5}$ samples the two main factors explained 79% of source contribution (Fig. 2.6b). Pb, Cu, Ni and Cr with 58% variance represented Factor 1 and household dust was identified as their main source. In PM_{10} Fe, Mn and Zn represented Factor 1 with 21% variance and was accredited to seeped outside dust.

2.6.4 Enrichment of Heavy Metals from Indoor Sources

Enrichment factor (EF) was calculated for the heavy metals to know their origin. Indoor metals may have originated from outdoor air or from indoor sources. A comparison between the ambient metallic concentrations with that of indoors, is of utmost significance which can be performed by calculating enrichment ratios for different

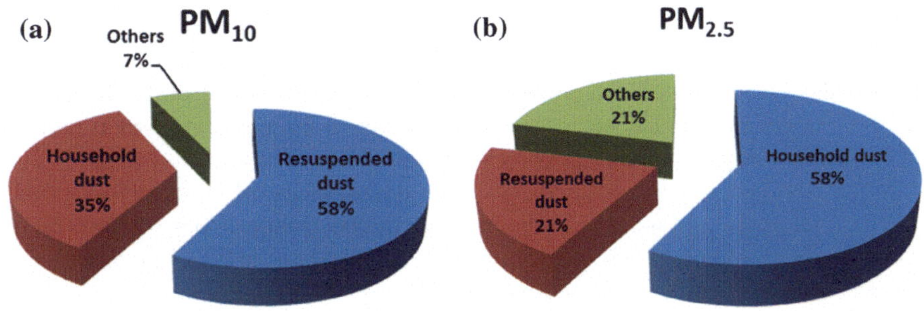

Fig. 2.6 **a–b** Pie diagram showing source distribution of particle constituent at Roadside microenvironment in **a** PM_{10}, **b** $PM_{2.5}$

elements taking into account, both indoor and outdoor concentrations. The concentrations are usually normalized to an element considered as being the most ambiguous indicator of source material. The enrichment factor was calculated by estimating indoor outdoor ratios of different metals which are usually normalized (Rahn 1976). In this study particulate concentration was taken as reference on an assumption that all the particulate matter in the air was of outdoor origin. The enrichment factor can be calculated as:

$$EF_{INDOOR} = \{X_{INDOOR}/PM_{INDOOR}\}/\{X_{OUTDOOR}/PM_{OUTDOOR}\} \quad (2.3)$$

Where, X is the concentration of the metal under investigation. Elements with EFs close to unity are likely to have originated from outdoor sources and are considered to be non-enriched. Elements with enrichment factors significantly greater than one are considered as enriched and likely have their origin other than outdoor air.

Well planned microenvironment—At well planned microenvironment only Ni was found to be enriched in PM_{10} samples with enrichment factor (EF) 2.59. While in $PM_{2.5}$ samples Pb, Ni and Zn were found to have enrichment factor greater than one i.e. 1.14, 1.10 and 1.21 respectively and therefore have some indoors sources as well. Presence of Pb may be due to wall dust indoors at these sites which contains paint. Ni and Zn are found to have additional sources indoors at all sites. This might be due to the use of cosmetics, household utensils and burning cigarettes (Kulshrestha et al. 2014).

Densely populated microenvironment—At the densely populated microenvironment no metal in PM_{10} sample showed EF < 1 indicating their origin from outdoor sources. In $PM_{2.5}$ samples Cu, Ni, Cr and Zn were found to have EFs greater than one viz. 4.39, 3.26, 1.21 and 1.43 respectively depicting their origin within the household. Pb associated with both sizes of particulate matter was likely to be emitted from outdoor sources. Pb was enriched at the roadside and well planned microenvironments whose major source was wall dust containing paint. Pigments in paints include oxides of Fe, Mn, Cr and Sb along with lead antiminate, cadmium selenide, lead, zinc and barium chromates. Pb is also used as a corrosion inhibitor in "lead-free" paints. Ni and Zn were emitted from indoor sources. Their probable sources are cosmetics, household utensils and cigarette smoke. High levels of indoor Zn have also been reported in a previous study (Feng and Barratt 1993). Goyal et al. (2012) Cr concentration has been associated with Ni and Zn indicating their mutual indoor source (Goyal et al. 2012). Cr, enriched in $PM_{2.5}$ samples, was found to have additional sources which were believed to be due to usage of kerosene/oil. In the other two microenvironments, Cr was mainly emitted from outdoor sources. Fe in all the microenvironments associated with PM_{10} and $PM_{2.5}$ had EF less than one and believed to be originated from outdoor sources.

Roadside microenvironment—At roadside microenvironment in PM_{10} samples Pb, Cu and Zn were found to be enriched with EFs 1.60, 1.70 and 1.12 respectively, while in $PM_{2.5}$ samples Pb, Mn, Cu, Ni and Zn were enriched having EFs 1.27, 2.79, 1.06, 1.28, and 3.69 respectively and hence it was concluded that the metals originated from some indoors sources in addition to outdoor infiltration.

2.7 Recommendations

Air pollution is a major environmental risk to human beings. By adopting suitable measures, nation can decrease the risk of stroke, heart disease, lung cancer, and both chronic and acute respiratory diseases which are associated with air pollution. Policies and investments supporting cleaner transport, energy-efficient homes, industry and improved l waste management can play key roles in curbing ambient pollution which directly affects the level of indoor pollution. Outdoor pollution control is possible through a concerted action taken by the public, policy makers working in sectors like transport, energy, waste management and urban planning and academicians. A few instances adopted in transport, urban planning, power generation and industry that reduce air pollution:

1. Facilitation of clean energy sources at affordable prices for cooking, heating and lighting.
2. To improve indoor air quality focus should be on increasing energy efficiency of buildings.
3. Use of low-emission fuels and renewable power means co-generation of heat and power and distributed energy production like solar panels should be promoted.

Indoor air pollution is a direct consequence of ambient pollution levels. The problem can be dealt with by a synchronized effort made by the government, industries, research and academia. The government should carve stringent policies and work in close coordination with various agencies like CPCB to chalk down a robust plan leading to the development of indoor air quality policies. Creating mass awareness and sensitizing both rural and urban areas of the country and it should be taken up as a national campaign. The focal theme of the awareness can be centred on the use of better quality fuels which are affordable at the same time. Promotions of clean energy options like solar energy lit lamps than the indigenously built kerosene lamps

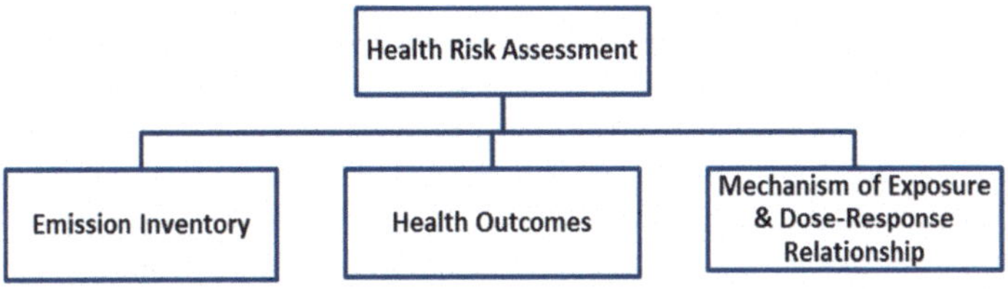

Fig. 2.7 Summary of an efficient health risk assessment study

should be done in masses. Usage of biogas and improved cooking stoves should be made available for users at a nominal price. Government should consult with stakeholders including environment scientists, engineers and public health experts to set guidelines and standards. The role of construction industries is also crucial for adopting measures to check indoor air pollution from the word go. Better building material like low VOC emitting materials, radon-resistant building practices can lessen the risks of lung cancer. Finally identification of the gaps in literature and its thorough assessment can be taken up by researchers and academicians. Long term exposure assessment studies and their health consequences should be worked upon to come up with concrete conclusions about the extent of damage done by the poor quality of indoor air to the human health. Research programmes highlighting gaps in research and adoption of novel measures to address the challenge are needed to be studied. Collaborations should be made with international agencies where IAQ guidelines are already in exercise. Researchers may help government in creating mass public awareness operations to apprise public about the indoor air quality allied issues and appropriate actions to control them (Goyal et al. 2012) (Fig. 2.7).

2.7.1 Modeling of Indoor Air Pollution

Air pollution modeling is used to assess the precision to which air pollutants can be measured in accurate quantities which is required for the accurate prediction of the level of pollution. An indoor air quality prediction model generally relates indoor pollutant concentration with factors like ventilation sources, source origin of the pollutants, various geometric parameters and sink. Through pollution modeling, accurate and desired function of concentration, such as peak concentration for a particular set of condition can be predicted through mathematical simulation rather than prediction by manual monitoring. Indoor air quality models are helpful in prediction of pollutants present indoors and their association with ambient pollutant concentrations, indoor-outdoor air-exchange rates, and sources of indoor pollutants. Pollutants may trespass indoors through infiltration or ventilation. Whereas indoor pollutants arise from point or diffuse sources. The pollutants may be transported and dispersed though a wide region. The first step in indoor modeling involves establishing a statement of the mass balance concerning the pollutant under consideration.

When air passes through a filter at a rate q_0 through a structure of volume V, some of the air is recirculated through another filter at a rate q_1, and air infiltrates the structure at a rate q_2. Each filter is characterized by a factor $F \equiv (C_{\text{inlet}} - C_{\text{outlet}})/C_{\text{inlet}}$. The pollutant level is anticipated to be constant all over. The indoor and outdoor pollutant concentrations at time t are C and C_0. The rate at which the pollutant is added to the indoor air is S. The rate at which the pollutant is removed from the air owing to internal sinks is R. In this case, a suitable initial equation is:

$$V\frac{dC}{dt} = q_0C_0(1 - F_0) + q_1C(1 - F_1) + qC_0 - (Q_0 + q_1 + q_2)C + S - R \tag{2.4}$$

The decay rate is a function of C, however, in modeling indoor air quality, the sink rate is often assumed to be constant. Accurate values of the ventilation parameters are usually more difficult to determine. Without forced ventilation, air exchange is primarily due to infiltration; in most forced-ventilation systems (which are balanced so that negative pressures are not created inside the building), the rate of infiltration is negligible, as compared to the forced-ventilation rate. In view of the uncertainties associated with many of the parameter values and the difficulty of doing otherwise, compartments have been widely used in modeling indoor air quality. A compartment is a region within which spatial variations in pollutant concentrations can be neglected. The main motive of indoor air quality modeling is to study the association between the indoor and outdoor pollutant concentration and the effect of geometric and ventilation characteristics of a building. To have an overall picture of the indoor environmental conditions, an efficient study must include the corresponding outside concentrations, geometric parameters, ventilation parameters sink and strength of the source. These factors are vital for a mass balance based model.

2.8 Conclusion

Air in urban cities has become highly polluted with pollutants crossing the standards set by the World Health Organization (WHO). Among all the air pollutants, particulate matter can cause significant health damage, specifically particles that are smaller than 10 μm. The present study conducted in three urban microenvironments of Lucknow city concluded that indoor air quality was worse in roadside houses having highest particulate concentrations. At all the three location Fe was found to be dominating element followed by Zn and Pb. Contaminated household dust was identified as one of the main sources of toxic heavy metals. From the PCA it was fond that vehicular emission, re-suspended dust and solid waste incineration were some of the common sources contributing the particulate matter. Crude fuel combustion was identified as another important factor. The metals were found to be enriched in both sizes of particulate matter. There is a need to assess the heavy metal contamination in different environmental components. Studies have reported the origin of heavy metals through source apportionment of particulate matter, however nothing substantial has been proposed that enables optimal control and regulation of emission of these heavy metals in the atmosphere to reduce impact on health. Economic factors and cost effective methods for monitor and control of these heavy metals need consideration.

Acknowledgements The authors greatly acknowledge Dr. (Mrs.) V. Prakash, Principal, Isabella Thoburn College and Prof. A.R. Khan, Head, Department of Chemistry, Integral University, Lucknow for their support.

References

Adams K, Greenbaum DS, Shaikh R, van Erp AM, Russell AG (2015) Particulate matter components, sources, and health: systematic approaches to testing effects. J Air Waste Manage Ass 65:544–558

Air Pollution in Delhi: An Analysis, ENVIS Centre on Control of Pollution (Water, Air, & Noise), 2016

Ajmani GS, Suh HH, Pinto JM (2016) Effects of ambient air pollution exposure on olfaction: a review. Environ Health Pers 124:1683–1693

Albalak R, Frisancho AR, Keeler GJ (1999) Domestic biomass fuel combustion and chronic bronchitis in two rural Bolivian villages. Thorax 54:1004–1008

Ambient (outdoor) air quality and health WHO, 2nd May 2018

Anderson JO, Thundiyil JG, Stolbach A (2012) Clearing the air: a review of the effects of particulate matter air pollution on human health. J Med Toxicol 8(2):166–175

Assem FL, Levy LS (2009) A review of current toxicological concerns on vanadium pentoxide and other vanadium compounds: gaps in knowledge and directions for future research. J Toxicol Environ Health B Crit Rev 12(4):289–306

Assessment of ambient Air Quality of Lucknow City, Post monsoon 2017. CSIR-IITR, Lucknow

Atkinson RW, Anderson HR, Sunyer J, Ayres J, Baccini M, Vonk JM et al (2001) Acute effects of particulate air pollution on respiratory admissions: results from APHEA 2 project. Air pollution and health: a European approach. Am J Respir Crit Care Med 164:1860–1866

Avila DS, Puntel RL, Aschner M (2013) Manganese in health and disease. Metal Ions Life Sci 13:199–227

Awasthi S, Glick HA, Fletcher RH (1996) Effect of cooking fuels on respiratory diseases in preschool children in Lucknow, India. Am J Trop Med Hyg 55:48–51

Balakrishnan K, Sankar S, Parikh J, Padmavathi R, Srividya K et al (2002) Daily average exposures to respirable particulates matter from combustion of biomass fuels in rural households of Southern India. Environ Health Perspect 110:1069–1075

Balakrishnan K, Ramaswamy P, Sambandam S, Thangavel G, Ghosh S, Johnson P et al (2011) Air pollution from household solid fuel combustion in India: an overview of exposure and health related information to inform health research priorities. Glob Health Action 4. https://doi.org/10.3402/gha.v4i0.5638

Balakrishnan K, Cohen A, Smith KR (2014) Addressing the burden of disease attributable to air pollution in India: the need to integrate across household and ambient air pollution exposures. Environ Health Perspect 122(1):A6–A7

Baranski B, Sitarek K (1987) Effect of oral and inhalation exposure to cadmium on the oestrous cycle in rats. Toxicol Lett 36(3):267–273

Barceloux DG (1999) Vanadium. J Toxicol Clin Toxicol 37(2):265–278

Basu R, Harris M, Sie L, Malig B, Broadwin R, Green R (2014) Effects of fine particulate matter and its constituents on low birth weight among full-term infants in California. Environ Res 128:42–51

Behera D, Dash S, Malik SK (1988) Blood carboxyhaemoglobin levels following acute exposure to smoke of biomass fuel. Indian J Med Res 88:522–524

Bhardawaj A, Tyagi R, Sharma BK et al (2013) A review of biofuel policy in India: current status and perspectives. Int J Appl Eng Res 8:1907–1912

Bhardawaj A, Habib G, Padhi et al (2016) Deteriorating air quality and increased health risks in Delhi: the decisions being delayed. IIOAB J 7:10–15

Block ML, Calderón-Garcidueñas L (2009) Air pollution: mechanisms of neuroinflammation and CNS disease. Trends Neurosci 32:506–516

Bollati V, Marinell B, Apostoli P, Bonzini M, Nordio F, Hoxha M, Pegoraro V, Motta V, Tarantini L, Cantone L, Schwartz J, Bertazzi PA, Baccarelli A (2010) Exposure to metal-rich particulate matter modifies the expression of candidate microRNAs in peripheral blood leukocytes. Environ Health Perspect 118(6):763–768. https://doi.org/10.1289/ehp.0901300

Braidy N, Poljak A, Marjo C, Rutlidge H, Rich A, Jayasena T et al (2014) Metal and complementary molecular bioimaging in Alzheimer's disease. Front Aging Neurosci 6:138. https://doi.org/10.3389/fnagi.2014.00138

Bruce N, Perez-Padilla R, Albalak R (2000) Indoor air pollution in developing countries: a major environmental and public health challenge. Bull World Health Organ 78:1078–1092

Cakmak S, Dales R, Kauri LM, Mahmud M, Van Ryswyk K, Vanos J et al (2014) Metal composition of fine particulate air pollution and acute changes in cardiorespiratory physiology. Environ Pollut 189:208–214

Caserta D, Graziano A, Lo Monte G, Bordi G, Moscarini M (2013) Heavy metals and placental fetal-maternal barrier: a mini-review on the major concerns. Eur Rev Med Pharmacol Sci 17(16):2198–2206

Census of India (2011) Census of India, Government of India, Office of the Registrar General and Census Commissioner, New Delhi

Cervantes-Yépez S, López-Zepeda LS, Fortoul TI (2018) Vanadium inhalation induces retinal Müller glial cell (MGC) alterations in a murine model. Cut Ocu Toxocol 37(2). https://doi.org/10.1080/15569527.2017.1392560

Chafe ZA, Brauer M, Klimont Z, Van Dingenen R, Mehta S, Rao S, Riahi K, Dentener F, Smith KR (2014) Household cooking with solid fuels contributes to ambient $PM_{2.5}$ air pollution and the burden of disease. Environ Health Perspect 122(12):1314–1320. http://doi.org/10.1289/ehp.1206340

Chang JW, Chen HL, Su HJ, Liao PC, Guo HR, Lee CC (2011) Simultaneous exposure of non-diabetics to high levels of dioxins and mercury increases their risk of insulin resistance. J Hazard Mater 185(2–3):749–755

Chorvatovicova D, Kovacikova Z (1992) Inhalation exposure of rats to metal aerosol. II. Study of mutagenic effect on alveolar macrophages. J Appl Toxicol 12(1):67–78

De Rosis F, Anastasio SP, Selvaggi L, Beltrame A, Moriani G (1985) Female reproductive health in two lamp factories: effects of exposure to inorganic mercury vapour and stress factors. Br J Ind Med 42(7):488–494

Dreher KL (2000) Particulate matter physicochemistry and toxicology. In search of causality—a critical perspective. Inhal Toxicol 12:45–57

Duffus JH (2002) Heavy metals: a meaningless term? Pure Appl Chem 74(5):793–807

Enamorado-Báez SM, Gómez-Guzmán JM, Chamizo E, Abril JM (2015) Levels of 25 trace elements in high-volume air filter samples from Seville (2001–2002): sources, enrichment factors and temporal variations. Atmos Res 155:118–129

Erisman JW, van Elzakker BG, Mennen MG, Hogenkamp J, Zwart E, van den Beld et al (1994) The Elspeetsche Veld experiment on surface exchange of trace gases: summary of results. Atmos Environ 28(3):487–496

Ezz WN, Mazaheri M, Robinson P et al (2015) Ultrafine Particles from Traffic Emissions and Children's Health(UPTECH) in Brisbane, Queensland (Australia): study design and implementation. Int J Environ Res Public Health 12:1687–1702

Feng Y, Barratt R (1993) An assessment of data of trace elements in indoor and outdoor dusts. Int J Environ Health Res 3:18–31

Fernandez-Real JM, Lopez-Bermejo A, Ricart W (2002) Cross-talk between iron metabolism and diabetes. Diabetes 51(8):2348–2354

Ferrannini E (2000) Insulin resistance, iron, and the liver. Lancet 355(9222):2181–2182

Fortoul TI, Salgado RC, Moncada SG, Sanchez IG, Lopez IE, Espejel G, Calderon NL, Saldivar L (1999) Ultrastructural findings in the murine nonciliated bronchiolar cells (NCBC) after subacute inhalation of lead acetate. Acta Vet Brno 68:51–55

Fortoul TI, Lara VR, Gonzalez-Villalva A, Rojas-Lemus M, Colin-Barenque, Bizzaro-Nevares P (2015) Health effects of metals in particulate matter. http://doi.org/10.5772/59749

Fowler BA (2009) Monitoring of human populations for early markers of cadmium toxicity: a review. Toxicol Appl Pharmacol 238(3):294–300

Fusco D, Forastiere F, Michelozzi P, Spadea T, Ostro B, Arca M et al (2001) Air pollution and hospital admissions for respiratory conditions in Rome, Italy. Eur Respir J 17:1143–1150

Galanis A, Karapetsas A, Sandaltzopoulos R (2009) Metal-induced carcinogenesis, oxidative stress and hypoxia signalling. Mutat Res 674(1–2):31–35

Garaga R, Sahu SK, Kota SH (2018) A review of air quality modeling studies in India: local and regional scale. Curr Pollut Rep 4(2):59–73. http://doi.org/10.1007/s40726-018-0081-0

Garcia-Leston J, Mendez J, Pasaro E, Laffon B (2010) Genotoxic effects of lead: an updated review. Environ Int 36(6):623–636

Genc S, Zadeoglulari Z, Fuss SH et al (2012) The adverse effects of air pollution on the nervous system. J Toxicol 2012:1–23

Gerber GB, Leonard A, Hantson P (2002) Carcinogenicity, mutagenicity and teratogenicity of manganese compounds. Crit Rev Oncol Hematol 42(1):25–34

Global Burden of Disease, 2016

Gordona T, Balakrishnanb K, Deyc S, Rajagopaland S, Thornburge J, Thurstona G, Agrawal A, Collmang G, Guleriah R, Limayei S, Salvii S, Kilaruj V, Nadadurg S (2018) Air pollution health research priorities for India: Perspectives of the Indo-U.S. Communities of Researchers. Environ Int 119:100–108

Goyal R, Khare M, Kumar P (2012a) Indoor air quality: current status, missing links and future road map for India. J Civil Environ Eng 2:4. https://doi.org/10.4172/2165-784X.1000118

Goyal R, Khare M, Kumar P (2012b) Indoor air quality: current status, missing links and future road map for India. J Civil Environ Eng 2:118. https://doi.org/10.4172/2165-784X.1000118

Gramotnev G, Ristovski Z (2004) Experimental investigation of ultra-fine particle size distribution near a busy road. Atmos Environ 38:1767–1776

Guo H, Kota SH, Sahu SK, Hu J, Ying Q, Gao A, Zhang H (2017) Source apportionment of PM2.5 in North India using source-oriented air quality models. Environ Pollut 231:426–436

Hartwig A (2013) Metal interaction with redox regulation: an integrating concept in metal carcinogenesis? Free Radic Biol Med 55:63–72

Henry RC, Lewis CW, Hopke PK, Williamson HJ (1984) Review of receptor model fundamentals. Atmos Environ 18(8):1507–1515

Hidy GM, Venkataraman C (1996) The chemical mass balance method for estimating atmospheric particle sources in Southern California. Chem Eng Commun 151:187–209. https://doi.org/10.1080/00986449608936548

Hinds WC (1999) Aerosol technology: properties, behavior, and measurement of airborne particles, 2nd edn. Wiley, New York

Homa D, Haile E, Washe AP (2016) Determination of spatial Chromium contamination of the environment around industrial zones. Int J Anal Chem Volume 2016, Article ID 7214932, 7 page

http://timesofindia.indiatimes.com/city

http://ueppcb.uk.gov.in/files/Ambient_Air_Quality_2015__(2).pdf

http://www.imd.gov.in/section/climate/extreme/

Hu X, Zhang Y, Ding ZH, Wang TJ, Lian HZ, Sun YY, Wu JC (2012) Bioaccessibility and health risk of arsenic and heavy metals (Cd Co, Cr, Cu, Ni, Pb, Zn and Mn) in TSP and PM2.5 in Nanjing, China. Atmos Environ 57:146–152

Imhof D, Weingartner E, Ordóñez C et al (2005) Real-world emission factors of fine and ultrafine aerosol particles for different traffic situations in Switzerland. Environ Sci Technol 39:8341–8350

Ingrid PS, Araújo Dayana B, Costa DB, de Moraes Rita JB (2014) Identification and characterization of particulate matter concentrations at construction jobsites. Sustainability 6:7666–7688. https://doi.org/10.3390/su6117666

Ishida S, Andreux P, Poitry-Yamate C, Auwerx J, Hanahan D (2013) Bioavailable copper modulates oxidative phosphorylation and growth of tumors. Proc Natl Acad Sci USA 110(48):19507–19512

Jain SK, Sahni YP, Rajput N, Gautam V (2011) Nanotoxicology: an emerging discipline. Vet World 4(1):35–40

Jayanthi AP, Beumer K, Bhattacharya S (2012) Nanotechnology: risk governance in India. Econ Polit Week 47(2012):34–40

Jomova K, Valko M (2011) Advances in metal-induced oxidative stress and human disease. Toxicology 283(2–3):65–87

Joshi C, Sharma N, Singh R, Ajay (2017) Biosorption: a review on heavy metal toxicity and advances of biosorption on conventional methods. J Chem Chem Sci 7:714–724

Karagulian F, Belis CA, Dora CFC, Prüss-Ustün AM, Bonjour S, Adair-Rohani H, Amann M (2015) Contributions to cities' ambient particulate matter (PM): a systematic review of local source contributions at global level. Atmos Environ 120:475–483

Keogh DU, Ferreira L, Morawska L (2009) Development of a particle number and particle mass vehicle emissions inventory for an urban fleet. Environ Model Soft 24:1323–1331

Koedrith P, Seo YR (2011) Advances in carcinogenic metal toxicity and potential molecular markers. Int J Mol Sci 12(12):9576–9595

Kulkarni MM, Patil RS (1999) Monitoring of daily integrated exposure of outdoor workers to respirable particulate matter in an urban region of India. Environ Monitor Assess 56:129–146

Kulmala M, Laaksonen A (1990) Binary nucleation of water sulfuric acid system: comparison of classical theories with different H_2SO_4 saturation vapor pressures. J Chem Phys 93:696–701

Kulmala M, Vehkamäki H, Petäjä T, Dal Maso M, Lauri A, Kerminen VM et al (2004) Formation and growth rates of ultrafine atmospheric particles: a review of observations. J Aerosol Sci 35:143–176

Kulshreshtha P, Khare M, Seetharaman P (2008) Indoor air quality assessment in and around urban slums of Delhi city, India. Indoor Air 18:488–498

Kulshrestha A, Massey DD, Masih J, Taneja A (2014) Source characterization of trace elements in indoor environments at urban, rural and roadside sites in a semi arid region of India. Aerosol Air Qual Res 14:1738–1751

Kumar P (2011) Footprints of airborne ultrafine particles on urban air quality and public health. J Civ Environ Eng 1:101

Kumar P, Kumar A, Lead JR (2012) Nanoparticles in the Indian environment: known, unknowns and awareness. Environ Sci Technol 46:7071–7072

Kumar P, Morawska L, Birmili W et al (2014) Ultrafine particles in cities. Environ Int 66:1–10

Lapuerta M, Armas O, Rodriguez-Fernandez J (2008) Effect of biodiesel fuels on diesel engine emissions. Prog Ene Comb Sci 34:198–223

Leung DYC (2015) Outdoor-indoor air pollution in urban environment: challenges and opportunity. Front Environ Sci 2:1–7. https://doi.org/10.3389/fenvs.2014.00069

Lin YY, Hwang YH, Chen PC, Chen BY, Wen HJ, Liu JH et al (2012) Contribution of gestational exposure to ambient traffic air pollutants to fetal cord blood manganese. Environ Res 112:1–7

Loo BW, Adachi RS, Cork CP, Goulding FS, Jaklevic JM, Landis DA, Searles WL (1979) A second generation dichotomous sampler for large-scale monitoring of airborne particulate matter. Lawrence Berkeley Laboratory Report, Lawrence Berkeley Laboratory

Loomis D, Grosse Y, Lauby-Secretan B, El Ghissassi F, Bouvard V, Benbrahim-Tallaa L et al (2014) The carcinogenicity of outdoor air pollution. Lancet Oncol 13:1262–1263

Magos L, Clarkson TW (2006) Overview of the clinical toxicity of mercury. Ann Clin Biochem 43(Pt 4):257–268

Mahish PK, Tiwari KL, Jadhav SK (2015) Biodiversity of fungi from lead contaminated industrial waste water and tolerance of lead metal ion by dominant fungi. Res J Environ Sci 9(4):159–168

Martinelli N, Olivieri O, Girelli D (2013) Air particulate matter and cardiovascular disease: a narrative review. Eur J Intern Med 24(4):295–302

Massey DD, Kulshrestha A, Taneja A (2013) Particulate matter concentrations and their related metal toxicity in rural residential environment of semi-arid region of India. Atmos Environ 67:278–286. https://doi.org/10.1016/j.atmosenv.2012.11.002

Mauderly JL, Chow JC (2008) Health effects of organic aerosols. Inhal Toxicol 20:257–288

Mertz W (1993) Chromium in human nutrition: a review. J Nutr 123(4):626–633

Migliaretti G, Cadum E, Migliore E, Cavallo F (2005) Traffic air pollution and hospital admission for asthma: a case-control approach in a Turin (Italy) population. Int Arch Occup Environ Health 78:164–169

Miller MR, Shaw CA, Langrish JP (2012) From particles to patients: oxidative stress and the cardiovascular effects of air pollution. Future Cardiol 8(4):577–602

Minoura H, Takekawa H, Terada S (2009) Roadside nanoparticles corresponding to vehicle emissions during one signal cycle. Atmos Environ 43:546–556

Mitra S, Keswani T, Dey M, Bhattacharya S, Sarkar S, Goswami S et al (2012) Copper-induced immunotoxicity involves cell cycle arrest and cell death in the spleen and thymus. Toxicology 293(1–3):78–88

Modgil S, Lahiri DK, Sharma VL, Anand A (2014) Role of early life exposure and environment on neurodegeneration: implications on brain disorders. Transl Neurodegener 3:9

Morawska L, Bofinger ND, Kocis L, Nwankwoala A (1998) Submicrometer and super micrometer particles from diesel vehicle emissions. Environ Sci Technol 32:2033–2042

Moschandreas DJ, Pelton DJ, Sibbett DJ, Stark JWC, McFadden JE (1978) Comparison of indoor-outdoor concentrations of atmospheric pollutants. Field monitoring protocol. Scientific report, GEOMET Report No. E-721

Nairz M, Haschka D, Demetz E, Weiss G (2014) Iron at the interface of immunity and infection. Front Pharmacol 5:152. https://doi.org/10.3389/fphar.2014.00152

National Research Council (US) Committee on Indoor Pollutants. https://www.ncbi.nlm.nih.gov/books/NBK234059/

Niu J, Liberda EN, Qu S, Guo X, Li X, Zhang J et al (2013) The role of metal components in the cardiovascular effects of $PM_{2.5}$. PLoS One 8(12):e83782. https://doi.org/10.1371/journal.pone.0083782

Oberdörster G, Sharp Z, Atudorei V, Elder A, Gelein R, Lunts A et al (2002) Extrapulmonary translocation of ultrafine carbon particles following whole-body inhalation exposure of rats. J Toxicol Environ Health A 65:1531–1543

Oberdörster G, Oberdörster E, Oberdörster J (2005) Nanotoxicology: an emerging discipline evolving from studies of ultrafine particles. Environ Health Perspect 113:823–839

Oucher N, Kerbachi R, Ghezloum A, Merabet H (2015) Magnitude of air pollution by heavy metals associated with aerosols particles in Algiers. Ene Proce 74:51–58

Pandey PK, Patel KS, Subrt P (1998) Trace elemental composition of atmospheric particulate at Bhilai in Central-East India. Sci Total Environ 215:123–134

Pant P, Guttikundab SK, Peltier RE (2016) Exposure to particulate matter in India: a synthesis of findings and future directions. Environ Res 147:480–496

Park K, Dam HD (2010) Characterization of metal aerosols in PM_{10} from urban, industrial, and Asian dust sources. Environ Monitor Assess 160:289–300

Pekkanen J, Timonen KL, Ruuskanen J, Reponen A, Mirme A (1997) Effects of ultrafine and fine particles in urban air on peak expiratory flow among children with asthmatic symptoms. Environ Res 74:24–33. https://doi.org/10.1006/enrs.1997.3750

Pennington MR, Johnston MV (2012) Trapping charged nanoparticles in the nano aerosol mass spectrometer(NAMS). Int J Mass Spectro 311:64–71

Penttinen P, Timonen KL, Tiittanen P, Mirme A, Ruuskanen J, Pekkanen J (2001) Number concentration and size of particles in urban air: effects on spirometric lung function in adult asthmatic subjects. Environ Health Perspect 109:319–323
Pérez N, Pey J, Cusack M et al (2010) Variability of particle number, black carbon, and PM_{10}, $PM_{2.5}$, and PM1levels and speciation: influence of road traffic emissions on urban air quality. Aero Sci Tech 44:487–499
Peters A, Wichmann HE, Tuch T, Heinrich J, Heyder J (1997) Respiratory effects are associated with the number of ultrafine particles. Am J Respir Crit Care Med 155:1376–1383
Pey J, Querol X, Alastuey A et al (2009) Source apportionment of urban fine and ultra-fine particle number concentration in a Western Mediterranean city. Atmos Environ 43:4407–4415
Pietrangelo A (1996) Metals, oxidative stress, and hepatic fibrogenesis. Semin Liver Dis 16(1):13–30
Pio F, Sun X, Liu S, Yamauchi T (2008) Concentrations of toxic heavy metals in ambient particulate matter in an industrial area of northeastern China Fron. Med China 2(2):207–210
Pradhan A, Waseem M, Dogra S, Khanna AK, Kaw JL (2004) Trends of metals in the respirable particulates: a comparative seasonal study in Lucknow city. Poll Res 23(3):445–450
Praveena S, Pasula S, Sameera S (2013) Trace elements in diabetes mellitus. J Clin Diagn Res 7(9):1863–1865
Qiao H, Liu W, Gu H et al (2015) The transport and deposition of nanoparticles in respiratory system by inhalation. J Nanomat 2015:1–8
Qin X, Wang S (2006) Filtration properties of electrospinning nanofibers. J Appl Polym Sci 102:1285–1290. https://doi.org/10.1002/app.24361
Rahn KA (1976) The chemical composition of the atmospheric aerosol. Technical Report, Graduate School of Oceanography, University of Rhode Island, Kingston
Rahn A (1976) The chemical composition of atmospheric Aerosol, Technical report. Graduate School of Oceanography, University of Rhode Island, USA
Ristovski ZD, Morawska L, Bofinger ND, Hitchins J (1998) Submicrometer and supermicrometer particles from spark ignition vehicles. Environ Sci Technol 32:3845–3852
Saadeh R, Klaunig J (2015) Children's inter-individual variability and asthma development. Int J Health Sci 9(4):456–467
Saksena S, Singh PB, Prasad RK, Malhotra P, Joshi V, Patil RS (2002) Exposure of infants to outdoor and indoor air pollution in low income urban areas: a case study of Delhi. East-West Center Working Papers 54:1–49
Salma I, Balásházy I, Winkler-Heil R, Hofmann W, Záray G (2002) Effect of particle mass size distribution on the deposition of aerosols in the human respiratory system. J Aerosol Sci 33:119–132. https://doi.org/10.1016/S0021-8502(01)00154-9
Saluja G (2017) Assessment of air pollution in Lucknow. Res Rev J Ecol Environ Sci 5(3):1–5
Sammut ML, Noack Y, Rose J, Hazemann JL, Proux O, Depoux M et al (2010) Speciation of Cd and Pb in dust emitted from sinter plant. Chemosphere 78:445–450
Schroeder WH, Dohson M, Kane DM, Johnson ND (1987) Toxic trace elements associated with air borne particulate matter: a review. J Air Pollut Control Assoc 33:1267–1285
Schuette FJ (1967) Plastic bags for collection of gas samples. Atmos Environ 1:515–519
Shi JP, Evans DE, Khan AA et al (2001) Sources and concentration of nanoparticles (< 10 nm diameter) in the urban atmosphere. Atmos Environ 35:1193–1202
Shrivastav R (2001) Atmospheric heavy metal pollution: development of chronological records and geochemical monitoring. Resonance 2:62–68
Silbergeld EK (2003) Facilitative mechanisms of lead as a carcinogen. Mutat Res 533(1–2):121–133
Singh AL, Jamal S (2012) A study of risk factors associated with indoor air pollution in the low income households in Aligarh city, India. J Environ Res Manag 3:1–8
Smith KR, Mehta S (2003) The burden of disease from indoor air pollution in developing countries: comparison of estimates. Int J Hyg Environ Health 206:279–289
Sorsa M (2011) Biological monitoring. In: La Ferla F, Lauwerys RR, Stellman JM (eds) Encyclopedia of occupational health and safety. International Labor Organization, Genova

Soto-Jimenez MF, Flegal AR (2011) Childhood lead poisoning from the smelter in Torreon, Mexico. Environ Res 111(4):590–596

Spix C, Anderson HR, Schwartz J, Vigotti MA, LeTertre A, Vonk JM et al (1998) Short-term effects of air pollution on hospital admissions of respiratory diseases in Europe: a quantitative summary of APHEA study results. Air pollution and health: a European approach. Arch Environ Health 53:54–64

Sun HL, Chou MC, Lue KH (2006) The relationship of air pollution to ED visits for asthma differ between children and adults. Am J Emerg Med 24:709–713

Theophanides T, Anastassopoulou J (2002) Copper and carcinogenesis. Crit Rev Oncol Hematol 42(1):57–64

Tolis E, Saraga D, Ammari G, Gkanas E, Gougoulas T, Papaioannou C et al (2014) Chemical characterization of particulate matter (PM) and source apportionment study during winter and summer period for the city of Kozani, Greece. Open Chem 12(6). https://doi.org/10.2478/s11532-014-0531-5

U.S. Environmental Protection Agency (1980) Environmental monitoring systems laboratory. List of designated reference and equivalent methods. Research Triangle Park: U.S. Environmental Protection Agency, p 22

U.S. Environmental Protection Agency. http://cfpub.epa.gov/ncea/isa/recordisplay.cfm?deid=149923

U.S. Environmental Protection Agency. http://cfpub.epa.gov/ncea/isa/recordisplay.cfm?deid=158823

U.S. Environmental Protection Agency. http://cfpub.epa.gov/ncea/isa/recordisplay.cfm?deid=198843

Vaio PD, Magli E, Caliendo G, Corvino A, Fiorino F, Frecentese F, Saccone I, Santagada V, Severino B, Onorati G, D'Onofrio Freda G, Manzo C, Perissutti E (2018) Heavy metals size distribution in PM_{10} and environmental-sanitary risk analysis in Acerra (Italy). Atmosphere 9(58):1–15. https://doi.org/10.3390/atmos9020058

Valko M, Morris H, Cronin MT (2005) Metals, toxicity and oxidative stress. Curr Med Chem 12(10):1161–1208

Varghese SK, Gangamma S, Patil RS, Sethi V (2005) Particulate respiratory dose to Indian Women from domestic cooking. Aer Sci Technol 39(12):1201–1207. https://doi.org/10.1080/02786820500444838

Verma A, Singh SN, Shukla MK (2003) Air quality of transgomti area of Lucknow city, India. Bull Environ Contam Toxicol 70(1):166–173

Verma MK, Chauhan LK, Sultana S et al (2014) The traffic linked urban ambient air superfine and ultrafine PM1mass concentration, contents of pro-oxidant chemicals, and their seasonal drifts in Lucknow, India. Atmos Poll Res 5:677–685

Verma AK, Saxena A, Khan AH, Sharma GD (2015) Air pollution problems in Lucknow City, India: a review. J Environ Res Dev 9(4):1176–1188

Wang J, Guo X, Zhu J, Reinert T, Heitmann J, Spemann D, Vogt J, Flagmeyer RH, Butz T (2000) Source identification of lead pollution in the atmosphere of Shanghai City by Analyzing Single Aerosol Particles (SAP). Environ Sci Technol 34(10):1900–1905. https://doi.org/10.1021/es9907818

Watson JG, Zhu T, Chow JC, Engelbrecht J, Fujita EM, Wilson WE (2002) Receptor modeling application framework for particle source apportionment. Chemosphere 49:1093–1136

Weinberg ED (2010) Can iron be teratogenic? Biometals 23(2):181–184

Welz B, Sperling M (1999) Atomic absorption spectrometry. Willey-VCH, Weinheim

Wichmann HE, Spix C, Tuch T, Wolke G, Peters A, Heinrich J et al (2000) Daily mortality and fine and ultrafine particles in Erfurt, Germany. Part I: role of particle number and particle mass. Res Rep Health Eff Inst 98:5–86

Wiseman CL, Zereini F (2009) Airborne particulate matter, platinum group elements and human health: a review of recent evidence. Sci Tot Environ 407(8):2493–2500

World Health Organization (2002) World Health Report, Geneva

World Health Organization (WHO) (2006) Health risks of particulate matter from long-range transboundary air pollution. WHO Regional Office for Europe

U.S. Environmental Protection Agency. http://www.epa.gov/air/caa/

Yang J, Teng Y, Song L, Zuo R (2016) Tracing sources and contamination assessments of heavy metals in road and foliar dusts in a typical Mining city, China. PLoS ONE 11(12):e0168528. https://doi.org/10.1371/journal.pone.0168528

Yu F, Turco RP (2000) Ultrafine aerosol formation via ion-mediated nucleation. Geophys Res Lett 27:883–886

Zhang H, He PJ, Shao LM(2008) Fate of heavy metals during municipal solid waste incineration in Shanghai. J Hazard Mater 156(—3):365–373

Zhang X, Zhang W, Yi M, Wang Y, Wang P, Xu J, Niu F, Lin F (2018) High-performance inertial impaction filters for particulate matter removal. Sci Rep 8:4757

Chapter 3
In-situ Measurements of Aerosols from the High-Altitude Location in the Central Himalayas

Hema Joshi, Manish Naja and Tarun Gupta

Abstract Aerosols, both natural and anthropogenic, affect the Earth's climate directly due to the absorption and scattering of solar radiation and indirectly by modifying the cloud microphysics. Due to the short lifetime of these aerosols, their distribution is non-uniform and large uncertainties exist in their estimates at global and regional scale. The characteristics of atmospheric aerosols vary largely from one region to another due to spatial and temporal variations in the emission sources, transport, atmospheric transformation and removal of aerosol particles. The aerosol measurements over the Himalayan region are of crucial importance in order to provide a far-field picture quite away from potential sources. The ground-based measurements of aerosols are utilized along with satellite data to explain various aerosol characteristics over the Himalayan region. The roles of different processes such as boundary layer dynamics, meteorology, regional and long-range transport are assessed. In addition, the aerosol variation over the foothills of the Himalayas in the Indo-Gangetic Plain region has also been studied and the role of the boundary layer dynamics and updraft/downdraft of aerosols is elaborated. The high-altitude location of Himalayas is characterized by the low aerosol loading specially in winter, while significant aerosol abundance is observed in the spring. However, the significant aerosol abundance is observed over the foothills location throughout the year. The strong confinement of aerosols in the foothill region is evident, which leads to the significant enhancement in the surface concentration of aerosols. Interestingly, in the spring season, significant aerosol abundance is seen over the Himalayan region as well. The investigation of the mixing layer depth and the vertical distribution of aerosols over this region in spring reveals the transport and buildup of aerosols from the foothills region to the Himalayan region. The role of absorbing aerosols in the radiation budget over the central Himalaya region is also discussed.

H. Joshi (✉) · T. Gupta
Indian Institute of Technology, Kanpur 208016, Uttar Pradesh, India
e-mail: joshihm@iitk.ac.in

M. Naja
Aryabhatta Research Institute of Observational Sciences, Nainital 263001, Uttarakhand, India

T. Gupta et al. (eds.), *Measurement, Analysis and Remediation of Environmental Pollutants*, Energy, Environment, and Sustainability,
https://doi.org/10.1007/978-981-15-0540-9_3

Keywords Aerosol optical depth · Black carbon · Mixing layer depth · Vertical distribution · Himalayas · Indo-Gangetic Plain

3.1 Introduction

The global climate system depends upon the balance between the incoming solar radiation and outgoing long-wave (infrared) terrestrial radiation (Charlson et al. 1991; Hansen et al. 1998; Seinfeld et al. 2004) and an interaction between its subcomponents. The changes in the aerosol content and greenhouses gases present in atmosphere alters the radiative balance of the climate system (Haywood and Ramaswamy 1998; Myhre et al. 1998). The quantification of radiative forcing by aerosols is more complicated and uncertain than greenhouse gases due to high variability in aerosol mass and particle number concentrations in space and time owing to their short lifetime (about a week) as compared to the greenhouse gases (which have longer lifetime). Aerosols are known to affect the radiation budget through direct effect (through scattering and absorption of the incoming radiation) and indirect effect (modification of the microphysical properties of clouds). Aerosols thus have substantial influence on the Earth climate at the regional and global scales by affecting the radiation budget, cloud microphysics, hydrological cycle, and monsoon patterns (Lelieveld et al. 2001; Jacobson 2001). Interestingly, the aerosols of anthropogenic origins are found to be associated with the climate change via their interaction with radiation and clouds (Boucher et al. 2013).

The lack of observations of aerosols characteristics and spatio-temporal distribution coupled with the multi-faceted impacts on climate makes aerosols one of the least understood components of the Earth's climate (IPCC 2007; Boucher et al. 2013). There exist large uncertainties in their global and regional estimates (Ramanathan et al. 2001a) due to non-uniform distribution over the globe. Although the model predictions have significantly improved in recent times, still the most of the available models possess high degree of uncertainty associated with their use and capturing the actual magnitude of aerosol. The ground based actual measurements of aerosols and their properties thus are of crucial importance for realistic impact assessments and model evaluations.

Some of the recent studies have indicated the significant aerosol abundance over the South Asian region (Ramanathan et al. 2001b; Lelieveld et al. 2001) and the possible impacts of aerosols. This region is also marked with the high levels of absorbing aerosols (Koch and Hansen 2005; Lawrence and Lelieveld 2010). In recent times, the increase in anthropogenic activity has been witnessed in terms of accelerated population density, industrialization and urbanization leading to aerosol abundance in the atmosphere. The enhancement in the aerosol emission is reported over the Asian region (Zhang and Reid 2010; Lu et al. 2012; Boucher et al. 2013) including the Indian region (Porch et al. 2007; Streets et al. 2009; Moorthy et al. 2013a). The Indo-Gangetic Plain (IGP) region is marked with the high abundance of aerosol and gaseous pollutants throughout the year, which not only affects significant pollution load over this region (Dey et al. 2004; Singh et al. 2004; Gautam et al. 2010, 2011; Giles et al. 2011; Joshi et al. 2016), but also the nearby elevated slopes of Himalayas (Pant et al. 2006; Dumka et al. 2010; Gobbi et al. 2010; Bonasoni et al. 2010; Marinoni et al. 2010; Kumar et al. 2011; Srivastava et al. 2012). Aerosols from IGP can be transported to high altitude regions by enhanced convection and uplifting, and can have myriads of effects, once reached the Himalayas. The close association between the pollution load of IGP is found to be associated to the Himalayan climate, which thereafter affects the regional climate significantly by affecting the dynamics, hydrological cycle and radiation budget of the region (Menon et al. 2002; Lau and Kim 2006; Lau et al. 2010).

3.1.1 Atmospheric Aerosols

Atmospheric aerosols are defined as particles in solid, liquid or in mixed phase dispersed in the atmosphere (Seinfeld and Pandis 1998). The term aerosol was coined in the year 1918 by the physical chemist E. G. Donnan. It was considered as analogous to the term hydrosol, which is a stable liquid suspension of solid particles (Hinds 1999). A. Schmauss got the term integrated into the meteorological literature in the early 1920s. Atmospheric aerosols are two/three-phase systems (solid-gas, liquid-gas and a combination of both), the common examples of which are dust, fume, smoke, fog, haze, smoke, smog and clouds (Seinfeld and Pandis 1998). The size of aerosols varies from ~10^{-3} to 100 μm (Junge 1963; Prospero et al. 1983).

3.1.2 Aerosol Sources and Production

Atmospheric aerosols can be of natural as well as of anthropogenic origin (Seinfeld and Pandis 1998) (Fig. 3.1). The natural sources include windborne dust, sea spray, volcanic activities and biomass burning, while anthropogenic activities mainly include fuel combustion, industrial emissions, transportation and biomass burning due to crop harvest. There are also common sources i.e., biomass burning, soil dust emissions etc., which can be originated from both the natural and anthropogenic sources. The natural aerosols are reported to be present in the higher amount ~4–5 times larger than anthropogenic ones on the global scale, but regional variations in anthropogenic aerosols are reported to change this ratio significantly in the industrialized Northern Hemisphere (Seinfeld and Pandis 1998). Aerosols, when directly emitted from their sources are called primary aerosols (e.g., black carbon, soil dust, sea salt). Aerosols can also be formed in atmosphere by the oxidation of the precursor gases (i.e., nitrogen oxides, sulphur dioxide, and volatile organic compounds), the resultant oxidation product then nucleate to form new particles or condense on pre-existing ones forming secondary aerosols (Seinfeld and Pandis 1998).

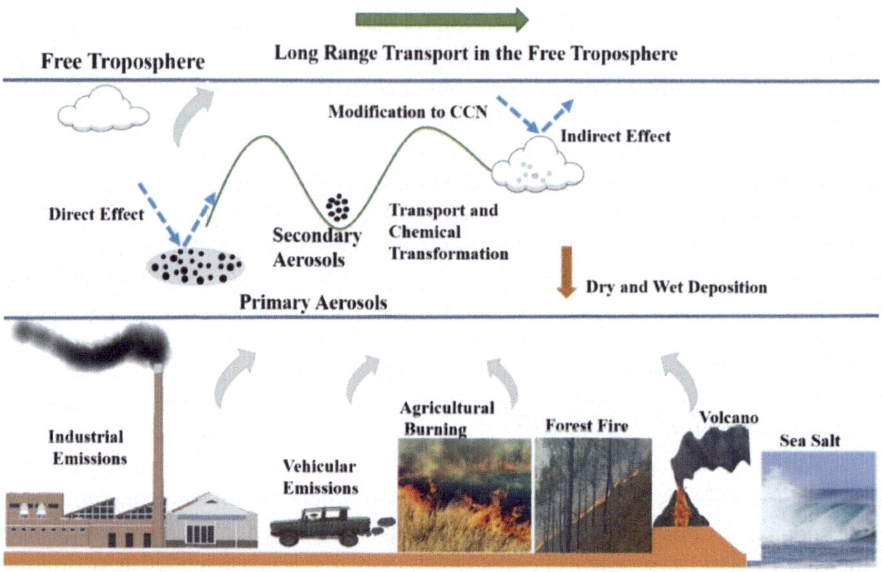

Fig. 3.1 Schematic representation of aerosol emission sources, deposition, transport and transformation processes, and aerosol effects (direct and indirect)

Aerosols are mainly produced by bulk to particle conversion process also called as "mechanical disintegration", which is the most common and major process of aerosol production (Prospero et al. 1983). The gas-to-particle production of aerosols however is a secondary process of aerosol production. The aerosols formation occurs as a result of chemical reactions between various gaseous species in the atmosphere, followed by nucleation and condensation. The gases or vapors with low vapor pressure in the atmosphere condense on to existing particles and increase the mass of aerosols or condense to form new particles in the air. The particles produced from gas-to-particle process are hygroscopic in nature and are efficient cloud condensation nuclei (CCN). Atmospheric aerosols are broadly classified in two main categories, fine and coarse mode aerosols (Fig. 3.2). The fine fraction of aerosols is comprised of the Aitken mode and the accumulation mode of particles. The particles in the accumulation mode are of crucial importance as they are most stable, characterized by larger lifetime and have the significantly high impact on climate. The most common types of the aerosols found in the atmosphere are mentioned below.

3.1.2.1 Dust Aerosols

Dust aerosols are one of the major type of natural aerosols (Alfaro et al. 1997) originated predominantly from desert regions in the Northern Hemisphere. The dust source regions mainly include deserts, semi-arid desert fringes, dry lakebeds, and the arid regions with low vegetation. The origin of dust can be both the natural as

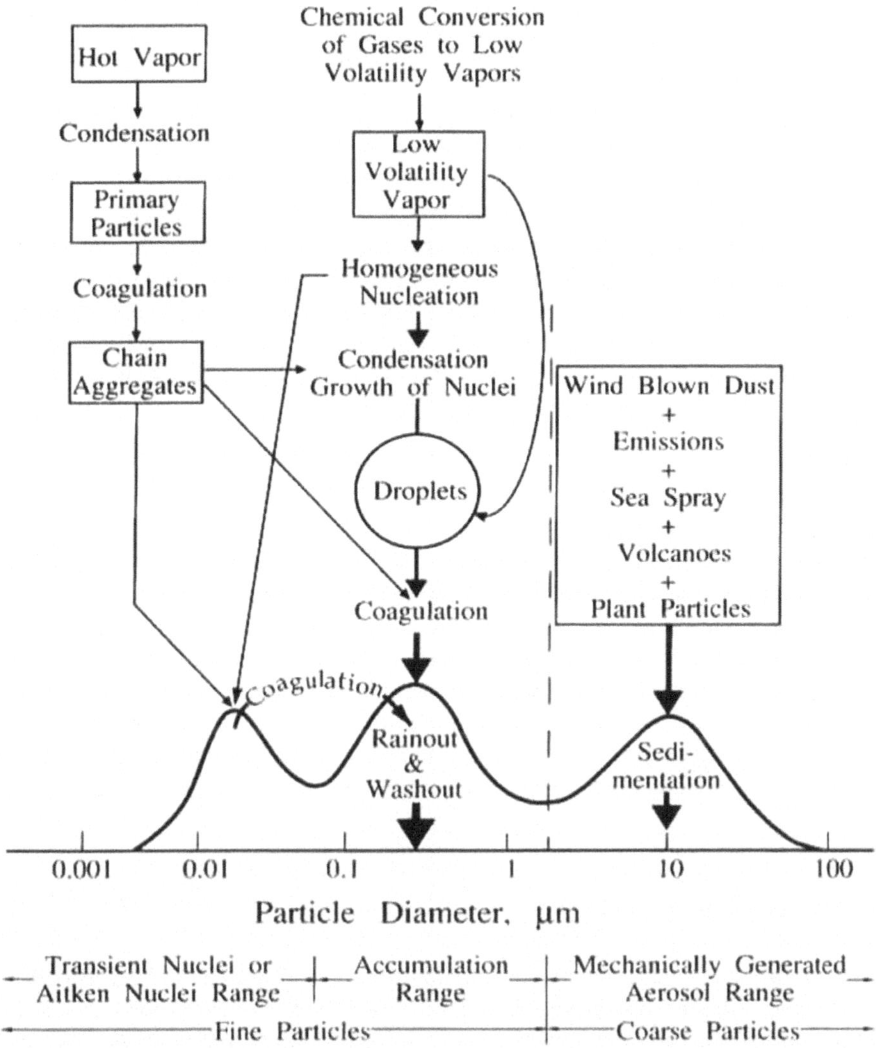

Fig. 3.2 Idealized representation of the size distribution (surface area) of the atmospheric aerosol (Whitby 1978). The common formation mechanism, sources and removal processes of aerosol are shown

well as anthropogenic in origin. The larger size dust particles are found near the source regions while the small size dust can easily transport to much larger distances (Prospero et al. 2002).

3.1.2.2 Sea Salt

Sea salt aerosols are of natural origin contributing to about ~30% of global aerosol burden (Fitzgerald 1991). These aerosols are generated by a wide range of physical processes, the most effective of which is bursting of air bubbles entrained in the ocean surface during white cap formation (Blanchard 1985; Monahan 1986), which strongly depends upon the wind speed. These aerosols are hygroscopic in nature and are very efficient cloud condensation nuclei (CCN).

3.1.2.3 Black Carbon

Black carbon also called as 'BC' is the most absorbing primary aerosol with strong and wide absorption in the visible and Infrared (IR) range. It is emitted from the incomplete combustion processes such as fossil fuel and biomass burning, thus much of the atmospheric BC is of anthropogenic origin. BC is known to retain its basic form at very high temperatures, with a vaporization temperature near 4000 K (Bond et al. 2013). It serves as an excellent tracer (due to its inert nature and longer lifetime) to study the anthropogenic influence at remote sites. BC can undergo the regional and intercontinental transport even to the remote and clean locations in Arctic (Stohl 2006).

3.1.2.4 Organic Carbon

Organic carbon (OC) is produced from both primary emissions as well as secondary transformations. The OC aerosols scatter the solar radiation and are known as scattering aerosols. OC aerosols are comprised of myriads of organic compounds with varying volatility and polarity, which makes the study of the effect of individual organic compounds quite challenging. Due to the presence of polar functional groups, for example carboxylic and dicarboxylic acids, OC aerosols also participate in cloud droplet nucleation (Saxena and Hildemann 1996).

3.1.2.5 Sulphate and Nitrate Aerosols

Sulphate aerosols are produced via chemical reactions in the atmosphere from gaseous precursors (i.e., SO_2 from anthropogenic sources and volcanoes, and dimethyl sulphide (DMS) from biogenic sources, especially marine planktons) and are efficient CCN. The volcanic eruption also adds large amount of SO_2 into the stratosphere, where it is converted to sulphate aerosols. The industrial activities also contribute to sulphate aerosols (Charlson et al. 1992). Nitrate aerosols are also produced by chemical reactions in the atmosphere from gaseous precursors, produced either naturally or anthropogenically and are also efficient CCN.

3.1.3 Aerosol Transport, Transformation and Removal

Atmospheric aerosols, once airborne can undergo the change in their size and composition by coagulation with other species, condensation of vapor, by evaporation, chemical reactions, and by activation in presence of water supersaturation to form clouds and fog. Atmospheric aerosols can also exist in different mixing state (externally or internally) depending upon how the constituent chemicals are distributed among the particles. The particles from diverse sources remain separated in an external mixture, while diverse chemical components get mixed to form a single particle in an internally mixed aerosols. The nature of mixing affects optical and physical properties of aerosols, and plays a crucial role in the aerosol radiative forcing. The changes in the aerosols number density, size distribution and hence in their optical and radiative properties can also occur during the transformation process. The main aerosol transformation processes are known to be condensational growth, coagulation, and the processing of aerosols by non-precipitating cloud cycling. The transformation processes are not the removal process as aerosol mass is not removed from the atmosphere during this process.

Aerosols are known to be removed from the atmosphere by two main processes: dry and wet removal processes. The dry removal processes, also called as dry deposition or sedimentation, involves sedimentation of particles due to gravity and impaction onto surfaces, while the wet processes involve the rain out and wash out processes (Pruppacher and Klett 1978). The sedimentation process is significant only for aerosols with diameters greater than a few micrometers. Wet removal however is the main removal process for smaller particles. The wash out scavenging is the below cloud scavenging and involves removal of aerosols from the atmosphere by the falling rain drops. The rainout process mainly describes the removal of a cloud condensation nuclei and takes place within the cloud. Aerosols also undergo transport at regional, intercontinental and global scales from their source region to distant locations by means of prevailing winds. The synoptic scale air masses influence the aerosol concentration significantly (Smirnov et al. 1995). Aerosols when get transported to the free troposphere, can stay there for a longer time and can undergo long range transport from there.

3.1.4 Brief Background and the Study Region

The systematic measurements of aerosols have been initiated over the Indian region since the 1980s at several distinct geographical regions under I-MAP (Indian Middle Atmosphere Programme) and ISRO-GBP (Indian Space Research Organization, Geosphere Biosphere Programme) (Moorthy et al. 1999). Several campaigns were also conducted i.e., The Indian Ocean Experiment (INDOEX) (Satheesh and Ramanathan 2000; Ramanathan et al. 2001b), Integrated Campaign for Aerosols, gases and Radiation Budget (ICARB) (Moorthy et al. 2008), and other land campaigns (LC-I, LC-II) in order to better understand the role of aerosols over various regions (mainland and oceans). The initiation of establishing several permanent aerosol monitoring stations from the ground based diverse locations started under Aerosol Radiative Forcing over India (ARFI) project of ISRO-GBP. Initially few monitoring stations were established, followed by more over the time in order to study aerosol impacts, to generate regional aerosol database, and radiative forcing maps for climate impact assessment over the Indian region. Realizing the importance of the aerosol measurements from the Himalayan region, aerosols observations were initiated at the high-altitude site, Aryabhatta Research Institute of Observational Sciences (ARIES) (earlier known as state observatory) (29.4° N, 79.5° E, 1958 m msl), Nainital in year 2002 under the ISRO-GBP program. The actual ground-based observations of aerosols are quite limited and nearly non-existing over the Himalayas considering the diversity of the region and topographic changes. In the later years, aerosols observations were also setup at Kullu (31.9° N, 77.1° E, 1154 m msl) (since year 2007), and more recently at Hanle (32.7° N, 78.9° E, 4520 m msl) (since year 2009).

Aerosol observation and reported studies from Nainital site had shown that the site is characterized by very low aerosol optical depth (AOD) comparable to the Antarctic environment in winter, while in summer typical to the continental location (Sagar et al. 2004). The observations of black carbon mass concentration however (started later in November 2004 at the site) indicated that the BC contributes about ~5.0% to the composite aerosol mass at the site in the winter season (Pant et al. 2006). The site also experiences the dust storm, which typically originates from the Thar Desert. The enhancement in the aerosol columnar loading due to the dust storm was reported to be two to four times higher particularly at longer wavelengths (Hegde et al. 2007). Aerosol optical and physical properties and surface concentration of BC was studied and reported from the site (Dumka et al. 2008, 2010). The possibility of influence of the aerosols from the nearby valley region to the observation site was also indicated, however the actual simultaneous observations at the foothills locations were not present for the comparison. Aerosol observation of BC mass concentration was then initiated at the low altitude foothills site, Pantnagar (29.0° N, 79.5° E, 231 m msl) (since year 2009) in the Indo-Gangetic Plain region in order to study the aerosol sources, the role of boundary layer dynamics and updraft/downdraft of aerosols from IGP region to the nearby Himalayan region.

In the present study, an attempt has been made to study the aerosol climatology from the high-altitude site in the central Himalayan region. The long-term datasets of aerosols have been utilized for the trend estimations in aerosols over this region. Further, aerosol characterization from foothills location in the Indo-Gangetic Plain region has also been made and reported and the associated processes have been discussed. The possible sources of aerosols, their seasonality and processes governing the observed concentration or loading have been discussed. The columnar, surface and vertical variation of aerosols have been studied and discussed here along with the aerosol measurement from the space borne sensors. These ground-based datasets have also been utilized for the comparison with the model simulations over this region.

3.2 Measurement Sites, Observations and Database

The ground-based observations of aerosols have been used from a high altitude, regional representative site, Nainital (29.4° N, 79.5° E, 1958 m msl) (Fig. 3.3) in the central Himalayas. It is an ideal site for studying the far-field impact of anthropogenic aerosols, by virtue of being situated far away from the major aerosol sources. In addition, aerosol observations from the low altitude, foothill site, Pantnagar (29.0° N, 79.5° E, 231 m msl) in the Indo-Gangetic Plain (IGP) region has also been used (Fig. 3.3), which represents as an ideal site for studying the aerosol characteristic near the source region. Although, these two sites are located at an aerial distance of ~30 km, they are characterized by different topography, vegetation type and meteorological parameters. The site, Nainital is located at the mountain top, the north and north-east of which is the high-altitude Himalayan

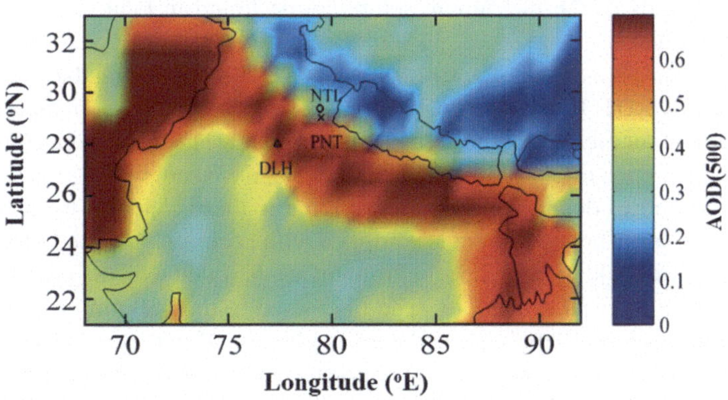

Fig. 3.3 Spatial variation of aerosol optical depth (AOD 550 nm) obtained from MODIS (Terra) satellite. The location of the high-altitude observation site, Nainital (NTL: 29.4° N, 79.5° E, 1958 m msl) in the central Himalayas and the foothill site, Pantnagar (PNT: 29.0° N, 79.5° E, 231 m msl) in the Indo Gangetic Plain region is shown. The country capital Delhi (DLH) is also shown

mountain range, while towards the south and southwest lies the low altitude plain merging with the IGP region. The mean temperature experienced at the Nainital site reaches maximum ~22.3 °C (T_{max} ~ 30 °C) in June, while minimum monthly mean in January is ~5.2 °C ($T_{min} \leq 0$), while at Pantnagar site, the maximum monthly mean temperature observed in June reaches ~30 °C (T_{max} ~ 42 °C), with minimum in January is ~10 °C ($T_{min} \geq 3$ °C) (Joshi et al. 2016). The additional details regarding the present high-altitude site have been reported in the earlier studies (Pant et al. 2006; Dumka et al. 2010; Naja et al. 2016).

The measurements of black carbon (BC) mass concentration has been made using the optical measurements from seven-channel (0.37–0.95 μm) Aethalometer (Magee Scientific, USA). The light attenuation through a quartz filter paper has been measured in order to compute the absorption coefficients by measuring the light transmission through the particle laden and particle free reference spot of filter. The absorption coefficients were then converted into the BC mass concentration by utilizing the calibration factors of corresponding wavelengths as mentioned by manufacturer (Hansen et al. 1984). This instrument utilizes the 0.88 μm wavelength for determination of BC mass concentration. BC is reported to be the principal absorber of light at this wavelength (Bodhaine 1995) with negligible contribution from other components. BC measurements at 0.88 μm from Aethalometer has been extensively used in literature (Hansen et al. 1984; Moorthy et al. 2004; Pant et al. 2006; Satheesh et al. 2008 and many others). Although the optical attenuation technique used in this instrument is widely used, it is found to suffer from various systematic errors as being the filter-based absorption technique and needs to be corrected. The instrument suffers from the uncertainty arising due to the multiple scattering in the quartz fibre matrix of the tape (also called C-factor) and another for 'shadowing effects', arising due to deposition of scattering material along with BC in filter tape, which needs to be corrected (Weingartner et al. 2003). Additional details regarding instrumentation, methodology and associated uncertainty are quite well documented (Hansen et al. 1984). The details of the instrument setup can be found in Pant et al. (2006) and Dumka et al. (2010).

The columnar abundance of aerosols has been studied using the hand held, Sun-photometer MICROTOPS-II (Solar Light Company, USA), a very popular and widely used instrument for measuring the aerosol optical depth. The instrument contains five different interference filters at 380, 440, 500, 675 and 870 nm wavelengths, and provides aerosol optical depth (AOD) corresponding to these channels using Bouguer-Lambert-Beer law. The instrument is equipped with accurately aligned optical collimators, with a full field view of 2.5°. Instrument is integrated with the internal baffles, which eliminate internal reflections. The filter used in the 380 nm channel has the wavelength precision of ±0.4 nm, and a full width at half maximum (FWHM) band pass of 4 nm, while other channels are characterized by peak wavelength precision of ±1.5 nm, and FWHM band pass of 10 nm. The instrument is operated only when the region of sky (~10°) around the Sun is free from clouds in order to avoid cloud contamination in the data (Pant et al. 2006). The details of the instrument have been well documented (Morys et al. 2001; Ichoku et al. 2002). The instrument setup at the site has been same as described in Pant et al. (2006) and Kumar et al. (2011). AOD observations at the foothills region have been studied using data from AERONET (Aerosol Robotic NETwork) (Holben et al. 1998, 2001). The direct Sun and diffuse sky radiances at eight spectral channels (340, 380, 440, 500, 675, 870, 940, and 1020 nm) are measured from the CIMEL sun/sky radiometer of AERONET (Holben et al. 2001; Eck et al. 1999). The direct Sun measurements were made once every 15 min at 340, 380, 440, 500, 675, 870, 940, and 1020 nm. The uncertainty in AOD retrieval under cloud free conditions is less than less than ± 0.02 for shorter wavelengths and ± 0.01 for $\lambda > 440$ nm. The quality assured datasets which are pre and post field calibrated, cloud cleared and manually inspected were used.

The vertical variation of aerosols has been studied using the Cloud-Aerosol Lidar and Infrared Pathfinder Satellite Observation (CALIPSO) (Winker et al. 2009) datasets. CALIPSO was launched in April 2006 in the A-train constellation of satellites (Stephens et al. 2002) as part of the NASA Earth System Science Pathfinder (ESSP) program in collaboration with the French space agency CNES. The main aim of CALIPSO was to understand and fill the existing gaps in the global distribution and properties of aerosols and clouds. The satellites of the A-train are in a 705 km sun synchronous polar orbit, with a16-day repeat cycle characterized by an equatorial passing time of about 13:30 local solar time. CALIOP is the first polarization lidar to provide global atmospheric measurements of aerosol and clouds around a solid-state Nd:YAG laser, which produces simultaneous, coaligned, pulses at 1064 and 532 nm. The elastic backscatter observations are utilized at 532 and 1064 nm channels (Winker et al. 2009). The CALIPSO profiles of aerosol extinction coefficient (version-3, 5-km horizontal resolution datasets) have been studied. The Hybrid Extinction Retrieval Algorithm (HERA; Young and Vaughan 2009) has been used to derive the extinction coefficients. The six aerosol types are represented by aerosol models with a prescribed bi-modal size distribution (fine and coarse mode) and a characteristic complex refractive index for each mode and wavelength (Omar et al. 2009). The discrimination between ice clouds and water clouds and the identification of non-spherical aerosol particles is done by depolarization measurements. Additionally, the other information like the aerosol size and type of aerosols as smoke, dust, polluted dust etc. has also been obtained from the ratios of the signals obtained at the two wavelengths.

The model simulations have also been used in order to evaluate the model performance over the region. The Weather Research and Forecasting Model (version 3.6) (Skamarock et al. 2008) coupled with chemistry (Grell et al. 2005) has been used to simulate BC mass concentration. In order to avoid influence due to topography, the observations from the low altitude site, Pantnagar has been compared. The model domain has been centered at Pantnagar (79.5° E, 29.0° N) and has (100, 100, 37) grid points in (longitude, latitude, vertical) directions. Aerosol processes have been taken from GOCART bulk aerosol scheme (Chin et al. 2002). Anthropogenic emissions of BC, OC and SO_2, PM2.5 and PM10 have been taken from Emission Database for Global Atmospheric Research (EDGAR) available at a spatial resolution of $0.1° \times 0.1°$ (Janssens-Maenhout et al. 2012). The complete model setup and other details have been reported in Kumar et al. (2012) and Joshi et al. (2016).

3.3 Aerosols Climatology and Trends from a High-Altitude Site

The high-altitude regions of Himalayas are far away from any major anthropogenic activities and are best suited to study the regional enhancement in aerosols and its impacts. The long-term ground-based datasets of aerosol black carbon (2005–2016) and aerosol optical depth (2005–2014) from the regional representative site, Nainital in the central Himalayas, has been utilized to infer aerosol climatology and trends in aerosols over this region. The ground-based estimation of aerosol trends over the Himalayan region are of crucial importance considering the complex topography of the region, limited availability of observations and close proximity of the Himalayan region from the highly polluted Indo Gangetic Plain (IGP) region.

The climatology in BC was examined. The minimum BC concentration was observed during the July–August due to scavenging during monsoon, while high BC was observed from March till May season each year. The monthly mean BC concentration during the observation period showed as high as ~2.49 $\mu g\ m^{-3}$ in April month, while monthly BC as low as ~0.29 $\mu g\ m^{-3}$ was observed in September month. The influence of the background level enhancement in the BC has been explored by examining the daytime and the nighttime levels in BC. As the site is located at the mountain top surrounded by the valley region, so the diurnal variations in BC are indicative of the evolution of the aerosols from nearby valley regions and other low altitude regions to the site. The amplitude in BC has been found to be maximum in winter season, while nearly flat pattern in BC was observed in spring. The close examination of data reveals that during winter (December to February) the background (night time) levels of BC were quite low, while BC during daytime showed a significant enhancement due to transport of BC from nearby valley regions to this site. The major contribution of BC in winter has been observed to be coming from the valley regions during the daytime, while during nighttime, BC aerosols along with other aerosols subside in the valley region. Quite interestingly, significant enhancement in background (night time) BC is seen in the spring (March to May) each year due to the biomass burning around the nearby regions. The long-term data of BC were then utilized for examining the trend over this region. The linear trends were estimated in the monthly mean BC, which shows a positive slope of ~1.511 (ng m^{-3} per year) with standard error (σ_{ω}) of ~0.975. The positive slope in BC indicates that the black carbon aerosols have been increasing at the site although at a slow rate. The reported BC trends are very less and nearly non-existing over the Indian region, specially over the Himalayan region. The observed trends might be useful to study the actual impact of aerosols over this region in future. The observed increasing trends in BC were consistent with some of the reported model based studies, which had reported increased warming, accompanied by earlier snow melt and retreat of high mountain glaciers over the Himalayas (Jain 2008; Kulkarn et al. 2007) and Tibetan Plateau (TP) regions (Ramanathan et al. 2007; Flanner et al. 2009; Menon et al. 2010). Based on the back trajectory analysis, Lu et al. (2012) also reported that the amount of BC received by the Himalayas and Tibetan Plateau (HTP) region has increased by 41% from 1996 to 2010, the main contributing source regions were reported to be South Asia and East Asia accounting for 67 and 17% of BC transported to the HTP on an annual basis.

In addition, the climatology in the columnar abundance of aerosols was examined, which revealed that the climatological daily mean AOD (500 nm) varied from 0.019 (January) to 0.820 (April) while monthly mean climatological mean AOD was found to vary from ~0.189 to ~0.515, with an annual mean ~0.189 (500 nm). The aerosol loading was found to be minimum in winter and maximum in spring season at Nainital. It is interesting that minimum AOD does not occur in the monsoon because of the hygroscopic growth of aerosols, but happens in winter. The enhancement in the aerosol due to the northern Indian biomass burning was reported at the site in spring season. The enhancement in the BC and AOD (0.5 mm) due to biomass burning was found to be ~1.8 $\mu g\ m^{-3}$ (~145%) and 0.3 (~150%) respectively at the site, leading to the additional atmospheric warming of 19 $W\ m^{-2}$, with the enhancement in the lower atmospheric heating rate by 0.8 K day^{-1} (Kumar et al. 2011). The northern Indian region experiences the influence of the dust transported from the northwestern arid regions to the IGP (Prospero et al. 2002; Agnihotri et al. 2011) in each spring (Sikka 1997; Dey et al. 2004). The influence of the transported dust from Thar Desert has been reported at the Nainital site (Hegde et al. 2007). Dust, when mixed with BC, is reported to cause enhanced absorption even greater than what is observed solely due to BC in IGP (Gautam et al. 2011).

AOD at the site also showed increasing trend. The monthly mean AOD data (500 nm) when examined for trend showed a positive slope of ~0.000774 $year^{-1}$ with a standard error (σ_{ω}) of ~0.000415. The multi-wavelength data of AOD was used to examine the aerosol size by investigating the 'angstrom exponent' (AE) by the method mentioned by Angstrom (1964). The slope of the monthly mean AE (440–870 nm) was found positive ~0.00374 $year^{-1}$ with a standard error (σ_{ω}) of ~0.00144. The increasing trends, observed over the site, indicate that the aerosol loading might have increased over the site in the recent times. The observed increase in the aerosol loading is in accordance to what is reported at the other locations, where AOD is reported to increase. Over the Indian region, Moorthy et al. (2013a) had utilized the AOD data of various ground-based networks of observatories (under ARFI project of Indian Space Research Organization) to deduce the regional nature of aerosol loading over this region. It was reported that AOD over the Indian region increased at a rate of 2.3% (of its value in 1985) per year, while more rapidly with enhancement of ~4% in the recent decade (Moorthy et al. 2013a). The major fraction of this enhancement was also found to be of anthropogenic origin.

The observed trends at the present site (~0.0007 $year^{-1}$) have been compared with the AOD trends reported by Babu et al. (2013). The observed trends at the present site were less (as the site is situated far away from any major anthropogenic source region) than what reported at Port Blair (an island location in the Bay of Bengal) ~0.0097 $year^{-1}$, and Minicoy (location in Indian Ocean) ~0.0208 $year^{-1}$, Trivandrum (a coastal location in south India) ~0.011 $year^{-1}$, Visakhapatnam (industrialized coastal location in east coast of India) ~0.0266 $year^{-1}$, Hyderabad (Industrialized, urban location close to central India) ~0.0127 $year^{-1}$, and Dibrugarh (North-Eastern

India) ~0.0135 year^{-1}. Interestingly, AOD trend over the two nearby IGP locations Patiala (semi urban, western part of Northern India) and Kanpur (Industrialized, urban location in IGP) were also found to be increasing at a rate of ~0.0127 year^{-1} and ~0.0089 year^{-1} respectively. The increased aerosol abundance at these two IGP locations indicated that there is a possibility of the regional level enhancement of aerosols over the IGP region, which might lead to the background levels enhancement of aerosols at nearby Himalayan regions in future. The combined effect of the observed enhanced aerosol loading and BC over this region indicates the possibility of enhanced atmospheric warming over this region in future as well. The atmospheric warming strongly depends upon the contribution of absorbing aerosols. The increasing trend in aerosol loading with increasing trend in BC contributes to the net warming, while decreasing trend in BC suggests decreased warming or less warming potential of aerosols.

3.4 Aerosol Variation at the Adjacent Foothills Location

In order to see the aerosol variation near the source region in the foothill locations in IGP region and its influence to the nearby Himalayas, the aerosol measurements were initiated at the foothills location, Pantnagar (29.0° N, 79.5° E, 231 m amsl) shown in Fig. 3.3. BC serves as an excellent tracer of anthropogenic influence and has been studied extensively over this foothill region. BC datasets (2009–2012) when examined showed that the BC concentrations in the foothills regions of the Himalayas were quite high as shown in Fig. 3.4, typical to that of the semi urban/urban site. BC levels were observed to show different diurnal variation (with two peaks; morning and evening peak with the daytime minimum BC) and seasonal variations with winter maximum at the site. This is in contrast to what was observed at the nearby site Nainital, where the diurnal variations result in a single afternoon peak in BC with maximum BC observed in the spring season. BC at the foothill site, was found to start increasing about an hour before sunrise, attaining the morning peak (linked with fumigation effect arising due to entrainment of BC aerosols from the nocturnal residual layer) within an hour or two and decreased afterwards during daytime. This resulted in lowest levels in the afternoon (due to turbulent mixing that dilutes the surface BC), followed by an increase after sunset attaining secondary peak in the evening, which then slowly decreases in the nighttime. The reduced mixing after sunset and rapid cooling of land leads to the formation of nocturnal layer resulting in the trapping of pollutants near surface during nighttime. The diurnal variations were clear and sharp in all the seasons at the foothill region showing maximum diurnal amplitudes in the winter season followed by autumn and minimum amplitude in summer-monsoon.

The impact of the local anthropogenic activities was also studied in terms of the weekly cycles in BC, which revealed a clear enhancement of ~88% in BC levels (in January) during weekdays, when compared with weekends levels. Higher BC in weekdays has been associated with traffic patterns and industrial activity (in nearby

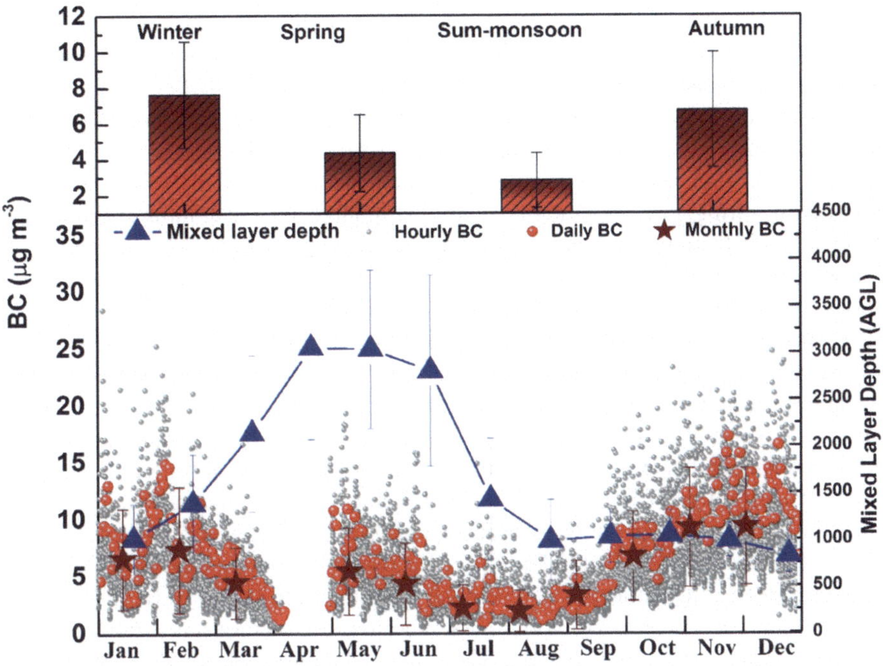

Fig. 3.4 Seasonal variations in BC mass concentrations (2009–2012), and seasonal mean noontime (1200–1400 h) mixing layer depth

region, within ~20 km) during weekdays, as compared to weekends with reduced industrial activity. The weekly variations in aerosols, however can be altered or completely masked by other factors like wind speed, wind direction, rainfall and long-range transport.

Investigation of the seasonal variation in BC when examined showed a maximum monthly mean BC in December ($9.3 \pm 5.1\ \mu g\ m^{-3}$) and minimum in August ($1.9 \pm 1.8\ \mu g\ m^{-3}$) (Fig. 3.4). BC was found to be maximum in winter ($7.9 \pm 5.2\ \mu g\ m^{-3}$), the biomass/biofuel burning for cooking and heating purpose in winter was the possible major source of BC during this season. The minimum BC was reported in summer-monsoon ($2.8 \pm 2.8\ \mu g\ m^{-3}$) (Fig. 3.4). The seasonal variation in BC shows reduction in the spring season at the foothill site. The reduction in the surface concentration of BC in spring season at the foothill region, despite the enhanced aerosol abundance in the spring season, was due to the extensive mixing of aerosols, as indicated by the mixing layer depth variation shown in Fig. 3.4. The aerosols near surface might have advected to the higher altitudes in spring. The observed BC concentration over the foothill site was then compared to what was observed at Nainital. BC was observed to be low at the Himalayan site (annual mean ~$1.0\ \mu g\ m^{-3}$), while significant high BC (~$5.50\ \mu g\ m^{-3}$) was observed over the foothills site with different diurnal and seasonal variations. Seasonal mean BC was found to be maximum at the Himalayan site in spring, while at IGP, maximum BC was observed in winter. BC was observed to be significantly high at foothills site in winter (~569%), followed by autumn (~543%), summer-monsoon (~304%) and spring (~240%) as compared to what observed at the Himalayan site. A clear evidence of the direct influence of aerosols from IGP to Himalayan region was observed in spring when the day to day variation in BC was in good association at these two sites. The possibility of direct transport was observed to be minimum in winter. The observed differences in BC, in particular in spring, provided the strong motivation to study the columnar abundance of the aerosols from the site and the spatial variation of aerosols over this region.

The columnar abundance of aerosols at the foothill site Pantnagar showed significant high AOD values with maximum columnar abundance of aerosol observed in the spring season (>0.6), which suggested that the surface level aerosols might have advected to the higher altitude regions in the spring leading to the low surface concentration of BC (along with other aerosols) while high total columnar abundance of aerosols. Monthly average AOD was maximum (>0.6) during late spring. The aerosol abundance was significant at the Pantnagar site throughout the year (annual mean AOD ~ 0.56 ± 0.32). The significant aerosol abundance was observed in the foothills regions in the winter season (AOD > 0.4), suggesting a typical semi-urban/urban nature of the site. The observed different seasonal cycle of AOD, as compared to BC, at Pantnagar indicates the smaller contribution of BC in the total columnar abundance of aerosols by mass. The investigation of angstrom exponent revealed the presence of the coarse mode of aerosols during March–June at the site, as the site receives the significant amount of dust aerosols during this period. The columnar abundance of aerosol over the Himalayan site reveals very low AOD in fine mode (typical to clean free tropospheric site) in winter with significant aerosol load observed in spring dominated by coarse mode of aerosols. The spatial variation of AOD from MODIS and absorbing aerosols (retrieved aerosol index) from the Ozone Monitoring Instrument (OMI) also showed maximum loading in the spring season.

3.5 Role of Mixing Layer Depth and BC-CO Relation

The influence and variations in planetary boundary layer height on the surface evolution of BC was explored by investigating the association between BC and mixing layer depth (MLD). The MLD plays an important role in determining the concentration of pollutants near surface and their distribution above the surface. It represents the volume available for dispersion of pollutants due to convection or mechanical turbulence. Due to lack of the ground-based observations of the planetary boundary layer height at the site, the MLD height was calculated using HYSPLIT model. The influence of the MLD in BC was examined at the low altitude foothills site, which also helped in understanding the uplifting of aerosols to the nearby higher altitude regions. The noon-time (1200–1400 h, IST) MLD estimation was carried out around Pantnagar ($1° \times 1°$), which showed MLD to be maximum in April–June (higher solar radiation and surface temperature), which then decreased to about 1000 m during August–January with a minimum value of ~832 ± 193 m in December as shown in

Fig. 3.4. The reduction in the solar radiation, cloudy conditional and rainfall causes reduction in MLD during monsoon. Interestingly, the higher BC level in winter (7.9 ± 5.2 μg m^{-3}) and autumn (6.5 ± 4.9 μg m^{-3}) was found to coincide with lower values of MLD during these seasons (about 1050 m) indicating that the mixing of BC emitted from sources is limited to quite a less volume as compared to spring and summer seasons (Fig. 3.5).

In order to better understand the evolution of BC with MLD, the daytime (1100–1700 h) variation in BC and MLD was studied and is shown in Fig. 3.5. Interestingly, an inverse association, with highest anti-correlation ($R^2 = 0.89$) is observed in winter followed by autumn ($R^2 = 0.76$) and spring ($R^2 = 0.61$) was found while minimum association was found in monsoon (July–August) ($R^2 = 0.33$). It was found that the variations in MLD had a strong role in governing wintertime BC, while MLD played a weak role during the monsoon period when the observation site is under the influence of the rainfall, and receives intense air-masses from oceanic regions.

The variation in BC and Carbon monoxide (CO) was then explored, both of which were found to be governed by similar evolution with MLD. The common sources of combustion were identified by utilizing the BC-CO relation for the first time over the Indian region. Both BC and CO have been linked closely with each other via combustion process, as the type and amount of combustion changes their emission ratio in the combustion process (Seinfeld and Pandis 1998). The CO observations were made using non-dispersive infrared (4.6 μm) absorption technique. The useful information regarding the influence of fresh and aged air masses at the site was accessed by utilizing the BC-CO association. A strong correlation indicates common combustion source with influence of fresh emission, while poor correlation indicates emission from diverse sources and aged emissions. BC and CO reveals a good correlation in winter ($R^2 = 0.75$) and late autumn seasons ($R^2 = 0.72$) indicating the presence of similar types of combustion sources (Fig. 3.5c). The positive correlation ($R^2 = 0.62$) was observed in spring, however a poor correlation ($R^2 = 0.05$) was observed in summer-monsoon. The marine air masses arrive from diverse directions in summer-monsoon, which cause scavenging of aerosols and gases over the region, but these air masses sometimes carry aerosols (from diverse regions) along with them to the site, which leads to the observed poor correlation between BC and CO during this season. In addition, due to different lifetime of BC and CO (CO possesses much longer lifetime than BC), and due to different processes of scavenging BC than CO showed poor correlation. The dominant BC emissions were then indicated from the average slope of BC-CO (BC versus CO) over this region. The observed slope was found to be ~10.4 μg $ppmv^{-1}$, which was even higher than what was observed over Mexico and Germany (Baumgardner et al. 2002). The influence from cooking/domestic use of fuel was indicated as the main contributor.

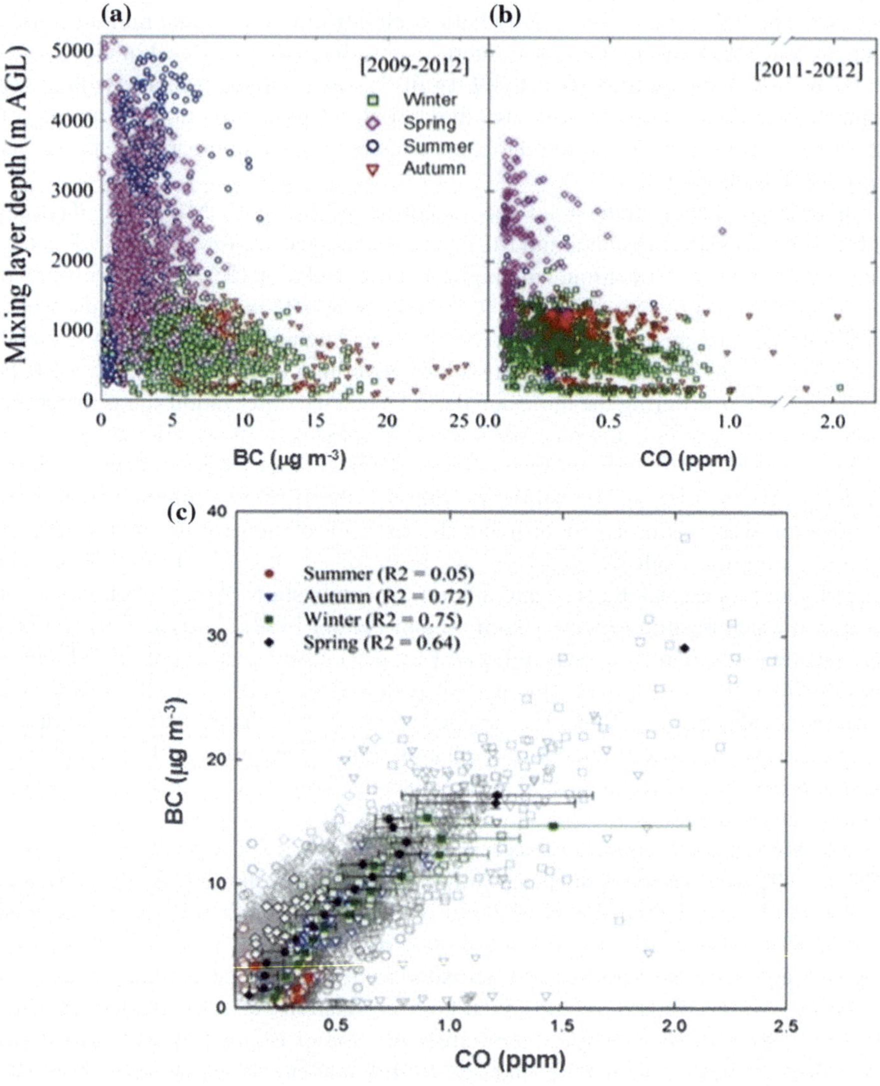

Fig. 3.5 **a, b** Correlation of hourly daytime (1100–1700 h) BC (µg m^{-3}) and carbon monoxide (CO) with mixed layer depth (m AGL) for different seasons at Pantnagar. **c** The hourly average correlation between surface CO and BC is shown by open grey symbols. The daily average (color open symbols) and binned averages (color filled symbols) correlation between CO and BC is also shown

3.6 Association of BC with Large-Scale Wind Patterns: Source Distributions

The influence of large-scale winds in the observed BC mass concentration was studied at the foothill site. The seven-days back-air trajectories were calculated utilizing the Hybrid Single-Particle Lagrangian Integrated Trajectory (HYSPLIT; Draxler and Rolph 2003) model and using the NCEP/NCAR global reanalysis data at 1000 m AGL at Pantnagar. The trajectories were calculated and clustered according to the ward's Hierarchical method by combining the nearest trajectories according to angular distance in order to determine the direction of arrival of air masses to the site. The calculated trajectories were then clustered as shown in Fig. 3.6. The trajectories in different seasons were then clustered in five different groups west (W), north-west (NW), north-east (NE), south-west (SW) and south-east (SE), based on the direction from which they were originating (Fig. 3.6). It was found that the site was influenced by the air masses originating from W and NW with a larger contribution from the NW group (~63%) in winter and from W group (55%) in spring. The marine air

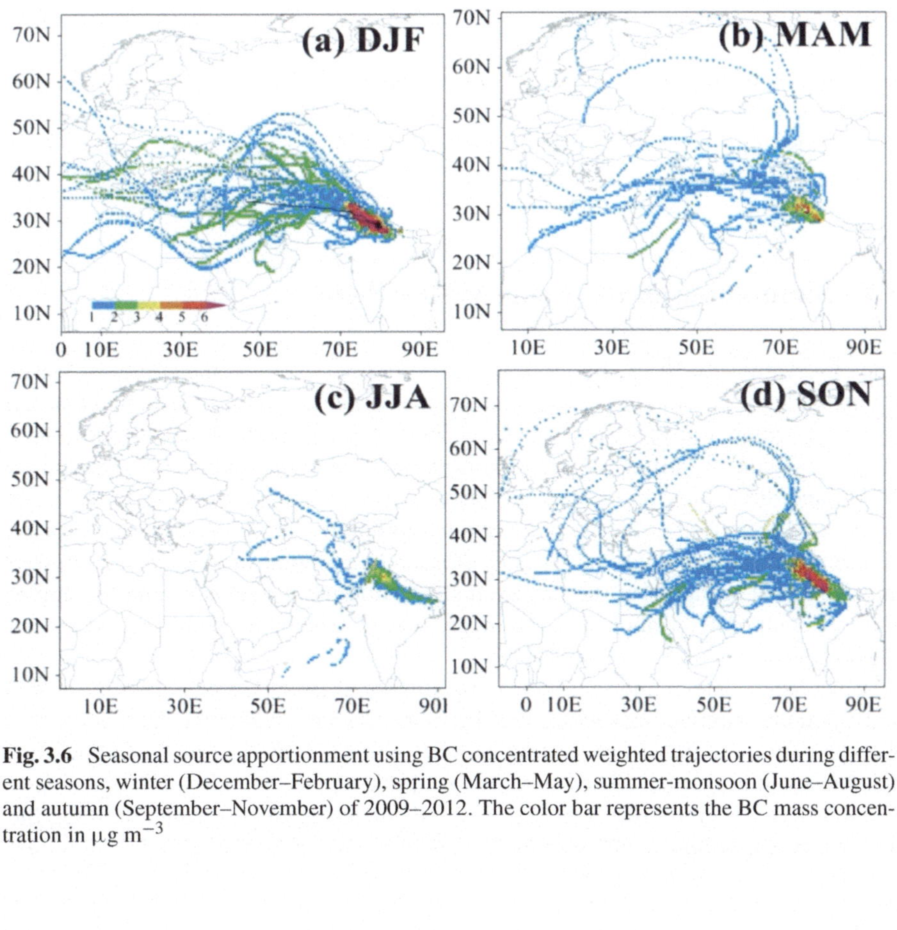

Fig. 3.6 Seasonal source apportionment using BC concentrated weighted trajectories during different seasons, winter (December–February), spring (March–May), summer-monsoon (June–August) and autumn (September–November) of 2009–2012. The color bar represents the BC mass concentration in $\mu g\ m^{-3}$

masses contribute ~63% (both SW and SE) from the oceanic regions of the Bay of Bengal (SE) and the Arabian Sea (SW), while ~37% of air masses of continental origin from NW directions. The change in the air mass pattern was again observed in autumn. The dominance of air masses originating from continental origin from W/NW was observed, while air masses arriving from marine origin from SE were reduced due to the withdrawal of the southwest monsoon from the Indian region.

The possible source regions contributing to the observed BC at the foothill site was investigated by the means of Concentration Weighted Trajectory (CWT) analysis (Stohl 1996). In this method, a grid was superimposed over the domain of all the trajectories computed and the weighted value of observed concentration was assigned to the trajectories that crossed the corresponding grid cell. The study revealed maximum BC contribution from NW and W ~ 7–8 μg m^{-3} in both winter and autumn. In winter, air masses arriving from the NW direction (~62% of air masses) contribute significant BC of ~8.54 μg m^{-3}. The change in the air masses was observed in the spring season when air masses shift from NW (45%) to W (55%). The air-masses from the W sector contributes BC of ~4.89 μg m^{-3}. In summer-monsoon, influence of SE (38%) and SW (24%) air masses were seen, with BC levels dropping to 2.21–2.59 μg m^{-3}. Soon after the withdrawal of the southwest monsoon, the dominant contribution from NW direction was again seen in autumn, with a consequent rise in BC levels (7.89 μg m^{-3}). The air masses that passed through the IGP region with higher amount of BC, before arriving at the foothill site have been reported to contribute to the observed BC at the site. The observed levels at the foothill site were found to be influenced and augmented by the synoptic air masses arriving via IGP, especially due north-west (NW) of the site.

3.7 Simulations of BC Using WRF-Chem

The observations of BC at the foothills site were used to assess the ability of WRF-Chem model in simulating seasonal and diurnal variability in BC. The model domain was kept centered at Pantnagar (79.5° E, 29.0° N) with (100, 100, 37) grid points in (x longitude, y latitude, z vertical) directions with horizontal grid spacing of 10 km and the model top located at 50 hPa. The model simulations were made for January, May and July (due to computational limitation) representing winter, spring and summer/monsoon seasons. The comparison of temporal variation in BC showed high temporal variation in the observed BC, while the model was able to capture some of temporal variation, but the magnitude was highly underestimated by the model (Fig. 3.7).

It was quite interesting that the model simulated BC at 29 N (representing the site by model simulations) was quite low with low episodic enhancements, while BC simulations at 28 N (representing the IGP regions far away from foothill region) was higher in magnitude. These simulations indicated the presence of large spatial variability in BC emissions around the foothill site especially in winter. Further, the day to day variability in BC was highly underestimated by the model. The model

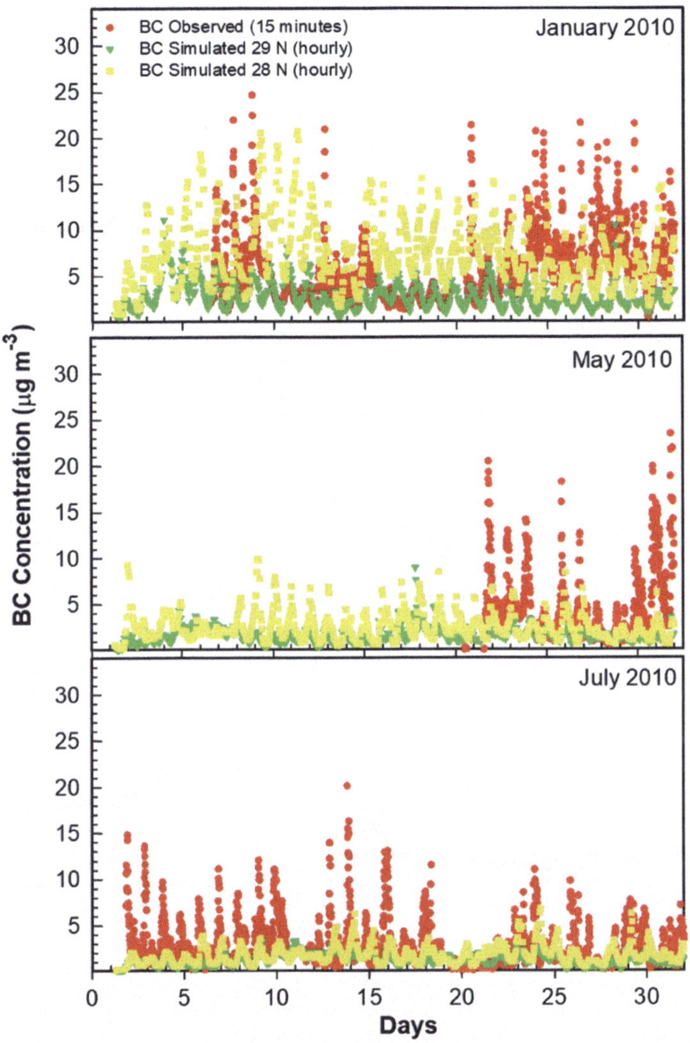

Fig. 3.7 The comparison of temporal variation in observed BC mass concentration with model simulated BC at Pantnagar for year 2010

performance in simulating the diurnal variability in BC was also investigated as the studies on diurnal pattern simulations of BC over the Indian region were sparse.

The model showed a mixed capability in simulating BC diurnal variation as shown in Fig. 3.8. The model was found to capture some features, while missed out on a few others. The high observed BC during nighttime and low BC during daytime as a result of diurnal changes in the planetary boundary layer height was captured by the model. The morning and evening peaks in observed BC were not captured by the model possibly due to the inadequacy of the ABL schemes in the model (as has been

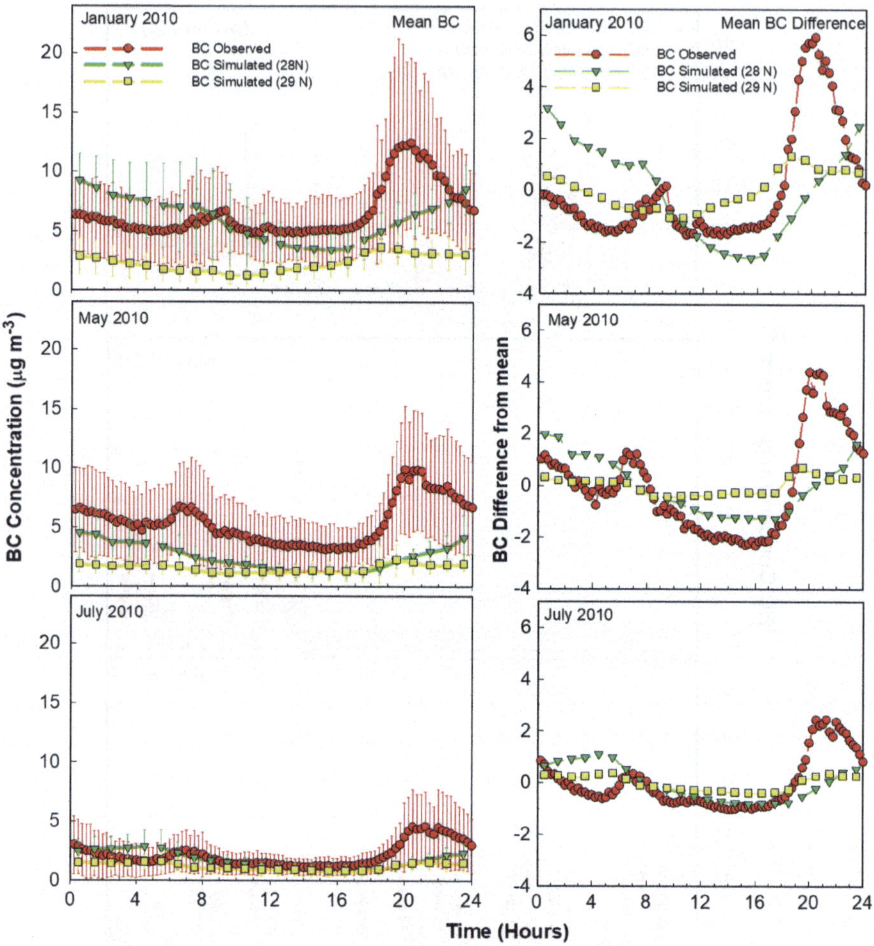

Fig. 3.8 Diurnal variations in observed BC and WRF-Chem model simulated BC during January, May and July (left panel) for year 2010. The difference in the diurnal variation in BC for observed and simulated BC from their respective mean BC (right panel)

pointed out by some earlier reports; (Nair et al. 2012; Moorthy et al. 2013b) using RegCM and GOCART models). Average observed and modeled BC at Pantnagar during January, May and July were 6.6 ± 2.2 and 2.4 ± 1.2 μg m^{-3}, 5.5 ± 1.8 and 1.6 ± 0.7 μg m^{-3}, and 2.2 ± 1.0 and 1.2 ± 0.5 μg m^{-3}, respectively. The BC levels simulated at 28 N were found to be 6.1 ± 3.1 μg m^{-3}, 2.5 ± 1.6 μg m^{-3} and 1.8 ± 1.0 μg m^{-3} for January, May and July, respectively. These average levels at 28 N were found to be in a good agreement with observations.

The model was able to capture the observed decrease in BC from January to July indicating the capability of the model to qualitatively capture the seasonal variations in BC. The model however failed to capture BC levels, thus the decrease in the magnitude of seasonal and diurnal variation in BC was underestimated. Additionally, the role of meteorology in controlling seasonal and diurnal variation in BC at Pantnagar was also highlighted as BC emissions provided as input to the model do not have a seasonal cycle. The WRF-Chem model was found to capture some of the observed features but in order to capture the observed magnitude in good agreement, improvement in the diurnal and seasonal variability in the BC emission sources in and around the foothills region was strongly suggested. In addition, need of co-located observations of planetary boundary layer height were suggested for better evaluation of model performance over the region.

3.8 Vertical Distribution of Aerosols Over the Region

In order to understand the aerosols variation at the foothill location and its vertical distribution above the site, aerosol vertical profiles from CALIPSO were examined. The aerosol profiles in different seasons were examined, which showed the confinement of aerosol over the low altitude regions in winter and autumn, while significant presence of aerosols to greater vertical extension (~about 4–5 km altitude) in spring and early summer (Fig. 3.9). The seasonal variation in aerosols was then explored in detail for which the aerosol extinction profiles were investigated at the foothill site. The cloud free and quality controlled CALIPSO extinction profiles (532 nm) were examined at 2 × 2 degree around Pantnagar.

The daytime and nighttime profiles of aerosols extinction coefficient were examined for days when the surface BC showed enhancement. The higher extinction coefficient near ground was observed in the nighttime profiles, which sharply decreased above ~1 km as compared to the daytime profile (Fig. 3.10). The surface BC, although constituting a minor fraction of columnar AOD, was reported to significantly affect the aerosol properties, especially in winter when most of the aerosols (including BC) are strongly confined near surface usually below ~1 km. The daytime levels in BC, mixing layer depth and daytime profile of extinction coefficient at the Pantnagar site when showed the possibility of uplifting and mixing of aerosols (including BC) to higher heights.

The seasonal variation in aerosols extinction coefficient at 532 nm as shown in Fig. 3.10 revealed high aerosol extinction coefficient (more than 2 km^{-1}) near surface in winter, which decreased sharply to less than 0.25 km^{-1} at about 1.5 km indicating the strong confinement of aerosols below 2 km. The vertical profile of aerosols extinction coefficient showed similar variation as that in winter but the aerosol extinction coefficient value near surface are found to be low as compared to the winter when quite high extinction coefficient was observed near surface. The vertical gradient in the aerosol extinction coefficient was also smaller in spring indicating mixing of aerosols between surface and ~3 km. Aerosol profiles in spring season also revealed reduction in the aerosol extinction coefficient near surface, while enhancement at the higher altitude regions.

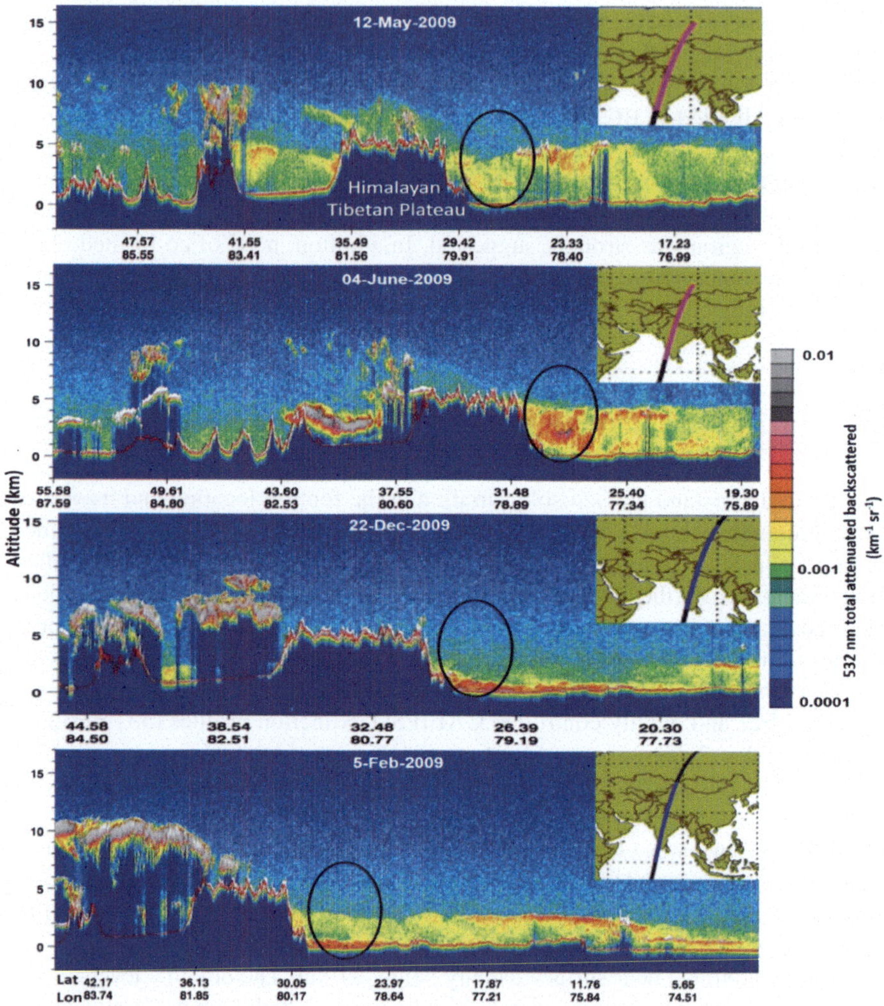

Fig. 3.9 CALIPSO, 532 nm total attenuated backscattered profiles (km^{-1} sr^{-1}) from southern India to the Himalayan Tibetan Plateau region (HTP). The orbital track of each profile is also shown. The profiles were investigated for 12 May 2009, representing spring, 4 June 2009 representing early summer, and 22 Dec 2009 and 5 Feb 2009 representing winter season

3.9 Conclusions

The variability of aerosols over the high-altitude region in the central Himalayas and its adjacent foothills location was studied. The sources and the processes governing the distribution of aerosols at these sites was explored. The decadal scale ground-based aerosol observations were used to present the climatology and trends of aerosols at the high altitude, regional representative site in the central Himalayas. The

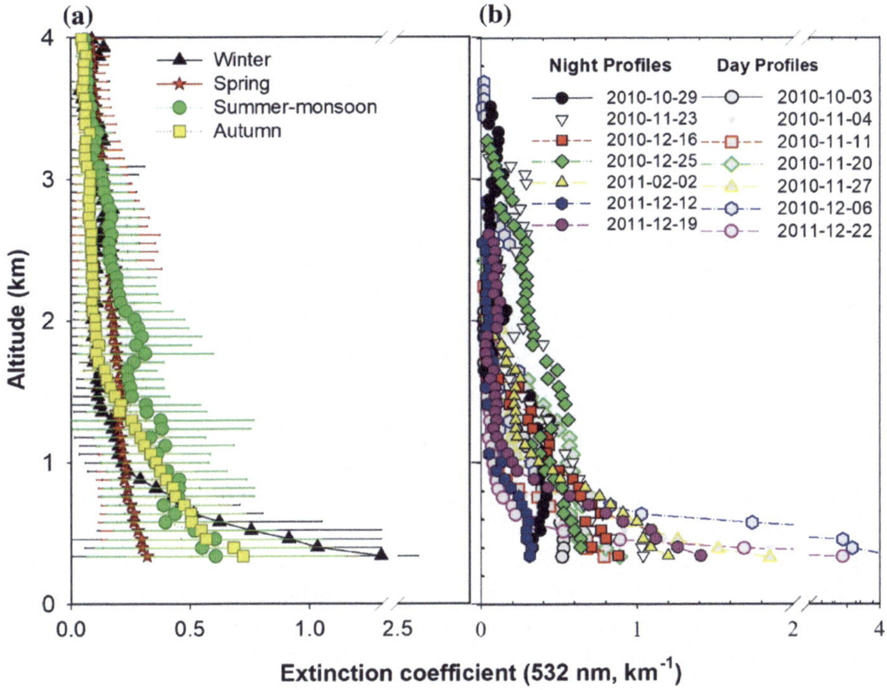

Fig. 3.10 **a** Seasonal profiles of aerosol extinction coefficients at 532 nm from CALIPSO during 2010–2011 for different seasons. **b** Variation in the aerosol extinction coefficient corresponding to high BC days

enhancement in the black carbon aerosols and columnar abundance in the aerosols was observed at the site. The observed positive trend in black carbon (BC) was found to be ~1.511 ng m^{-3} per year with a standard error (σ_ω) of ~0.975, indicating increase in the absorbing aerosols over this region. The columnar abundance of aerosols (AOD at 500 nm) also showed an increasing trend of ~0.000774 per year with a standard error (σ_ω) of ~0.000415. The increasing trends in aerosol abundance and BC indicate the more warming potential of aerosol over this region. The regional characterization of aerosols was made from the adjacent foothill site in the Indo-Gangetic Plain region in order to understand the sources of aerosols, and aerosol characterization near the source region. The influence of aerosol sources from the IGP region to the nearby Himalayan region were studied. The foothill region was found to be marked with high levels of BC mass concentration as well as the columnar loading of aerosols. Large variations were observed in BC at diurnal and seasonal scales, closely associated with the local/regional anthropogenic influences, mesoscale and synoptic meteorological processes.

The influence of the local emissions and the atmospheric boundary layer dynamics was examined in the diurnal variations in BC. The mixing layer depth was observed to play a dominant role in governing the surface distribution of BC in winter. BC exhibited nearly an inverse relation with MLD in all seasons, which was found to be strongest in winter and weakest in monsoon. The influence of fresh emissions with common combustion sources were observed in winter and autumn, while the influence of aged air mass and emissions from different combustion sources were witnessed in summer-monsoon.

Unlike BC, the co-located study of aerosol optical depths (AOD) and aerosol absorption reveals maximum in spring over IGP, possibly due to the presence of absorbing aerosols (including BC, dust) above the above the ABL (in the free troposphere). The aerosol optical depth (AOD 500 nm) variation shows annual peak in May–June. The dominance of coarse mode of aerosols was found in spring and early summer, while dominance of fine mode of aerosols was found in late autumn and early winter.

The influence of the large-scale winds was also studied at the foothill site, which revealed that the air masses arriving through the IGP region contribute to the BC observed at the site. The vertical profile of aerosol from CALIPSO satellite showed sharp confinement of aerosols near surface in winter followed by autumn in the IGP region, while aerosols were observed to reach the high-altitude Himalayan regions in spring season. The aerosol extinction coefficient values were found to be quite high near surface in winter over the IGP region. In spring, the reduction in aerosol extinction coefficient was observed near surface (over IGP region) while it showed enhancement in the higher altitudes (altitudes >2 km).

The performance of the WRF-Chem model was also evaluated in simulating the observed ground based black carbon aerosols over the foothill regions in IGP. The improvement in the emission inventory with finer temporal and spatial resolution in the IGP region, along with the boundary layer measurements was strongly suggested in order to capture and match the actual levels of aerosols observed over this region by the present-day models, which can then be used for accurate impact assessment of aerosols over this region.

Acknowledgements The present work was carried out under ISRO-GBP (ARFI) project. The authors highly acknowledge the ISRO-GBP (ARFI) and ISRO-GBP (ATCTM) projects. The author HJ wants to acknowledge Dr. Rajesh Kumar for providing the WRF model simulations. The CO data was obtained from Piyush Bhardwaj. The help and support received from Prof. K. P. Singh during measurement at Pantnagar site is highly acknowledged. Special thanks to the PI of AERONET site at Pantnagar for providing quality assured data. The satellite datasets of aerosols were obtained from MODIS via Giovanni site, and CALIPSO from NASA Langley Research Center Atmospheric Science Data Center. The HYSPLIT model made available from NOAA Air Resources Laboratory was also used and is acknowledged.

References

Agnihotri R, Mandal TK, Karapurkar SG, Naja M, Gadi R, Ahammmed YN, Kumar A, Saud T, Saxena M (2011) Stable carbon and nitrogen isotopic composition of bulk aerosols over India and northern Indian Ocean. Atmos Environ 45:2828–2835. https://doi.org/10.1016/j.atmosenv.2011.03.003

Alfaro SC, Gaudichet A, Gomes L, Maillé M (1997) Modeling the size distribution of a soil aerosol produced by sandblasting. J Geophys Res 102(D10):11239–11249. https://doi.org/10.1029/97JD00403

Angstrom, A. (1964) The parameters of atmospheric turbidity. Tellus 16(1):64–75

Babu SS, Manoj M, Moorthy KK, Gogoi M, Nair V, Kompalli S, Satheesh SK, Niranjan K, Ramagopal K, Bhuyan PK, Singh D (2013) Trends in aerosol optical depth over Indian region: potential causes and impact indicators. J Geophys Res: Atmos 118:11794–11806

Baumgardner D, Raga GB, Peralta O, Rosas I, Castro T, Kuhlbusch T, John A, Petzold A (2002) Diagnosing black carbon trends in large urban areas using carbon monoxide measurements. J Geophys Res: Atmos 107(D21):8342. https://doi.org/10.1029/2001jd000626

Blanchard DC (1985) The oceanic production of atmospheric sea salt. J Geophys Res 90:961–963

Bodhaine BA (1995) Aerosol absorption measurements at Barrow, Mauna Loa and the south pole. J Geophys Res 100(D5):8967–8975. https://doi.org/10.1029/95JD00513

Bonasoni P, Laj P, Marinoni A, Sprenger M, Angelini F, Arduini J, Bonafe U, Calzolari F, Colombo T, Decesari S, Di Biagio C, di Sarra AG, Evangelisti F, Duchi R, Facchini MC, Fuzzi S, Gobbi GP, Maione M, Panday A, Roccato F, Sellegri K, Venzac H, Verza GP, Villani P, Vuillermoz E, Cristofanelli P (2010) Atmospheric Brown Clouds in the Himalayas: first two years of continuous observations at the Nepal-Climate Observatory at Pyramid (5079 m). Atmos Chem Phys Discuss 10:4823–4885

Bond TC, Doherty SJ, Fahey DW, Forster PM, Berntsen T, DeAngelo BJ, Flanner MG, Ghan S, Kärcher B, Koch D, Kinne S, Kondo Y, Quinn PK, Sarofim MC, Schultz MG, Schulz M, Venkataraman C, Zhang H, Zhang S, Bellouin N, Guttikunda SK, Hopke PK, Jacobson MZ, Kaiser JW, Klimont Z, Lohmann U, Schwarz JP, Shindell D, Storelvmo T, Warren SG, Zender CS (2013) Bounding the role of black carbon in the climate system: a scientific assessment. J Geophys Res Atmos 118:5380–5552. https://doi.org/10.1002/jgrd.50171

Boucher O, Randall D, Artaxo P, Bretherton C, Feingold, G, Forster P, Kerminen V-M, Kondo Y, Liao H, Lohmann U, Rasch P, Satheesh SK, Sherwood S, Stevens B, Zhang XY (2013) Clouds and aerosols. In: Climate change 2013: the physical science basis. Contribution of Working Group I to the fifth assessment report of the Intergovernmental Panel on Climate Change. Cambridge University Press, USA

Charlson RJ, Langner J, Rodhe H, Leovy CB, Warren SG (1991) Perturbation of the northern hemisphere radiative balance by backscattering from anthropogenic sulfate aerosols. Tellus 43AB:152–163

Charlson RJ, Schwartz SE, Hales JM, Cess RD, Coakley JA, Hansen JE, Hofmann DJ (1992) Climate forcing by anthropogenic aerosols. Science 255:423–430

Chin M, Ginoux P, Kinne S, Torres O, Holben BN, Duncan BN, Martin RV, Logan JA, Higurashi A, Nakajima T (2002) Tropospheric aerosol optical thickness from the GOCART model and comparisons with satellite and sunphotometer measurements. J Atmos Sci 59:461–483

Dey S, Tripathi SN, Singh RP (2004) Influence of dust storms on the aerosol optical properties over the Indo-Gangetic basin. J Geophys Res 109:D20211. https://doi.org/10.1029/2004JD004924

Draxler RR, Rolph GD (2003) Real-time environmental applications and display system (READY). NOAA Air Resources Laboratory, Silver Spring, MD. http://www.arl.noaa.gov/ready/hysplit4.html

Dumka UC, Moorthy KK, Pant P, Hegde P, Sagar R, Pandey K (2008) Physical and optical characteristics of atmospheric aerosols during ICARB at Manora Peak, Nainital: a sparsely inhabited, high-altitude location in the Himalayas. J Earth Syst Sci 117(S1):399–405

Dumka UC, Moorthy KK, Kumar R, Hegde P, Sagar R, Pant P, Singh N, Babu SS (2010) Characteristics of aerosol black carbon mass concentration over a high altitude location in the Central Himalayas from multi-year measurements. Atmos Res 96:510–521. https://doi.org/10.1016/j.atmosres.2009.12.010

Eck TF, Holben BN, Reid JS, Dubovik O, Kinne S, Smirnov A, O'Neill NT, Slutsker I (1999) The wavelength dependence of the optical depth of biomass burning, urban and desert dust aerosols. J Geophys Res 104:31333–31350

Fitzgerald JW (1991) Marine aerosols: a review. Atmos Environ 25A:533–545

Flanner MG, Zender CS, Hess PG, Mahowald NM, Painter TH, Ramanathan V, Rasch PJ (2009) Springtime warming and reduced snow cover from carbonaceous particles. Atmos Chem Phys 9(7):2481–2497. https://doi.org/10.5194/acp-9-2481-2009

Gautam R, Hsu NC, Tsay S-C, Lau WK, Holben B, Bell S, Smirnov A, Li C, Hansell R, Ji Q, Payra S, Aryal D, Kayastha R, Kim KM (2011) Accumulation of aerosols over the Indo-Gangetic plains and southern slopes of the Himalayas: distribution, properties and radiative effects during the 2009 pre-monsoon season. Atmos Chem Phys 11:12841–12863. https://doi.org/10.5194/acp-11-12841-2011

Gautam R, Hsu NC, Lau K-M (2010) Premonsoon aerosol characterization and radiative effects over the Indo-Gangetic Plains: implications for regional climate warming. J Geophys Res 115:D17208. https://doi.org/10.1029/2010JD013819

Giles DM, Holben BN, Tripathi SN, Eck TF, Newcomb WW, Slutsker I, Dickerson RR, Thompson AM, Mattoo S, Wang S-H, Singh RP, Sinyuk A, Schafer JS (2011) Aerosol properties over the Indo-Gangetic Plain: a mesoscale perspective from the TIGERZ experiment. J Geophys Res 116:D18203. https://doi.org/10.1029/2011JD015809

Gobbi GP, Angelini F, Bonasoni P, Verza GP, Marinoni A, Barnaba F (2010) Sunphotometry of the 2006–2007 aerosol optical/radiative properties at the Himalayan Nepal Climate Observatory—Pyramid (5079 m a.s.l.). Atmos Chem Phys Discuss 10:1193–1220

Grell GA et al (2005) Fully coupled "online" chemistry within the WRF model. Atmos Environ 39:6957–6975

Hansen AD, Rosen AH, Novakov T (1984) The Aethalometer—an instrument for the real-time measurement of optical absorption by aerosol particles. Sci Total Environ 36:191–196. https://doi.org/10.1016/0048-9697(84)90265-1

Hansen JE, Sato M, Lacis A, Ruedy R, Tegen I, Matthews E (1998). Climate forcings in the industrial era. Proc Natl Acad Sci 12753–12758

Haywood JM, Ramaswamy V (1998) Global sensitivity studies of the direct radiative forcing due to anthropogenic sulfate and black carbon aerosols. J Geophys Res 103:6043–6058

Hegde P, Pant P, Naja M, Dumka UC, Sagar R (2007). South Asian dust episode in June 2006: aerosol observations in the Central Himalayas. Geophys Res Lett 34:L23802. https://doi.org/10.1029/2007gl030692

Hinds W (1999) Aerosol technology: properties, behavior, and measurement of airborne particles. Wiley, New York

Holben BN, Tanre D, Smirnov A, Eck TF, Slutsker I, Abuhassan N, Newcomb WW, Schafer JS, Chatenet B, Lavenu F, Kaufman YJ, Vande Castle J, Setzer A, Markham B, Clark D, Frouin R, Halthore R, Karneli A, O'Neill NT, Pietras C, Pinker RT, Voss K, Zibordi G (2001) An emerging ground-based aerosol climatology: aerosol optical depth from AERONET. J Geophys Res 106(D11):12067–12097

Holben BN, Eck TF, Slutsker I, Tanré D, Buis JP, Setzer A, Vermote E, Reagan JA, Kaufman YJ, Nakajima T, Lavenu F, Jankowiak I, Smirnov A (1998) AERONET—a federated instrument network and data archive for aerosol characterization. Remote Sens Environ 66:1–16

Ichoku C, Levy L, Kaufman YJ, Remer LA, Rong-Rong L, Martins VJ, Holben BN, Abuhassan N, Slutsker I, Eck TF, Pietras C (2002) Analysis of the performance characteristics of the five-channel Microtops II sun photometer for measuring aerosol optical thickness and precipitable water. J Geophys Res 107:D13. https://doi.org/10.1029/2001JD001302

IPCC (2007) Summary for policymakers. In: Solomon S, Qin D, Manning M, Chen Z, Marquis M, Averyt KB, Tignor M, Miller HL (eds) Climate change 2007: the physical science basis. Contribution of Working Group I to the fourth assessment report of the Intergovernmental Panel on Climate Change. Cambridge University Press, Cambridge, United Kingdom/New York, NY, USA

Jacobson MZ (2001) Strong radiative heating due to the mixing state of black carbon in atmospheric aerosols. Nature 409:695–697

Jain SK (2008) Impact of retreat of Gangotri glacier on the flow of Ganga River. Curr Sci 95:1012–1014

Janssens-Maenhout G et al (2012) EDGAR-HTAP: a harmonized gridded air pollution emission dataset based on national inventories. EUR Report No. EUR 25229, European Commission Publications Office, Ispra, Italy, p 40

Joshi H, Naja M, Singh KP, Kumar R, Bhardwaj P, Babu SS, Satheesh SK, Moorthy KK, Chandola HC (2016) Investigations of aerosol black carbon from a semi-urban site in the Indo-Gangetic Plain region. Atmos Environ. https://doi.org/10.1016/j.atmosenv.2015.04.007

Junge CE (1963) Air chemistry and radioactivity. Academic Press, New York

Koch D, Hansen J (2005) Distant origins of Arctic black carbon: a Goddard Institute for Space Studies ModelE experiment. J Geophys Res 110:D04204. https://doi.org/10.1029/2004JD005296

Kulkarn AV, Bahuguna IM, Rathore BP, Singh SK, Randhawa SS, Sood RK, Dhar S (2007) Glacial retreat in Himalaya using Indian Remote Sensing Satellite data. Curr Sci 92(1):69–74

Kumar R, Naja M, Satheesh SK, Ojha N, Joshi H, Sarangi T, Pant P, Dumka UC, Hegde P, Venkataramani S (2011) Influences of the springtime northern Indian biomass burning over the central Himalayas. J Geophys Res 116:D19302. https://doi.org/10.1029/2010JD015509

Kumar R, Naja M, Pfister GG, Barth MG, Brasseur GP (2012) Simulations over South Asia using the Weather Research and Forecasting model with Chemistry (WRF-Chem): set-up and meteorological evaluation. Geosci Model Dev 5:321–343. https://doi.org/10.5194/gmd-5-321-2012

Lau K-M, Kim K-M (2006) Observational relationships between aerosol and Asian monsoon rainfall, and circulation. Geophys Res Lett 33:L21810. https://doi.org/10.1029/2006GL027546

Lau KM, Kim M, Kim K, Lee W (2010) Enhanced surface warming and accelerated snow melt in the Himalayas and Tibetan Plateau induced by absorbing aerosols. Environ Res Lett 5:025204. https://doi.org/10.1088/1748-9326/5/2/025204

Lawrence MG, Lelieveld J (2010) Atmospheric pollutant outflow from southern Asia: a review. Atmos Chem Phys 10:11017–11096. https://doi.org/10.5194/acp-10-11017-2010

Lelieveld J, Crutzen PJ, Ramanathan V, Andreae MO, Brenninkmeijer CAM, Campos T, Cass GR, Dickerson RR, Fischer H, de Gouw JA, Hansel A, Jefferson A, Kley D, de Laat ATJ, Lal S, Lawrence MG, Lobert JM, Mayol-Bracero OL, Mitra AP, Novakov T, Oltmans SJ, Prather KA, Reiner T, Rodhe H, Scheeren HA, Sikka D, Williams J (2001) The Indian Ocean experiment: widespread air pollution from South and Southeast Asia. Science 291(5506):1031–1036. https://doi.org/10.1126/science.1057103

Lu Z, Streets DG, Zhang Q, Wang S (2012) A novel back-trajectory analysis of the origin of black carbon transported to the Himalayas and Tibetan Plateau during 1996–2010. Geophys Res Lett 39:L01809. https://doi.org/10.1029/2011GL049903

Marinoni A, Cristofanelli P, Laj P, Duchi R, Calzolari F, Decesari S, Sellegri K, Vuillermoz E, Verza GP, Villani P, Bonasoni P (2010) Aerosol mass and black carbon concentrations, a two year record at NCO-P (5079 m, Southern Himalayas). Atmos Chem Phys 10:8551–8562. https://doi.org/10.5194/acp-10-8551-2010

Menon S, Hansen J, Nazarenko L, Luo Y (2002) Climate effects of black carbon aerosols in China and India. Science 297:2250–2253

Menon S, Koch D, Beig G, Sahu S, Fasullo J, Orlikowski D (2010) Black carbon aerosols and the third polar ice cap. Atmos Chem Phys 10:4559–4571. https://doi.org/10.5194/acp-10-4559-2010

Monahan EC (1986) The ocean as a source for atmospheric particles. In: Buat-Menard P (ed) The role of air-sea exchange in geochemical cycling. D. Reidel, Hingham, MA, pp 129–163

Moorthy et al (1999) Aerosol climatology over India. 1-ISRO GBP MWR network and database. ISRO/GBP, SR-03-99

Moorthy KK, Babu SS, Sunilkumar SV, Gupta PK, Gera BS (2004) Altitude profiles of aerosol BC, derived from aircraft measurements over an inland urban location in India. Geophys Res Lett 31:L22103. https://doi.org/10.1029/2004gl021336

Moorthy KK, Satheesh SK, Babu SS, Dutt CBS (2008) Integrated Campaign for Aerosols, gases and Radiation Budget (ICARB): an overview. J Earth Syst Sci 117:243–262

Moorthy KK, Babu SS, Manoj MR, Satheesh SK (2013a) Buildup of aerosols over the Indian region. Geophys Res Lett 40:1011–1014. https://doi.org/10.1002/grl.50165

Moorthy KK, Beegum SN, Srivastava N, Satheesh SK, Chin M, Blond N, Babu SS, Singh S (2013b) Performance evaluation of chemistry transport models over India. Atmos Environ 71:210–225

Morys M, Mims FM III, Hagerup S, Anderson SE, Backer A, Kia J, Walkup T (2001) Design, calibration and performance of Microtops II hand-held ozone monitor and sun photometer. J Geophys Res 106:14573–14582

Myhre G, Stordal F, Restad K, Isaksen I (1998) Estimates of the direct radiative forcing due to sulfate and soot aerosols. Tellus Ser B 50:463–477

Nair VS, Solmon F, Giorgi F, Mariotti L, Babu SS, Moorthy KK (2012) Simulation of South Asian aerosols for regional climate studies. J Geophys Res 117:D04209. https://doi.org/10.1029/2011JD016711

Naja M, Bhardwaj P, Singh N, Kumar P, Kumar R, Ojha N, Sagar R, Satheesh SK, Moorthy KK, Kotamarthi VR (2016) High-frequency vertical profiling of meteorological parameters using AMF1 facility during RAWEX–GVAX at ARIES Nainital. Curr Sci 111(1):132–140. ISSN 0011-3891

Omar AH, Winker DM, Vaughan MA, Hu Y, Trepte CR, Ferrare RA, Lee KP, Hostetler CA, Kittaka C, Rogers RR, Kuehn RE, Liu Z (2009) The CALIPSO automated aerosol classification and Lidar ratio selection algorithm. J Atmos Ocean Technol 26:1994–2014. https://doi.org/10.1175/2009JTECHA1231.1

Pant P, Hegde P, Dumka UC, Sagar R, Satheesh SK, Moorthy KK, Saha A, Srivastava MK (2006) Aerosol characteristics at a high-altitude location in central Himalayas: optical properties and radiative forcing. J Geophys Res 111:D17206. https://doi.org/10.1029/2005JD006768

Porch W, Chylek P, Dubey M, Massie S (2007) Trends in aerosol optical depth for cities in India. Atmos Environ 41:7524–7532. https://doi.org/10.1016/j.atmosenv.2007.05.055

Prospero JM, Ginoux P, Torres O, Nicholson SE, Gill TE (2002) Environmental characterization of global sources of atmospheric soil dust identified with the Nimbus 7 Total ozone Mapping Spectrometer (TOMS) absorbing aerosol product. Rev Geophys 40:1002. https://doi.org/10.1029/2000RG000095

Prospero JM, Charlson RJ, Mohnen B, Jaencke R, Delany AC, Mayers J, Zoller W, Rahn K (1983) The atmospheric aerosol system—an overview. Rev Geophys 21:1607–1629

Pruppacher HR, Klett JD (1978) Microphysics of clouds and precipitation. D. Reidel Publishing Company, Holland

Ramanathan V, Crutzen PJ, Kiehl JT, Rosenfeld D (2001a) Aerosols, climate and the hydrological cycle. Science 294:2119–2124

Ramanathan V et al (2001b) Indian Ocean experiment: an integrated analysis of the climate forcing and effects of the great Indo-Asian haze. J Geophys Res 106:28371–28398

Ramanathan V, Raman MV, Roberts G, Kim D, Corrigan C, Chung C, Winker D (2007) Warming trends in Asia amplified by brown cloud solar absorption. Nature 448:575–578. https://doi.org/10.1038/nature06019

Sagar R, Kumar B, Dumka UC, Moorthy KK, Pant P (2004) Characteristics of aerosol optical depths over Manora Peak: a high altitude station in the central Himalayas. J Geophys Res 109:D06207. https://doi.org/10.1029/2003JD003954

Satheesh SK, Ramanathan V (2000) Large differences in the tropical aerosol forcing at the top of the atmosphere and Earth's surface. Nature 405:60–63. https://doi.org/10.1038/35011039

Satheesh SK et al (2008) Climate implications of large warming by elevated aerosol over India. Geophys Res Lett 35:L19809. https://doi.org/10.1029/2008GL034944

Saxena P, Hildemann LM (1996) Water soluble organics in atmospheric particles: a critical review of the literature and application of thermodynamics to identify candidate compounds. J Atmos Chem 24:57–109

Seinfeld JH et al (2004) ACE-Asia—regional climatic and atmospheric chemical effects of Asian dust and pollution. Bull Am Meteorol Soc 85(3):367–380

Seinfeld JH, Pandis SN (1998) Atmospheric chemistry and physics—from air pollution to climate change. Wiley, New York, USA

Sikka DR (1997) Desert climate and its dynamics. Curr Sci India 72:35–46

Singh RP, Dey S, Tripathi SN, Tare V (2004) Variability of aerosol parameters over Kanpur, northern India. J Geophys Res 109:D23206. https://doi.org/10.1029/2004JD004966

Skamarock WC et al (2008) A description of the advanced research WRF version 3. NCAR Tech. Note NCAR/TN-475+STR, Natl. Cent. for Atmos. Res., Boulder, CO, pp 125

Smirnov A, Villevalde Y, O'Neill NT, Royer A, Tarussov A (1995) Aerosol optical depth over the oceans: analysis in terms of synoptic air mass types. J Geophys Res 16:639–650, 24513

Srivastava AK, Ram K, Pant P, Hegde P, Joshi H (2012) Black carbon aerosols over Manora Peak in the Indian Himalayan foothills: implications for climate forcing. Environ Res Lett 7:014002. https://doi.org/10.1088/1748-9326/7/1/014002

Stephens GL et al (2002) The CloudSat mission and the A-Train: a new dimension of space-based observations of clouds and precipitation. Bull Am Meteorol Soc 83:1771–1790

Stohl A (1996) Trajectory statistics—a new method to establish source-receptor relationships of air pollutants and its application to the transport of particulate sulfate in Europe. Atmos Environ 30(4):579–587. https://doi.org/10.1016/1352-2310(95)00314-2

Stohl A (2006) Characteristics of atmospheric transport into the Arctic troposphere. J Geophys Res 111(11):D11306. https://doi.org/10.1029/2005JD006888

Streets DG, Yan F, Chin M, Diehl T, Mahowald N, Schultz M, Wild M, Wu Y, Yu C (2009) Anthropogenic and natural contributions to regional trends in aerosol optical depth, 1980–2006. J Geophys Res 114:D00D18. https://doi.org/10.1029/2008jd011624

Weingartner E, Saathof H, Schnaiter M, Streit N, Bitnar B, Baltensperger U (2003) Absorption of light by soot particles: determination of the absorption co-efficient by means of Aethalometers. J Aerosol Sci 34(10):1445–1463. https://doi.org/10.1016/S0021-8502(03)00359-8

Whitby KT (1978) The physical characteristics of sulfur aerosols. Atmos Environ 12:135–159

Winker DM, Vaughan MA, Omar AH, Hu Y, Powell KA, Liu Z, Hunt WH, Young SA (2009) Overview of the CALIPSO mission and CALIOP data processing algorithms. J Atmos Ocean Technol 26:2310–2323. https://doi.org/10.1175/2009JTECHA1281.1

Young SA, Vaughan MA (2009) The retrieval of profiles of particulate extinction from Cloud Aerosol Lidar Infrared Pathfinder Satellite Observations (CALIPSO) data: algorithm description. J Atmos Ocean Technol 26:1105–1119. https://doi.org/10.1175/2008JTECHA1221.1

Zhang J, Reid JS (2010) A decadal regional and global trend analysis of the aerosol optical depth using a data-assimilation grade over-water MODIS and Level 2 MISR aerosol products. Atmos Chem Phys 10(22):10949–10963. https://doi.org/10.5194/acp-10-10949-2010

Chapter 4
Analysis of Atmospheric Pollutants During Fireworks Festival 'Diwali' at a Residential Site Delhi in India

Pallavi Saxena, Anju Srivastava, Shivangi Verma, Shweta, Lakhwinder Singh and Saurabh Sonwani

Abstract Globally, a number of firework events have been celebrated on a large scale in the names of different festivals. Diwali is one among the popular Indian festival held during October or November every year with huge fireworks. In the present study, various air pollutants like nitrogen oxides (NO_x), particulate matter ($PM_{2.5}$ and PM_{10}) and ozone (O_3) were analyzed in pre, during and post Diwali in two consecutive years i.e. October 2016 and October 2017 in capital city of India, Delhi. The results showed that the background values of particulate matter are exceeding 5–6 times in 2016 and 7–8 times in 2017 than permissible limits set by National Ambient Air Quality Standards (NAAQS), India. In Diwali-2016, the highest PM_{10} and $PM_{2.5}$ concentrations were about 8 times and 7 times higher than NAAQS limits respectively. For Diwali-2017, there was rapid increase in PM_{10} and $PM_{2.5}$ concentrations that were about 10 times and 13 times higher than NAAQS threshold value respectively. Moreover, PM_{10} and $PM_{2.5}$ concentrations in 2017, higher than 2016 were found to be 5–8 times more as compared to background concentrations. However, the concentrations of NO_x and O_3 look similar during background event, pre Diwali, Diwali and post Diwali periods in both the years of 2016 and 2017. The huge Diwali induced air pollution is influenced by transboundary air mass movements from nearby regions of Delhi and adjoining countries in both the selected years of 2016 and 2017 particularly in case of particulate matter in Diwali and Post Diwali. The study concludes that during background and in and around Diwali period receives the air masses containing the emissions from biomass burning which significantly increases the air pollution load.

Keywords Air pollution · Fireworks · Diwali · Delhi

P. Saxena (✉)
Department of Environmental Sciences, Hindu College, University of Delhi, Delhi 110007, India
e-mail: pallavienvironment@gmail.com

A. Srivastava · S. Verma · Shweta · L. Singh
Department of Chemistry, Hindu College, University of Delhi, Delhi 110007, India

S. Sonwani
School of Environmental Sciences, Jawaharlal Nehru University, New Delhi 110067, India

T. Gupta et al. (eds.), *Measurement, Analysis and Remediation of Environmental Pollutants*, Energy, Environment, and Sustainability,
https://doi.org/10.1007/978-981-15-0540-9_4

4.1 Introduction

India is considered as the land of festivals due to its diverse culture and traditions in its different states. It is also observed to have rapid increase in industrialization and urbanization since 19th century onwards. The global organizations like World Health Organization (WHO) and the World Bank have observed that Delhi is one of the metropolis of India which is second most polluted in the world particularly in case of one of the primary air pollutant i.e. particulate pollution (World Bank 2004). The main sources of these air pollutants are high traffic density, high emissions from industries, biomass burning, transboundary mass air movements from adjoining states and domestic activities (Saxena et al. 2012; WHO 2014). Delhi is the 7th highest populated metropolis at global scale. The population of Delhi is 16.75 million during 2001–2011. The total population of Delhi has been ascended to 20.26% during 2001–2011 period with annual growth rate of 1.92% per annum. Upraise in population is mainly due to the migration of people from other states and it is expected to rise by 40% by 2021 (Census of India 2011). Delhi's drastic increase, on the basis of economic aspect and population growth would give rise to an increasing need for transport, imposing excessive pressure on the city's transport infrastructure. Delhi has the highest number of vehicles (~7.4 millions) among the Indian cities (GNCT of Delhi report 2011–2012).

Diwali is famous fireworks festival that is celebrated with great joy and passion all over India every year in October/November. Firing crackers is a significant part of the celebrations in Diwali. Normally the crackers consists of 75% potassium nitrate, 15% charcoal, 10% sulphur, potassium and trace elements, that highly affects the environment as well as human health (Kulshrestha et al. 2004; Agrawal et al. 2011). On burning, fire crackers release air pollutants like CO, NO_x, SO_2 and O_3, Black carbon and particulates are responsible for the production of dense clouds of smoke (Khaiwal et al. 2003; Swamy et al. 2012). A large quantity of crackers are burnt mostly on the onset of festival (Diwali day), on the day before (pre-Diwali day) and after (post-Diwali day) Diwali. Burning of firecrackers produces toxic substances which release toxic gases and particulate matter of fine size to the atmosphere leading to severe health and environmental hazards (Perrino et al. 2011; Rao et al. 2012).

Studies have been made in several parts of the world to determine the effects of firework activities on the ambient air quality. Fleischer et al. (1999) revealed that highly toxic pollutants like furans and polychlorinated dioxins are produced during the display of fireworks like "Fountains" and "Blue lightning rockets". Firecracker activities on New Year's Eve on Oahu (USA) was contributing for an enhancement of total suspended particulates with an average of 300% at 14 locations and with approximately 700% in the lung-penetrating size ranges at a particular location (Bach et al. 1975). An increase of about 57, 25 and 123% in SO_2, NO_2 and PM_{10} levels respectively has been reported during Lantern festival in Beijing by the fireworks display over the previous day. The effect of firework activities on aerosol concentrations in the air during the celebration of New Year's Eve 2005 in Germany has been shown by some studies (Liu et al. 1997; Drewnick et al. 2006). During the FIFA World Cup

2006 celebration, fireworks increased the concentration of metals like Mg, Sr, Ba, Cu and K by 22, 120, 12, 6 and 11 times respectively in the ambient air of Italy (Vecchi et al. 2008). A number of studies have also reported with a rise in trace gases and particulate matter pollution during Diwali period in Delhi, Lucknow, Hisar and Howrah (Attri et al. 2001; Barman et al. 2009; CPCB 2010; Ganguly 2009; Perrino et al. 2011; Khaiwal et al. 2003; Thakur et al. 2010; Kulshrestha et al. 2004; Singh et al. 2010).

In the present study, the variations of trace gases and particulate matter at one of the residential and representative site of Delhi during Diwali festival in 2016 and 2017 were reported. Moreover, the role of transboundary movements was also discussed in and around Diwali festival during selected years.

4.2 Methodology

4.2.1 Monitoring Sites and Data Sources

Measurements of air pollutants viz. nitrogen oxides (NO_x), particulate matter ($PM_{2.5}$ and PM_{10}) and ozone (O_3) in 2016 and 2017 in background, pre, during and post Diwali at one of the residential and representative site in Delhi were analyzed. The location of the site is shown in Fig. 4.1. Hourly concentrations of PM_{10}, $PM_{2.5}$, NO_x and O_3 at selected site were obtained from the Central Pollution Control Board

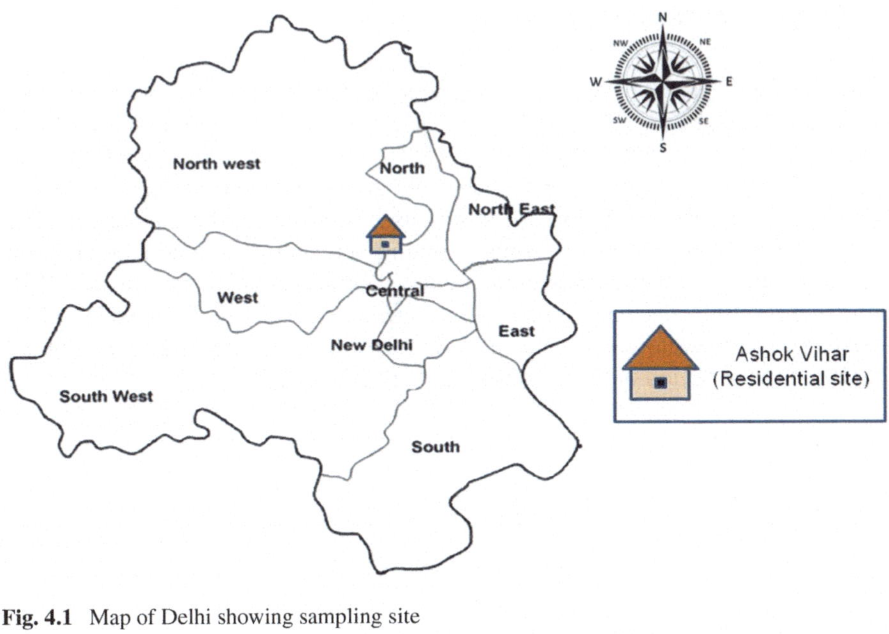

Fig. 4.1 Map of Delhi showing sampling site

(CPCB) (http://cpcb.nic.in/real-time-air-quality-data/) website. O_3 and NO_x were measured using Chemiluminescence method and PM_{10} and $PM_{2.5}$ using tapered element oscillating microbalance (http://cpcb.nic.in/air-quality-standard/). The selected air pollutants were analyzed during the given periods as: (a) background (control days i.e. 25th October, 2016 and 14th October, 2017), (b) Pre Diwali (before the onset of Diwali festival i.e. 26–29th October, 2016 and 15–18th October, 2017), (c) Diwali (on the day of festival i.e. 30th October, 2016 and 19th October, 2017) and Post Diwali (after the Diwali festival i.e. 31st October to 3rd November, 2016 and 20th to 23rd October, 2017).

4.2.2 Description of Study Site

Ashok Vihar (latitude: 28.6910° N, longitude: 77.1765° E) is one of residential site of Delhi surrounded by commercial complexes and nearby to an industrial area. There are around four major market places and 2 major commercial complexes. It is also nearby to a metro station, Keshavpuram and 3 major arterial roads. It has also various traffic intersections which are well connected with ring roads and outer ring roads. The passenger per unit (PCU) at Ashok Vihar has been registered 3,267 per hour. One of the famous industrial area of Delhi is located nearby Ashok Vihar i.e. Wazirpur Industrial area which is a hub of many large scale and small scale industries.

4.2.3 Back Trajectory Analysis

Back trajectory analysis have been widely used for the identification of the air mass movement and long-range transport in the Asian as well as in western countries (Stein et al. 2015; Su et al. 2015; Deka et al. 2015; Escudero et al. 2006; Langford et al. 2018). The online Hybrid Single Particle Lagrangian Integrated Trajectory Model (HYSPLIT) (Draxler and Rolph 2014) was used to calculate the trajectories. The transport of pollutants like trace gases or atmospheric aerosols generally analyzed by Lagrangian-based particle trajectory models (White et al. 2007; Stein et al. 2015). In the present study, five day air mass back trajectories using the NOAA-HYSPLIT model over Delhi during the selected period of Diwali and nearby Diwali days along with background period in the year 2016 and 2017 is used.

4.3 Results and Discussion

4.3.1 Air Mass Back Trajectory Analysis

Figures 4.2a–j and 4.3a–j show the five day air mass back trajectories using the NOAA-HYSPLIT model over Delhi at elevation of 500, 1000, 1500 m above mean sea level (AMSL) during the selected period of Background, Pre Diwali, Diwali and post Diwali days in years 2016 and 2017 both. Figure 4.2a–j depicts the selected

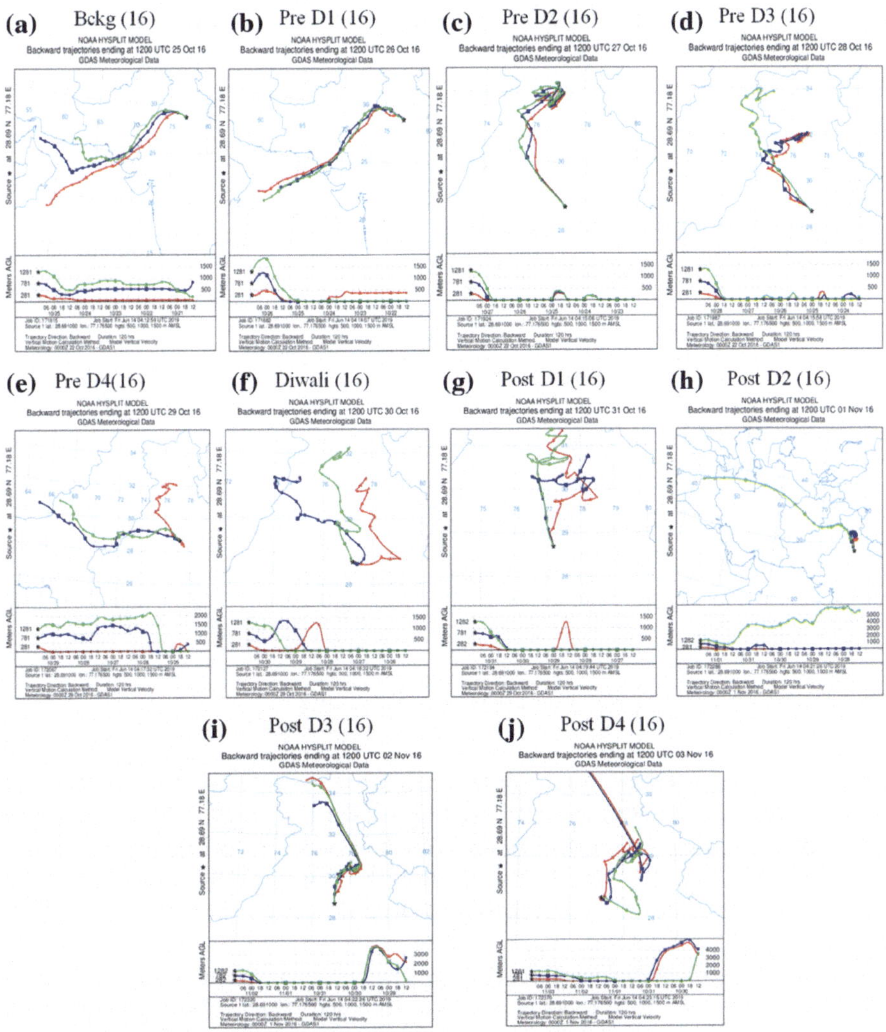

Fig. 4.2 **a–j** Back trajectory plots during selected period in 2016

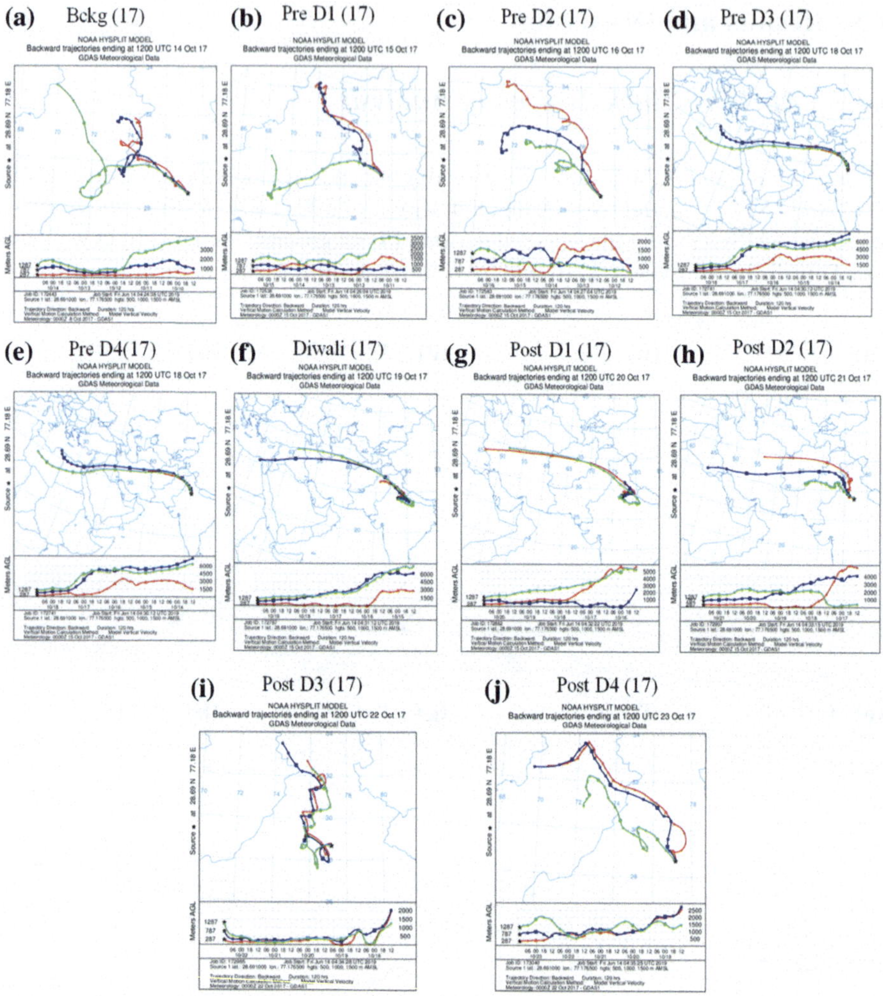

Fig. 4.3 **a–j** Back trajectory plots during selected period in 2017

periods as (a) Background Day, 25th Oct 2016, (b–e) Pre-Diwali day, 26–29th Oct 2016, (f) Diwali Day, 30th Oct 2016 and (g–j) Post Diwali Day, 31st to 3rd Nov 2016. Likewise, Fig. 4.3a–j depicts the selected periods as (a) Background Day, 14th Oct 2017, (b–e) Pre-Diwali day, 15–18th Oct 2017, (f) Diwali Day, 19th Oct 2017 and (g–j) Post Diwali Day, 20th to 23rd Oct 2017. Hysplit back trajectory (ARL Laboratory, NOAA, USA) have been widely used for the identification of the air mass movement and long-range transport in the Asian as well as in Western countries (Stein et al. 2015; Su et al. 2015; Deka et al. 2015; Escudero et al. 2006; Langford et al. 2018). A detailed analysis of these trajectories depicted in Fig. 4.2a. Background Day, 25th Oct 2016, the trajectories are originating from Pakistan and Arabian Sea during 2016, whereas neighboring state of Delhi such as Punjab and Haryana and country like Pakistan were identified as origin of air mass during Background Day, 14th Oct 2017 as plotted in Fig. 4.3a. During Pre Diwali period (Fig. 4.2b–e) 26–29th Oct 2016, it was observed that most of the air masses originated from the neighboring states such as Punjab, Himachal Pradesh, Jammu & Kashmir, apart from this some of the air masses were also originated from Pakistan and some part of Arabian Sea. Moreover, during the Pre Diwali days (Fig. 4.3b–e), 15–18th Oct 2017 the air masses were coming from Punjab and some part of Pakistan, whereas it was also noticed that some of the air mass originated from distant countries in Middle East Countries, and Africa. These regions contain high emission sources of pollutants which may carry forward with the help of winds over Delhi region. On the day of Diwali (30th Oct 2016), apart from the local sources air masses originating from neighboring states such as Punjab, Uttar Pradesh, Uttarakhand, it is also clear from the trajectory that some of the air masses also coming from the Pakistan at same day in 2016. In Diwali Day, 19th Oct 2017 (Fig. 4.3f) the major air masses were coming from neighboring states such as Punjab, Uttar Pradesh and countries like Pakistan and some Middle East countries, which may increase the pollutant load during Diwali day in Delhi region. In the analysis of the trajectories during Post Diwali Day, 31st to 3rd Nov 2016 (Fig. 4.2g–j) the most of the air masses were originated from Punjab, Haryana, Uttarakhand and several part of the Middle East countries including Pakistan. Moreover, the Post Diwali trajectories of 20th to 23rd Oct 2017 (Fig. 4.3g–j) showed similar source origin like in 2016.

Overall, the analysis of the air mass back trajectory conclude that the study period in 2016 and 2017 in and around Diwali period was found under the influence of the air masses coming from the nearby states and countries, which may increase the pollution load during the study period in Diwali in both the years.

4.3.2 Annual Variation in Trace Gases and Particulate Matter Concentrations

Figures 4.4a, b and 4.5a, b show the annual variations of trace gases and particulate matter concentrations at selected residential site which is assumed as typical a representative of emission scenario in Delhi city. To analyze the role of emissions in and around Diwali period, the following days are divided per se in two selected years: (a) background (control days i.e. 25th October, 2016 and 14th October, 2017), (b) Pre Diwali (before the onset of Diwali festival i.e. 26–29th October, 2016 and 15–18th October, 2017), (c) Diwali (on the day of festival i.e. 30th October, 2016 and 19th October, 2017) and Post Diwali (after the Diwali festival i.e. 31st October to 3rd November, 2016 and 20th to 23rd October, 2017). The variations among different periods in different years and varies as per different air pollutants are shown in Figs. 4.4a, b and 4.5a, b.

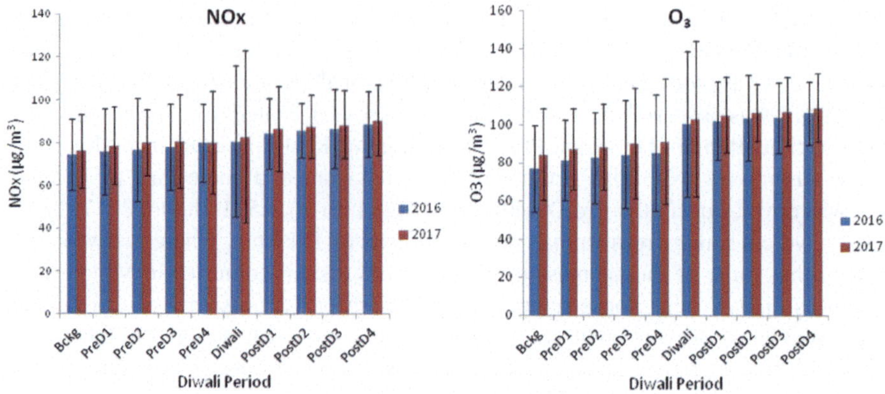

(a) NOx during Diwali period (b) O_3 during Diwali period

Fig. 4.4 **a**, **b** Annual variation of trace gases during selected period of 2016 and 2017

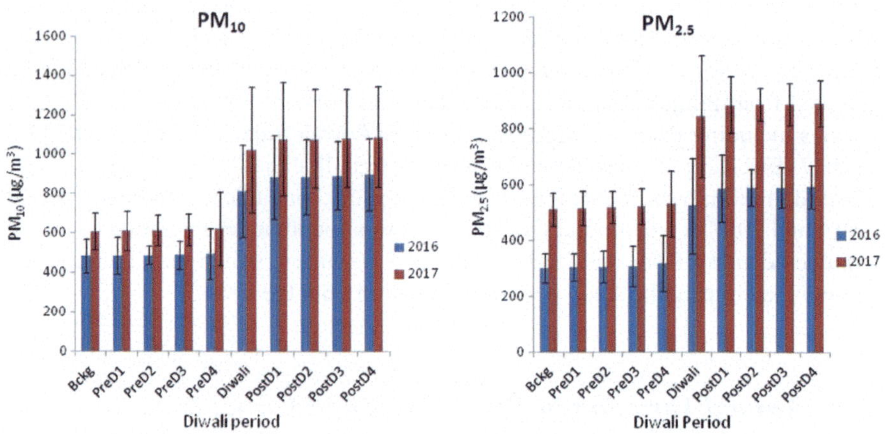

(a) PM_{10} during Diwali period (b) $PM_{2.5}$ during Diwali period

Fig. 4.5 **a**, **b** Annual variation of particulate matter during selected period of 2016 and 2017

4.3.3 Trace Gases

Annual average variations of NO_X and O_3 of selected residential site during all the selected periods viz. Background, Pre Diwali, Diwali and Post Diwali are shown in Fig. 4.4a, b. The concentrations found during different selected in and around festive periods were also being compared with National Ambient Air Quality Standards (NAAQS). In case of NO_X, during the year 2016, in background and Pre-Diwali days, the concentrations were found in the range of 74.3 ± 16.6 to $79.8 \pm 18.2\ \mu g/m^3$ which is slightly near to NAAQS permissible limit. However, the background NO_X concentration of $74.3 \pm 16.6\ \mu g/m^3$ is showing significantly lesser than NAAQS permissible limit but the last and fourth day of Pre-Diwali (PreD4), insignificant difference of NO_x concentration (79.8 ± 18.2 $\mu g/m^3$) was found as compared to NAAQS permissible limit. This is may be due to the reason that, in general, just before the day of Pre-Diwali (PreD4), the firework activities have relatively increased than other days of Pre Diwali (PreD1, PreD2 and PreD3). In addition to that, the selected site is residential site, having about more than 40 major colonies, which will also contribute the high concentrations of NO_x on fourth and last day of Pre Diwali period (PreD4). Interestingly, as shown in Fig. 4.4a, b, there is insignificant difference between the concentrations of background (74.3 ± 16.6 $\mu g/m^3$) and first 3 days of Pre-Diwali (75.5 ± 20.1; 76.4 ± 24.1 and 77.8 ± 20.3 $\mu g/m^3$). This is may be due to the fact that fireworks activities generally becomes higher just before one day of Diwali as compared to previous days of Pre Diwali and since Ashok Vihar is a residential area and connected to ring roads, are responsible for high concentrations of NO_x just before one day of Diwali, as more of shopping and transportation activities becomes relatively higher at that day as compared to other previous days of Pre-Diwali period. Hence, insignificant change was observed during background and first three days of Pre Diwali period. On Diwali day, 80.65 ± 35.5 $\mu g/m^3$ of NO_x was found which is insignificantly different from NAAQS permissible limit, may be due to, firework activities were started from the last day of Pre Diwali, therefore, there is no significant difference between just before one day of Diwali and on the day of Diwali. However, significant difference was found between average value of Pre Diwali period (77.37 ± 15.67 $\mu g/m^3$) and the day of Diwali (80.65 ± 35.5 $\mu g/m^3$). During Post Diwali period, in all the days (PostD1–PostD4), higher concentrations of NO_x (84.32 ± 16.5–88.76 ± 15.3 $\mu g/m^3$) were found that are higher than NAAQS permissible limit. This may be due to the lower boundary layer and low temperature during October month which trapped the air pollutants and resulted in higher concentrations (Srinivas et al. 2015; Sonwani and Saxena 2016; Saxena and Naik 2018). Higher concentrations of NO_x were found in the year 2017 as compared to 2016 during all the selected days of Diwali period. Similar trend was also reported in 2017, in background and Pre-Diwali days, that the concentrations were found in range of 75.9 ± 17.2 to 80.08 ± 24.2 $\mu g/m^3$ which is also slightly near to or at the threshold level of NAAQS limit. In this year too, the background NO_x concentration of 75.9 ± 17.2 $\mu g/m^3$ is showing significantly lesser than NAAQS permissible limit while the 3 days of Pre-Diwali i.e. PreD2, PreD3 and PreD4 were at the threshold or very near to NAAQS permissible limit. On Diwali day, higher concentration of NO_x was found to be 82.65 ± 40.5 $\mu g/m^3$ in 2017 as compared to 2016 which is just crossing the NAAQS permissible limit. Moreover, significant difference was observed between mean value of Pre Diwali period (79.70 ± 22.4 $\mu g/m^3$) and the day of Diwali (82.65 ± 40.5 $\mu g/m^3$). During Post Diwali period, in all the days (PostD1–PostD4), higher concentrations of NO_x (86.43 ± 20.1–90.54 ± 16.7 $\mu g/m^3$) were found that are higher than NAAQS permissible limit and also reported to higher than 2016. This is may be due to the transboundary movements of air masses from adjoining states and neighboring countries for the increasing pattern of air pollutants found to be more prevalent than 2016 as already shown in Figs. 4.2a–j and 4.3a–j.

In case of O_3, significantly lower concentrations were found during background and Pre Diwali period in both the years of 2016 (77.13 ± 22.8–85.43 ± 30.5 μg/m^3) and 2017 (84.34 ± 24.1–91.22 ± 33.0 μg/m^3) as compared to NAAQS permissible limit. This may be due to the fact that, ozone formation is a non linear process, therefore its precursor-product relationship is not dependent on only precursor (e.g. NO_x, VOCs or CO) rather other precursors too (Sonwani et al. 2016). On the day of Diwali, in both the years, ozone concentrations in 2016 (100.54 ± 38.2 μg/m^3) as well as in 2017 (103.23 ± 41.0 μg/m^3) were just crossing the threshold limit of NAAQS. This may be due to that firework activities are found to be more on the day of Diwali as compared to other previous days. Moreover, vehicular activities just before the day of Diwali are generally high and due to lower boundary layer and stable atmospheric conditions, ozone concentrations were found to be high at the day of Diwali (Thakur et al. 2010; Singh et al. 2010; Saxena et al. 2019). During post Diwali period, ozone concentrations were found to be significantly higher than NAAQS permissible limit in both 2016 (102.43 ± 20.0 to 106.54 ± 16.5 μg/m^3) and 2017 (105.32 ± 20.1 to 109.34 ± 18.1 μg/m^3). This may be due to the lower boundary layer and low temperature during October month which trapped the air pollutants and resulted in higher concentrations (Srinivas et al. 2015; Sonwani and Saxena 2016). In case of O_3 too, higher concentrations were found to be more in 2017 as compared to 2016 due to the transboundary movements of air masses from adjoining states and neighboring countries for the increasing pattern of air pollutants found to be more prevalent than 2016 as already shown in Figs. 4.2a–j and 4.3a–j.

4.3.4 Particulate Matter

Figure 4.5a, b shows the changes in the annual mean $PM_{2.5}$ and PM_{10} concentrations at the selected residential site during in and around Diwali period. The concentrations found during different selected in and around Diwali periods were also being compared with National Ambient Air Quality Standards (NAAQS), India. In case of $PM_{2.5}$, the background (301.22 ± 50.30 μg/m^3) and Pre Diwali period concentrations (304.55 ± 48.60 to 318.76 ± 100.50 μg/m^3) are 5 times higher than the permissible limit of NAAQS in the year 2016. Moreover, significant difference was found between background concentration (301.22 ± 50.30 μg/m^3) and all days of Pre Diwali period (PreD1 to PreD4). This may be due to the fact that Ashok Vihar is very near to one of the main industrial area of Delhi i.e. Wazirpur Industrial area which is a hub of construction as well as electroplating industries. These industries may emit a large number of particulate matter which got suspended in the atmosphere. Moreover, in late October month, low temperature and lower boundary layer trapped the particulate matter more firmly as compared to trace gases like NO_x and O_3 (Nishanth et al. 2012; Sonwani and Kulshrestha 2019). Hence, very high concentrations of $PM_{2.5}$ were found during the background and Pre Diwali period. On the day of Diwali, alarmingly high concentration of $PM_{2.5}$ (524.23 ± 170.30 μg/m^3) was reported which is 7 times higher than NAAQS permissible limit. Firework activities and lower boundary layer and low temperature can be the major reason for such high concentrations. During Post Diwali, $PM_{2.5}$ concentrations were found to be 8–9 times higher than NAAQS permissible limit in all the four days of Post Diwali period (589.32 ± 120.50 to 594.22 ± 78.30 $\mu g/m^3$). The high loading of atmospheric particulate matter particularly during winter months due to low temperature leads to the low mixing height and stable atmospheric conditions and transportation of emission coming from biomass burning or crop residue burning activities in nearby states and countries as already shown in Figs. 4.2a–j and 4.3a–j. During the year 2017, $PM_{2.5}$ concentrations were also found to very high and about 8–13 times higher than NAAQS permissible value, which is higher than the values reported in year 2016. Significant difference was found between background concentration (511.23 ± 60.20 $\mu g/m^3$) and all days of Pre Diwali period (516.22 ± 63.0 to 531.28 ± 120.50 $\mu g/m^3$). On the day of Diwali, 845.34 ± 220.50 $\mu g/m^3$ of concentration was found which is almost 13 times higher than NAAQS permissible limit. Moreover, during post Diwali period, in all the days, concentrations (886.12 ± 102.70 to 891.23 ± 83.90 $\mu g/m^3$) were found be 13–14 times higher than NAAQS permissible value. Higher concentrations of $PM_{2.5}$ was found in all the days of background and in and around Diwali period in the year 2017 as compared to 2016 may be due to increased firework activities, transboundary movements of air masses from neighboring states and countries as already discussed in Figs. 4.2a–j and 4.3a–j, increased biomass burning and crop residue burning (Ambade and Ghosh 2013; Sonwani et al. 2016; Sonwani and Kulshreshtha 2016; Sonwani and Kulshrestha 2018).

Similarly, in case of PM_{10}, during the year 2016, concentrations were found to be 5–8 times than NAAQS permissible limit in all the days of Diwali period and background day and 6 to 10 times higher in the year 2017. During the year 2016, significant difference was found between background day (484.56 ± 85.2 $\mu g/m^3$) and all the days of Pre Diwali period (486.55 ± 91.5 to 497.65 ± 126.2 $\mu g/m^3$). In the year 2017 too, very high concentrations of PM_{10} during the period of background (609.76 ± 92.50 $\mu g/m^3$) as well as Pre Diwali period (612.33 ± 99.3 to 623.24 ± 185.80 $\mu g/m^3$) was found which clearly depicted the significant difference between them. This may be due to the fact that likely in the case of $PM_{2.5}$ too, industrial emissions from nearby industrial area of Ashok Vihar emitted a large amount of particulate matter which got suspended in the atmosphere. Moreover, in October month, low temperature and lower boundary layer trapped the particulate matter more firmly as compared to trace gases like NO_x and O_3 (Nishanth et al. 2012). Hence, very high concentrations of $PM_{2.5}$ were found during the background and Pre Diwali period. On the day of Diwali, 813.23 ± 233.70 $\mu g/m^3$ in 2016 and 1023.13 ± 320.90 $\mu g/m^3$ in 2017 PM_{10} concentrations were found which are 8 and 10 times higher than NAAQS standard. Moreover, during Post Diwali period concentrations were found in the range of 885.43 ± 211.78 $\mu g/m^3$ to 899.65 ± 182.80 $\mu g/m^3$ in 2016 and 1077.43 ± 287.20 $\mu g/m^3$ to 1089.23 ± 255.80 $\mu g/m^3$ in 2017. These were also found to be 8–11 times higher NAAQS permissible limit. This is also due to the similar reason as mentioned in the above section of $PM_{2.5}$.

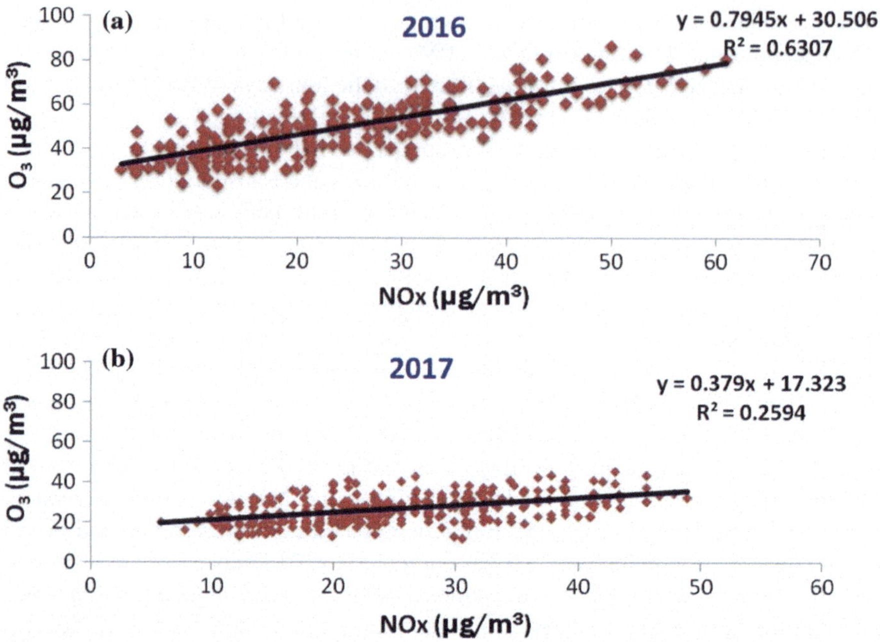

Fig. 4.6 **a, b** Correlation plots of NO_x and O_3 during selected period of 2016 and 2017

4.3.5 Relationship Between NO_x and O_3 in and Around Fireworks Festival

Figure 4.6a shows a positive moderate correlation with a correlation coefficient of 0.63 in 2016. This shows that O_3 was formed from the photolysis of NO_x during the selected period of Diwali but can be produced from other sources too. Similar observation was found in the study of Nishanth et al. (2012). While a weak correlation was found between O_3 and NO_x as shown in Fig. 4.6b with a correlation coefficient of 0.25 in 2017. This depicts that O_3 and NO_x are not well correlated and the O_3 may be produced from other sources during Diwali period.

4.4 Conclusion

The present study concludes that the levels of particulate matter as well as trace gases stay near or above the NAAQS threshold level for the "very healthy" and in some cases "severe" category at one of the residential site of Delhi in 2016 and 2017 for most of the selected period of in and around Diwali. Generally, a significant rise in the concentrations of selected air pollutants (trace gases and particulate matter) from 4 days back from Diwali, considering during Pre Diwali period onwards reveals the impact of high background emissions and transported component behind their high concentrations. Moreover, the transboundary movements of air masses from adjoining states of Delhi along with nearby neighboring countries like Pakistan and Middle East countries have also contributed to a larger extent in the form of biomass burning. The concentrations of trace gases and particulate matter also showed an increasing trend year after year. Therefore, higher concentrations of trace gases and particulate matter were found in 2017 as compared to 2016. In Diwali-2016, the highest PM_{10} and $PM_{2.5}$ concentrations were about 8 times and 7 times higher than National Ambient Air Quality Standards (NAAQS) limits respectively. For Diwali-2017, there was rapid increase in PM_{10} and $PM_{2.5}$ concentrations that were about 10 times and 13 times higher than NAAQS threshold value respectively. Moreover, PM_{10} and $PM_{2.5}$ concentrations in 2017, higher than 2016 were found to be 5–8 times more as compared to background concentrations. Moreover, particulate matter (PM_{10} and $PM_{2.5}$) seems similar during Diwali and post Diwali period. The relatively similar concentration (no significant difference) of PM even after Diwali period pointed towards some transported component along with the air-masses coming from the neighboring states and countries, which contribute high load of PM during and after Diwali period in Delhi. Thus, the heavy background emissions due to biomass burning/agricultural-waste burning may contribute significantly to the total pollution load (especially PM) in and around firework festival in Delhi.

References

Agrawal A, Upadhyayl VK, Sachdeva K (2011) Study of aerosol behavior on the basis of morphological characteristics during festival events in India. Atmos Environ 45(21):3640–3644

Ambade B, Ghosh S (2013) Characterization of PM10 in the ambient air during Deepawali festival of Rajnandgaon district, India. Nat Hazards 69:589–598

Attri AK, Kumar U, Jain VK (2001) Microclimate: formation of O_3 by fireworks. Nature 411(6841):1015

Bach W, Daniels A, Dickinson L, Hertlein F, Morrow J, Margolis S, Dinh V (1975) Fireworks pollution and health. Pol J Environ Stud 7:183–192

Barman SC, Singh R, Negi MPS, Bhargava SK (2009) Fine particles (PM2.5) in ambient air of Lucknow city due to fireworks on Diwali festival. J Environ Biol 30(5):625–632

Census of India (2011). Available at: http://censusindia.gov.in/2011-prov-results/provdata_productsdelhi.html

CPCB (Central Pollution Control Board) (2010). Ambient air and noise pollution levels—Deepawali 2010. Press Release, Delhi, pp 1–8

Deka JP, Baruah B, Singh S, Chaudhury R, Prakash A, Bhattacharyya P, Kumar M (2015) Tracing phosphorous distributions in the surficial sediments of two eastern Himalayan high altitude lakes through sequential extraction, multivariate and HYSPLIT back trajectory analyses. Environ Earth Sci 73(11):7617–7629

Draxler R, Rolph G (2014) HYSPLIT (HYbrid Single-Particle Lagrangian Integrated Trajectory) Model access via NOAA. NOAA Air Resources Laboratory ARL READY Website http://ready.arl.noaa.gov.HYSPLIT.php

Drewnick F, Hings SS, Curtius J, Eerdekens G, Williams J (2006) Measurement of fine particulate and gasphase species during the New Year's fireworks 2005 in Mainz, Germany. Atmos Environ 40:4316–4327

Escudero M, Stein A, Draxler RR, Querol X, Alastuey A, Castillo S, Avila A (2006) Determination of the contribution of northern Africa dust source areas to PM10 concentrations over the central Iberian Peninsula using the Hybrid Single-Particle Lagrangian Integrated Trajectory model (HYSPLIT) model. J Geophys Res: Atmos 111(D6)

Fleischer O, Wichmann H, Lorenz W (1999) Release of polychlorinated dibenzo-p-dioxins and dibenzofurans by setting off fireworks. Chemosphere 39:925–932

Ganguly ND (2009) Surface ozone pollution during the festival of Diwali, New Delhi, India. J Earth Sci India 2(4):224–229

GNCT of Delhi report, 2011–2012, Transport Department, pp 1–175

Khaiwal R, Mor S, Kaushik CP (2003) Short-term variation in air quality associated with firework events: a case study. J Environ Monit 5:260–264

Kulshrestha UC, Rao TN, Azhaguvel S, Kulshrestha MJ (2004) Emissions and accumulation of metals in the atmosphere due to crackers and sparkles during Diwali festival in India. Atmos Environ 38(27):4421–4425

Langford AO, Alvarez RJ II, Brioude J, Evan S, Iraci LT, Kirgis G, Senff CJ (2018) Coordinated profiling of stratospheric intrusions and transported pollution by the Tropospheric Ozone Lidar Network (TOLNet) and NASA Alpha Jet experiment (AJAX): observations and comparison to HYSPLIT, RAQMS, and FLEXPART. Atmos Environ 174:1–14

Liu DY, Rutherford D, Kinsey M, Prather KA (1997) Real-time monitoring of pyrotechnically derived aerosol particles in the troposphere. Anal Chem 69(10):1808–1814

Nishanth T, Praseed KM, Rathnakaran K, Kumar MKS, Krishna RR, Valsaraj KT (2012) Atmospheric pollution in a semi-urban, coastal region in India following festival seasons. Atmos Environ 47:295–306

Perrino C, Tiwari S, Catrambone M, Torre SD, Rantica E, Canepari S (2011) Chemical characterization of atmospheric PM in Delhi, India, during different periods of the year including Diwali festival. Atmos Pollut Res 2(4):418–427

Rao PS, Gajghate DG, Gavane AG, Suryawanshi P, Chauhan C, Mishra S, Gupta N, Rao CVC, Wate SR (2012) Air quality status during Diwali Festival of India: a case study. Bull Environ Contam Toxicol 89:376–379

Saxena P, Bhardwaj R, Ghosh C (2012) Status of air pollutants after implementation of CNG in Delhi. Curr World Environ 7(1):109

Saxena P, Naik V (eds) (2018) Air Pollution: Sources, Impacts and Controls. CABI

Saxena P, Srivastava A, Tyagi M, Kaur S (2019) Impact of tropospheric ozone on plant metabolism – a review. Pollut Res 38(1):175–180

Singh DP, Gadi R, Mandal TK, Dixit CK, Singh KST, Singh N et al (2010) Study of temporal variation in ambient air quality during Diwali festival in India. Environ Monit Assess 169(1):1–13

Sonwani S, Amreen H, Khillare PS (2016, July) Polycyclic Aromatic Hydrocarbons (PAHs) in urban atmospheric particulate of NCR, Delhi, India. In 41st COSPAR Scientific Assembly (Vol. 41)

Sonwani S, Kulshreshtha U (2016) Particulate matter levels and it's associated health risks in East Delhi. Proceedings of Indian aerosol science and technology association conference on aerosol and climate change: insight and challenges. IASTA Bull 22(1–2). ISSN 09714510

Sonwani S, Kulshrestha U (2018) Morphology, elemental composition and source identification of airborne particles in Delhi, India. J Indian Geophys Union 22(6):607–620

Sonwani S, Kulshrestha, UC (2019) PM 10 carbonaceous aerosols and their real-time wet scavenging during monsoon and non-monsoon seasons at Delhi, India. J Atmos Chem 1–30

Sonwani S, Saxena P (2016) Identifying the sources of primary air pollutants and their impact on environmental health: a review. Int J Eng Tech Res 6(2):111–130

Srinivas R, Panicker AS, Parkhi NS, Peshin SK, Beig G (2015) Sensitivity of online coupled model to extreme pollution event over a mega city Delhi. Atmos Pollut Res. APR-D-14-00373

Stein AF, Draxler RR, Rolph GD, Stunder BJ, Cohen MD, Ngan F (2015) NOAA's HYSPLIT atmospheric transport and dispersion modeling system. Bull Am Met Soc 96(12):2059–2077

Su L, Yuan Z, Fung JC, Lau AK (2015) A comparison of HYSPLIT backward trajectories generated from two GDAS datasets. Sci Total Environ 506:527–537

Swamy YV, Venkanna R, Nikhil GN, Chitanya DNSK, Sinha PR, Ramakrishna M et al (2012) Impact of oxides of nitrogen, volatile organic carbons and black carbon emissions on ozone weekend/weekday variations at a semi arid urban site in Hyderabad. Aerosol Air Qual Res 12:662–671

Thakur B, Chakraborty S, Debsarkar A, Chakrabarty S, Srivastava RC (2010) Air pollution from fireworks during festival of lights (Deepawali) in Howrah, India—a case study. Atmósfera 23(4):347–365

Vecchi R, Bernardoni V, Cricchio D, D'Alessandro A, Fermo P, Lucarelli F, Nava S, Piazzalunga A, Valli G (2008) The impact of fireworks on airborne particles. Atmos Environ 42:1121–1132

White AB, Darby LS, Sneff CJ, King CW, Banta RM, Koermer J, Wilczak JM, Neiman PJ, Angevine WM, Talbot R (2007) Comparing the impact of meteorological variability on surface ozone during the NEAQS (2002) and ICARTT (2004) field campaigns. J Geophys Res 112:D10S14

WHO (2014). http://www.who.int/phe/health_topics/outdoorair/databases/cities-2014/en/

World Bank (2004) Most dangerous polluted cities: particulate matter air pollution 2004. Available at: http://www.allcountries.org/air_pollution.html

Chapter 5
Organic Air Pollutants: Measurement, Properties & Control

Abhishek Chakraborty

Abstract In last few decades, air pollution has emerged as a major threat to human health and different gaseous and particulates air pollutants are found to be influencing global climate directly or indirectly via altering radiative forcing or of cloud micro-physical properties. Organic air pollutants contribute a substantial portion to the total pollutants load and usually, in polluted location, their contribution can go up to 90% of the total particulate phase air pollutants or aerosols. Organic particulate pollutants or aerosols can be both primary or secondary in nature and generally changes their characteristics significantly upon reacting with different atmospheric oxidants like ozone or hydroxyl radicals. Sources and characteristics of organic air pollutants generally display a wide range of spatiotemporal variability. Several techniques are available the detection and measurement of organic air pollutants but time resolution and types of organic pollutants detected by different techniques vary significantly. However, since the characteristics of the organic pollutants can change relatively quickly via atmospheric processing it is desirable to measure these pollutants in real time. Recent advances in mass spectrometric techniques have enabled the scientists to understand the evolution of these organic pollutants in real-time and that led to more accurate source apportionment and identification of factors that influence the formation of secondary organic aerosols. In this chapter readers will get a comprehensive overview of organic air pollutants characteristics, their sources, evolution in the atmosphere and possible ways to control their abundance.

Keywords Organic aerosols · Measurement · Secondary formation · Control

5.1 Introduction

Air pollution has emerged as a global threat to climate and human health (Arnold et al. 2009), and the situation is especially troublesome in developing countries like India, China, Brazil etc. Air pollutants can be broadly divided into two categories

A. Chakraborty (✉)
ESED, IIT Bombay, Mumbai, India
e-mail: abhishekc@iitb.ac.in

T. Gupta et al. (eds.), *Measurement, Analysis and Remediation of Environmental Pollutants*, Energy, Environment, and Sustainability,
https://doi.org/10.1007/978-981-15-0540-9_5

depending upon their physical form; gaseous and particulate (PM). Gaseous pollutants are those which exist in gaseous form under normal ambient conditions and include methane, CFCs (Chlorofluorocarbons), ground level ozone (O_3), oxides of nitrogen and sulfur (NO_x & SO_x), and VOCs (Volatile organic compounds). Several gaseous and particulate (or aerosol) pollutants can interact with incoming or outgoing solar radiation either directly or indirectly and can alter the temperature profile of Earth's atmosphere. Gaseous pollutants like CH_4 can absorb outgoing long-wave radiation and heat up the atmosphere. On the other hand, CFCs are mainly responsible for the destruction of stratospheric ozone, while NO_x, SO_x & VOCs can produce several secondary pollutants including ground level ozone, which in turn damage materials and cause irritation in eyes and respiratory system. Particulate phase air pollutants or particulate matters (PM) are also widely known as aerosols; liquid droplets and/or solid particles suspended in a gaseous medium (which in this context; air). In terms of origin, aerosols can directly enter the atmosphere via natural and anthropogenic (human-made) sources and termed as primary aerosols. Natural sources include volcanic eruptions, forest fires, seas-salt, resuspension of mineral dust, etc. while anthropogenic sources can be combustion activities (biomass burning, engine emission), mining, construction, power generation, etc. Secondary aerosols can be formed via gas phase chemical reactions and subsequent nucleation and condensation of gas phase products. NO_x & SO_x can be converted into nitrate and sulfate aerosols via gas phase oxidation and gas to particle conversion. Composition wise aerosols can be divided into two major fractions; organic and inorganic. Organic aerosols (OA) or organic carbon (OC) is a subset of carbonaceous aerosols, which includes black carbon (BC) and elemental carbons (EC) as well. Black and elemental carbon are the soot particles, made up of pure carbon, refractory and have a significant impact on the atmospheric radiation budget. BC & EC both are a refractory fraction of the carbonaceous aerosols but measured by different techniques; BC by optical absorption method and EC by thermal oxidation followed by optical detection of the produced CO_2. OA on the other hand, mainly composed of non-refractory/organic carbon (OC), hydrogen and oxygen along with some hetero atoms like sulfur and nitrogen (as minor components and not always present). For this chapter, we will use OC & OA interchangeably as they can be easily converted from one value to another (OC to OA and vice versa) using some well characterized regional ratios for ambient aerosols. Inorganic aerosols are mostly a combination of sulfur or nitrogen with oxygen and/or hydrogen, sometimes chlorine and bromine are also found in inorganic aerosols in lesser proportions. Black carbon, a form of particulate pollution can absorb the incoming solar radiation thus increases the temperature of the atmosphere. On the other hand, sulfate and nitrate aerosols can scatter the incoming solar radiation to impart a cooling effect on the atmosphere. Aerosols can indirectly influence the incoming solar radiation by acting as condensation nuclei for cloud (and fog), which in turn will reflect a substantial portion of incoming solar radiation. Aerosols can also affect the micro-structure of the cloud; it can stay longer and delay the precipitation or can influence the intensity of the rainfall. Aerosols have been linked to diseases related to human respiratory and cardiovascular systems

(Atkinson and Arey 2003; Aumont et al. 2005). Global burden of disease study (Baltensperger et al. 2005) conducted by WHO (World Health Organization) revealed that outdoor air pollution is the second largest killer globally behind only to the heart attack. Two countries; India and China contribute more than half of the total premature deaths caused by air pollution. The formation mechanism, properties and impacts of inorganic aerosols are well characterized, however, a significant amount of uncertainty remains on the impact of organic aerosols on human health and global climate due to less understanding of the formation, evolution and characteristics of organic aerosols.

5.2 Budget and Properties of OA

Organic aerosols contribute significantly to the total aerosol budget of the troposphere, depending upon geographical area, local emission sources it can contribute between 20 and 90% of the total fine aerosols (aerosol with <2.5 μm diameter) (Bertram et al. 2001; Birch and Cary 1996; Bond et al. 2004). Organic aerosols (OA) exhibit a wide variation in terms of sources, evolution via atmospheric processing and physico-chemical characteristics. Broadly, OA can be divided into primary (POA) and secondary OA (SOA), depending on whether emitted directly or formed via atmospheric processing.

5.2.1 Primary Organic Aerosols (POA)

Emission budget

Primary OA (or OC) emitted directly by different natural and anthropogenic sources; mostly due to incomplete combustion activities like forest fires, biomass burning, vehicular emission cooking, and industrial processes. Some other natural non-combustion sources of POA are marine aerosols and biological aerosols (cellular materials like fungal spores, and pollen etc.) (Canagaratna et al. 2007; Cappiello et al. 2003; Chakraborty et al. 2015). Estimated mean global annual emission of primary OC (POC) from combustion is around 34 Tg (1 Tg = 10^9 kg) (Chakraborty et al. 2016). POC from natural combustion events like forest fires or burning of Savana grassland (an example of open biomass burning) contributes around 74% (maximum) to total combustion related OC, followed by OC emissions from anthropogenic combustion like biofuel (19%) and fossil fuel burning (7%). However, this estimate has a high degree of uncertainty with a range of OC emission varying from 17 to 77 Tg/year, the uncertainty mostly stemmed from the level of accuracy associated with the emission inventory and different volatility profiles of different POAs, which dictates the gas to particle phase portioning. So, part of the volatile organic compounds (VOCs) that partitioned into the particle phase to produce POA may again evaporate back to the gas phase when dilution takes place after being released to the atmosphere. This partitioning between gas and particle phase can be instantaneous and very sensitive to measuring conditions so this could add significant variations in the estimation of emission inventory. Marine POA emission rates are estimated to be around 10 (±5) Tg/year (Cappiello et al. 2003; Chakraborty et al. 2016), but these estimations are again quite uncertain with some studies estimating marine POA emission rates as high as 40 Tg/year (Chirico et al. 2010). Marine POA generally enters into the atmosphere via a burst of sea bubbles; these bubbles form via breaking waves which scavenge surface active and other types of organic materials during their rise to ocean surface (Cappiello et al. 2003). Marine POA emission generally shows a strong seasonality with maximum concentrations during spring/summer and minimum during winter, possibly due to higher spring time biological activity in the ocean. Biological materials also contribute to ambient organic aerosols and POA emission rates associated with primary biological aerosol particles (PBAP) from fungi estimated to be around 28 Tg/year (Chakraborty et al. 2015), some studies (Crippa et al. 2013; Cubison et al. 2011) reported total emission rates of 50–164 Tg/year as well with major contribution (~25%) from fungal spores.

Characteristics of POA

Primary OA characteristics mostly determined by the source from which it is being emitted. OA emitted from fossil fuel combustion is generally reside in the Aitken to accumulation mode range, typically with a mass median diameter of around 90 nm. More than 90% of the traffic related POA (vehicular emission, fossil fuel combustion) reside in ultrafine mode (<100 nm), while most of the POA emitted from biomass burning associated with particles of diameter 200 nm or less (Decesari et al. 2000). POA emitted from cooking activities generally display dual mode size distribution with a lower size mode around 200 nm and higher size mode around 500 nm. Most of the marine POA is found to be associated with accumulation mode particles with diameters between 50 and 300 nm (Cappiello et al. 2003). The contribution of marine POA (MOA) to total sea-salt mass in accumulation mode can be as high as 60%, especially during summer/spring time when ocean biological activities are at peak (Docherty et al. 2008). Marine POA in the submicron size range is mostly associated with phytoplankton residues and polysaccharides, while in super-micron size range, it is dominated by carboxylic acid additives and sea-salt aerosols (Dockery and Pope 1994; Donahue et al. 2006). Fossil fuel POA (FOA) is generally considered to be non-hygroscopic (has no water affinity and will not grow under high humidity) with hygroscopic growth factor (GF) of 1 (GF = ratio of wet to dry aerosol diameter at 90% RH, unless stated otherwise) (Donahue et al. 2012). POA from biomass burning activity (BBOA) is more hygroscopic than FOA and can act as cloud condensation nuclei (CCN) for cloud/fog formation. Hygroscopic GF of marine POA is found to be 1.25, which is on the lower side. However, it's CCN activity is quite high (Elbert et al. 2007), which means a high percentage of marine POA will grow into droplets when RH is sufficiently high. This apparent contradiction is mostly due to the reduction of water surface tension by surface active organics, which facilitates the droplet growth (Elbert et al. 2007). In terms of volatility, most of the POA is found to be semi to intermediate-volatile in nature, with saturation vapor pressure ranging from 1 to 1000 $\mu g/m^3$ (1.7×10^{-7} to 1.7×10^{-4} mbar) (Ervens and Volkamer 2010). However, volatility characteristics of POA emitted from both cooking and biomass burning can vary greatly depending upon type of fuel/food/meat and burning/cooking temperatures along with ambient conditions (Ervens et al. 2003). POA, in general, is less oxidized compared to SOA with O/C rarely exceeding 0.30, though POA emitted from biomass burning activities found to be reasonably oxidized (~0.30) even at the source (Ervens et al. 2011). Mixing state of different aerosol constituents can influence optical and other physicochemical properties of the overall aerosol population; thus understanding of mixing state is essential. Marine POA associated with smaller aerosols (<200 nm) is more likely to be externally mixed with other components of marine aerosols like sea-salt, ammonium sulfate while primary OA associated with larger aerosols (~1000 nm) is more likely to be internally mixed with other chemical constituents (Donahue et al. 2006; Forouzanfar et al. 2015). Some part of the BBOA is also light absorbing and shows a significant dependence on the wavelength (at which absorption is measured) and burning conditions, while the type of biomass has less influence on absorbing characteristics of BBOA (Gantt and Meskhidze 2013). POA emitted from fossil fuel combustion is relatively less absorbing than BBOA, mostly due to compositional differences.

5.2.2 Secondary Organic Aerosol (SOA)

Secondary OA in the atmosphere can arise out of oxidation of gas phase VOCs and subsequent gas to particle partitioning of those oxidized products and heterogeneous, multiphase oxidation of primary organic compounds. Properties of these SOA are entirely different than their precursors and SOA plays a vital role in influencing overall OA and their characteristics relevant to climatic and health impacts.

SOA budget

SOA budget estimation is considerably more difficult than POA estimation as several other factors like type and extent of oxidation reaction, reaction yield and partitioning of reaction products come into play apart from the accuracy of primary OA and VOC emission inventories. Reported global SOA budget varies significantly depending upon the chosen approach and type of precursor inventory. Global SOA budget using traditional 'Bottom-up' approach (using known or assumed flux rates of dominant VOCs and utilizing SOA production rate from those VOCs obtained from laboratory smog chamber studies) ranges between 14 and 82 Tg/year (Arnold et al. 2009). More than 80% of the budgeted SOA is biogenic in nature, originating from the oxidation of biogenic VOCs like Isoprene, α-pinene, limonene, etc. However, VOC emission rates are mostly taken for terrestrial eco-system, biogenic VOCs of marine origin have mostly been neglected in global SOA estimation schemes, and this could result in the severe underestimation of global SOA production as shown by a few studies (Ge et al. 2012; George and Abbatt 2010). Global SOA budget estimated using a relatively new 'top-down' approach by constraining the final fate of the precursors based on laboratory (field observations) leads to 200–1300 Tg/year, values which are an order of magnitude higher than obtained from 'Bottom-up' approach. Higher end values of this 'Bottom-up' approach are quite unrealistic & unlikely as SOA yield data obtained from laboratory smog chamber experiments for different precursors are not in line with the suggested higher values (Arnold et al. 2009). This large discrepancy between two approaches is indicative of serious gaps in our understanding of SOA formation and lacunas in the way actual ambient processing is represented in smog chamber experiments. A 'Hybrid' SOA budget estimate method combining some features of the other two previously mentioned approaches have been developed recently, and the best estimate for SOA is now stood at 190 Tg/year (converting SOC of 136 Tg/year by assuming an OM (or OA)/OC ratio of 1.40), with an upper and lower limit of 7 & 400 Tg/year. Oxidation products of biogenic VOCs (emitted by trees) contribute around 60% of the total estimated SOA, with the rest being contributed by biomass burning and anthropogenic activities. A wide range of estimated upper and lower bound SOA values along with huge discrepancies between different approaches indicate that uncertainty associated with these estimation methods is quite significant. Gaps in the characterization of VOC and other precursors, reaction mechanisms and SOA yield, along with dynamic and continuously evolving characteristics of SOA are some of the primary reasons behind the uncertainties associated with global SOA budget estimation.

SOA formation

Gas phase oxidation of VOCs

SOA forms mainly through gas phase oxidation of VOCs by hydroxyl (OH·) radicals, nitrate ($NO_3\cdot$) radicals, ozone (O_3) or via photolysis and subsequent gas to particle phase partitioning of the oxidation products. VOC oxidation (especially biogenic VOCs) occurs mostly by following major mechanisms: (a) Addition of functional groups to $C = C$ bonds by the atmospheric oxidants (b) Abstraction of H-atom from C–H bonds (and rarely from O–H bonds) by nitrate and OH radicals. Ozone mostly reacts with VOCs only via addition to $C = C$ bonds mechanism (Gilardoni et al. 2009). The most effective oxidant and the dominant reaction pathway will be determined by the structure of the precursor VOCs and ambient conditions (Arnold et al. 2009; Gilardoni et al. 2009). Initial reactions VOC s and oxidants generally produce first generation oxygenated products with one or more oxygen containing functional groups like ketone (–C(=O)–), alcohol (–OH), aldehyde (–C(=O)H), nitrate ($-ONO_2$), carboxylic acid (–C(=O)OH), peroxyacyl nitrate ($-C(=O)OONO_2$), percarboxylic acid (–C(=O)OOH) and hydroperoxide (–OOH). Not all the VOCs emitted have same SOA formation potential, depending upon their reactivity, vapor pressure and other properties it is found that some classes of VOCs have more SOA formation capacity than others. Cyclic VOCs (like cycloalkenes, aromatic hydrocarbons and the majority of the terpenes) are generally efficient in producing SOA as their oxidation predominantly occurs via ring opening and addition of oxygenated functional groups to the existing carbon backbone (Arnold et al. 2009). The majority of the understanding of the mechanism of VOC oxidation comes from chamber experiments though only a fraction of VOCs has been tested in those experiments. However, the application of structure-reactivity relationship obtained from those studies can provide insights into the mechanism of oxidation for many other VOCs (Goldstein et al. 2008; Hallquist et al. 2009).

Gas to particle partitioning of organics

Secondary organic compounds formed via gas phase oxidation of VOCs often partition into condensed phase (dry or wet particulates and droplets; jointly known as 'aerosols') and form SOA depending upon their volatility and ambient conditions (temperature, humidity, existing aerosol loading condition etc.). It is assumed that SOA is made up of semi-volatile compounds which partition between gas and particle phases depending upon the existing conditions. Partitioning theory was first proposed by Pankow in 1990s (Hamilton et al. 2004; Hawkins and Russell 2010) and partitioning of each organic compound can be expressed by the following model (Source: Donahue et al. 2006; Hallquist et al. 2009):

$$\frac{C_i^p}{C_i^g} = K_{p,i}C_{OA} = \frac{C_{OA}}{C_i^*}$$

where C_i^g is the mass concentration of species *i* per unit volume of air (μg/m^3) in gas phase, C_i^p is the mass concentration of species i per unit volume of air (μg/m^3) in the particulate phase, $K_{p,i}$ (m^3/μg) is equilibrium partitioning coefficient. $C_i^*\left(\mu g/\mathrm{m}^3\right)$ is the saturation vapor concentration (inverse of $K_{p,i}$), and C_{OA} is the mass concentration of the total absorbing particle phase per unit volume of air (μg/m^3).

This theory is an absorptive partitioning model where gas phase semi-volatile organics absorbed into the existing particulate or aqueous phase organics. Some semi-volatile organics will always partition into the particulate phase as long as some OA is available irrespective of whether the saturation vapor pressure for a particular semi-volatile compound has been reached or not. The fraction of organics (F_i) that will be partitioned into the particle phase can be obtained from the following equation (Source: Donahue et al. 2006; Hallquist et al. 2009):

$$F_i = \frac{C_i^p}{C_i^p + C_i^g} = \frac{C_{OA}K_{p,i}}{1 + C_{OA}K_{p,i}} = \frac{1}{1 + C_i^*/C_{OA}}$$

As the existing OA concentration (absorbing phase) increases, even the compounds with greater volatility (with higher C_i^* values) begin to partitioned into the particulate phase. This indicates that in polluted locations (like India, China) the gas phase products will be partitioned into particle phase more efficiently and contribute further to already existing higher aerosol loadings. Using this model Odum et al. (1996) proposed a method to calculate SOA yield of different organic precursors via the following equation (Source: Hallquist et al. 2009; Odum et al. 1996):

$$F_{OA} = \frac{\Delta M_{OA}}{\Delta M_{HC}} = C_{OA} \sum_i \frac{\alpha_i K_{p,i}}{1 + C_{OA} K_{p,i}} \equiv \sum_i \frac{\alpha_i}{1 + C_i^* / C_{OA}}$$

α_i is the stoichiometric yield (mass based) of species i. Two parameters represent volatility distribution of the products; α_i and C_i^*, thus this model is known as two product model and widely used for SOA yield estimation. However, limitations of the above model become evident as additional knowledge about complexities SOA formation and wide variations in their physicochemical properties obtained through field and chamber studies. Mainly the vast range of C_{OA} values and ongoing oxidation of organic even after partitioning to particle phase makes it difficult to apply the simple absorption based model with confidence to estimate SOA concentrations. To address these shortcomings, Donahue and co-workers (Hawkins et al. 2010; Heal et al. 2012) proposed a "Volatility basis set" (VBS) approach. In this approach, a group of compounds with very similar C_i^* values are lumped together to form a 'bin' and a total of 9 such 'bins' are constructed with one order of magnitude separation among the 'bins' in C_i^* values at 300 K. Using the VBS, different SOA formation reactions and their end products with different volatilities can be mapped in one or several sets of bins depending upon existing ambient OA levels. So, using this model, changes in SOA production due to aging (continuous degradation/evolution in OA characteristics due to ambient oxidation and other processing) can also be mapped, provided kinetics and volatilities of the end products is somewhat known.

Condensed phase formation and evolution of SOA

Particle phase

Condensed phase reactions of organic aerosols have recently gained considerable attention from the aerosol science community. This interest mostly stems from the realization that understanding of condensed phase OA processing may help to explain some of the discrepancies between observed and model-estimated SOA production and characteristics. Particle phase OA processing may also offer some insights into the formation of high molecular weight organic compounds (like HULIS) detected in atmospheric aerosols. A significant number of laboratory studies have been conducted to understand the mechanisms, yields, influencing parameters and impact of condensed phase reactions on overall OA characteristics and lifetime. Condensed phase processing of organics involve several other physicochemical processes such as diffusion of gas phase organics to the particle/droplet surface, mass accommodation coefficient (transfer efficiency of organics from gas phase to the condensed phase upon collision with a particle/droplet), subsequent reactions on surface and bulk followed by the diffusion/dissolution in the aerosol/droplet phase (Fig. 5.1) resulting in formation of various products with varying volatilities. The radicals need to be first taken up by the condensed phase and represented by reactive uptake coefficient (γ). The reactive uptake coefficient indicates the probability that a reaction will occur after gas-surface (of the condensed phase) collision.

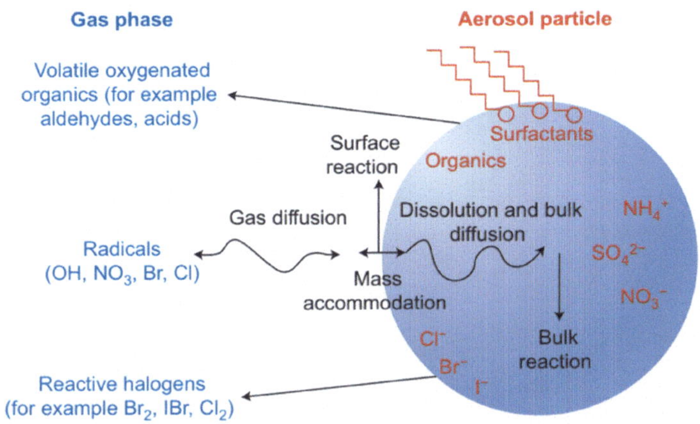

Fig. 5.1 Processes involved in condensed phase evolution of OA. *Image source* George and Abbatt (2010)

Laboratory studies carried out in flow-tubes using thin organic films suggested $\gamma > 0.1$ (range 0.1–1) for OH radical uptake onto condensed phase organics (Heald and Spracklen 2009; Hearn and Smith 2006; Hennigan et al. 2009). The values of the same coefficient for NO_3 radicals are reported to be within the range of 0.0002–0.1 depending on the condition and type of organics present on the condensed phase surface (Herrmann 2003). Reacto-diffusive length is another important parameter which can indicate whether a radical will mostly interact with surface organics or mainly participate in bulk reactions. For highly reactive radicals like OH, this length is often in the order of manometer, indicating that OH· will mostly take part in surface reactions before going into bulk phase (if at all). Radicals can collide and react directly with the surface organics (Eley–Rideal or ER kinetic mechanism), or it can be first absorbed into the condensed phase followed by reactions with surface organics (Langmuir–Hinshelwood or LH kinetic mechanism) (Heald and Spracklen 2009). For OH radical and condensed phase organics interactions, laboratory observations suggest that LH mechanism could be the dominating one (Hildemann and Saxena 1996). The generalized radical initiated condensed phase OA processing is shown in Fig. 5.2. Like gas-phase oxidation, initial condensed phase oxidation also involves hydrogen abstraction followed by the formation of alkylperoxy radical (RO_2). RO_2 radical can then produce RO (Alkoxy radical) through self-reaction or by reacting with NOx and/or HO_2 radical. Alkoxy radicals can produce both low to high volatility stable oxygenated species and additional organic radicals (R·) via a number of reaction pathways: (a) Isomerization (via reaction with O_2) and subsequent chain propagation reactions leading to secondary loss of RH (b) RO decomposition via scission of C–C bond which generally leads to the formation of aldehyde and alkyl radicals along with other compounds of lower molecular weight and higher volatility (Heald and Spracklen 2009).

End products of OH initiated condensed phase OA oxidation are often found to be carbonyls, alcohols and organic acids. Carbonyls found to be consistently higher than

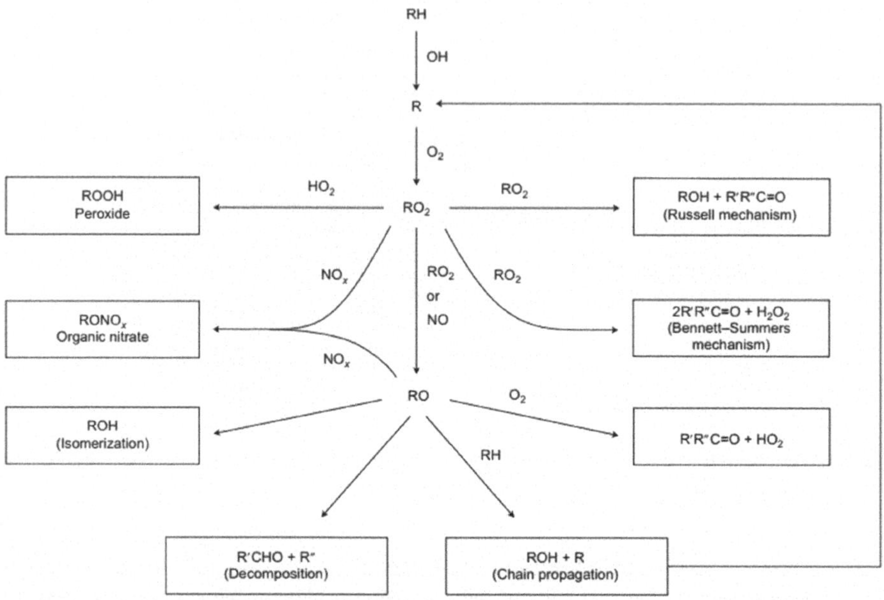

Fig. 5.2 General reaction pathways for condensed phase OA reactions. *Image source* George and Abbatt (2010), Molina et al. (2004)

alcohols indicating $RO + O_2$ and RO_2 self-reactions are more dominant than other reaction pathways (lower right half of Fig. 5.2). Another vital aspect of condensed phase OA processing is the formation of high molecular weight (MW) compounds or oligomers (MW > 200). These species generally have higher MW compared to other individual end products of gas or condensed phase reactions and generally of low to very-low volatility. Some oligomers have very high oxygen content as indicated by their high O/C ratios (0.40–0.80), especially when formed from biogenic precursors (Arnold et al. 2009). Oligomer formation can happen via several pathways and an enhanced rate of formation observed with increasing particle acidity, possibly due to acid catalyzed reactions. However, oligomers can form even in the absence of particle acidity and most likely influenced by water content (or water activity) of the existing aerosols as well (Hinz et al. 1996; Huffman et al. 2009). Esterification reactions between carboxylic acid group and alcoholic moieties could lead to oligomer formation, especially under high-NO_X condition. Some studies (Iinuma et al. 2007) have reported that depending upon NO_X levels, the type oligomers formed can vary significantly with organic peroxides and hemiacetals dominating in low-NO_X regime and oligoesters in high-NO_X conditions. Other moieties like epoxides and anhydrides can also participate in esterification under some circumstances. Carbonyls like aldehydes can also form oligomers in the form of hemiacetals/acetals via aldol reactions. Although, several oligomeric compounds have been identified in atmospheric aerosols in significant quantity (Iinuma et al. 2007; Jang et al. 2002), the actual physico-chemical impacts of oligomer formation on the existing aerosols are not well characterized and should be an active area for future research. Thus, particle phase reactions can significantly influence the OA characteristics by producing both more oxidized high MW compounds with low volatility and small organic fragments with high volatility depending upon particle phase characteristics, ambient conditions, and nature of gas phase SOA and/or other volatile precursors.

Aqueous phase

Several studies (Jayne et al. 2000; Kalberer et al. 2004, 2006; Kanakidou et al. 2005) have also focused on OA chemistry of atmospheric aqueous phase as condensed have significant water content either as aerosol water (in aerosol pores or as a surface layer) or in the form of cloud/fog droplets (Li et al. 2016). Radicals like OH, NO_3, HO_2 can initiate reactions in the aqueous phase as well along with non-radical oxidants like H_2O_2 & O_3. Gas phase oxidation products, VOCs can dissolve into the aqueous phase according to their solubility and affinity towards water whereas some part of OA from the CCN (condensation nuclei on which the cloud/fog droplets form) can also enter the solution. After dissolution chemical transformation of the organics takes place by radical/non-radical initiated reactions and some portion of the end products, become a part of the condensed phase giving rise to aqueous SOA (aqSOA) (Fig. 5.3). Among all the possible aqueous phase oxidants, OH radical is found to be the most potent oxidizing agent (Liu and Wang 2010) for ambient aerosols. OH radical can enter aqueous phase via gas to droplet partitioning or they can be produced inside the droplets via nitrite photolysis and/or Fenton reactions (assisted by transition metals like Cu & Fe) (Arnold et al. 2009).

Some modeling studies (Liu et al. 2008; Lu et al. 2015) have estimated OH concentrations of 10^{-13} and 10^{-12} M for cloud and aerosol water respectively, while first order rate constants for OH radicals in cloud and aerosol water are estimated to be 10^{-3} and 10^{-2} s^{-1}, respectively. For gaseous SOA formation, pre-existing OA will determine how much gas phase oxidation products can be partitioned into particulate phase, while for aqueous SOA (aqSOA), amount of water present in wet aerosols/cloud/fog droplets is extremely important. Amount of partitioned aqSOA fraction (F_p) can be obtained from following equation (very similar to gas phase SOA partitioning equations) (Source: Ervens et al. 2011)

$$F_p(\text{aqSOA}) = \frac{\text{LWC}/C^*_{\text{aq}}}{1 + \text{LWC}/C^*_{\text{aq}}}$$

C_{aq} is the equilibrium concentration of aqueous phase analogous to saturation vapor pressure of C^* for gas phase organics. C^*_{aq} and C^*_{aq} are related by Henry's law constant, K_H. $K_H = C^*_{aq}/C^*$, pre-existing OA mass from the gas phase partitioning theory has been replaced with LWC (liquid water content) for aqSOA calculation. Usually, compounds with K_H values $>10^3$ M/atm generally partitioned into the atmospheric aqueous phase and produce aqSOA. However, some compounds (especially carbonyls) with much smaller K_H values have also detected in the aqueous phase. Hydration reactions between different compounds and the aqueous phase are one of

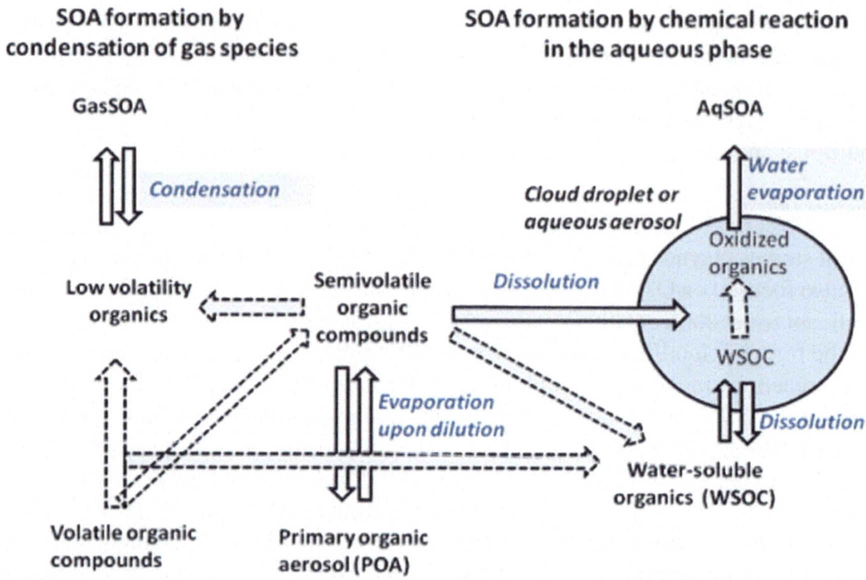

Fig. 5.3 Processes associated with aqueous phase SOA (aqSOA) formation. *Image source* Ervens et al. (2011)

the reasons to find compounds with lower K_H values in aerosol/cloud/fog water. OA dissolved in aerosol water generally represents a much concentrated solution than cloud/fog droplets, due to several order of magnitudes higher LWC associated with the later ($\mu g/m^3$ vs. mg/m^3 of LWC for aerosol water & cloud/fog droplets, respectively). The ionic strength of aerosol water is much higher than fog/cloud droplets and found to be an important factor which can influence the aqueous phase reaction kinetics (Liu and Wang 2010). The highly concentrated solution of OA favors the oligomerization (and accretion) reaction in aerosol water and lead to the production of high MW compounds. However, these kinds of reactions are negligible in cloud/fog water with much more diluted organics solution. The rate of aqSOA formation inside the aerosol water is generally higher than in fog/cloud droplets, but due to high LWC of the later, absolute aqSOA production in aerosol water and droplets is comparable. Apart from LWC, for cloud/fog processes, amount and characteristics of aqSOA also influenced by droplet size distribution and presence of transition metals (Fe, Cu) (Mahowald et al. 2008). Aqueous SOA contains some unique, organic compounds like organosulfates, oxalates, and di/oxocarboxylic acids, which have no known gas phase sources and thus can be used as a tracer for aqSOA formation. RH (humidity) and NO_x can influence the aqSOA production by influencing the aerosol/droplet water contents, volatility and water solubility of the precursor compounds. Several field studies reported enhanced Fp, higher WSOC (water soluble organic carbon) and O/C ratios with increasing RH (Maria et al. 2003; Meskhidze and Nenes 2006), possibly due to more gas to aqueous phase partitioning of organics due to a higher amount of aerosol water/larger droplets at elevated RH levels. RH & NO_x levels can also alter the gas phase oxidation products distribution making them more suitable for aqueous partition and reactions. It is observed form laboratory studies that aqSOA formation enhanced under high NOx regime (opposite the trend observed gas phase SOA), mostly due to formation of water soluble gas phase oxidation products like carbonyls (Li et al. 2016). Under High NOx condition, aqueous phase processing can yield a comparable or higher amount of SOA (aqSOA) than gas phase oxidation (gaseous SOA) from volatile organic precursors, especially of biogenic origin. In regions with high NO_x, humidity and biogenic VOC emissions, aqSOA can contribute equally to the total SOA mass.

SOA characteristics

SOA characteristics are very different from POA and evolving continuously with their aging processes in the atmosphere. In general, SOA is more oxidized, hygroscopic, and less volatile than POA; however, depending upon the precursors and their level of atmospheric processing SOA properties can also vary significantly. Properties of gaseous and aqueous SOA also differ considerably from each other as well. AqSOA is generally more oxidized than gasSOA (SOA produced via gas phase oxidation & subsequent partitioning), with gasSOA O/C ranging from 0.3 to 0.5, while aqSOA can vary from 0.5 to 2 depending upon precursors and atmospheric conditions. Often more oxidized & water soluble fraction of OA (including gasSOA) act as a precursor for aqSOA, leading to even higher O/C ratios (Li et al. 2016). Higher O/C ratios generally make OA more hygroscopic, and laboratory studies found that aqSOA is more hygroscopic with higher GF than gasSOA (Michaud et al. 2009). Optical properties of SOA vary, and both types of SOA can absorb light at different wavelengths and collectively can be termed as "brown carbon". GasSOA generally absorbs lights more efficiently at lower wavelengths (~300 nm) while aqSOA absorbs more at longer wavelengths (~350–600 nm). This difference can be attributed to the higher oxidation levels and presence of HULIS like substances in the composition of aqSOA (Li et al. 2016). SOA formation also alters the size distribution of the particles, while POA is mostly associated with smaller sizes, SOA is found in mid to higher range of accumulation mode particles, generally, aqSOA (~0.3 μm) is associated with larger particles than gasSOA (~0.5–1 μm) (Li et al. 2016). In terms of volatility, aqSOA is much less volatile than gas SOA & POA due to higher oxygenation and presence of high MW products. This indicate that even after cloud/fog droplets evaporation a substantial aqSOA will remain in the particle phase under ambient conditions (Li et al. 2016; Middlebrook et al. 1998).

5.3 OA Measurement

One of the major issue aswith comprehensive OA characterization is the sheer number of individual species involved (in millions). So, any realistic OA measurement technique can't accomplish individual species level characterization in all practical purposes. This individual species characterization may not be required as well

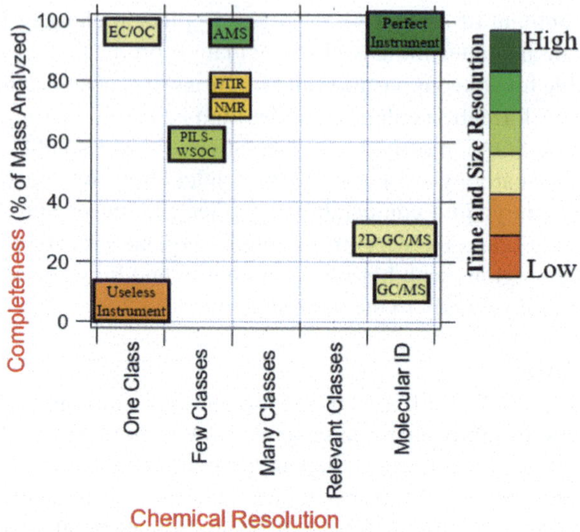

Fig. 5.4 A three-dimensional representation of currently available popular OA measurement and characterization techniques according to their mass closure, levels of chemical speciation, time and size resolution capabilities. *Image source* Hallquist et al. (2009)

if some key parameters controlling overall OA characteristics can be identified and constrained. OA measurement can be broadly divided into three categories: (a) online techniques (b) offline techniques (c) indirect estimation (mostly for SOA). SOA can be indirectly estimated by subtracting measured POA from total OA. Figure 5.4 represents an overview of the currently available popular OA characterization instruments in terms of their strengths and shortcomings. An ideal instrument should reside at the top right corner of the Fig. 5.4, offers high size & time resolution along with complete mass closure & molecular level species characterization. However, most of the currently available OA characterization instruments generally lack one or several characteristics of an ideal device, so, only a combination of several complementary/supplementary measurement techniques can provide the desired level of information.

5.3.1 Offline Techniques

Among offline sophisticated OA characterization techniques, GC/MS is one of the most widely applied technique to identify and quantify individual OA species. However, the vast complexity of OA composition with millions of primary and secondary species means that only a handful of individual OA species can be reliably identified and quantified while most of the remaining OA mass is poorly represented in overlapping and unidentified heave of peaks. The recent development of two dimensional

GC combined with time of flight MS (GCxGC-TOFMS) along with some other modifications now allows better separation of unknown OA species with a relatively better temporal resolution ($\sim$1 h) (Mohr et al. 2009; Moise et al. 2002). LC/MS techniques are also becoming quite popular among aerosol researchers, especially for the separation of polar organics with high water solubility. LC/MS techniques are especially suited for the detection of polar high-MW compounds or oligomers without any further derivatization or pre-concentration as often the case with GC/MS. LC/MS analysis has also been used to detect nitrate and sulfate groups incorporated into SOA, and helped in understanding their formation processes (Molina et al. 2004; Nobel and Prather 1996; Ng et al. 2008). Most LC/MS deploys soft electrospray ionization (ESI) process for OA speciation and quantification, so, most of the parent OA molecules remain intact (unlike in hard electron impact (EI) ionization procedure) thus structural changes in OA due to atmospheric processes and oligomer formation can be studied. A two dimensional LC coupled with ESI-TOF MS had successfully quantified organic (carboxylic) acids from ambient aerosol samples (Ng et al. 2010), indicating the potential of the technique. Further developments should be directed to enhance the mass spectral resolution along with lowering the analysis time.

To reduce the complexity associated with identification of individual OA species, researchers have focused on groups or types of species based on functional groups, polarity, etc. Ion chromatography (IC) and proton NMR (Nuclear magnetic resonance spectroscopy) techniques can provide those information relatively easily compared to other very complex offline techniques used for individual species identification (O'Dowd et al. 2004; Orsini et al. 2003). Using NMR or IC, researchers have segregated water soluble organics (WSOC) into three major categories based on polarity: (1) mono/di-carboxylic acids, (2) poly-carboxylic acids (3) Neutral/basic compounds. NMR technique has been successful in the identification of fulvic acids (a part of humic like substances or HULIS) form ambient aerosols (Ovadnevaite et al. 2011). However, quantitative analysis of OA in ambient aerosols by IC & NMR remains a challenge. Another type of widely used offline OA characterization technique is FTIR (Fourier-transform infrared) spectroscopy. FTIR technique has been used to identify and quantify several organic functional groups like aliphatic (saturated & unsaturated), aromatic, carbonyl ($C = O$), organosulfur (C–O–S), etc. (Pankow 1994; Pathak et al. 2007). FTIR is particularly useful for the quantification of amines and organosulfur species, which are otherwise quite difficult to measure by other techniques. Some elemental ratios like OM/OC ratios can also be obtained from this technique, which can provide some insights into OA evolution (Pol et al. 2006). FTIR has been successfully applied to many filter samples for OA characterization (Pankow 1994; Pathak et al. 2007; Pol et al. 2006; Polidori et al. 2008) and showed good agreement with other more advanced & costly OA measurement techniques. Offline or semi-continuous OC/EC (elemental carbon) determination technique is one of the most popular methods to determine total OC in ambient aerosols. Ambient aerosols are collected on quartz filters either by a separate PM sampler (offline) or directly by the EC/OC instrument (semi-continuous) itself. EC/OC instrument operates on thermo-optical principle using a combination of high temperatures and laser transmittance/reflectance from the sampled filters to find EC & OC (Putaud et al. 2004; Ridley et al. 2018). Using OC/EC ratio, indirect estimation of SOC (and SOA) is also possible, as it is assumed that primary OC (POC) will exhibit minimum OC/EC ratio. So, regression analysis of EC & OC data during periods less impacted by SOC (peak traffic hours early morning or evening etc.) can reveal the minimum OC/EC ratio which is then can be used to estimate SOC (Roelofs 2008).

$$\mathrm{POC} = (\mathrm{OC/EC})_\mathrm{p} * \mathrm{EC};\ \ \mathrm{SOC} = \mathrm{OC} - \mathrm{POC}$$
$$(\mathrm{OC/EC})_\mathrm{p} = \mathrm{OC/EC\ ratio\ for\ primary\ emitted\ organics}$$

However, the selection of this primary OC/EC ratio is tricky and often leads to biased estimation of SOC. Also, this EC/OC method provides only total EC & OC amount but fails to provide any information on different types of primary and secondary OC present.

5.3.2 Online Techniques

Offline techniques of OA characterization generally involves filter sample collection and post-analysis of those collected filters using different techniques. Some of the major disadvantages of offline OA characterization are low time resolution, sample storage requirement, volatilization, adsorption & reactions during sampling and/or storage, manual labour and handling error. The advent of real-time measurement techniques has revolutionized the research related to OA in terms of analysis time and the information obtained. In the last two decades, the number of online MS techniques has grown many folds and continuous improvement is happening in terms of resolution, information detail etc. The general operating principle of online OA characterization involves three major steps: (1) introduction of air borne particles to the instrument (2) vaporization & ionization of organics in those particles (3) detection of ions by MS. Configuration of different online techniques depend on the choice of particle introduction system (nozzles, capillaries and aerodynamic lenses), vaporization (thermal or laser induced vaporization) & ionization methods deployed (electron impact, electro spray or chemical ionization). The AMS instrument (Aerosol mass spectrometer, commercialized by Aerodyne Inc.) is one of the most popular OA measurement technique (Rogge et al. 1993; Russell et al. 2009) currently available in the market. The AMS uses thermal vaporization (flash evaporation at 600 °C) of aerosols and ionizes them via electron impact (EI with 70 eV energy), the ionized aerosols then detected by quadruple or time of flight (ToF) MS. It can provide real-time (1 min or less) concentration data of OA, sulfate, nitrate, chloride and ammonium but due to extensive OA fragmentation, molecular level characterization of OA is not possible. Another disadvantage of the AMS is the inability to identify non-refractory materials (anything that doesn't vaporize at 600 °C) like black/elemental carbon and metallic oxides etc. However, AMS can provide some useful information like elemental ratios (O/C, H/C, N/C & OM/OC) after high resolution data obtained from a ToF-AMS, which can provide insights into the real-time evolution of OA in the atmosphere. AMS data can also be utilized for source apportionment via PMF (Positive matrix factorization) (Saunders et al. 2003; Smith and Hearn 2006) leading to the identification of different types of primary and secondary OA present in the atmosphere. AMS-PMF combination has been successfully used to quantify POA contribution from vehicular exhaust, biomass burning, sea-salt spray and cooking activities (Sorooshian et al. 2007, 2013). Several types of oxidized OA (a subset of SOA, you can get total SOA by adding them all) with different oxidation levels (O/C) have also been identified using AMS in the ambient aerosols (Meskhidze and Nenes 2006; Saunders et al. 2003; Spracklen et al. 2008; Storey and Pankow 1992; Surratt et al. 2006), clearly showing the continuously evolving nature of OA/SOA. Several other online aerosol MS techniques have also come into existence in last decades or so with different vaporization and ionization techniques. Laser ablation MS can analyze individual particles by utilizing a laser to vaporize & ionizer aerosols, ATOFMS (aerosol ToF MS by TSI Inc.) is one such commercially available instrument (Ulbrich et al. 2009; Van Der Werf et al. 2006). Aerosol MS using soft ionization techniques like ESI or CI (chemical ionization) have also been developed (Vlasenko et al. 2008; Weber et al. 2001) but their application to ambient aerosol measurement is still somewhat limited (Arnold et al. 2009).

Particle into liquid sampler (PILS) facilitates aerosols rapid extraction with the water for further analysis (by IC) (Wu et al. 2012; Yao et al. 2003). This technique enables the continuous measurement of WSOC and some inorganic ions with a time resolution of minutes. WSOC mostly consists of SOA and sometimes biomass burning OA (BBOA), so, indirectly, WSOC concentrations can be used to infer information about SOA levels in regions least affected by biomass burning activities. Combining total OC measurement techniques (like EC/OC, AMS etc.) with PILS, researchers can get water insoluble OC (WIOC) as well. Recently, by applying some modifications, PILS combine with IC techniques was able to measure both gas and particle phase water soluble organics (Maria et al. 2003). One major disadvantage of PILS techniques is that size-resolved information can't be obtained and only bulk WSOC can be measured without any speciation information.

5.4 OA Control

Control of OA can be divided into two parts: (1) control of primary OA and (2) control of SOA. Primary OA control can be achieved by controlling emissions from mostly anthropogenic sources like vehicular exhaust, industrial emission, reduction in biomass burning activities etc. Reduction in emission can be achieved via a combination of public awareness and stringent emission regulation policy coupled with technological innovation. Affordable & effective emission control technologies coupled with more efficient combustion processes can reduce the POA emission significantly. OA levels in US are steadily declining for last few decades (at a rate of 0.32 $\mu g/m^3$ per decade) thanks to stringent emission control measures taken by

US EPA, especially targeting vehicular and power plant emissions (Yu et al. 2005). OA reduction resulted in a 31% reduction in $PM_{2.5}$ between 1990 and 2012; this translates into the prevention of 84000 premature deaths from air pollution (Yu et al. 2005). It is also shown that the majority of this OA reduction is due to a reduction in anthropogenic POA, rather than SOA. In China, a short term very stringent pollution control campaign was carried out during 2015 victory day parade (Zhang et al. 2007). During the control period, the number of on-road vehicles was reduced by half, close to 10,000 enterprises were either shut down or asked to reduce production, cooking in open charcoal grills was prohibited and major construction activities were banned. These measures resulted in 64 & 53% decrease in $PM_{2.5}$ & OA, respectively, compared to pre-control period. In OA fraction, POA emission from vehicular exhaust and cooking was reduced by more than 40%, clearly showing the impact of emission control on POA levels. Control technologies such as diesel particulate filter (DPF), diesel oxidation catalyst (DOC; a part of DPF) and selective catalytic reduction (SCR) system can reduce pollutants emission from vehicles. Modern vehicles equipped with DPF pollution control device could reduce POA by more than 90% (Zhang et al. 2011). Here, the POA implies actual particulate OA and other emitted semi-volatile organic gases which are in constant equilibrium with gas & particulate phase.

SOA control is far more complicated than POA as SOA can only be effectively controlled by eliminating its precursors such as VOCs. A significant part of the VOCs is biogenic and hence beyond the conventional control strategies. Anthropogenic VOCs; those are emitting from vehicular exhaust, solvent industries, and other combustion activities can be controlled. Several studies have also indicated that other inorganic precursors like NO_x, SO_x can also enhance SOA formation from biogenic VOCs, so controlling those inorganic gases also help in reducing SOA levels. A Laboratory study indicated that DPF equipped with DOC could reduce SOA formation from diesel vehicle exhaust by 90% via oxidation of gas phase precursors and altering their composition (Zhang et al. 2014). NOx emissions can be reduced by 50–90% by SCR depending upon type of catalyst used and engine loading conditions; this reduction may also reduce the SOA production from VOCs (Zhang et al. 2011). For gasoline vehicles, NMOG (non-methane organic gases): NO_x ratio is the key to reduce SOA formation, higher ratios could lead to enhance SOA formation from gasoline exhaust (Zhao et al. 2017). This indicates that the regulatory agency should aim for a synergic reduction of VOCs and NO_x to achieve the desired level of SOA reduction.

5.5 Conclusion

OA represents a major part of ambient fine aerosols and have both climatic and health impacts. However, complete understanding of ambient OA formation, evolution and removal processes still remain a challenge. Incredible complexity associated with OA characteristics, evolution, sources and composition makes it very difficult to fully characterize ambient OA. Several smog chamber studies have been carried out to understand the OA formation and evolution processes from different ambient precursors like VOCs. However, often those studies don't fully represent the actual complex ambient conditions leading to end products with different composition and characteristics than found in actual atmosphere. Chamber studies still remain useful as they can shed light on the kinetics and mechanisms of OA formation and evolution processes along with the influence of environmental parameters on the same. Ambient OA can be broadly divided into primary (POA) and secondary OA (SOA), with former being emitted directly into the atmosphere and latter forms in the atmosphere via atmospheric processing. POA can be emitted by natural like sea salt, volcanic eruptions, forest fire, dust storms and/or anthropogenic sources such as vehicular exhaust, industrial emission, cooking & biomass burning. SOA can be produced via gas phase oxidation of VOCs by oxidants like OH, NO_3 & O_3 and subsequent gas to particle partitioning of the oxidation products, SOA produced via this process is known as gaseous SOA. In the condensed phase (on aerosol surface or inside cloud/fog droplets), gaseous SOA undergoes further transformation with the production of more oxidized, low volatility organics. SOA can also be formed via aqueous phase (in particulate bound water or fog/cloud droplets) processing of gaseous SOA, POA and VOCs which is known as aqueous SOA. SOA generally more oxidized, hygroscopic (water attracting) and less volatile than POA and also shows more significant variations in characteristic than POA. Currently available OA measurement techniques can characterize important OA properties, however, no one technique alone can characterize all important OA parameters. Both online and offline OA measurement techniques are available, especially the online OA characterization techniques are improving much faster and becoming more popular due to their higher time & size resolution. However, much more work needed to be done to improve the details of the information available via online OA characterization techniques. Effective OA control should incorporate the reduction of precursors VOCs (mostly anthropogenic ones) for SOA, apart from controlling POA emission directly from the source. Technologies are available for control of POA and SOA precursors' emission; however, policy makers should ensure the proper implementation of those technologies. Researchers should also focus to improve the efficiency and affordability of the emission control technologies.

References

Arnold SR, Spracklen DV, Williams J et al (2009) Evaluation of the global oceanic isoprene source and its impacts on marine organic carbon aerosol. Atmos Chem Phys 9:1253–1262. https://doi.org/10.5194/acp-9-1253-2009

Atkinson R, Arey J (2003) Gas-phase tropospheric chemistry of biogenic volatile organic compounds: a review. Atmos Environ 37:197–219. https://doi.org/10.1016/S1352-2310(03)00391-1

Aumont B, Szopa S, Madronich S (2005) Modelling the evolution of organic carbon during its gas-phase tropospheric oxidation: development of an explicit model based on a self generating approach. Atmos Chem Phys 5:2497–2517. https://doi.org/10.5194/acp-5-2497-2005

Baltensperger U, Kalberer M, Dommen J et al (2005) Secondary organic aerosols from anthropogenic and biogenic precursors. Faraday Discuss 130:265–278. https://doi.org/10.1039/b417367h
Bertram AK, Ivanov AV, Hunter M et al (2001) The reaction probability of OH on organic surfaces of tropospheric interest. J Phys Chem A 105:9415–9421. https://doi.org/10.1021/jp0114034
Birch ME, Cary RA (1996) Elemental carbon-based method for occupational monitoring of particulate diesel exhaust: methodology and exposure issues. Analyst 121:1183–1190. https://doi.org/10.1039/an9962101183
Bond TC, Streets DG, Yarber KF et al (2004) A technology-based global inventory of black and organic carbon emissions from combustion. J Geophys Res D Atmos 109:1–43. https://doi.org/10.1029/2003JD003697
Canagaratna MR, Jayne JT, Jimenez JL et al (2007) Chemical and microphysical characterization of ambient aerosols with the aerodyne aerosol mass spectrometer. Mass Spectrom Rev 26:185–222. https://doi.org/10.1002/mas.20115
Cappiello A, De Simoni E, Fiorucci C et al (2003) Molecular characterization of the water-soluble organic compounds in fogwater by ESIMS/MS. Environ Sci Technol 37:1229–1240. https://doi.org/10.1021/es0259990
Chakraborty A, Bhattu D, Gupta T et al (2015) Real-time measurements of ambient aerosols in a polluted Indian city: sources, characteristics, and processing of organic aerosols during foggy and nonfoggy periods. J Geophys Res 120. https://doi.org/10.1002/2015JD023419
Chakraborty A, Ervens B, Gupta T, Tripathi SN (2016) Characterization of organic residues of size-resolved fog droplets and their atmospheric implications. J Geophys Res 121. https://doi.org/10.1002/2015JD024508
Chakraborty A, Ervens B, Gupta T, Tripathi SN (2016b) Characterization of organic residues of size-resolved fog droplets and their atmospheric implications. J Geophys Res Atmos 121:4317–4332. https://doi.org/10.1002/2015JD024508
Chirico R, DeCarlo PF, Heringa MF et al (2010) Impact of aftertreatment devices on primary emissions and secondary organic aerosol formation potential from in-use diesel vehicles: results from smog chamber experiments. Atmos Chem Phys 10:11545–11563. https://doi.org/10.5194/acp-10-11545-2010
Crippa M, Decarlo PF, Slowik JG et al (2013) Wintertime aerosol chemical composition and source apportionment of the organic fraction in the metropolitan area of Paris. Atmos Chem Phys 13:961–981. https://doi.org/10.5194/acp-13-961-2013
Cubison MJ, Ortega AM, Hayes PL et al (2011) Effects of aging on organic aerosol from open biomass burning smoke in aircraft and laboratory studies. Atmos Chem Phys 11:12049–12064. https://doi.org/10.5194/acp-11-12049-2011
Decesari S, Facchini MC, Fuzzi S, Tagliavini E (2000) Characterization of water-soluble organic compounds in atmospheric aerosol: a new approach. J Geophys Res Atmos 105:1481–1489. https://doi.org/10.1029/1999JD900950
Docherty KS, Stone EA, Ulbrich IM et al (2008) Apportionment of primary and secondary organic aerosols in southern california during the 2005 study of organic aerosols in riverside (SOAR-1). Environ Sci Technol 42:7655–7662. https://doi.org/10.1021/es8008166
Dockery DW, Pope CA (1994) Acute respiratory effects of particulate air pollution. Annu Rev Public Health 15:107–132. https://doi.org/10.1146/annurev.pu.15.050194.000543
Donahue NM, Robinson AL, Stanier CO, Pandis SN (2006) Coupled partitioning, dilution, and chemical aging of semivolatile organics. Environ Sci Technol 40:2635–2643. https://doi.org/10.1021/es052297c
Donahue NM, Robinson AL, Trump ER et al (2012) Volatility and aging of atmospheric organic aerosol. Springer, Berlin, pp 97–143
Elbert W, Taylor PE, Andreae MO, Pöschl U (2007) Contribution of fungi to primary biogenic aerosols in the atmosphere: wet and dry discharged spores, carbohydrates, and inorganic ions. Atmos Chem Phys 7:4569–4588. https://doi.org/10.5194/acp-7-4569-2007

Ervens B, Volkamer R (2010) Glyoxal processing by aerosol multiphase chemistry: towards a kinetic modeling framework of secondary organic aerosol formation in aqueous particles. Atmos Chem Phys 10:8219–8244. https://doi.org/10.5194/acp-10-8219-2010
Ervens B, George C, Williams JE et al (2003) CAPRAM 2.4 (MODAC mechanism): an extended and condensed tropospheric aqueous phase mechanism and its application. J Geophys Res Atmos 108:1–21. https://doi.org/10.1029/2002JD002202
Ervens B, Turpin BJ, Weber RJ (2011) Secondary organic aerosol formation in cloud droplets and aqueous particles (aqSOA): a review of laboratory, field and model studies. Atmos Chem Phys 11:11069–11102. https://doi.org/10.5194/acp-11-11069-2011
Forouzanfar MH, Alexander L, Bachman VF et al (2015) Global, regional, and national comparative risk assessment of 79 behavioural, environmental and occupational, and metabolic risks or clusters of risks in 188 countries, 1990–2013: a systematic analysis for the global burden of disease study 2013. Lancet 386:2287–2323. https://doi.org/10.1016/S0140-6736(15)00128-2
Gantt B, Meskhidze N (2013) The physical and chemical characteristics of marine primary organic aerosol: a review. Atmos Chem Phys 13:3979–3996. https://doi.org/10.5194/acp-13-3979-2013
Ge X, Setyan A, Sun Y, Zhang Q (2012) Primary and secondary organic aerosols in Fresno, California during wintertime: results from high resolution aerosol mass spectrometry. J Geophys Res 117. https://doi.org/10.1029/2012jd018026
George IJ, Abbatt JPD (2010) Heterogeneous oxidation of atmospheric aerosol particles by gas-phase radicals. Nat Chem 2:713–722. https://doi.org/10.1038/nchem.806
Gilardoni S, Liu S, Takahama S et al (2009) Characterization of organic ambient aerosol during MIRAGE 2006 on three platforms. Atmos Chem Phys 9:5417–5432. https://doi.org/10.5194/acp-9-5417-2009
Goldstein AH, Worton DR, Williams BJ et al (2008) Thermal desorption comprehensive two-dimensional gas chromatography for in-situ measurements of organic aerosols. J Chromatogr A 1186:340–347. https://doi.org/10.1016/j.chroma.2007.09.094
Hallquist M, Wenger JC, Baltensperger U et al (2009) The formation, properties and impact of secondary organic aerosol: current and emerging issues. Atmos Chem Phys 9:5155–5236
Hamilton JF, Webb PJ, Lewis AC et al (2004) Partially oxidised organic components in urban aerosol using GCXGC-TOF/MS. Atmos Chem Phys 4:1279–1290. https://doi.org/10.5194/acp-4-1279-2004
Hawkins LN, Russell LM (2010) Polysaccharides, proteins, and phytoplankton fragments: four chemically distinct types of marine primary organic aerosol classified by single particle spectromicroscopy. Adv Meteorol 2010:1–14. https://doi.org/10.1155/2010/612132
Hawkins LN, Russell LM, Covert DS et al (2010) Carboxylic acids, sulfates, and organosulfates in processed continental organic aerosol over the southeast Pacific Ocean during VOCALS-REx 2008. J Geophys Res D Atmos 115:D13201. https://doi.org/10.1029/2009JD013276
Heal MR, Kumar P, Harrison RM (2012) Particles, air quality, policy and health. Chem Soc Rev 41:6606–6630. https://doi.org/10.1039/c2cs35076a
Heald CL, Spracklen DV (2009) Atmospheric budget of primary biological aerosol particles from fungal spores. Geophys Res Lett 36:1–5. https://doi.org/10.1029/2009GL037493
Hearn JD, Smith GD (2006) Reactions and mass spectra of complex particles using aerosol CIMS. Int J Mass Spectrom 258:95–103. https://doi.org/10.1016/J.IJMS.2006.05.017
Hennigan CJ, Bergin MH, Russell AG et al (2009) Gas/particle partitioning of water-soluble organic aerosol in Atlanta. Atmos Chem Phys 9:3613–3628. https://doi.org/10.5194/acpd-9-635-2009
Herrmann H (2003) Kinetics of aqueous phase reactions relevant for atmospheric chemistry. Chem Rev 103(12):4691–4716. https://doi.org/10.1021/CR020658Q
Hildemann LM, Saxena P (1996) Hygroscopic characteristics of atmospheric organic aerosols. Abstr Pap Am Chem Soc 211:106-ANYL
Hinz K-P, Kaufmann R, Spengler B (1996) Simultaneous detection of positive and negative ions from single Airborne particles by real-time laser mass spectrometry. Aerosol Sci Technol 24:233–242. https://doi.org/10.1080/02786829608965368

Huffman JA, Docherty KS, Mohr C et al (2009) Chemically-resolved volatility measurements of organic aerosol from different sources. Environ Sci Technol 43:5351–5357. https://doi.org/10.1021/es803539d

Iinuma Y, Muller C, Berndt T et al (2007a) Evidence for the existence of organosulfates from beta-pinene ozonolysis in ambient secondary organic aerosol. Environ Sci Technol 41:6678–6683. https://doi.org/10.1021/es070938t

Iinuma Y, Mueller C, Boege O et al (2007b) The formation of organic sulfate esters in the limonene ozonolysis secondary organic aerosol (SOA) under acidic conditions. Atmos Environ 41:5571–5583. https://doi.org/10.1016/j.atmosenv.2007.03.007

Jang M, Czoschke NM, Lee S, Kamens RM (2002) Heterogeneous atmospheric aerosol production by acid-catalyzed particle-phase reactions. Science 298:814–817. https://doi.org/10.1126/science.1075798

Jayne JT, Leard DC, Zhang XF et al (2000) Development of an aerosol mass spectrometer for size and composition analysis of submicron particles. Aerosol Sci Technol 33:49–70. https://doi.org/10.1080/027868200410840

Kalberer M, Paulsen D, Sax M et al (2004) Identification of polymers as major components of atmospheric organic aerosols. Science 303:1659–1662. https://doi.org/10.1126/science.1092185

Kalberer M, Sax M, Samburova V (2006) Molecular size evolution of oligomers in organic aerosols collected in urban atmospheres and generated in a smog chamber. Environ Sci Technol 40:5917–5922. https://doi.org/10.1021/es0525760

Kanakidou M, Seinfeld JH, Pandis SN et al (2005) Organic aerosol and global climate modelling: a review. Atmos Chem Phys 5:1053–1123

Li H, Zhang Q, Duan F et al (2016) The "Parade Blue": effects of short-term emission control on aerosol chemistry. Faraday Discuss 189:317–335. https://doi.org/10.1039/C6FD00004E

Liu X, Wang J (2010) How important is organic aerosol hygroscopicity to aerosol indirect forcing? Environ Res Lett 5. https://doi.org/10.1088/1748-9326/5/4/044010

Liu ZG, Berg DR, Swor TA, Schauer JJ (2008) Comparative analysis on the effects of diesel particulate filter and selective catalytic reduction systems on a wide spectrum of chemical species emissions. Environ Sci Technol 42:6080–6085. https://doi.org/10.1021/es8004046

Lu Z, Streets DG, Winijkul E et al (2015) Light absorption properties and radiative effects of primary organic aerosol emissions. Environ Sci Technol 49:4868–4877. https://doi.org/10.1021/acs.est.5b00211

Mahowald N, Jickells TD, Baker AR et al (2008) Global distribution of atmospheric phosphorus sources, concentrations and deposition rates, and anthropogenic impacts. Global Biogeochem Cycles 22:1–19. https://doi.org/10.1029/2008GB003240

Maria SF, Russell LM, Turpin BJ et al (2003) Source signatures of carbon monoxide and organic functional groups in Asian Pacific Regional Aerosol Characterization Experiment (ACE-Asia) submicron aerosol types. J Geophys Res 108:8637. https://doi.org/10.1029/2003JD003703

Meskhidze N, Nenes A (2006) Phytoplankton and cloudiness in the Southern Ocean. Science 314:1419–1423. https://doi.org/10.1126/science.1131779

Michaud V, El Haddad I, Liu Y et al (2009) In-cloud processes of methacrolein under simulated conditions—Part 3: hygroscopic and volatility properties of the formed secondary organic aerosol. Atmos Chem Phys 9:5119–5130

Middlebrook AM, Murphy DM, Thomson DS (1998) Observations of organic material in individual marine particles at cape grim during the first aerosol characterization experiment (ACE 1). J Geophys Res Atmos 103:16475–16483. https://doi.org/10.1029/97JD03719

Mohr C, Huffman JA, Cubison MJ et al (2009) Characterization of primary organic aerosol emissions from meat cooking, trash burning, and motor vehicles with high-resolution aerosol mass spectrometry and comparison with ambient and chamber observations. Environ Sci Technol 43:2443–2449. https://doi.org/10.1021/es8011518

Moise T, Talukdar RK, Frost GJ et al (2002) Reactive uptake of NO_3 by liquid and frozen organics. J Geophys Res 107:4014. https://doi.org/10.1029/2001JD000334

Molina MJ, Ivanov AV, Trakhtenberg S, Molina LT (2004) Atmospheric evolution of organic aerosol. Geophys Res Lett 31. https://doi.org/10.1029/2004GL020910
Nobel CA, Prather KA (1996) Real-time measurement of correlated size and composition profiles of individual atmospheric aerosol particles. Environ Sci Technol 30(9):2667–2680. https://doi.org/10.1021/ES950669J
Ng NL, Kwan AJ, Surratt JD et al (2008) Secondary organic aerosol (SOA) formation from reaction of isoprene with nitrate radicals (NO_3). Atmos Chem Phys 8:4117–4140. https://doi.org/10.5194/acp-8-4117-2008
Ng NL, Canagaratna MR, Zhang Q et al (2010) Organic aerosol components observed in northern hemispheric datasets from aerosol mass spectrometry. Atmos Chem Phys 10:4625–4641. https://doi.org/10.5194/acp-10-4625-2010
O'Dowd CD, Facchini MC, Cavalli F et al (2004) Biogenically driven organic contribution to marine aerosol. Nature 431:676–680. https://doi.org/10.1038/nature02959
Odum JR, Hoffmann T, Bowman F, et al (1996) Gas/particle partitioning and secondary organic aerosol yields. Environ Sci Technol 30:2580–2585. https://doi.org/10.1021/es950943+
Orsini DA, Ma Y, Sullivan A et al (2003) Refinements to the particle-into-liquid sampler (PILS) for ground and airborne measurements of water soluble aerosol composition. Atmos Environ 37:1243–1259. https://doi.org/10.1016/S1352-2310(02)01015-4
Ovadnevaite J, Ceburnis D, Martucci G et al (2011) Primary marine organic aerosol: a dichotomy of low hygroscopicity and high CCN activity. Geophys Res Lett 38. https://doi.org/10.1029/2011GL048869
Pankow JF (1994) An absorption model of the gas/aerosol partitioning involved in the formation of secondary organic aerosol. Atmos Environ 28:189–193. https://doi.org/10.1016/1352-2310(94)90094-9
Pathak RK, Presto AA, Lane TE et al (2007) Ozonolysis of α-pinene: parameterization of secondary organic aerosol mass fraction. Atmos Chem Phys 7:3811–3821. https://doi.org/10.5194/acp-7-3811-2007
Pol J, Hohnova B, Jussila M, Hyotylainen T (2006) Comprehensive two-dimensional liquid chromatography-time-of-flight mass spectrometry in the analysis of acidic compounds in atmospheric aerosols. J Chromatogr A 1130:64–71. https://doi.org/10.1016/j.chroma.2006.04.050
Polidori A, Turpin BJ, Davidson CI et al (2008) Organic PM2.5: fractionation by polarity, FTIR spectroscopy, and OM/OC ratio for the Pittsburgh aerosol. Aerosol Sci Technol 42:233–246. https://doi.org/10.1080/02786820801958767
Putaud JP, Raes F, Van Dingenen R et al (2004) European aerosol phenomenology-2: chemical characteristics of particulate matter at kerbside, urban, rural and background sites in Europe. Atmos Environ 38:2579–2595. https://doi.org/10.1016/j.atmosenv.2004.01.041
Ridley DA, Heald CL, Ridley KJ, Kroll JH (2018) Causes and consequences of decreasing atmospheric organic aerosol in the United States. Proc Natl Acad Sci 115:290–295. https://doi.org/10.1073/PNAS.1700387115
Roelofs GJ (2008) A GCM study of organic matter in marine aerosol and its potential contribution to cloud drop activation. Atmos Chem Phys 8:709–719. https://doi.org/10.5194/acp-8-709-2008
Rogge WF, Mazurek MA, Hildemann LM et al (1993) Quantification of urban organic aerosols at a molecular-level—identification, abundance and seasonal-variation. Atmos Environ Part A-Gen Top 27:1309–1330. https://doi.org/10.1016/0960-1686(93)90257-y
Russell LM, Takahama S, Liu S et al (2009) Oxygenated fraction and mass of organic aerosol from direct emission and atmospheric processing measured on the R/V *Ronald Brown* during TEXAQS/GoMACCS 2006. J Geophys Res 114:D00F05. https://doi.org/10.1029/2008JD011275
Saunders SM, Jenkin ME, Derwent RG, Pilling MJ (2003) Protocol for the development of the master chemical mechanism, MCM v3 (Part A): tropospheric degradation of non-aromatic volatile organic compounds. Atmos Chem Phys 3:161–180. https://doi.org/10.5194/acp-3-161-2003
Smith GD, Hearn JD (2006) Studies of radical-initiated oxidation of organic particles using aerosol CIMS. Abstr Pap Am Chem Soc 231

Sorooshian A, Ng NL, Chan AWH et al (2007) Particulate organic acids and overall water-soluble aerosol composition measurements from the 2006 Gulf of Mexico atmospheric composition and climate study (GoMACCS). J Geophys Res 112. https://doi.org/10.1029/2007jd008537

Sorooshian A, Wang Z, Coggon MM et al (2013) Observations of sharp oxalate reductions in stratocumulus clouds at variable altitudes: organic acid and metal measurements during the 2011 E-PEACE campaign. Environ Sci Technol 47:7747–7756. https://doi.org/10.1021/es4012383

Spracklen DV, Arnold SR, Sciare J et al (2008) Globally significant oceanic source of organic carbon aerosol. Geophys Res Lett 35:1–5. https://doi.org/10.1029/2008GL033359

Storey JME, Pankow JF (1992) Gas particle partitioning of semivolatile organic-compounds to model atmospheric particulate materials.1. Sorption to graphite, sodium-chloride, alumina, and silica particles under low humidity conditions. Atmos Environ Part a-Gen Top 26:435–443. https://doi.org/10.1016/0960-1686(92)90328-i

Surratt JD, Murphy SM, Kroll JH et al (2006) Chemical composition of secondary organic aerosol formed from the photooxidation of isoprene. J Phys Chem A 110:9665–9690. https://doi.org/10.1021/jp061734m

Ulbrich IM, Canagaratna MR, Zhang Q et al (2009) Interpretation of organic components from positive matrix factorization of aerosol mass spectrometric data. Atmos Chem Phys 9:2891–2918

Van Der Werf GR, Randerson JT, Giglio L et al (2006) Interannual variability in global biomass burning emissions from 1997 to 2004. Atmos Chem Phys 6:3423–3441. https://doi.org/10.5194/acp-6-3423-2006

Vlasenko A, George IJ, Abbatt JPD (2008) Formation of volatile organic compounds in the heterogeneous oxidation of condensed-phase organic films by gas-phase OH. J Phys Chem A 112:1552–1560. https://doi.org/10.1021/jp0772979

Weber RJ, Orsini D, Daun Y et al (2001) A particle-into-liquid collector for rapid measurement of aerosol bulk chemical composition. Aerosol Sci Technol 35:718–727. https://doi.org/10.1080/02786820152546761

Wu C, Ng WM, Huang J et al (2012) Determination of elemental and organic carbon in PM2.5 in the Pearl River Delta Region: inter-instrument (sunset vs. DRI model 2001 thermal/optical carbon analyzer) and inter-protocol comparisons (improve vs. ACE-Asia protocol). Aerosol Sci Technol 46:610–621. https://doi.org/10.1080/02786826.2011.649313

Yao XH, Lau APS, Fang M et al (2003) Size distributions and formation of ionic species in atmospheric particulate pollutants in Beijing, China: 2-dicarboxylic acids. Atmos Environ 37:3001–3007. https://doi.org/10.1016/s1352-2310(03)00256-5

Yu LE, Shulman ML, Kopperud R, Hildemann LM (2005) Characterization of organic compounds collected during Southeastern Aerosol and Visibility Study: Water-soluble organic species. Environ Sci Technol 39:707–715. https://doi.org/10.1021/es0489700

Zhang Q, Jimenez JL, Canagaratna MR et al (2007) Ubiquity and dominance of oxygenated species in organic aerosols in anthropogenically-influenced Northern Hemisphere midlatitudes. Geophys Res Lett 34. https://doi.org/10.1029/2007gl029979

Zhang Q, Jimenez JL, Canagaratna MR et al (2011) Understanding atmospheric organic aerosols via factor analysis of aerosol mass spectrometry: a review. Anal Bioanal Chem 401:3045–3067. https://doi.org/10.1007/s00216-011-5355-y

Zhang JK, Sun Y, Liu ZR et al (2014) Characterization of submicron aerosols during a month of serious pollution in Beijing, 2013. Atmos Chem Phys 14:2887–2903. https://doi.org/10.5194/acp-14-2887-2014

Zhao Y, Saleh R, Saliba G et al (2017) Reducing secondary organic aerosol formation from gasoline vehicle exhaust. Proc Natl Acad Sci USA 114:6984–6989. https://doi.org/10.1073/pnas.1620911114

Chapter 6
Chemical Speciation and Source Apportionment of Airborne Coarse Particles at Kanpur

Pragati Rai and Tarun Gupta

Abstract The key objective of this study was to unravel the major sources of $PM_{10\text{-}2.5}$ (defined as PM_{10}-$PM_{2.5}$ or coarse particles) within and near the city of Kanpur. Airborne particulate matter (PM) samples were collected from 1st April to 15th July, 2011. The average mass concentration of coarse particles was found to be $64.3 \pm 51.16\ \mu g/m^3$. In addition to the mass concentrations of coarse particles their black carbon (BC), water soluble inorganic carbon (WSIC), water soluble organic carbon (WSOC), water-soluble ions and elemental composition were also determined. The size-resolved data set was processed via Positive Matrix Factorization (PMF) to carefully identify major sources and an attempt was made to quantify their respective source strengths. The contribution of these sources to the coarse PM mass concentration was also assessed. The identified sources and their contributions for the coarse particles were paved road dust (53%), vehicular emission (7%), coal combustion & brick kilns (2.5%), construction activities and incineration (0.5%), crustal dust (32%) and biomass burning and oil combustion (5%).

Keywords Source apportionment · PMF · WSOC · $PM_{10\text{-}2.5}$

6.1 Introduction

Increasing level of air pollution has become a major health concern now days worldwide. Various studies reported that atmospheric particulate matter appears to have an adverse effect on respiratory and cardiovascular systems (Nastos et al. 2010; WHO 2013). Exposure for different mortality and morbidity effects is highly dependent on individual particle size fractions such as $PM_{10\text{-}2.5}$, $PM_{2.5\text{-}1.0}$ and $PM_{1.0}$ (Rajput et al. 2019). Ambient particles originate through various sources. These sources are primarily categorized into industrial, biomass burning, transport, domestic and fugitive emissions (Chakraborty and Gupta 2010; Gupta and Mandariya 2013; Ghosh et al. 2014). Different size fraction elements typically serve as markers for different

P. Rai · T. Gupta (✉)
Department of Civil Engineering, Indian Institute of Technology, Kanpur, UP 208016, India
e-mail: tarun@iitk.ac.in

T. Gupta et al. (eds.), *Measurement, Analysis and Remediation of Environmental Pollutants*, Energy, Environment, and Sustainability,
https://doi.org/10.1007/978-981-15-0540-9_6

sources (Visser et al. 2015) such as $PM_{10-2.5}$ is mainly due to resuspended road dust (Kupiainen et al. 2005; Li et al. 2013) and tires and brake emissions (Councell et al. 2004; Chang et al. 2014).

The World Health Organization (WHO) 2011, states that Kanpur which is the central part of Indo Gangetic Plain (IGP) region is the second most polluted city in India. IGP is the dust prone region during pre-monsoon (April-June) season every year due to the long-range transport of mineral dust originating from Iran, Afghanistan, Pakistan, and the Thar Desert (western India) (Ram et al. 2012). Mineral dusts are the essential components of aerosol-climate system that enter into the atmosphere via wind erosion of desert, arid regions and disturbed soils (Ghosh et al. 2014). Several studies have been carried out focusing on dust characteristics in the IGP (Dey et al. 2004; Chinnam et al. 2006; Chakraborty and Gupta 2010; Ghosh et al. 2014; Mishra et al. 2014; Sharma et al. 2014). Despite of various studies in IGP region, no study has focused on size segregated coarse fraction of PM. Coarse particle fraction comprises of two components, a wind-suspended component and nonwind-suspended component (Harrison et al. 2000). Wind suspended components originated from natural sources such as sea spray and surface soils or dusts on paved areas while non-wind suspended components mainly from anthropogenic sources such as industrial processes, traffic-generated resuspension, construction activities, and biological particles. Positive matrix factorization (PMF) (Paatero and Tapper 1994) is a powerful source apportionment tool to quantify sources based on trace element measurements. Numerous studies have applied PMF on either elements only or in combination with other species, such as carbon species and inorganic ions (Hammond et al. 2008; Vedal et al. 2009; Gu et al. 2011; Amato et al. 2013; Yang et al. 2013; Zhang et al. 2013; Yang et al. 2015). However, these measurements are typically performed only for a single size fraction, preventing the study of short-term changes in air pollution exposure levels. Very few studies have been carried out so far on different particle size fraction source apportionment (Soluri et al. 2007; Cheng et al. 2014; Owoade et al. 2015; Visser et al. 2015). Therefore, this study focuses on the source apportionment of $PM_{10-2.5}$ during a dust campaign (April-July, 2011) at urban city Kanpur.

6.2 Experimental Methods and Data Analysis

$PM_{2.5}$ and PM_{10} were sampled using inertial impaction based high volume samplers for 8 h every 3rd day starting from 1st April, 2011 to 15th July, 2011. Two journal articles have already been published for the same dust campaign, one of the articles has highlighted the chemical characterization of crustal dust (Ghosh et al. 2014) while other article reported the optical, chemical, and physical properties of dust (Mishra et al. 2014). Detailed description of sampling site, meteorological conditions, samples collection and sample analysis procedures are given in our companion papers (Ghosh et al. 2014; Mishra et al. 2014).

6.3 Results and Discussion

6.3.1 Mass Concentration and Chemical Characteristics of $PM_{10\text{-}2.5}$

The average mass concentration of coarse particle ($PM_{10\text{-}2.5}$) was $64.3 \pm 51.16\ \mu g/m^3$ and it ranged from 5.7 to 212.71 $\mu g/m^3$ during campaign period. On average $PM_{10\text{-}2.5}$ know as coarse particle mass (CPM) accounted for nearly 63% of PM_{10}. The CPM concentration of this study was higher than near a Hong-Kong roadway (Cheng et al. 2014).

The average concentration of chemical species is shown in Fig. 6.1. The error bars in the figure depict variability in the elemental concentrations during this study period. The concentration of water soluble ionic species (NO_3^-, SO_4^{2-}, NH_4^+ and Cl^-), WSOC, WSIC, BC and crustal elements (Ca, Fe, Mg and Na) were higher in CPM. Very less concentration was observed for trace elements as compares to other species during this study.

This is likely due to better dispersion of pollutants in summer season because of higher wind speed, high temperature and elevated mixing height (Chakraborty and Gupta 2010). Enrichment factor (EF) for the crustal elements indicates their crustal nature from previous study (Ghosh et al. 2014); while it indicates non-crustal origin for other trace elements.

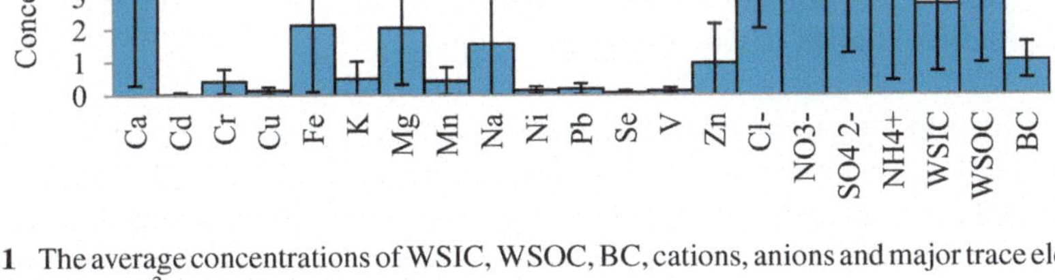

Fig. 6.1 The average concentrations of WSIC, WSOC, BC, cations, anions and major trace elements of $PM_{10\text{-}2.5}$ ($\mu g/m^3$) at Kanpur

6.3.2 $PM_{10\text{-}2.5}$ Source Apportionment by PMF and Data Analysis

EPA PMF v 5.0.5 model was employed to find the source profile and source contribution of $PM_{10\text{-}2.5}$. The measured concentrations and their related uncertainties were used as input to the model (Rajput et al. 2016). A similar approach to Polissar et al. (1998) has been attempted here to estimate the concentration values and their associated error estimates. The model performance in a base run for this study showed determination coefficient (R^2) between the modeled and experimental concentration of $PM_{10\text{-}2.5}$ and other chemical species. Some strong species including $PM_{10\text{-}2.5}$ were very well modeled ($r^2 > 0.7$). Scaled residuals for most of the species were within −3 to +3. Parameter Q is the goodness of model fit which was evaluated to identify the optimal number of factors and the optimal solution should lie in FPEAK range (Paatero et al. 2014). To decrease the influence of extreme values in the PMF solution, the model was run in the default robust mode. Different numbers of sources were explored by trial and error technique to find the optimal number solution. In this study, the theoretical Q_{th} value was 414 (Norris et al. 2014). Robust Q value was the value for which the impact of outliers was minimized, while true Q value was the value for which the influence of extreme values was not controlled. The minimized Q robust was 558 and the ratio of Q_{robust} versus Q_{th} was 1.25 which is similar to previous studies (Chan et al. 2011; Ethirajan and Mohan 2012). This ratio should be ~1 for the stable model solution. In the present study, the robust Q values were very close to the true Q values, implying that the model accounts for the outliers reasonably well. Based on the evaluation of the model results and the Q value variations in the factors, 20 random runs and random seeds were used and 3–10 factors were selected to find the optimum model which was a six-factor solution and it provided the most feasible (physically interpretable) results (with FPEAK = −0.5 based on G-space edges showing no correlations among the resolved sources). The description of the model has been explained in US-EPA PMF 5.0 user manual (Norris et al. 2014) and characterization of species for particular sources have been carried out in several previous studies (Chakraborty and Gupta 2010; Gu et al. 2011; Gupta and Mandariya 2013; Tao et al. 2013; Cheng et al. 2014; Visser et al. 2015). The factors resolved for $PM_{10\text{-}2.5}$ were, paved road dust, vehicular emission, coal combustion and brick kilns, construction activities and incineration, crustal dust and biomass burning and oil combustion. Using PMF analysis, source profiles and contributions of $PM_{10\text{-}2.5}$ mass concentration were identified (Figs. 6.2 and 6.3). Contribution from each factors to total $PM_{10\text{-}2.5}$ mass were also identified by PMF (Fig. 6.4).

Source 1: The species associated with factor 1 included high contribution from Mn, Cl^-, WSIC, NH_4^+ and NO_3^- and less contribution from BC and Na. This factor is classified as paved road dust. The major source of Mn is mineral dust (Querol et al. 2008; Rastogi and Sarin 2009; Ghosh et al. 2014) and it is also the marker of paved road dust (Watson et al. 2004). WSIC is a possible contributor for resuspended dust (Cesari et al. 2014). NH_4^+ and NO_3^- show the formation of ammonium nitrate and is the common marker of secondary nitrate but secondary nitrate contribution was

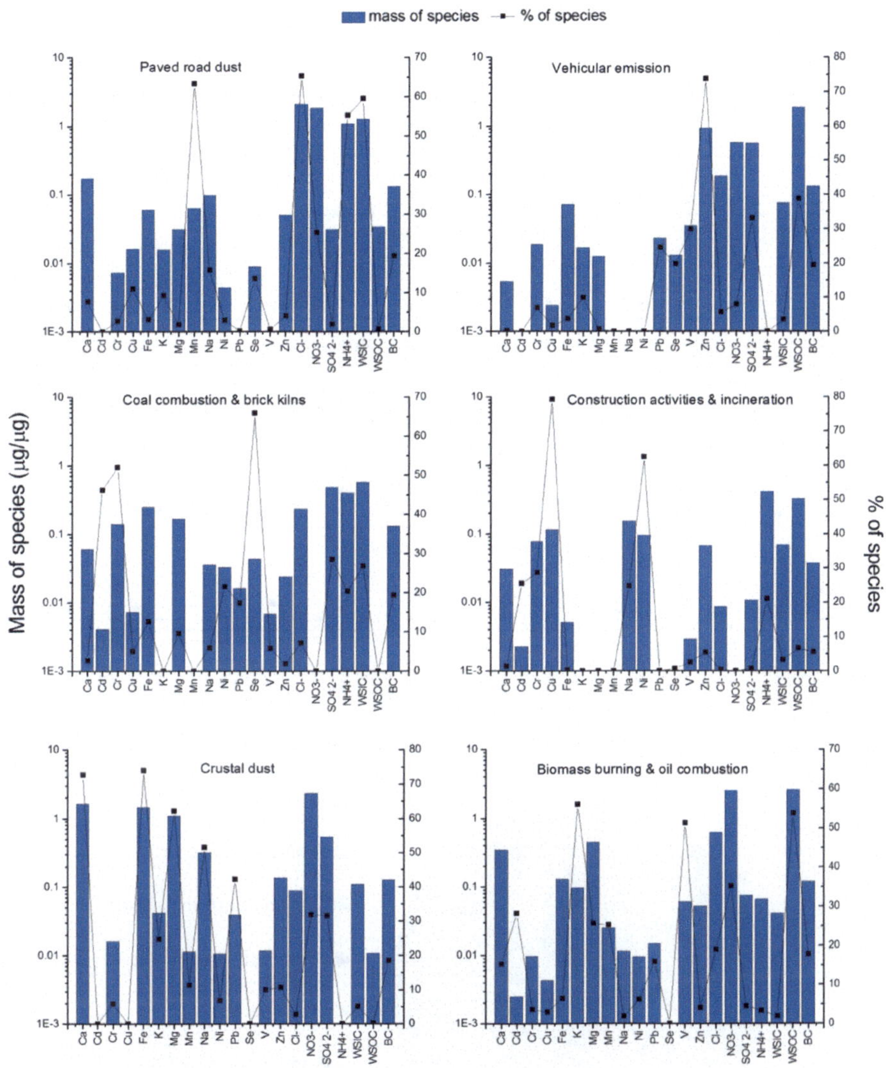

Fig. 6.2 PMF source profile of all six factors in Kanpur for $PM_{10\text{-}2.5}$ mass

lesser in comparison to road dust emission. The contribution of this factor was 53% of the total $PM_{10\text{-}2.5}$.

Source 2: this factor shows high loading on Zn, WSOC, SO_4^{2-}, V and Pb. These elements are in good agreement with vehicular emission. Zn can be emitted from tire wear and it is common component of fuel additives and SO_4^{2-}can be emitted from sulphur containing diesel exhaust (Chakraborty and Gupta 2010).

Pb has been phased out of gasoline from year 2000 in India. However, over the years it has mixed with the road dust and it still persists from earlier vehicular exhaust

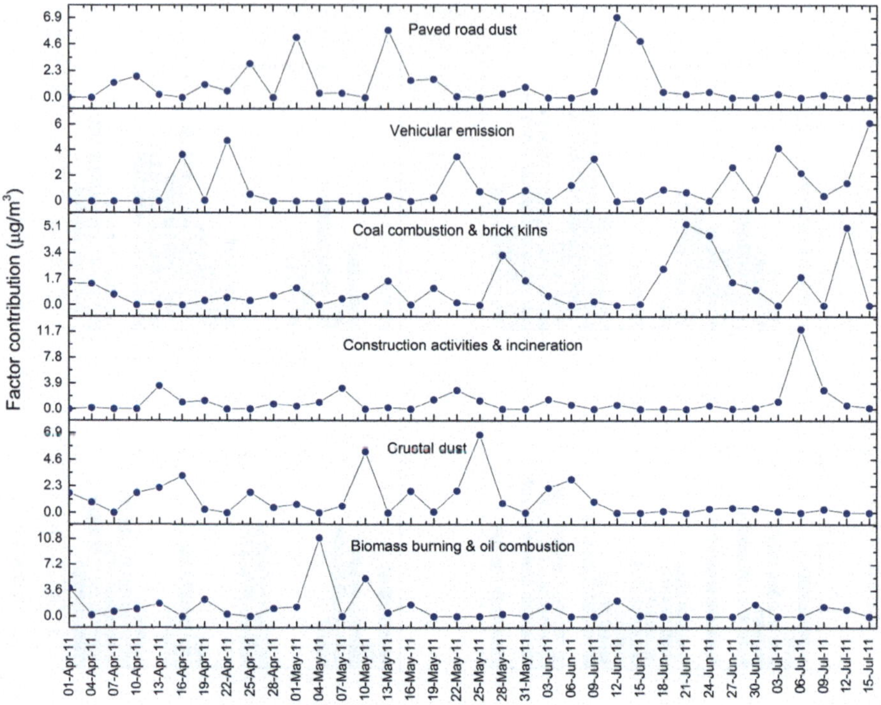

Fig. 6.3 $PM_{10-2.5}$ source contribution of all six factors from April-July, 2011

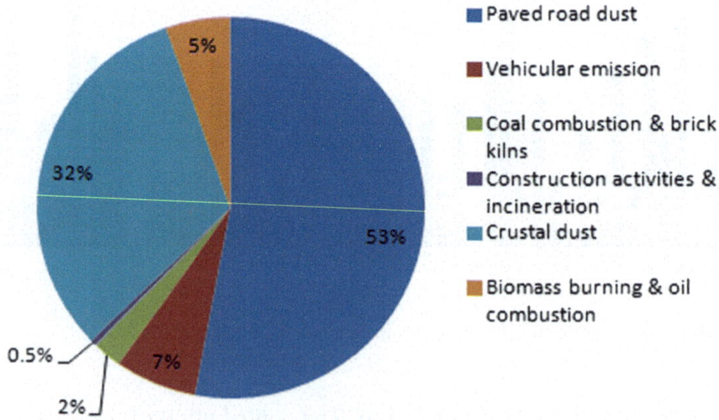

Fig. 6.4 Percentage source apportionment of $PM_{10-2.5}$ mass in Kanpur estimated by PMF

emission. Other likely source of Pb is tire wear and brake lining. WSOC is the tracer for vehicular emission and road dust (Yang et al. 2015). SO_4^{2-} and WSOC forms aqueous phase reactions from the vehicular exhaust. Therefore this factor indicates vehicular emission. Vehicular emissions have been found to be a major source of PM in many India cities. This factor contributed 7% of the total $PM_{10\text{-}2.5}$.

Source 3: this factor has major loading on Se, Cd and Cr while moderate loading on SO_4^{2-}, WSIC, Pb, NH_4^+, BC and Ni. Se, Cd, Pb, Ni and Cr are the major elements for coal combustion (Vincent and Passant 2006; Chakraborty and Gupta 2010; Tao et al. 2013). Cd and Pb are emitted via high temperature coal and oil combustion (Uberol and Shadman 1991). Coarse mode BC may originate from aerosolization of coal dust and other fugitive emissions from thermal power plants using coal (Owoade et al. 2015). BC, SO_4^{2-} and $NH_4^+$are common markers of brick kilns. This factor thus indicates coal combustion and brick kilns. This factor contributed 2.5% of the total coarse particles mass.

Source 4: this factor is highly dominated by Cu, Ni and Cr and is less dominated by Cd, Na and NH_4^+. Ni and Cu are found to have higher concentration in cement and clay (Kulshrestha et al. 2009). Cr is present in steel and cutting tools (Kulshrestha et al. 2009). Cd and NH_4^+ are mainly emitted due to burning of fossil fuels and the incineration of municipal solid waste (Ghosh et al. 2014). Incineration of municipal solid waste can be possible because there is one Asia's largest solid waste management plant A2Z in South-East direction about 8 km from IIT Kanpur. Some construction activities were going on within the premises of the sampling site. Therefore, this can be attributed to construction activity and incineration. The contribution of this factor was only 0.5% to the total coarse particles mass.

Source 5: this factor is highly dominated by Ca, Fe, Mg, Na, K, Pb, NO_3^-, SO_4^{2-} and BC. Ca, Fe, Mg, Na and K are the markers for crustal dust (Gu et al. 2011; Gupta and Mandariya 2013; Tao et al. 2013). Coarse mode SO_4^{2-} may be originated from biogenic decomposition processes or from soils which are rich in gypsum (Yuan et al. 2008; Ghosh et al. 2014). Ghosh et al. (2014) reported that enrichment factor for crustal elements were closed to 1 on all days. Therefore, this factor indicates crustal dust (Rajput et al. 2018). The contribution of this factor was 32% of the total $PM_{10\text{-}2.5}$ mass.

Source 6: this factor has high loading on K, V, WSOC, NO_3^- and Cd while low loading on Mg, Mn and Cl^-. K is the typical marker of biomass burning (Schleicher et al. 2011; Yang et al. 2015). Highly correlated WSOC and K are well known tracers for biomass combustion (Tao et al. 2013). V, Cd and NO_3^- are the tracers for oil combustion (Harrison et al. 1996; Watson 2004). Therefore, this factor is biomass burning and oil combustion. This might represent two distinct sources which due to poor statistical power in this study emerged as a mixed source. This factor contributed 5% of the total mass of $PM_{10\text{-}2.5}$.

From Fig. 6.3, paved road dust, crustal dust, construction activities and incineration and biomass burning and oil combustion contributions were high in pre-monsoon (April-May). A study done at various locations in Kanpur during June and December 2011 reported that major construction work was going on during June, 2011 (Kumar et al. 2014). Cement use and construction work release more coarse particles into the atmosphere. In addition, during April and May the forest fire was dominant in Nepal and nearby regions and might have resulted in mixed plumes on days when air masses originated from those regions. In the monsoon season, dust removal processes via wet scavenging are dominant and therefore their contribution was lower during monsoon.

Although vehicular emission and coal combustion and brick kilns contribution was dominated in monsoon (June-July), their contribution was significant throughout sampling period. Daily heavy fleet of vehicles movement on the Grand Trunk Road (a major national highway) passing through the middle of Kanpur city and terrible traffic congestion are responsible for severe vehicular emission in Kanpur. There is also a coal-based power plant (Panki) located in South-East direction about 8 km from sampling location and several brick kilns industries in North-East-South direction. Therefore, their contribution was dominated during whole sampling period. From the PMF model, paved road dust was the major source of $PM_{10\text{-}2.5}$ at Kanpur. In this study, the major source of $PM_{10\text{-}2.5}$ apportioned by PMF was similar to the studies done during 2012 in London (Visser et al. 2015) and during 2004–2005 in Hong Kong (Cheng et al. 2014).

6.4 Conclusion

In this study, coarse fraction of the airborne particulate matter $PM_{10\text{-}2.5}$ was sampled from April to July 2011 at IIT Kanpur. The sources have been identified using USEPA's PMF v5.0. Six sources were identified and average contribution from each of those sources was calculated. These are 53% from paved road dust, 7% from vehicular emission, 2.5% from coal combustion and brick kilns, 0.5% from construction activities and incineration, 32% from crustal dust and 5% from biomass burning and oil combustion. We conclude that paved road dust is the major source of $PM_{10\text{-}2.5}$ followed by crustal dust.

References

Amato F, Schaap M, Denier van der Gon HAC, Pandolfi M, Alastuey A, Keuken M, Querol X (2013) Short-term variability of mineral dust, metals and carbon emission from road dust resuspension. Atmos Environ 74:134–140

Cesari DA, Ielpo GP, Siciliano M, Mascolo G, Grasso FM, Contini D (2014) Source apportionment of $PM_{2.5}$ in the harbour–industrial area of Brindisi (Italy): identification and estimation of the contribution of in-port ship emissions. Sci Total Environ 497–498:392–400

Chakraborty A, Gupta T (2010) Chemical characterization and source apportionment of submicron (PM_1) aerosol in Kanpur Region. India Aeros Air Qual Res 10(5):433–445

Chan YC, Hawas O, Hawker D, Vowles P, Cohen DD, Stelcer E, Simpson R, Golding G, Christensen E (2011) Using multiple type composition data and wind data in PMF analysis to apportion and locate sources of air pollutants. Atmos Environ 45:439–449

Cheng Y, Lee S, Gu Z, Ho K, Zhanga Y, Huang Y, Chow JC, Watson JG, Cao J, Zhang R (2014) $PM_{2.5}$ and $PM_{10-2.5}$ chemical composition and source apportionment near a Hong Kong roadway. Particuology 30

Chinnam N, Dey S, Tripathi SN, Sharma M (2006) Dust events in Kanpur, northern India: chemical evidence for source and implications to radiative forcing. Geophys Res Lett 33:L08803

Councell TB, Duckenfield KU, Landa ER, Callender E (2004) Tire-wear particles as a source of zinc to the environment. Environ Sci & Technol 38:4206–4214

Dey S, Tripathi SN, Singh RP, Holben BN (2004) Influence of dust storms on the aerosol optical properties over the Indo-Gangetic basin. J Geophys Res Atmos 109:20211

Ethirajan R, Mohan S (2012) Comparative evaluation of VOC source profiles developed by PMF and UNMIX models. Int J Environ Sci Dev 3(5)

Ghosh S, Gupta T, Rastogi N, Gaur A, Misra A, Tripathi SN, Paul D, Tare V, Prakash O, Bhattu D, Dwivedi AK, Kaul DS, Dalai R, Mishra SK (2014) Chemical characterization of summertime dust events at Kanpur: insight into the sources and level of mixing with anthropogenic emissions. Aerosol Air Qual Res 14:879–891

Gu JW, Pitz M, Schnelle–Kreis J, Diemer J, Reller A, Zimmermann R, Soentgen J, Toelzel M, Wichmann HE, Peters A, Cyrys J (2011) Source apportionment of ambient particles: comparison of positive matrix factorization analysis applied to particle size distribution and chemical composition data. Atmos Environ 45:1849–1857

Gupta T, Mandariya A (2013) Sources of submicron aerosol during fog-dominated wintertime at Kanpur. Environ Sci Poll Res 20:5615–5629

Hammond DM, Dvonch JT, Keeler GJ, Parker EA, Kamal AS, Barres JA, Yip FY, Brakefield-Caldwell W (2008) Sources of ambient fine particulate matter at two community sites in Detroit, Michigan. Atmos Environ 42:720–732

Harrison RM, Smith DIT, Luhana L (1996) Source apportionment of atmospheric polycyclic aromatic hydrocarbons collected from an urban location in Birmingham, U.K. Environ Sci Tech 30(3):825–832

Harrison RM, Shi JP, Xi SH, Khan A, Mark D, Kinnersley R (2000) Measurement of number, mass and size distribution of particles in the atmosphere. Philos Trans R Soc Lond Ser A 358:2567–2580

Kulshrestha A, Satsangi PG, Masih J, Taneja A (2009) Metal concentration of $PM_{2.5}$ and PM_{10} particles and seasonal variations in urban and rural environment of Agra, India. Sci Total Environ 407:6196–6204

Kumar A, Srivastava D, Agrawal M, Goel A (2014) Snapshot of PM loads evaluated at major road and railway intersections in an urban locality. Int J Environ Prot 4(1):23–29

Kupiainen KJ, Tervahattu H, Raisanen M, Makela T, Aurela M, Hillamo R (2005) Size and composition of airborne particles from pavement wear, tires, and traction sanding. Environ Sci Technol 39:699–706

Li R, Wiedinmyer C, Hannigan MP (2013) Contrast and correlations between coarse and fine particulate matter in the United States. Sci Total Environ 456–457:346–358

Misra A, Gaur A, Bhattu D, Ghosh S, Dwivedi A, Dalai R, Paul D, Gupta T, Tare V, Singh S, Eck T, Welton E, Holben B, Tripathi SN, Mishra SK (2014) An overview of the physico-chemical characteristics of dust at Kanpur in the central Indo-Gangetic basin. Atmos Environ 97:386–396

Nastos T, Athanasios G, Michael B, Eleftheria SR, Kostas NP (2010) Outdoor particulate matter and childhood asthma admission in Athens, Greece: a time-series study. Environ Health 9:1–9

Norris GA, Duvall R, Brown SG, Bai S (2014) EPA positive matrix factorization (PMF) 5.0 fundamentals and user Guide prepared for the U.S. environmental protection agency office of research and development. Washington, DC. (EPA/600/R-14/108; STI-910511–5594-UG, April)

Owoade KO, Hopke PK, Olise FS, Ogundele LT, Fawole OG, Olaniyi BH, Jegede OO, Ayoola MA, Bashiru MI (2015) Chemical compositions and source identification of particulate matter ($PM_{2.5}$ and $PM_{2.5\text{-}10}$) from a scrap iron and steel smelting industry along the Ife-Ibadan highway. Nigeria Atmos Pollut Res 6:107–119

Paatero P, Tapper U (1994) Positive matrix factorization: a non-negative factor model with optimal utilization of error estimates of data values. Environmetrics 5:111–126

Paatero P, Eberly S, Brown SG, Norris GA (2014) Methods for estimating uncertainty in factor analytic solutions. Atmos Meas Tech 7:781–797

Polissar AV, Hopke PK, Paatero P (1998) Atmospheric aerosol over Alaska-2. Elemental composition and sources. J Geophys Res-Atmos 103:19045–19057

Querol X, Alastuey A, Moreno T, Viana MM, Castillo S, Pey J, Rodriguez S, Artinano B, Salvador P, Sanchez M, Garcia Dos Santos S, HerceGarraleta MD, Fernandez-Patier R, Moreno-Grau S, Negral L, Minguillon MC, Monfort E, Sanz MJ, Palomo-Marin R, Pinilla-Gil E, Cuevas E, de la Rosa J, Sanchez de la Campa A (2008) Spatial and temporal variations in airborne particulate matter (PM_{10} and $PM_{2.5}$) across Spain 1999–2005. Atmos Environ 42:3964–3979

Rajput P, Mandaria A, Kachawa L, Singh DK, Singh AK, Gupta T (2016) Chemical characterisation and source apportionment of PM_1 during massive loading at an urban location in Indo-Gangetic Plain: impact of local sources and long-range transport. Tellus B 68:30659

Rajput P, Singh DK, Singh AK, Gupta T (2018) Chemical composition and source-apportionment of sub-micron particles during wintertime over Northern India: new insights on influence of fog-processing. Environ Pollut 233:81–91

Rajput P, Izhar S, Gupta T (2019) Deposition modeling of ambient aerosols in human respiratory system: health implication of fine particles penetration into pulmonary region. Atmos Pollut Res 10:334–343

Ram K, Sarin MM (2012) Carbonaceous aerosols over Northern India: sources and spatio-temporal variability. Proc Indian natnSciAcad 78(3):523–533

Rastogi N, Sarin MM (2009) Quantitative chemical composition and characteristics of aerosols over western India: one-year record of temporal variability. Atmos Environ 43:3481–3488

Schleicher NJ, Norra S, Chai F, Chen Y, Wang S, Cen K, Yu Y, Stueben D (2011) Temporal variability of trace metal mobility of urban particulate matter from Beijing—A contribution to health impact assessments of aerosols. Atmos Environ 45:7248–7265

Sharma SK, Mandal TK, Saxena M, Rashmi Sharma A, Datta A, Saud T (2014). Variation of OC, EC, WSIC and trace metals of PM_{10} in Delhi, India. J Atmos Sol-Terr Phys 113:10–22

Soluri DS, Godoy MLDP, Godoy JM, Roldão LA (2007). Multi-site $PM_{2.5}$ and $PM_{2.5-10}$ aerosol source apportionment in Rio de Janeiro, Brazil. J Braz Chem Soc 18(4):838–845

Tao J, Zhang L, Engling G, Zhang R, Yang Y, Cao J, Zhu C, Wang Q, Luo L (2013) Chemical composition of $PM_{2.5}$ in an urban environment in Chengdu, China: importance of springtime dust storms and biomass burning. Atmos Res 122:270–283

Uberol M, Shadman F (1991) High–temperature removal of cadmium compounds using solid sorbents. Environ Sci Technol 25:1285–1289

Vedal S, Hannigan MP, Dutton SJ, Miller SL, Milford JB, Rabinovitch N, Kim SY, Sheppard L (2009) The denver aerosol sources and health (DASH) study: overview and early findings. Atmos Environ 43:1666–1673

Vincent K, Passant N (2006) Assessment of heavy metal concentrations in the United Kingdom, AEA technology, http://uk-air.defra.gov.uk/assets/documents/reports/cat16/0604041205_heavy_metal_issue1_final.pdf

Visser S, Slowik JG, Furger M, Zotter P, Bukowiecki N, Canonaco F, Flechsig U, Appel K, Green DC, Tremper AH, Young DE, Williams PI, Allan JD, Coe H, Williams LR, Mohr C, Xu L, Ng NL, Nemitz E, Barlow JF, Halios CH, Fleming ZL, Baltensperger U, Prévôt ASH (2015) Advanced source apportionment of size-resolved trace elements at multiple sites in London during winter. Atmos Chem Phys 15:11291–11309

Watson JG (2004) Desert research institute, protocol for applying and validating the CMB Model for $PM_{2.5}$ and VOC, US environmental protection agency, air quality modeling group

WHO (2013) Review of evidence on health aspects of air pollution–REVIHAAP Project, Report, WHO European Centre for Environment and Health, Bonn

Yang L, Cheng S, Wang X, Nie W, Xu P, Gao X, Yuan C, Wang W (2013) Source identification and health impact of $PM_{2.5}$ in a heavily polluted urban atmosphere in China. Atmos Environ 75:265–269

Yang H, Chen J, Wen J, Tian H, Liu X (2015) Composition and sources of $PM_{2.5}$ around the heating periods of 2013 and 2014 in Beijing: implications for efficient mitigation measures. Atmos Environ 30:1–9
Yuan H, Zhuang G, Li J, Wang Z (2008) Mixing of mineral with pollution aerosols in dust season in beijing: revealed by source apportionment study. Atmos Environ 42:2141–2157
Zhang R, Jing J, Tao J, Hsu S-C, Wang G, Cao J, Lee CSL, Zhu L, Chen Z, Zhao Y, Shen Z (2013) Chemical characterization and source apportionment of $PM_{2.5}$ in Beijing: seasonal perspective. Atmos Chem Phys 13:7053–7074

Chapter 7
Analysis of an Aerosol Environment in an Urban Region and Its Impact on Regional Meteorology

Shamitaksha Talukdar and Animesh Maitra

Abstract The aerosol environment over a region is a potential parameter to perturb the regional meteorology and it remains as a large source of uncertainties. This chapter presents an analysis of the aerosol environment over Kolkata, a densely-populated metropolitan area in the Eastern India, near to the land-ocean boundary of Bay of Bengal. Along with the natural sources of aerosols, anthropogenic aerosols constantly contribute to the total aerosol loading in this region. This analysis shows how the spectral dependence pattern of aerosol optical parameters estimates the dominant type of aerosols over a region. Furthermore, the investigation shows the impact of the dominant type of aerosols on the boundary layer meteorology utilizing ground based observations. In the course of this discussion, the role of atmospheric black carbon (BC) aerosols in controlling the aerosol environment is focused. This urban region experiences much high BC concentrations mostly originating from incomplete combustion of fossil fuel used in vehicles. The chapter also includes a discussion on back trajectory analysis showing the influence of transported air masses on the regional aerosol environment. The investigation presented in this chapter depicts significant perturbation in the boundary layer temperature profile in association with the abundance of pollutant aerosols which in turn can affect the normal atmospheric convective processes over a region. This chapter contributes to a better understanding of impact of pollutant aerosols on the regional meteorology and thermodynamic processes.

Keywords Urban pollutants · Boundary layer temperature · Lapse rate · Weakening of convection

---

S. Talukdar (✉)
National Atmospheric Research Laboratory, Gadanki, India
e-mail: shamitaksha@gmail.com

A. Maitra
Institute of Radio Physics and Electronics, University of Calcutta, Kolkata, India
e-mail: animesh.maitra@gmail.com

T. Gupta et al. (eds.), *Measurement, Analysis and Remediation of Environmental Pollutants*, Energy, Environment, and Sustainability,
https://doi.org/10.1007/978-981-15-0540-9_7

7.1 Introduction

An aerosol environment over a region is a resultant or a matured condition due to the followings: loading of aerosols in an urban, rural or any other environment, removal of aerosols from its atmosphere due to rainfall (rain scavenging or washout) or dispersal due to wind and transportation of aerosols by different air mass in different time. Thus an aerosol environment over a location consists of different types of scattering and absorbing aerosols. Scattering aerosols decrease the incoming solar radiation by scattering and exert cooling effect on the surface of earth and on the other hand absorbing aerosols, by trapping the shortwave solar radiation warms up the atmosphere above the earth surface (Charlson et al. 1992). It has been reported in recent studies that absorbing aerosol's heating efficiency is more in the upper boundary layer compared to surface layer (Ding et al. 2016). This is a very crucial aspect in the context of thermodynamic behaviour of the boundary layer and it may influence the normal upward movement of air-mass in the lower atmosphere. Depending on the possible aerosol sources a region may experience the dominance of a particular type of aerosol or a group of aerosol species which directly impact the aerosol forcing (direct and indirect) over that region. The spectral dependence pattern of aerosol optical parameters may give basic estimation about the dominant type of aerosols (fine mode or coarse mode particles) for a regional aerosol environment (Bergstrom et al. 2002; Dubovik et al. 2002). Aerosol optical depth (AOD) and single scattering albedo (SSA) are such two important aerosol optical parameters which have been used in the present analysis. AOD is the extinction of solar irradiance due to aerosols (Angstrom 1929) present in the air column from ground to the top of the atmosphere and SSA is the ratio of the scattering to the total extinction (absorption and scattering) (Bergstrom and Russell 1999; Ganguly et al. 2005; Haywood and Shine 1995, 1997). The logarithmic slope of AOD versus wavelength curve, known as Angström wavelength exponent (α), is an indicator of aerosol size (Eck et al. 1999). From the value of α we can have an estimation of the abundance of fine mode (anthropogenic aerosol associated with urban pollution)or coarse mode aerosol particles (natural aerosols like desert dust and sea salts) (Bergstrom et al. 2002; Sokolik and Toon 1997) in an aerosol environment. Simultaneously the wavelength dependence pattern of SSA also indicates the dominant type of aerosols over a region. Thus, both of these optical properties in combination can give a reliable estimation about the dominant aerosol type over a location (Bergstrom et al. 2007). The study location experiences both anthropogenic and natural aerosols loading in building up of the aerosol environment as it is a metro city near to the land-ocean boundary of Bay of Bengal. Very few studies have been carried out on the estimation of SSA and Angstrom exponent over the Gangetic West Bengal region so far (Goto et al. 2011; Lawrence and Lelieveld 2010). In the present analysis the spectral dependence pattern of AOD and SSA are used to classify the dominance aerosol types over Kolkata.

7.2 Description of the Site and Instruments Used

Kolkata is a densely populated urban region in the eastern India with a heavy daily traffic load and some industrial belts at the coast of river Ganges. The monitoring instrument set up that has been used to study the aerosol environment over this region is sited at the University of Calcutta which is in the midst of the city Kolkata. As a prevailing meteorological feature this region experiences around 72% of annual rainfall during South-west monsoon (June–September) season (www.imdkolkata.gov.in). But reportedly 15% of total rainfall over Kolkata takes place during pre-monsoon (March–May) season (Bhattacharya et al. 2011). Around 10–12% rainfall takes place during post-monsoon (October and November) and some rare rainfall events take place in winter (December–February). Convective rainfall is a typical feature in pre-monsoon and also sometimes in monsoon season. Apart from rain occurrences, a possibility of cloudiness is there in any season which may restrict optical aerosol measurement process on those days. Wind from different direction in different time of the year has influence on regional aerosol loading. The boundary layer height over this tropical urban region is observed maximum in pre-monsoon with a monthly average value of around 1.6 km and it becomes low in monsoon and winter months as shown in Fig. 7.1b. The yearly average lapse rate profile shown in Fig. 7.1a, has values around 6 K/km. Its vertical profile remains almost consistent up to 1 km height and above of this height, lapse rate decreases. AOD over the present location is monitored using a multi wavelength radiometer (MWR) with a spectral range of 340 nm to 1025 nm (at ten wavelengths 340, 400, 450, 500, 600, 650, 750, 850, 935, 1025 nm). MWR is capable of measuring the aerosol extinction of directly transmitted solar radiation as a function of solar zenith angle. To measure black carbon concentration, a 7-channel Aethalometer (AE-31) (Magee Scientific made) is constantly operated at the same location at seven wavelengths (370, 470, 520, 590,

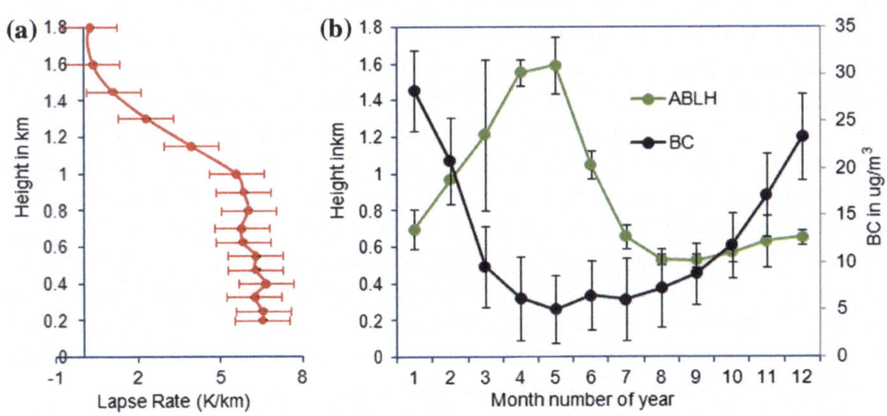

Fig. 7.1 **a** Mean lapse rate profile (2012–2014) within the height of surface boundary layer (the standard deviations of mean are shown by error bars). **b** Annual variation of atmospheric boundary layer height (ABLH) and near-surface BC concentration over Kolkata

660, 880, 950 nm). In accordance to that seven MWR wavelengths (400, 450, 500, 600, 650, 850, 935 nm) which are nearly equal to the aethalometer wavelengths are used to calculate SSA values. Temperature profile used in this analysis is monitored by a multi-frequency microwave radiometer (RPG-HATPRO) which is continuously operated at this site.

7.3 Data and Methodology

In this analysis, cloudy days (fully and partially) are not taken into account and only the cloud free days data are considered for further estimations. Cloudiness occurs very often which may restrict the measurement of AOD and that may cause limited data for some periods or months. MWR is not operated when it is a fully cloudy sky or a rainy day. However, the partially clouded day over the study location is determined using the MWR itself because in the presence of partial cloud coverage, MWR data logging software itself marks the data set as "Bad data" or "Doubtful data". Even in the case of cloud passing over the region, the MWR notes the dataset as "Flat or Zig-Zag" dataset. So the presence of cloudy dates is easily noticeable and is excluded from the analysis. The days marked as "Good data" by MWR (for cloud free days) are only considered for the study. To have sufficient dataset for each season the dataset of 2012 to 2014 have been combined. In order to have a coherent dataset analysis, all the dataset of AOD, BC and temperature profile have been collected during the day time [afternoon (AN) (1139 h (IST) to 1700 h (IST))] of cloud free days. The global reanalysis data product of NCEP/NCAR (National Centers for Environmental Prediction and National Center for Atmospheric Research) global reanalysis data product of NOAA (National Oceanic and Atmospheric Administration) (ftp://arlftp.arlhq.noaa.gov/pub/archives/reanalysis) is used in back trajectory analysis. Trajectories are drawn using TrajStat (Trajectory Statistics) which is geographic information system (GIS) based software. The trajectory calculation function in the TrajStat, appears from Hybrid Single Particle Lagrangian Integrated Trajectory (HYSPLIT) model. HYSPLIT is one of the most broadly used models describing atmospheric transport, deposition and dispersion calculations developed by NOAA's Air Resources Laboratory (Stein et al. 2015). Detailed descriptions of HYSPLIT modeling system development along with its historical background have been presented by Stein et al. (2015) in their recent article (Stein et al. 2015).

Atmospheric boundary layer height (ABLH) is not monitored at the study site. Three hourly gridded data of boundary layer height are obtained from NOAA Earth System Research Laboratory (ESRL) data base using the website (http://www.esrl.noaa.gov/psd/data/gridded/data.20thC_ReanV2.monolevel.mm.html) and monthly averaged boundary layer height has been plotted in Fig. 7.1b.

Aerosol single scattering albedo (ω) is the ratio of scattering to the total extinction due to atmospheric aerosols at particular wavelengths and is calculated by the expression

$$(\text{SSA})\omega = \frac{\tau_{sca}}{\tau_{sca} + \tau_{abs}} = \frac{\tau_{sca}}{\tau_{\lambda}} \quad (7.1)$$

In this equation τ_{abs} and τ_{sca} are absorption and scattering extinction and τ_{λ} is the AOD at the wavelength λ. The decreasing tendency of the spectral pattern of SSA indicates the dominance of fine mode aerosols in an aerosol environment and the increasing tendency of the spectral pattern indicates the abundance of dust aerosols (Bergstrom et al. 2007). Now, in order to separate out the terms τ_{abs} and τ_{sca} the absorption extinction is to be calculated. For the present analysis, the absorption of incoming solar radiation only due to BC aerosols over the study region is taken in account. Earlier researchers have reported that BC is the most effective absorbing component among the particulate matters (PM) (Heintzenberg et al. 1997; Ramana et al. 2010; Tiwari et al. 2013) and second largest heating agent next to CO_2 in warming up our atmosphere (Jacobson 2001; Ramanathan and Carmichael 2008). BC particles are emitted from incomplete combustion of fossil fuels (used in vehicles and industries) and also from biomass burning (Koelmans et al. 2006; Penner et al. 1993). It is reported in a study from Gadanki, India that 80–90% of net absorbed radiation within the atmosphere is contributed by BC aerosols (Gadhavi and Jayaraman 2010). The present site experiences notable high concentration of BC aerosol even in comparison with many other Indian urban stations as shown in Table 7.1. Moreover, the present location we are discussing is near to the Bay-of Bengal land-ocean boundary hence, the contribution of other absorbing aerosols has been found negligible compared to BC aerosols (Aruna et al. 2014). Considering all the mentioned

Table 7.1 BC concentration as reported from different urban locations in India (references are cited therein from where the values are listed in this table)

Station name	BC (μg m^{-3})	References
Goa (Urban)	3 (±0.07)	(Upadhyay and Singh 2010)
Ahmedabad (Urban and Industrial)	3.43 (dry season), 1.3 (wet season)	(Ramachandran and Rajesh 2007)
Pune (Urban)	3.58 (±01.55)	(Safai et al. 2013)
Bangalore (Urban)	4.2 (±0.2)	(Satheesh et al. 2011)
Udaipur (Urban)	5.6	(Vyas 2010)
Hyderabad (Urban)	6.67 (± 0–22) (dry season), 2.36 (±0.09) (wet season)	(Dumka et al. 2013)
Mumbai (Urban and Industrial)	12.5	(Venkatraman et al. 2002)
Varanasi (Urban)	14.5	(Upadhyay and Singh 2010)
Kanpur (Urban and Industrial)	6–20	(Tripathi et al. 2005)
Delhi (Urban and Industrial)	13.4 (±10)	(Tiwari et al. 2013)
Kolkata (Urban)	27 (±3) (dry season), 5 (±1.5) (wet season)	(Talukdar et al. 2015)

aspects it seems logical for the study region to account the absorption of incoming solar radiation solely by BC aerosols. The Absorption coefficient (β_{abs}) of BC can be calculated using the following equation (Aruna et al. 2014)

$$\beta_{abs} = \left[\left(\frac{\Delta ATN}{V}\right) \times A\right] C.R \tag{7.2}$$

Here '*C*' and '*R*' in this Eq. (7.2) are calibration factors introduced to aethalometer attenuation measurement in order to have authentic absorption coefficient values. 'C' is a factor which accounts the multiple scattering of transmitted light by the fibres of the quartz filter and its value is 1.9 (Bodhaine 1995; Weingartner et al. 2003). 'R' is an empirical factor introduced to account the loading of light scattering particles along with BC on the quartz filter and mentioned as 'shadowing effect'. The value of 'R' is considered as unity (Arnott et al. 2005; Weingartner et al. 2003). ΔATN is the change in attenuation of light through filter tape. A is the area of spot on filter tape where BC is deposited and V indicates the volume of air passed through filter.

Boundary layer height is a meteorological parameter having profound control on surface BC concentration. It is reported in earlier studies that most of the BC particles are confined within the atmospheric boundary layer height and are homogeneously mixed (Aruna et al. 2014; Gogoi et al. 2011; Stull 1989). Figure 7.1b shows the average boundary layer height over this tropical urban region during pre-monsoon months (March to May) is around 1.5 km, whereas in monsoon (June–Sep) it is around 600 m, and it is below 1000 m during winter (Dec–Feb). The figure also depicts that monthly mean near surface BC concentration decreases with an increase in the boundary layer height. It happens due to the strong control of regional boundary layer height on the near surface concentration of aerosol particles.

Diurnal and seasonal variations of atmospheric boundary layer height directly control the BC concentration capping the major fraction of total atmospheric BC content below it. Figure 7.2 depicts the diurnal variation of near surface BC concentration which is a boundary layer dependent variation. Figure 7.2 also presents the degree of near surface BC concentration more precisely over the year over this study region. Low surface BC in the month of Apr–May is associated with high boundary layer (shown in Fig. 7.1b) and the reverse case is evident in the winter months Dec–Feb. In monsoon BC is low mainly due to wet removal and the boundary layer height is also low in monsoon. Under these conditions, considering a homogenous mixing of BC within boundary layer height, extinction due to BC (τ_{abs}) has been calculated, by multiplying the derived value of absorption coefficient (β_{abs}) of BC by the atmospheric boundary layer height (ABLH) value (ABLH data used is not retrieved, it is NOAA ESRL gridded data).

$$\tau_{abs} = \text{ABLH} \times \beta_{abs} \tag{7.3}$$

It is to be noted here that a fraction of total content of BC is practically present there above ABL height (Babu et al. 2011) but there is a limitation of having BC profile information through the atmospheric column over the study location. Hence,

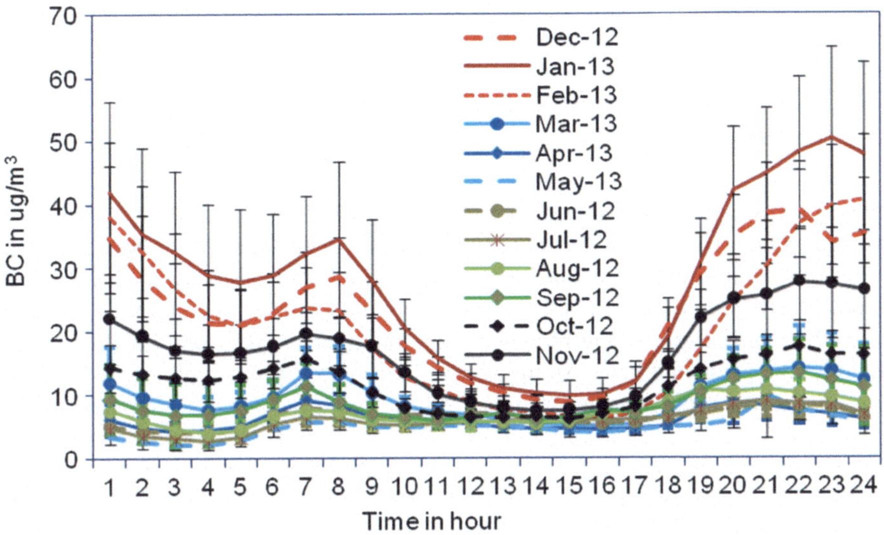

Fig. 7.2 Diurnal and annual variation of near surface black carbon concentration over Kolkata

the fraction of BC beyond Boundary layer height has not being taken into account. Therefore, the absorption extinction obtained in this calculation is presenting the absorption by BC within boundary layer height and possibly underestimating the total absorption due to BC.

The scattering extinction (τ_{sca}) is calculated by subtracting τ_{abs} values from τ_λ in different available wavelengths.

$$\tau_{sca} = \tau_\lambda - \tau_{abs} \tag{7.4}$$

and then SSA is derived by using Eq. (7.1).

The spectral pattern of AOD is used to calculate the Angström exponent (α). AOD is linked to Angström exponent by the equation

$$AOD_\lambda = \beta\lambda^{-\alpha} \tag{7.5}$$

α is an indicator of the particle size. Values of α > 1 indicates the dominance of fine mode pollutant and $\alpha < 1$ indicates an abundance of coarse mode aerosol particles and α can be calculated using the following equation

$$\ln AOD_\lambda = \alpha \ln \lambda + \ln \beta \tag{7.6}$$

Using linear regression in the Eq. (7.6), α is determined by the slope of the line. β is the coefficient of atmospheric turbidity (Eck et al. 1999).

7.4 Observations and Analysis

7.4.1 Seasonal Variability of Aerosol Parameters Over Kolkata

Overall aerosol loading over Kolkata is high as reflected through the high yearly average value (~0.8) of AOD (500 nm). In winter months it is ~1 or above and also in pre-monsoon months it persists above the mean. Comparatively it shows lower values (0.6–0.7) in monsoon and post-monsoon months as shown in Fig. 7.3a. Due to insufficient data set in the month of June the mean AOD value for this month could not be presented here. Along with it the BC absorption coefficient shown in same figure is observed to have higher values in winter months compared to the other months. The spectral variation of AOD shows a definite decreasing pattern with wavelength in every season and it does not show significant seasonal variation over the year, depicted in Fig. 7.3b. Figure 7.3c presents the spectral pattern of absorption coefficient in different seasons. In use, 880 nm is considered for standard BC measurement because it is the closest to 830 nm at which aerosols other than BC have negligible absorption (Bodhaine 1995). Using these observations the calculated SSA values have been plotted in Fig. 7.4a, b. Figure 7.4a shows that the monthly mean SSA varies between ~0.78 to ~0.96 with a yearly mean of 0.87 and the maximum is observed in monsoon season as depicted in Fig. 7.4b. It can be noted that SSA values along with absorption coefficient are low in pre-monsoon which apparently may appear inconsistent but boundary layer plays a significant role here. In the pre-monsoon season, atmospheric boundary layer height drastically increases which may cause a well mixing of absorbing particles throughout an extended boundary layer and this eventually dilutes the near surface concentration of absorbing particles. It is evident from Fig. 7.4b that the wavelength dependence of SSA observed in post-monsoon, winter and pre-monsoon season shows a definite decreasing spectral pattern, and, as mentioned before, it reveals the abundance of fine mode pollutant particles at the present location. On the other hand, the wavelength dependence of SSA in monsoon does not show a definite decreasing pattern with wavelengths unlike the other seasons which is indicating a less dominance of pollutant aerosols in monsoon.

BC is an important factor for the determination of the spectral tendency of SSA because of its absorption efficiency over a broad range of visible wavelengths (Bergstrom et al. 2002). For the present analysis over this highly populated urban region, significant amount of BC aerosol is present in its environment because of their continuous emission due to fossil fuel combustion in vehicle (Weingartner et al. 2003). It has been shown in Fig. 7.5 that BC concentrations have positive correlation with Angström exponent (α) values in every season indicating a strong influence of BC concentration in total loading of aerosols over this urban region. This also remains in good agreement with the consideration that in calculating the absorption extinction, the absorption of solar radiation only due to BC aerosols has been taken into account.

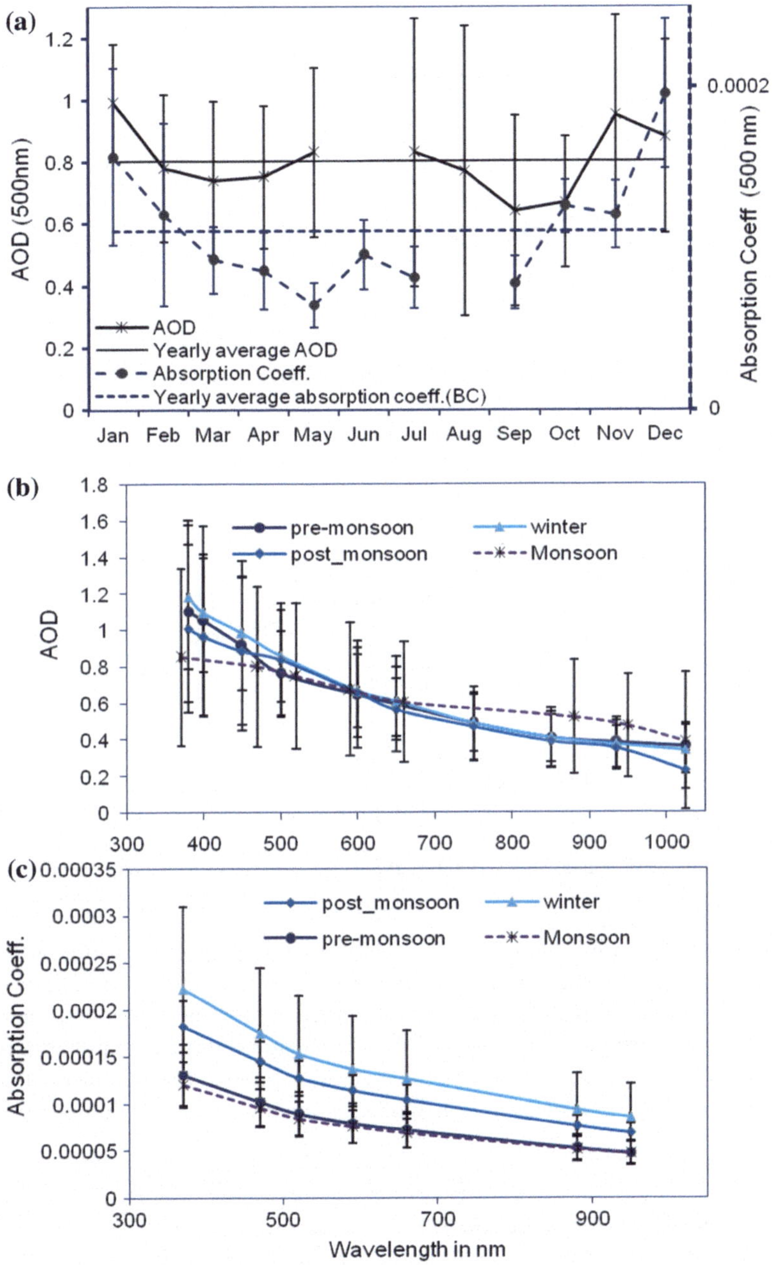

Fig. 7.3 **a** Annual variability of AOD and absorption coefficient at 500 nm. **b** Seasonal variability of the spectral patterns of AOD. **c** Seasonal spectral pattern of absorption coefficient

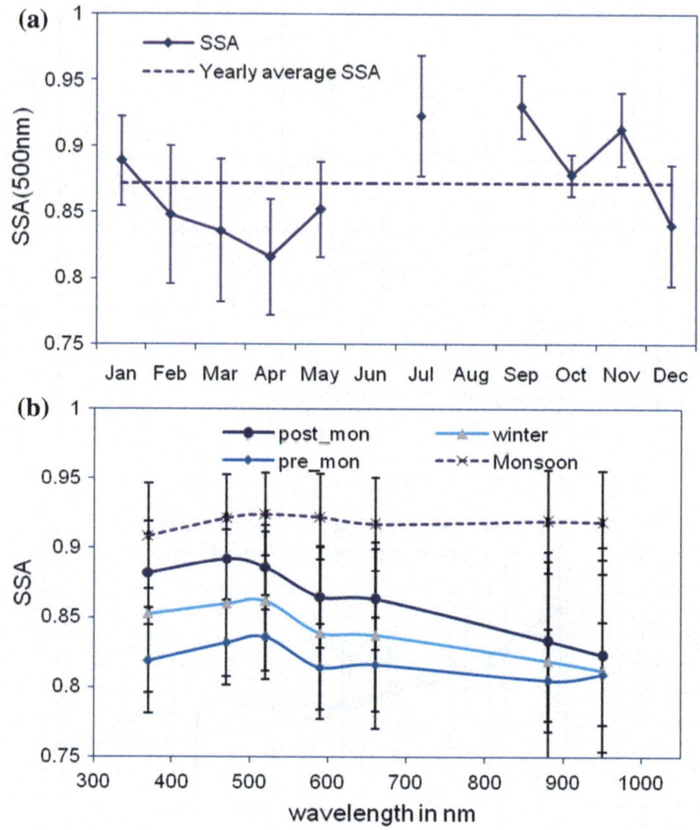

Fig. 7.4 **a** Annual variation of SSA at 500 nm. **b** Spectral patterns of SSA in different seasons

In most of the observations it is found that the spectral pattern of SSA has a decreasing tendency and corresponding α values are greater than unity, shown in Fig. 7.6a, b. It indicates the dominance of fine mode pollutant aerosols over the region. However, it must be noted that there are significant number of observations showing an increasing trend in the spectral variation of SSA, and α calculated for those days have values less than 1 indicating the dust aerosol's dominance.

7.4.2 Impact of the Abundance of Pollutant Aerosols on Regional Boundary Layer Temperature Profile

A major fraction of urban pollutants consists of light-absorbing carbonaceous aerosols which are fine particulate matter and efficiently warm up of the atmosphere (Bergstrom et al. 2007). Hence, in the present section, the impact of the abundance

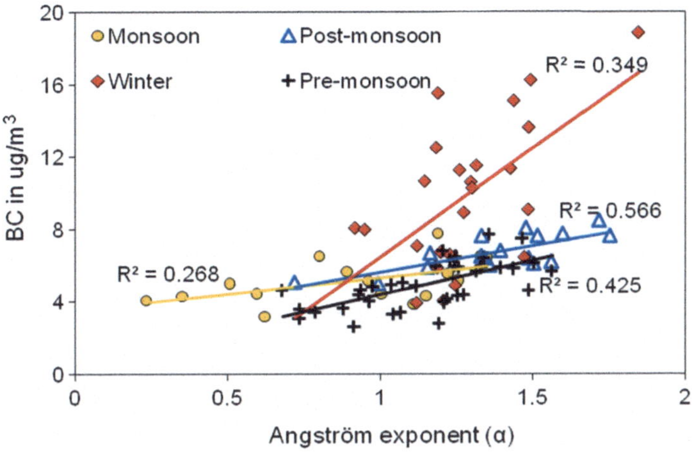

Fig. 7.5 Correlations between BC concentrations and Angström exponent (α) values in pre-monsoon, monsoon, post-monsoon and winter of 2013

of pollutant aerosols on temperature profile has been investigated in the following analysis.

The boundary layer temperature profile has been examined for the group of days having dominance of pollutant aerosols and separately for the group of days having dust aerosol dominance and that are also separated out for different seasons.

It is notable in Fig. 7.7 that the boundary layer temperature profile on pollutant dominating days has higher values than that observed on the dust dominating days. The disparity is most prominent during winter. It is mentioned in the previous section that the absorption coefficient value is maximum in winter and it is also reported that near surface BC shows maximum concentration during winter (Weingartner et al. 2003) when shallow boundary layer (around 760 m: average during December: 655 m, January: 702 m, February: 952 m) at the present location. Similar observation of boundary layer height over the Indo-Gangetic plain in winter month was also reported by earlier researchers (Niranjan et al. 2006). This shallowness of boundary layer in the presence of near surface temperature inversion in winter causes a quick increase in the concentration of pollutant particles near the ground and this may warm up the bottom level of the boundary layer resulting in a significant increase in near surface temperature profile. This very change in temperature profile is not prominent in monsoon and post-monsoon which may be due to the wet removal of major fraction of aerosols from the atmosphere (Kohler et al. 2001).

However, in pre-monsoon season a notable difference in temperature profile between these two groups of days is noticed which is different in nature that observed in winter. The extended boundary layer height in pre-monsoon allows a homogeneous mixing of pollutant aerosols within boundary layer (Saha et al. 2014; Tiwari et al. 2013) and significantly absorbs solar radiation all through this extended height. This may cause a notable fall in boundary layer lapse rate on pollutant dominating days

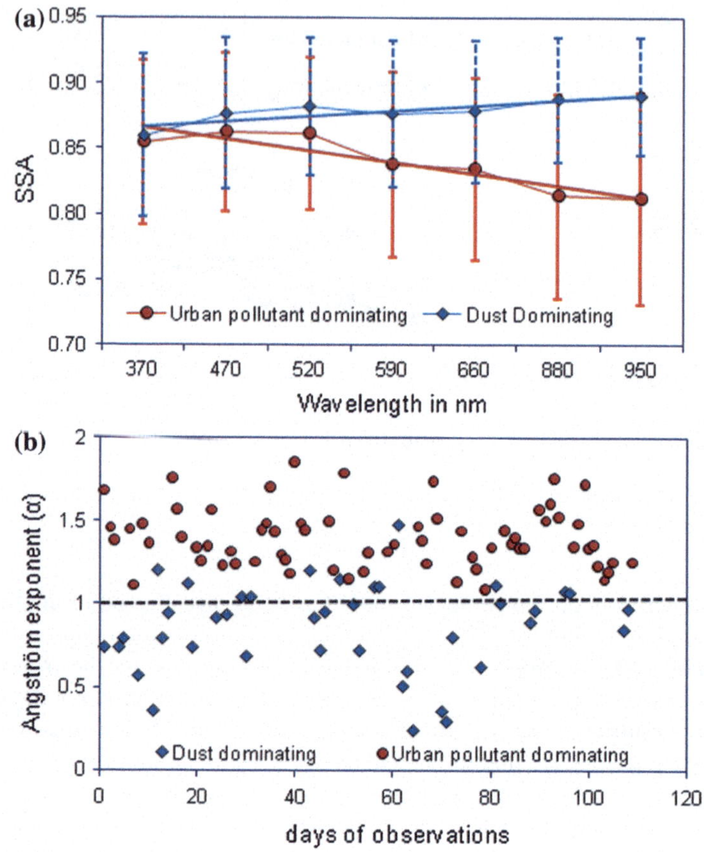

Fig. 7.6 **a** Angstrom exponent values for urban pollutant dominating days and dust dominating days. **b** Mean wavelength dependence of SSA for corresponding days (with standard deviation)

compared to the dust dominating days in the pre-monsoon months. Figure 7.8 depicts the comparison of the temperature profiles of pollutant dominating days with dust dominating days for a pre-monsoon month (May 2013). It is observed in Fig. 7.8 that the vertical gradient of temperature in the height range 200 m to 800 m on pollutant dominating days is distinguishably different than that of the dust dominating days.

The observed difference in temperature profile results in a sharp decrease in lapse rate profile on pollutant dominating days. To highlight the fall more evidently, the lapse rate within the boundary layer for pollutant dominating days are compared with dust dominating days for all the four seasons alone and plotted in Fig. 7.9. It shows that the temperature lapse rate within the boundary layer has a notable fall in pre-monsoon season for the pollutant dominating days compared to the dust dominating days. In other seasons this difference becomes not noticeable.

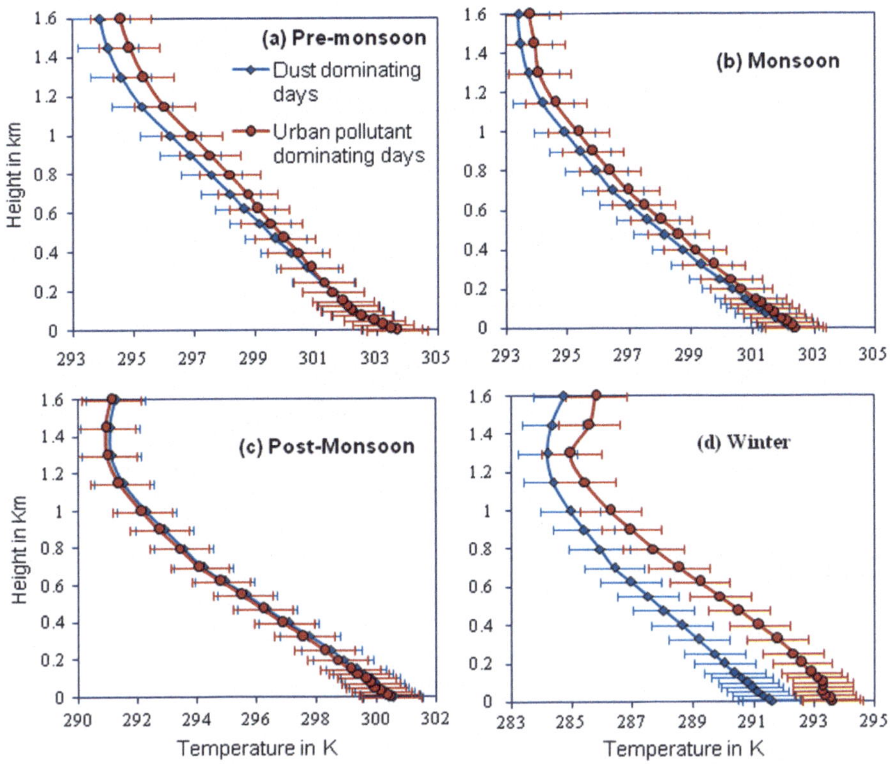

Fig. 7.7 Boundary layer temperature profiles compared on dust dominating days and urban pollutant dominating days in **a** pre-monsoon, **b** monsoon, **c** post-monsoon and **d** winter (standard deviations of mean value are marked by error bars) (Profiles are shown up to 1.6 km as ABLH is rarely found beyond this height at this location)

7.4.3 Influence of the Dominance of Pollutant Aerosols on Temperature Lapse Rate and Atmospheric Convection

The growth of convective processes strongly depends on lower atmospheric instability which is a function of the boundary layer temperature lapse rate. Convective available potential energy (CAPE) is an instability index to measure the development of convective process (Holton 2004). In the further analysis, the fall in lapse rate on pollutant dominating days and the corresponding CAPE values for those days in pre-monsoon has been intensively investigated. CAPE is calculated using the expression as follows

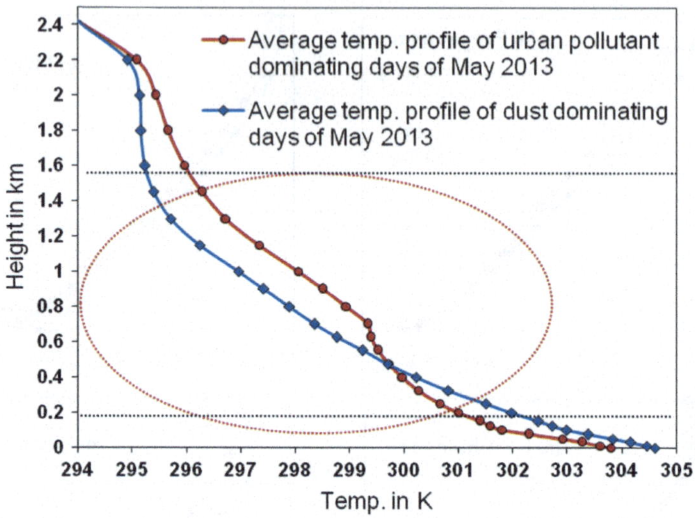

Fig. 7.8 Comparison between the average temperature profiles for pollutant dominating days and dust dominating days (May 2013)

$$\mathrm{CAPE} = \int_{Z_{LFC}}^{Z_{LNB}} \mathrm{g}\left(\frac{T_{parcel} - T_{evn}}{T_{evn}}\right) dz \tag{7.7}$$

T_{parcel} and T_{evn} are moist air parcel temperature and environment temperature respectively. The upper and lower limit of the integration is the level of neutral buoyancy Z_{LNB} and level of free convection Z_{LFC}. Equation (7.7) reveals that a small change in the temperature lapse rate profile can significantly affect the available potential for convection. It has been intended to observe the probable impact of the abundance of pollutant aerosols on the growth of convective energy over the present location in pre-monsoon season when the convective processes are likely to occur. A 10 days time span window in pre-monsoon is considered and within this span of time pollutant and dust dominating days are grouped separately.

The lapse rate profiles for the group of urban pollutant dominating days are compared with the group of dust particle dominating days. This 10 days time span window has been considered in order to avoid or minimize the effect of temporal variability in temperature profile and in the lapse rate as well. The disparity in lapse rate due to abundance of pollutants will be effectively more notable through this observation window avoiding the seasonal or temporal variations in temperature and lapse rate. Some of the lapse rate profiles comparisons shown in Fig. 7.10 depict the drastic fall in lapse rate on pollutant dominating days.

This notable decrease in lapse rate in association with the abundance of pollutant aerosols may cause reduction in updraft motion of wind and thus the convective growth processes may be affected (Niranjan et al. 2006; Parameswaran et al. 2004).

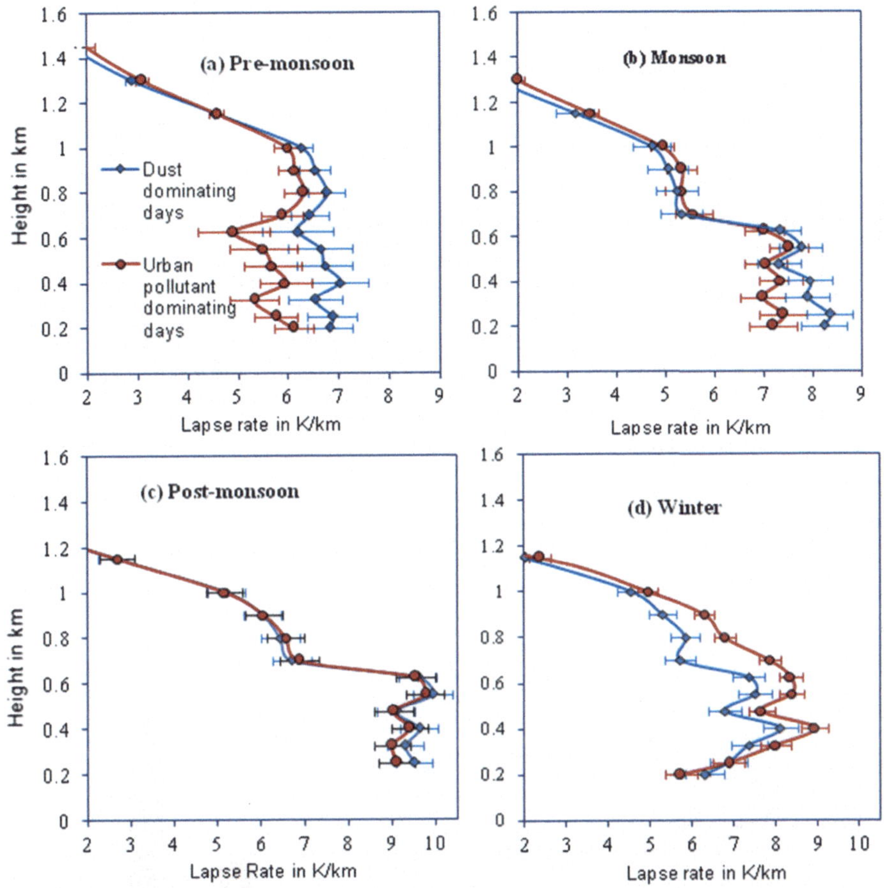

Fig. 7.9 Temperature lapse rate of dust dominating days and pollutant dominating days in **a** pre-monsoon, **b** monsoon, **c** post-monsoon and **d** winter

It has been noted from our investigations that daytime maximum CAPE values are much lower on the days which have dominance of urban pollutants compared to those days having dust aerosol's dominance. In Fig. 7.11a case study shows an instance which supports the above mentioned observation that the day time maximum CAPE value is much lower on a pollutant dominating day than that of a dust dominating day.

Figure 7.11a shows the temperature lapse rate profile on 23 May 2012 (a pollutant dominating day) has a negative value at around 600 m height which indicates the presence of temperature inversion at that height associated with a warming up of the atmosphere due to the dominance of pollutants. Temperature inversion directly affects the normal convection in the atmosphere which has been reflected in the drastic fall in the CAPE (maximum value 359 J/Kg) on that day. Whereas on the subsequent day, 24 May 2012 (which is a dust dominating day), the lapse rate is

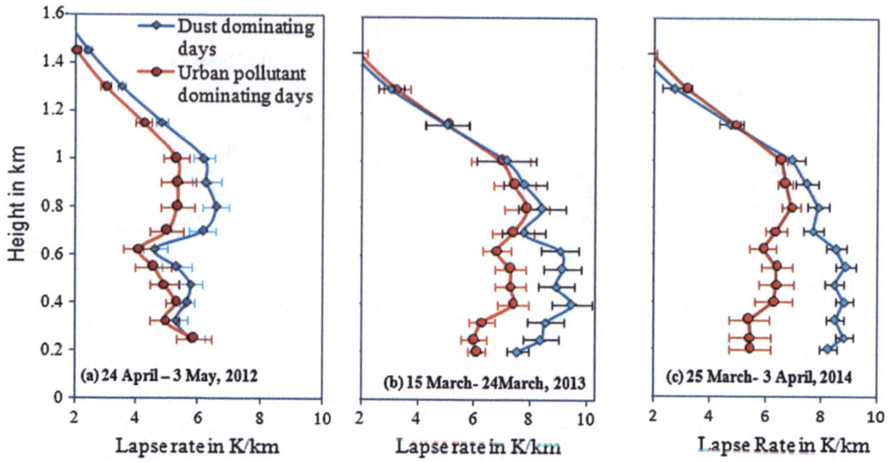

Fig. 7.10 Temperature lapse rate profile for groups of pollutant dominating days and dust dominating days in pre-monsoon season, Three sets of examples each group with 10 days span, are presented: **a** 24 April–3 May 2012 **b** 15 March–24 March 2013 and **c** 25 March–3 April 2014

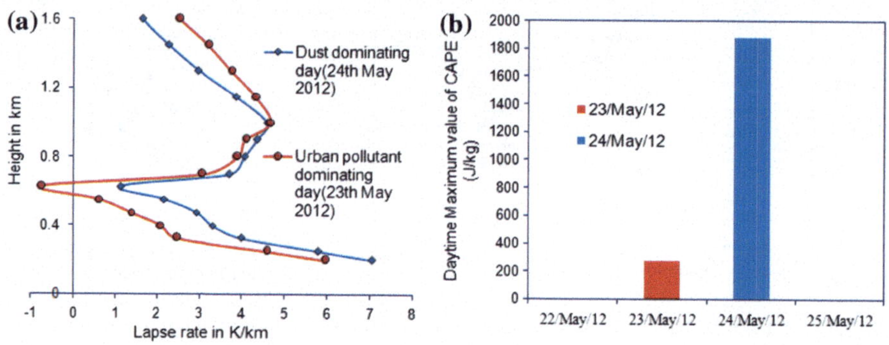

Fig. 7.11 Comparison of **a** lapse rate for a pollutant dominating day (23 May 2012) with a dust dominating day (24 May 2012) and **b** corresponding CAPE values

comparatively higher than the previous day, and the maximum value of CAPE is found around five times higher (1870 J/Kg).

The abundance of urban pollutants causes to decrease the temperature lapse rate which can potentially affect the convective processes mostly in pre-monsoon season over urban locations like the present one. The maximum values of CAPE during day time on pollutant dominating days are observed to have low values compared to the dust dominating days in most of the cases during pre-monsoon season. 15 sets of such available observation have been presented in Fig. 7.12. Each set of observations contains group of dust and pollutant dominating days within a window of 10 days span. The comparison of daytime maximum CAPE value for the group of dust dominating days and pollutant dominating days is shown in Fig. 7.12a and

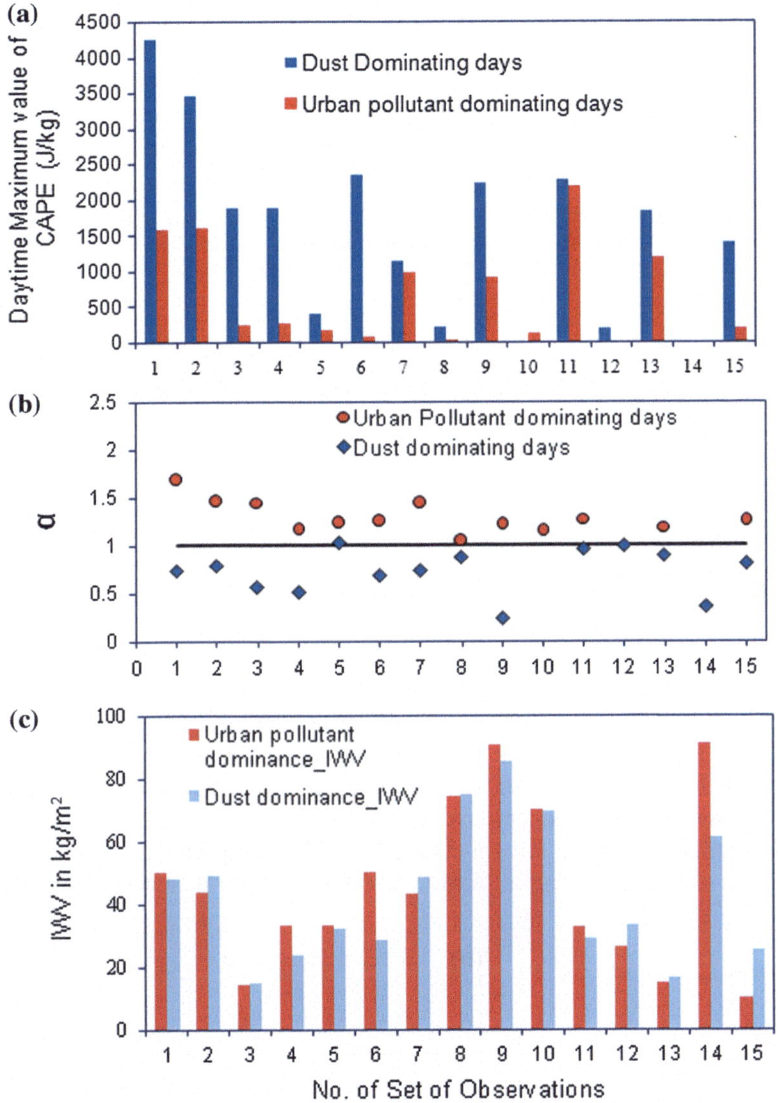

Fig. 7.12 **a** Maximum daytime value of CAPE compared between the group of pollutant dominating days and dust dominating days. **b** Corresponding Angstrom exponent values for and **c** Corresponding IWV values

corresponding sets of Angström exponent (α) values are presented in Fig. 7.12b. Moreover, to check the variation in the contribution of water vapour in building up CAPE values in these two cases, the integrated amount of water vapour (IWV) (from radiometric observations) for dust and pollutant dominating days are plotted in Fig. 7.12c. It shows that there are no significant differences in IWV contribution between these two cases but the CAPE values are significantly low for pollutant dominating days. This substantiates the fact that the decrease in lapse rate endorsed by the abundance of pollutant aerosols has mostly caused the weakening of the convective processes.

7.4.4 Influence of Transported Air Mass on Regional Aerosol Loading

In this analysis the back trajectories for all the dates within the observation span have been examined. A 5-day back trajectory analysis is presented using TrajStat for the pollutant dominating and dust dominating days separately.

Analyzing the back trajectories it is found that the majority of the trajectories (at 2000 m agl) for pollutant dominating days are coming from Indian sub-continental region whereas, trajectories for dust dominating days are found travelling mostly over the far arid region of Middle-East, shown in the Fig. 7.13a, b. This typical difference was not found while examining the trajectories at 1000 m agl (above ground level) because the turbulences present at this height sometime may restrict long range transport path (Garratt 1992). It was also reported by earlier researchers that only the trajectories beyond boundary layer or 2000 m are considerable to investigate the long range transport aspect (Gogoi et al. 2009). The transported air mass at 2000 m agl passing over the arid region of Middle-east may carry dust aerosols which can

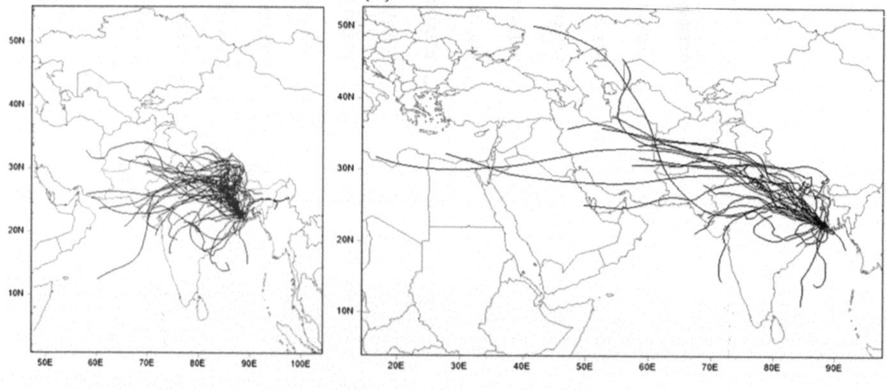

Fig. 7.13 Comparison between back trajectory analysis for **a** urban pollutant-dominating days and **b** dust-dominating days (at 2000 m agl)

turn the dominance of locally generated aerosols into the dominance of coarse dust particles. On the other hand the air mass advected from Indian subcontinent will mostly carry fine mode anthropogenic aerosols which may add up with the regional concentration of fine mode aerosols. It indicates that transported air mass has strong influence on the regional aerosol loading which can even control the spectral pattern of aerosol single scattering albedo (SSA) over a region.

In monsoon the spectral dependence pattern of SSA shows a feeble increasing trend with wavelength which indicates a less dominance of pollutant aerosols particles in this season. Unlike the other seasons monsoon experiences the back trajectories mostly coming from far distances even on those days when α value is less than 1 and SSA also increases with wavelengths. The reason may be the scavenging of aerosols due to rain fall in monsoon and the presence of marine aerosol in south westerly air mass which may have a stronger control on the total aerosol loading at this location in monsoon (Moorthy et al. 1997).

7.5 Summary

An analysis of the aerosol environment over Kolkata, an urban region in the eastern part of India, has been presented in this chapter. Primarily a set of ground-based measurement data are utilized in this analysis and the dominance between urban pollutant and dust aerosols are well discriminated using spectral dependence pattern of AOD and SSA in different seasons. The analysis of aerosol environment over a region like the present one is much significant considering its geographical location, type of aerosol loading and per capita emission of pollutant aerosols. It may give crucial inputs to future studies on atmospheric aerosols and also in different policy making issues. The main conclusions of this analysis are as follows:

I. The spectral dependence of aerosol single-scattering albedo shows a decreasing pattern with wavelengths pointing to the abundance of fine-mode urban pollutant aerosols for most of the observations over this urban region. The corresponding Ångström coefficients tested for the same have values greater than one.

II. Aerosol optical depth over Kolkata is overall high and shows maximum mean value (~1) in winter and comparatively low value (0.6–0.7) in monsoon and post-monsoon months with a yearly average of 0.8. SSA shows minimum value in pre-monsoon months with its mean values at around 0.8–0.85 and maximum in monsoon months with mean values around 0.9–0.95.

III. The concentration of black carbon (BC) aerosols over this region is distinguishably high and it shows significant positive correlation with Angström exponent (α) values in every season indicating a strong influence of the amount of BC aerosol in total regional aerosol loading over this urban location.

IV. The abundance of pollutant aerosols is observed to affect the temperature lapse rate within the boundary layer. The lapse rate for the group of urban pollutant dominating days experiences prominent decrease compared to that of the dust aerosol dominating days and this causes the weakening of the localized convective processes.
V. In monsoon the spectral dependence pattern of SSA does not show a decreasing trend, unlike the other seasons. It is possibly due to the relatively low concentration of fine-mode pollutants in monsoon season.
VI. Air masses transported from the continent typically carry anthropogenic aerosols which additionally contribute to the local loading of fine mode pollutant aerosol particles. On the other hand the long range transport of dust aerosols (desert dust) from the arid region of Middle-East can potentially influence the regional aerosol environment and its signature is observed in the spectral pattern of SSA over the present location.

Acknowledgements The main work was supported by the grants from the Indian Space Research Organization (ISRO) under the Project "Studies on Aerosol Environment at Kolkata Located Near the Land-Ocean Boundary as a Part of ARFI Network under ISRO-GBP" and was carried out at Institute of Radio Physics and Electronics, University of Calcutta, Kolkata.

References

Angstrom A (1929) On the atmospheric transmission of Sun radiation and on dust in the air. Geogr Ann 11:156–166
Arnott WP, Hamasha K, Moosmuller H, Sheridan PJ, Ogren JA (2005) Towards aerosol light absorption measurements with a 7-wavelength aethalometer: evaluation with a photoacoustic instrument and 3-wavelength nephelometer. Aerosol Sci Technol 39:17–29
Aruna K, LakshmiKumar TV, NarayanaRao D, KrishnaMurthy BV, Babu SS, Krishnamoorthy K (2014) Scattering and absorption characteristics of atmospheric aerosols over a semi-urban coastal environment. J Atmos Solar Terr Phys 119:211–222
Babu SS, Moorthy KK, Manchanda RK, Sinha PR, Satheesh SK, Vajja DP, Srinivasan S, Kumar VHA (2011) Free tropospheric black carbon aerosol measurements using high altitude balloon: do BC layers build "their own homes" up in the atmosphere? Geophys Res Lett 38:L08803. https://doi.org/10.1029/2011GL046654
Bergstrom RW, Russell PB (1999) Estimation of aerosol direct radiative effects over the mid-latitude North Atlantic from satellite and in situ measurements. Geophys Res Lett 26:1731–1734
Bergstrom RW, Russell PB, Hignett P (2002) Wavelength dependence of the absorption of black carbon particles: predictions and results from the TARFOX experiment and implications for the Aerosol Single Scattering Albedo. J Atmos Sci 59:567–577
Bergstrom R, Pilewskie WP, Russell PB, Redemann J, Bond TC, Quinn PK, Sierau B (2007) Spectral absorption properties of atmospheric aerosols. Atmos Chem Phys 7:5937–5943
Bhattacharya R, Bhattacharya A, Das R, Pal S, Mali P (2011) Pre-monsoon climate and rainfall activity over West Bengal: a survey, International. J. Phys 4:141–148
Bodhaine BA (1995) Aerosol absorption measurements at Barrow, Mauna Loa and the South Pole. J Geophys Res 100:8967–8975
Charlson RJ, Schwartz SE, Hales JM, Cess RD, Coakley JA, Hansen JE, Hofmann DJ (1992) Climate forcing by anthropogenic aerosols. Science 255:423–430

Ding AJ, Huang X, Nie W et al (2016) Enhanced haze pollution by black carbon in megacities in China. Geophys Res Lett 43(6):2873–2879

Dubovik O, Holben BN, Eck TF, Smirnov A, Kaufman YJ, King MD, Tanre D, Slutsker I (2002) Climatology of atmospheric aerosol absorption and optical properties in key locations. J Atmos Sci 59:590–608

Dumka UC, Manchanda RK, Shinha PR, Srenivasan S, Moorthy KK, Suresh Babu S (2013) Temporal variability and radiative impact of black carbon aerosols over tropical urban station Hyderabad. J Atmos Sol Terr Phys 105–106:81–90

Eck TF, Holben BN, Reid JS, Dubovik O, Smirnov A, O'Neill NT, Slutsker I, Kinne S (1999) Wavelength dependence of the optical depth of biomass burning, urban, and desert dust aerosols. J Geophys Res 104:31333–31349

Gadhavi H, Jayaraman A (2010) Absorbing aerosols: contribution of biomass burning and implications for radiative forcing. Ann Geophys 28:103–111

Ganguly D, Gadhavi H, Jayaraman A, Rajesh TA, Misra A (2005) Single scattering albedo of aerosols over the central India: implications for the regional aerosol radiative forcing. Geophys Res Lett 32:L18803. https://doi.org/10.1029/2005GL023903

Garratt JR (1992) The atmospheric boundary layer. Cambridge University Press, p 168

Gogoi MM, Moorthy KK, Babu SS, Bhuyan PK (2009) Climatology of columnar aerosol properties and the influence of synoptic conditions: first-time results from the northeastern region of India. J Geophys Res 114:D08202. https://doi.org/10.1029/2008JD010765

Gogoi MM, Pathak B, Moorthy KK, Bhuyan PK, Babu SS, Bhuyan K, Kalita G (2011) Multi-year investigations of near surface and columnar aerosols over Dibrugarh, Northeastern location of India: heterogeneity in source impacts. Atmos Env 45:1714–1724

Goto D, Takemura T, Nakajima T, Badarinath KVS (2011) Global aerosol model-derived black carbon concentration and single scattering albedo over Indian region and its comparison with ground observations. Atmos Env 45:3277–3285

Haywood JM, Shine KP (1995) The effect of anthropogenic sulphate and soot aerosols on clear sky planetary radiation budget. Geophys Res Lett 22:603–605

Haywood JM, Shine KP (1997) Multi-spectral calculations of the radiative forcing of tropospheric sulphate and soot aerosols using a column model. Q J R Meteorol Soc 123:1907–1930

Heintzenberg JR, Clarke AD, Charlson RJ, Liousse C, Ramaswamy V, Shine KP, Wendisch M (1997) Measurement and modelling of aerosol single-scattering albedo: progress, problems and prospects. Contrib Atmos Phys 70:249–264

Holton JR (2004) An introduction to dynamic meteorology, 4th edn. Elevier Academic Press, California

Jacobson MZ (2001) Strong radiative heating due to the mixing state of black carbon on atmospheric aerosols. Nature 409:695–697

Koelmans AA, Jonker MTO, Cornelissen G, Bucheli TD, Van Noort PCM, Gustafsson Ö (2006) Black carbon: the reverse of its dark side. Chemosphere 63:365–377

Kohler I, Dameris M, Ackermann I, Hass H (2001) Contribution of road traffic emissions to the atmospheric black carbon burden in the mid-1990s. J Geophys Res 106(D16):17997–18014

Lawrence MG, Lelieveld J (2010) Atmospheric pollutant outflow from southern Asia: a review. Atmos Chem Phys 10:11017–11096

Moorthy KK, Satheesh SK, Murthy BVK (1997) Investigations of marine aerosols over the tropical Indian Ocean. J Geophys Res 102:18827–18842

Niranjan K, Sreekanth V, Madhavan BL, Krishna Moorthy K (2006) Wintertime aerosol characteristics at a north Indian site Kharagpur in the Indo-Gangetic plains located at the outflow region into Bay of Bengal. J Geophys Res 111:D24209. https://doi.org/10.1029/2006JD007635

Parameswaran K, Sunilkumar SV, Rajeev K, Nair PR, MoorthyK Krishna (2004) Boundary layer aerosols at Trivandrum tropical coast. Adv Space Res 34:838–844

Penner JE, Eddleman H, Novakov T (1993) Towards the development of a global inventory for black carbon emissions. Atmos Env 27:1277–1295

Ramachandran S, Rajesh TA (2007) Black carbon aerosol mass concentrations over Ahmedabad, an urban location in western India: comparison with urban sites in Asia, Europe, Canada, and the United States. J Geophys Res 112 (D06211). https://doi.org/10.1029/2006jd007488

Ramana MV, Ramanathan V, Feng Y, Yoon SC, Kim SW, Carmichael GR, Schauer JJ (2010) Warming influenced by the ratio of black carbon to sulphate and the black-carbon source. Geosci, Nat. https://doi.org/10.1038/NGEO918

Ramanathan V, Carmichael G (2008) Global and regional climate changes due to black carbon. Nat Geosci 1:221–227

Safai PD, Raju MP, Budhavant KB, Rao PSP, Momin GA, Ali K, Devara PCS (2013) Long term studies on the characteristics of black carbon aerosols over tropical urban station Pune, India. Atmos Res 132–133:178–184

Saha U, Talukdar S, Jana S, Maitra A (2014) Effects of air pollution on meteorological parameters during Deepawali festival over an Indian urban metropolis. Atmos Env 98:510–519

Satheesh SK, Vinoj V, Moorthy KK (2011) Weekly periodicities of aerosol properties observed at an urban location in India. Atmos Res 101:307–313

Sokolik I, Toon OB (1997) Direct radiative forcing by anthropogenic airborne mineral aerosols. Nature 381:681–683

Stein AF, Draxler RR, Rolph GD, Stunder BJB, Cohen MD, Ngan F (2015) NOAA's HYSPLIT atmospheric transport and dispersion modeling system. Bull Amer Meteor Soc 96:2059–2077

Stull RB (1989) An introduction to boundary layer meteorology. Kluwer Academic Publishers, Dordrecht, p 620

Talukdar S, Jana S, Maitra A, Gogoi M (2015) Characteristics of black carbon concentration at a metropolitan city located near land–ocean boundary in Eastern India. Atmos Res 151:526–534

Tiwari S, Srivastava AK, Bisht DS, Parmita P, Srivastava MK, Attri SD (2013) Diurnal and seasonal variations of black carbon and PM2.5 over New Delhi, India: influence of meteorology. Atmos Res 125–126:50–62

Tripathi SN, Dey S, Tare V, Satheesh SK (2005) Aerosol black carbon radiative forcing at an industrial city in northern India. Geophys Res Lett 32(L08802). https://doi.org/10.1029/2005gl022515

Upadhyay SN, Singh RS (2010) Investigation of atmospheric aerosols from Varanasi in the Indo-Gangetic plane. ARFI & ICARB Scientific Progress Report ISRO-GBP, India, pp 56–60

Venkatraman C, Reddy K, Jonson K, Reddy MS (2002) Aerosol size and chemical characteristics at Mumbai, India, during INDOEX-IFP (1999). Atmos Env 36:1979–1991

Vyas BM (2010) Studies of regional features of atmospheric aerosol, total carbonaceous aerosols and their role in the atmospheric radiative forcing effect over the tropical semi-arid location i.e. Udaipur, western region part of India. ARFI & ICARB Scientific Progress Report ISRO-GBP, India, pp 67–70

Weingartner E, Saathoff H, Schnaiter M, Streit N, NBitnar B, Baltensperger U (2003) Absorption of light by soot particles: determination of the absorption coefficient by means of aethalometers. Aerosol Sci 34:1445–1463

Chapter 8
Vertical Profiling of Aerosol and Aerosol Types Using Space-Borne Lidar

Alaa Mhawish, K. S. Vinjamuri, Nandita Singh, Manish Kumar and Tirthankar Banerjee

Abstract Aerosol remote sensing has become a powerful tool to characterize the optical and microphysical properties of aerosols. Several satellite sensors such as MODIS, MISR, OMI, Tropomi and PARASOL utilize the solar electromagnetic radiation for retrieving aerosol properties from space. These instruments have high spatial coverage and can provide aerosol properties globally on a repeated basis. However, these passive sensors mostly lack the information regarding vertical distribution of aerosol and its types. Using active remote sensing technique however, provide valuable information to understand the vertical distribution of aerosols which is very useful to predict the lifetime of atmospheric aerosols, long-range transport and subsequent interaction with cloud droplets. CALIOP is an active sensor flying on board CALIPSO satellite provide height-dependent aerosol extinction repeatedly on a global basis. CALIOP aerosol retrieval algorithm retrieves aerosol information in 5 km horizontal resolution and 30–60 m vertical resolution. The latest updated CALIOP aerosol retrieval algorithm version 4 (V4) has the ability to identify ten aerosol subtypes; six for tropospheric aerosols and four for stratospheric aerosols. In this context, the annual, seasonal and diurnal variation of smoke aerosol have been investigated over central Indo-Gangetic Plain (IGP), South Asia using ten years V4 CALIOP profile data. We noted that for all the seasons, the highest smoke aerosol extinction observed near surface and contributed 40–60% to the total aerosol extinction during winter (DJF) and postmonsoon seasons (ON). In premonsoon (MAM) and monsoon (JJAS) seasons the highest contribution of smoke to the total extinction coefficient found at relatively higher altitude (premonsoon: 60% at 7–9 km, monsoon: 75% at 5–8 km). The day-night occurrence frequency of smoke aerosol found higher during the day time in winter at 4 km, while during monsoon the occurrence of the smoke was found higher at night time.

A. Mhawish · N. Singh · M. Kumar · T. Banerjee (✉)
Institute of Environment and Sustainable Development, Banaras Hindu University, Varanasi, India
e-mail: tb.iesd@bhu.ac.in

K. S. Vinjamuri
DST-Mahamana Centre of Excellence in Climate Change Research, Banaras Hindu University, Varanasi, India

T. Gupta et al. (eds.), *Measurement, Analysis and Remediation of Environmental Pollutants*, Energy, Environment, and Sustainability,
https://doi.org/10.1007/978-981-15-0540-9_8

Keywords Aerosol · Smoke · CALIPSO · Vertical profile · South Asia

8.1 Introduction

Aerosols are one of the major components of the atmosphere and play crucial role in Earth's radiation budget, atmospheric visibility, climate change, air quality and human health (Schulz et al. 2006; Evans et al. 2013; Kumar et al. 2016; Burney and Ramanathan 2014; Banerjee et al. 2018; Singh et al. 2019). The aerosol optical properties are mainly function of its chemical composition, morphology, and the size of the particles. The atmospheric aerosols interact with the incoming solar radiation by absorbing or scattering the light. The sum of the attenuated light (scattering + absorption) due to the aerosol particles called extinction and can be measured using ground-based, airborne and space-borne instruments. However, here we will explicitly consider the space-based remote sensing (satellite remote sensing) technique to explore vertical profile of aerosols.

Retrieval of aerosol optical and microphysical properties through satellite based measurements has become one of the major tool as it provides near real-time information with high spatial coverage. Numerous studies have used the satellite data to study the aerosol climatology (Kumar et al. 2018; Soni et al. 2015), estimation of the ground-level particulate matter (PM) concentration (Sorek-Hamer et al. 2013a; Chowdhury et al. 2019), air quality (Kumar et al. 2017a, b; Singh et al. 2018), tracking the pollution episodes (Kumar et al. 2016; Sorek-Hamer et al. 2013b) and for health-related studies (Van Donkelaar et al. 2010). Many satellites measurements are used for retrieving atmospheric aerosol properties, but each sensor has some strengths and limitations. Moreover, different satellites provide different information about the atmospheric aerosols at multiple spatiotemporal resolutions. National Aeronautics and Space Agency (NASA) with other international collaborators have several Earth-Observing Satellites (EOS) flying around the Earth collecting Earth and atmosphere information for research purpose. A-Train constellation is a NASA's group of sun synchronic satellites flying close to each other around the Earth (Fig. 8.1). Afternoon constellation satellites cross the equator around 13:30 local time with few minutes difference from each other. The small passing differences allow us to retrieve different aerosol properties utilizing the strength of each satellite and the type of information given by each satellite. Sensors like MODIS, MISR, PARASOL and OMI provide useful information of the aerosol optical depth and the microphysical properties of aerosols on a daily basis, while it has minimal information about the vertical distribution of aerosols. However, the A-Train active sensor CALIOP onboard of CALIPSO satellite provide valuable information about the vertical distribution of aerosols and aerosol types with limited spatial coverage. Therefore, integration of the data from different satellites provides the opportunity to use comprehensive information on aerosol properties for understanding 3-dimensional distribution of the aerosol on a global basis.

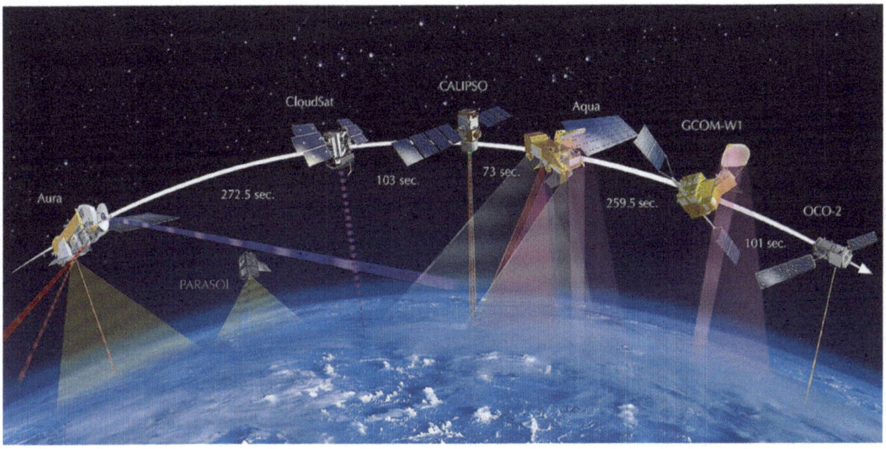

Fig. 8.1 The A-Train satellite constellation (*Source* https://atrain.gsfc.nasa.gov)

The Indo-Gangetic Plain (IGP) at South Asia, demarcated with red border in Fig. 8.2 is considered one of the global aerosol hotspots with high and diverse aerosol loading throughout the year (Mhawish et al. 2017, 2018, 2019; Singh et al. 2017a, b; Gautam et al. 2010, 2011). Figure 8.2 shows the spatial variation of

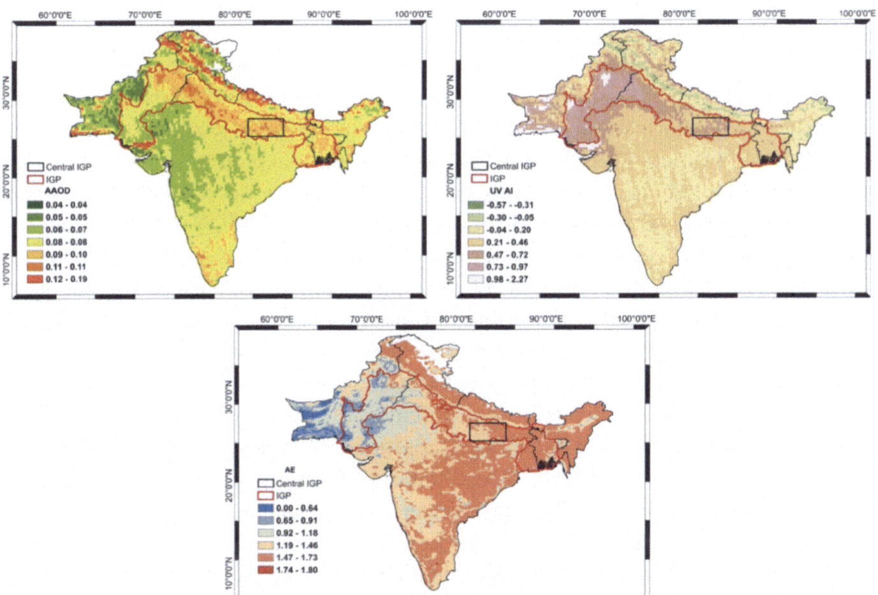

Fig. 8.2 Satellite retrieval of absorbing aerosol optical depth (AAOD), aerosol index (AI) and Angstrom Exponent (AE) from A-Train Satellites over South Asia, from 2008 to 2017

absorbing aerosol loading (in terms of AAOD) and aerosol types (in terms of Aerosol index and Angstrom exponent) retrieved by A-Train satellite sensors like MODIS and OMI. As indicated in the figure, the IGP region is mostly dominated by highly absorbing fine aerosol particles. The higher AAOD (>0.1), AI (0.21–2.27), and AE (0.92–1.8) indicate the dominance of fine absorbing particles across the region with specifically higher absorbing over the upper IGP. Several studies have investigated the horizontal distribution of aerosol and aerosol type across the region using passive sensors such as MODIS (Kumar et al. 2018; Mhawish et al. 2017), MISR (Dey and Di Girolamo 2010), and OMI (Jethva et al. 2018; Vinjamuri et al. 2019). However, vertical profile of aerosol across IGP is still limited and consist multiple uncertainties mainly due to the lack of coordinated ground experiments and non-availability of systematic investigation using space-borne lidar. We thereby, explore the space-borne active lidar sensor CALIOP for vertical profiling of aerosol and aerosol types over central IGP which will be useful to reduce uncertainties in aerosol-cloud-climate interaction over this highly polluted region.

8.2 CALIOP Sensor and Aerosol Retrieval Algorithm

CALIPSO was launched on April 28, 2006, and was a joint venture between the NASA and the French Space Agency (CNES) to explore the role of aerosols and clouds in the Earth's climate system. It is ascending node equatorial crossing time is at 13:30 h. It used to fly at an altitude of 705 km in the "A-Train" constellation till September 13, 2018, before lowering to 688 km to join the "C-Train" with CloudSat. The CALIPSO mainly carries three instruments, the Cloud-Aerosol Lidar with Orthogonal Polarization (CALIOP), the Imaging Infrared Radiometer (IIR), and the Wide Field Camera (WFC). However, we will only explore the CALIOP Lidar, a frequency doubled laser which provides the information on vertical structures of aerosols and clouds and their classifications (Winker et al. 2009; Young and Vaughan 2009). The Nd: YAG lasers transmits the light simultaneously at 532 and 1064 nm with a pulse repetition rate of 20.16 Hz and a pulse length of 20 nsec. The energy of the transmitted laser pulse is 110 mJ and produces a beam diameter of 70 m on the Earth's surface using beam expanders. The horizontal and vertical resolutions are 333 m and 30 m respectively (Winker et al. 2009; Young and Vaughan 2009).

CALIPSO products are available at different levels. At Level 1, the algorithm produces geo-located attenuated backscatter coefficient profiles for 532 nm, 1064 nm and the 532 nm cross-polarized. These datasets are used in Level 2 algorithms, which perform three major steps in producing science data products. The first step is detecting aerosol or cloud using Selective Iterated BoundarY Locator (SIBYL) module (Winker et al. 2009; Young and Vaughan 2009). This can be achieved by taking advantage of significant variations in backscattered coefficients. The profiles are then averaged based on the assigned resolutions. The individuality of the CALIPSO algorithm is the variable vertical resolution at different heights. For example, in the 5 km Level 2 profiles which we have used for this study, the vertical resolution is

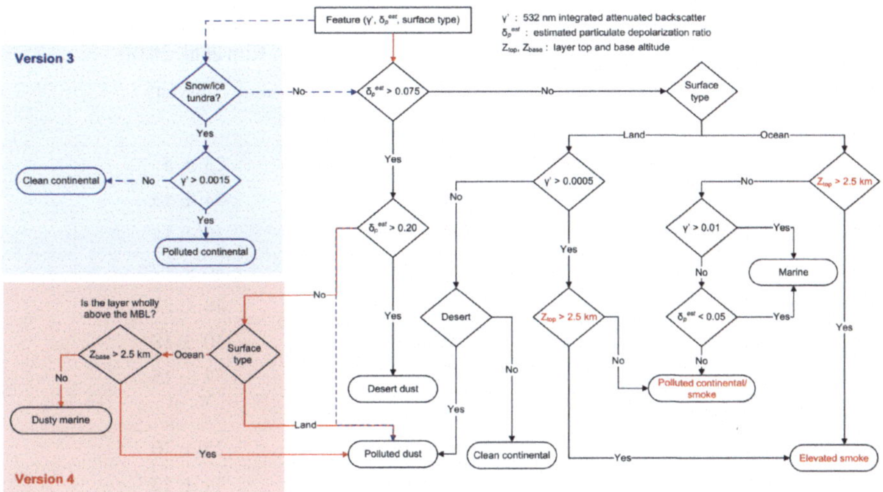

Fig. 8.3 CALIPSO flow chart of aerosol subtype selection scheme for tropospheric aerosols. *Note* The blue shaded-regions and arrows removed in the latest version 4 (V4) which have been used in V3. (Adopted from Kim et al. 2018, under CC BY 4.0 License)

60 m from −0.5 to 20.2 km and 180 m from 20.2 to 30.1 km. The second step is a Scene Classification Algorithm (SCA), which uses a combination of observed and CALIPSO aerosol model information in choosing an appropriate lidar ratio (extinction-to-backscatter ratio) and multiple scattering functions. These CALIPSO aerosol models were made by analyzing long-term AERONET observations. The lidar ratios for different aerosol types and the flow chart in selecting different aerosol types in the current 4.10 version is included in Fig. 8.3.

Based on the obtained parameters from SCA (Table 8.1), the final extinction profiles are obtained by the final step Hybrid Extinction Retrieval Algorithms (HERA). The HERA is executed based on the complexity involved in retrieving the extinctions. In order to enhance the quality and accuracy, the produced data is combined with ancillaries like meteorological data obtained from NASA's Global Modelling and Assimilation Office (GMAO), instrument status and elevation statistics. The highest quality CALIPSO products i.e., standard data products are however, only recommended for research studies.

8.3 Applications of Lidar Profile to Identify Biomass Burning Aerosols Across IGP

Vertical profiling of aerosol and its sub-type is extremely crucial over the IGP due to presence of variety of emission sources and their potential impacts on environment and human health. Different forms of biomass burning i.e. open crop residue burning,

Table 8.1 Aerosol lidar ratio with expected uncertainties for tropospheric and stratospheric aerosol subtypes at 532 and 1064 nm in CALPSO version 4 (Modified from Kim et al. 2018)

Aerosol subtype	S_{532}(sr)	S_{1064}(sr)
Tropospheric aerosols		
Clean marine	23 ± 5	23 ± 5
Dust	44 ± 9	44 ± 13
Polluted continental/smoke	70 ± 25	30 ± 14
Clean continental	53 ± 24	30 ± 17
Polluted dust	55 ± 22	48 ± 24
Elevated smoke	70 ± 16	30 ± 18
Dusty marine	37 ± 15	37 ± 15
Stratosphere aerosols		
Polar stratosphere aerosol	50 ± 20	25 ± 10
Volcanic ash	44 ± 9	44 ± 13
Sulfate/other	50 ± 18	30 ± 14
Smoke	70 ± 16	30 ± 18

waste incineration, household solid fuel combustion are frequently practiced across the region which result into emission of a huge amount of smoke aerosols. Despite contributing major fraction of ambient aerosols, limited investigation was made for the vertical distribution of smoke aerosols over IGP. Biomass burning is one of the prime emission sources of atmospheric particulate and for radiatively and chemically active gases which induce serious threats to human health and climate, both at the regional and global scale (Wan et al. 2017; Rajput et al. 2014; Singh et al. 2018; Vadrevu et al. 2015). Biomass burning is a distinguished global event including forest fire, grassland fires and agricultural residue burning. The agricultural residue burning is the second largest contributor after the forest fire to total fire emissions in South and Southeast Asia (Chang and Song 2010). IGP is extremely densely populated region and encompasses nearly 20% of the Indian geographical area and produces 50% of total food grains (Pal et al. 2009). Paddy and wheat are principally cultivated crops in the north-western part of IGP and mechanized harvesting generates enormous crop residues (Haryana and Punjab: 48% of total straw; Sharma et al. 2010). The farmers usually practice in situ burning of these agricultural residues to clear the fields mainly because of the minimum time gap between the paddy harvesting and wheat sowing period, and associated additional expenses for labour-based crop residue removal from the fields.

The agriculture residue burning during post-monsoon (October-November) and summer (April-May) over the Punjab and Haryana states result in reduced visibility, increased concentration of fine particulates and trace gases across the region in down-wind direction (Rajput et al. 2014; Jethva et al. 2018; Cusworth et al. 2018; Singh et al. 2017b). The north-westerly prevailing winds during post-monsoon carries the smoke and other pollutants throughout IGP, increase the smoke-exposed population and health risk. Various studies using ground and satellite-based observations have shown that the agriculture fires over the upper IGP significantly influence the aerosol composition and trace gases throughout IGP (Kaskaoutis et al. 2014; Singh et al. 2018; Vinjamuri et al. 2019; Vadrevu et al. 2015). The reduced visibility, frequent haze and fog formation due to stable meteorological conditions are very common over IGP during post-monsoon and winter seasons (Kumar et al. 2015). It is therefore, absolutely required to estimate the biomass burning emissions at the source level. Further, the biomass plume injection height is basic input in aerosol transport models and it primarily determines its travel distance, when and where it will impact the nearby community.

In India, investigation of aerosol extinction at different altitudes was initiated during the early 2000s. Gadahvi and Jayaraman (2006) used aircraft observations from a low power micro pulse lidar to characterize the vertical aerosol distribution over an urban location. Mehta et al. (2018) recently used decadal observations from CALIOP to characterize the vertical mixing of smoke aerosols. CALIOP based observations distinctly identified the confinement of smoke at higher altitude in comparison to other aerosol types. The IGP region is reported to have high smoke aerosols, varying considerably across the seasons and height (Vinjamuri et al. 2019) while the Eastern USA, South America, South Africa and most of the European countries exhibit a decline in smoke aerosols due to stringent environmental regulations. The incremental trends in the accumulation of smoke over IGP are prominent during winter (DJF) and post-monsoon (ON) seasons due to increased emission source strengths (Kumar et al. 2015; Kumar et al. 2017a, b; Rajput et al. 2010). The accumulation of smoke at an altitude as high as 5 km during the winter was reported by Kumar et al. (2015) using CALIPSO vertical feature mask indicating its wide distribution over the IGP region.

Few attempts of aerosol vertical profiling over IGP were made during pre-monsoon season, using both site-specific aircraft measurements (Vaishya et al. 2018) and by CALIPSO lidar profile (Gautam et al. 2010). These experiments suggest a clear shift in dominant aerosol types from super-micron mode (over western India) to sub-micron mode (over IGP), and aerosol accumulation at a comparatively high altitude, mainly influenced by prevailing westerly. The sub-micron mode aerosol mixtures over the IGP region were mostly emitted from anthropogenic sources, i.e. industries, fossil fuel combustion and biomass burning. The vertical profiles of single scattering albedo (SSA) reflected sharp heterogeneity within the IGP region. The SSA over central IGP region were found consistently increasing from the surface to the higher altitudes indicating the accumulation of smoke at lower altitudes in contrast to the eastern IGP with more smoke at higher altitude. The presence of such a huge quantity of smoke during the pre-monsoon season makes the aerosol composition highly absorbing over the IGP region. The key roles of smoke in contributing to overall aerosol burden and its role in climate dynamics are therefore, required with precise sounding techniques.

8.4 Vertical Profile of Aerosols and Aerosol Types Over Central IGP

The variable atmospheric lifetime of different aerosol types in free troposphere leads to their widespread transport over a large spatial domain. Further existing meteorology also influences the advection process and vertical structure. The importance of understanding the vertical distribution of aerosols is therefore, important to trace their transport, sources and other climatic implications. In this section, we explored the vertical distribution of smoke aerosols and its diurnal variation in different seasons over the central IGP region.

A typical CALIPSO V4 vertical feature mask and aerosol subtype retrieved over IGP on May 16, 2018 is included in Fig. 8.4. The vertical feature mask clearly

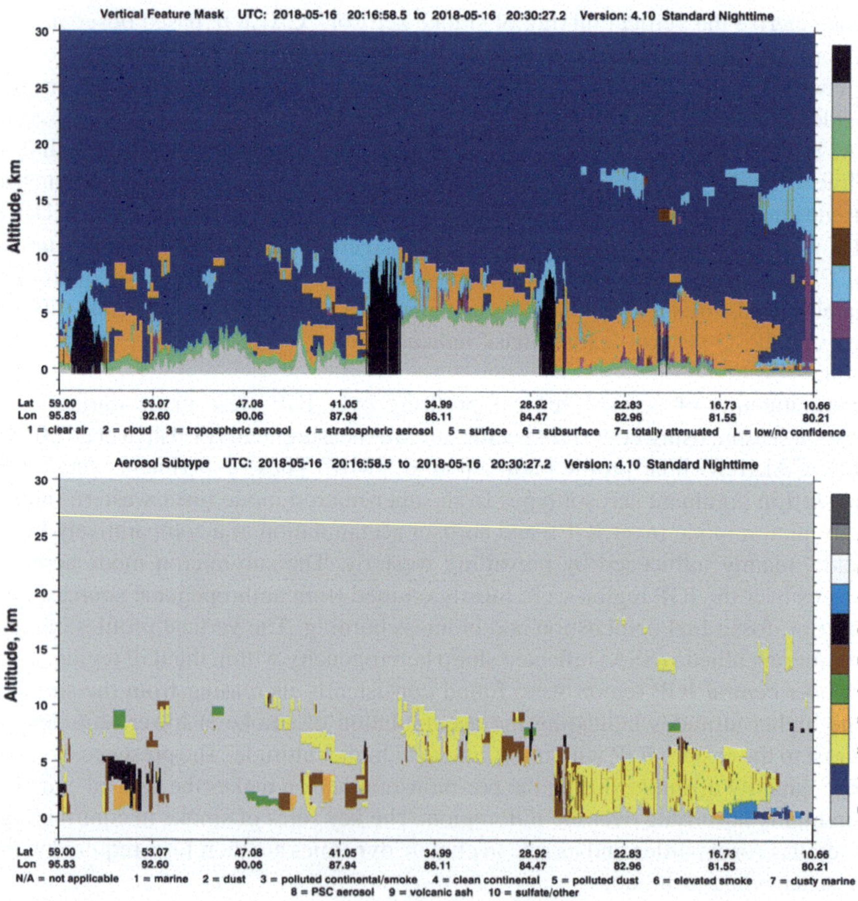

Fig. 8.4 Typical CALIPSO vertical feature mask and aerosol subtype over the IGP (Retrieved from https://www-calipso.larc.nasa.gov/products/lidar, on April 20, 2019)

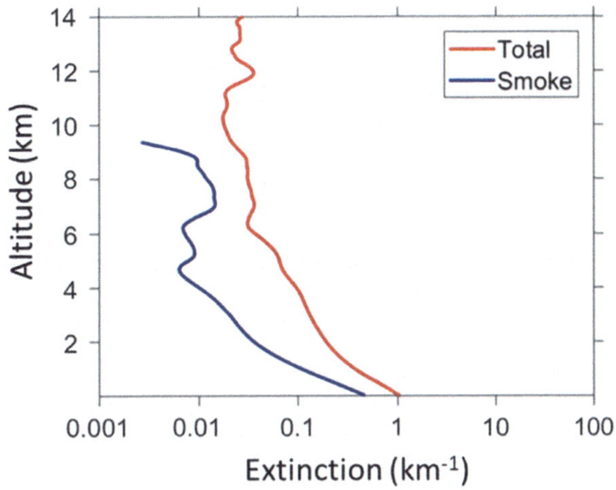

Fig. 8.5 CALIPSO annual mean extinction coefficient for total and smoke aerosols over central IGP

indicates the presence of huge quantity of aerosols with varying altitude across IGP. Likewise, from 28.92° N, 84.47° E to 22.83° N, 82.96° E, there is a clear dominance of tropospheric aerosols spread from near-surface to a height of 5 km. Further, the aerosol subtype clearly distinguishes the prevailing aerosol type as pre-monsoon specific dust aerosols with mixing of polluted dust. In order to investigate the vertical distribution of aerosols, a total of 53,710 CALIPSO Level 2 attenuated backscatter profiles were retrieved from 2008 to 2017 over central IGP and classified smoke aerosol extinction coefficient against the total aerosol extinction. Figure 8.5 shows the vertical distribution of total aerosol and smoke aerosol type over the central IGP on an annual scale. Both total and smoke aerosol extinction coefficient was found higher near the surface, before reducing with increasing height. We note that the smoke extinction decreases with increasing the altitude, before getting accumulated at 7–8 km. Retrieval of comparably high smoke extinction at 7–8 km height may indicate the convective transfer of smoke aerosols from the ground and/-or transport of aerosols by prevailing wind (Vinjamuri et al. 2019). However, on an annual scale, the total aerosol extinction did not show any sign of accumulation at a higher height which need to be explored further.

The central IGP is dominated by smoke aerosols mainly during DJF and ON (Singh et al. 2017a). Therefore, in order to investigate the contribution of smoke aerosol to the total aerosols, we have also explored the seasonal and annual contribution of smoke extinction to the total extinction over the central IGP. In Fig. 8.6, the CALIPSO vertical profile showed significant seasonal variations with maximum smoke contribution near the surface observed during winter months (DJF, 60%) and in post-monsoon (ON, 40%). The high smoke contribution during DJF and ON is mainly associated with open biomass burning (Jethva et al. 2018; Vadrevu et al. 2015). During MAM and JJAS, the higher smoke aerosols observed at greater altitude (monsoon: ~70%, Premonsoon: ~60%) indicate vertical transport of air masses and/-or smoke injection at higher altitude (Vinjamuri et al. 2019). This information is critical in

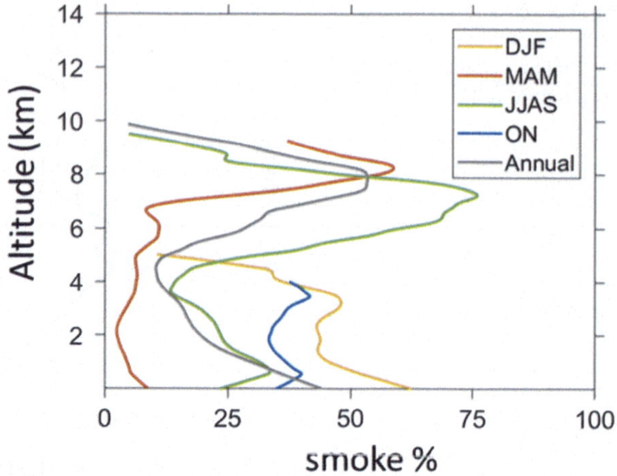

Fig. 8.6 Relative contribution of smoke extinction to total extinction over central IGP

several aerosol-climate models to predict the residential time of the aloft smoke as well as the fate of smoke aerosol transport. This information can also improve the chemical transport models for more accurate simulation of smoke aerosols.

Since the CALIOP is an active sensor, the lidar profile can be retrieved both during day and night which is useful to study the diurnal variation of vertically distributed aerosols. In Fig. 8.7, the diurnal variation of smoke aerosols has been plotted for different seasons. During MAM, there was minimum difference between day and night time smoke profile except at very high altitude (>6 km) where smoke aerosols were more prevalent during nighttime. Day-night occurrence frequency remains almost stable during ON while in winter, there was greater possibility of accounting smoke aerosols at 3–5 km during daytime which was found to dissipate at night. It is worth to mention that the sensitivity of CALIPSO to detect smoke

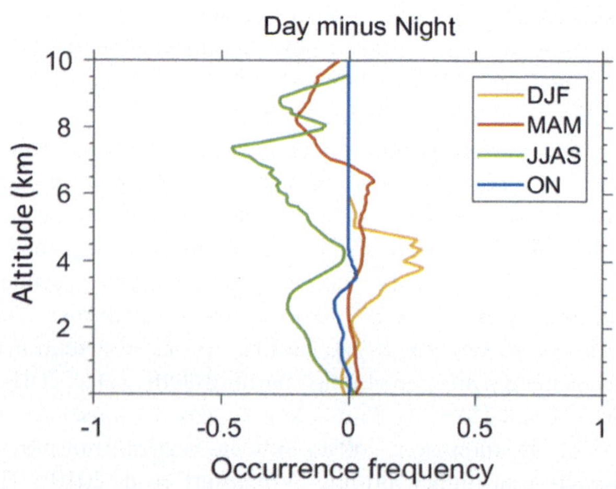

Fig. 8.7 Seasonal variations of smoke aerosols day-night occurrence frequency

aerosols is less during daytime because of lower signal-to-noise ratio (SNR) due to solar radiation. During night time, the higher SNR enhanced the sensitivity of CALIPSO to detect even the low concentration of smoke aerosols at higher altitude, as shown in Fig. 8.7.

8.5 Conclusion

The information about the vertical distribution of aerosols retrieved by CALIPSO attenuated backscatter lidar profile is crucial to understand the aerosols sources, transport and climatic implications. Over the IGP, the diverse biomass burning mainly of seasonal crop residue over north-western India enhance the high aerosol loading, mostly fine absorbing smoke aerosols. Here, we have explored the retrieval capability of CALIPSO lidar to classify the vertical distribution of smoke aerosol against total aerosols over the central IGP. CALIPSO attenuated backscatter profile showed a maximum extinction of smoke aerosols at very near to the surface indicating main contribution of local sources. The contribution of smoke aerosols to total aerosols was maximum during winter (~60%) and in post-monsoon (~40%). This was in contrast to the premonsoon (~60%) and monsoon (~70%) season when maximum contribution was noted at relatively higher altitude. Further, the day-night observations of smoke aerosols indicate significant diurnal variation. Our analysis has implication in regional air quality model to reduce uncertainty in aerosol transport, aerosol-cloud interaction and in AOD-PM modelling.

Acknowledgements The research is ASEAN-India Science and Technology Development Fund (CRD/2018/000011) under ASEAN-India Collaborative R&D Scheme, Government of India support from University Grants Commission (UGC) under UGC-Israel Science Foundation bilateral project (6-11/2018 IC).

References

Banerjee T, Kumar M, Singh N (2018) Aerosol, climate, and Sustainability. In: Della Sala DA, Goldstein MI (eds) The encyclopedia of the anthropocene, vol 2. Elsevier, Oxford, pp 419–428

Burney J, Ramanathan V (2014) Recent climate and air pollution impacts on Indian agriculture. Proc Natl Acad Sci 111(46):16319–16324

Chang D, Song Y (2010) Estimates of biomass burning emissions in tropical Asia based on satellite-derived data. Atmos Chem Phys 10(5):2335–2351

Chowdhury S, Dey S, Di Girolamo L, Smith KR, Pillarisetti A, Lyapustin A (2019) Tracking ambient $PM_{2.5}$ build-up in Delhi national capital region during the dry season over 15 years using a high-resolution (1 km) satellite aerosol dataset. Atmos Environ 204:142–150

Cusworth DH, Mickley LJ, Sulprizio MP, Liu T, Marlier ME, DeFries RS, Guttikunda SK, Gupta P (2018) Quantifying the influence of agricultural fires in northwest India on urban air pollution in Delhi India. Environ Res Lett 13(4):044018

Dey S, Di Girolamo L (2010) A climatology of aerosol optical and microphysical properties over the Indian subcontinent from 9 years (2000–2008) of Multiangle Imaging Spectroradiometer (MISR) data. J Geophys Res 115:D15204

Evans J, van Donkelaar A, Martin RV, Burnett R, Rainham DG, Birkett NJ, Krewski D (2013) Estimates of global mortality attributable to particulate air pollution using satellite imagery. Environ Res 120:33–42

Gadhavi H, Jayaraman A (2006) Airborne lidar study of the vertical distribution of aerosols over Hyderabad, an urban site in central India, and its implication for radiative forcing calculations. Ann Geophys 24(10):2461–2470

Gautam R, Hsu NC, Lau, KM (2010) Premonsoon aerosol characterization and radiative effects over the Indo-Gangetic Plains: implications for regional climate warming. J Geophys Res 115(D17)

Gautam R, Hsu NC, Tsay SC, Lau KM, Holben B, Bell S, Payra S et al (2011) Accumulation of aerosols over the Indo-Gangetic plains and southern slopes of the Himalayas: distribution, properties and radiative effects during the 2009 pre-monsoon season. Atmos Chem Phys 11(24):12841–12863

Jethva H, Chand D, Torres O, Gupta P, Lyapustin A, Patadia F (2018) Agricultural burning and air quality over northern India: a synergistic analysis using NASA's A-train satellite data and ground measurements. Aerosol Air Qual Res 18:1756–1773

Kaskaoutis DG, Kumar S, Sharma D, Singh RP, Kharol SK, Sharma M, Singh AK, Singh S, Singh A, Singh D (2014) Effects of crop residue burning on aerosol properties, plume characteristics, and long-range transport over northern India. J Geophys Res: Atmos 119(9):5424–5444

Kim M-H, Omar AH, Tackett JL, Vaughan MA, Winker DM, Trepte CR, Hu Y, Liu Z, Poole LR, Pitts MC, Kar J, Magill BE (2018) The CALIPSO version 4 automated aerosol classification and lidar ratio selection algorithm. Atmos Meas Tech 11:6107–6135. https://doi.org/10.5194/amt-11-6107-2018

Kumar M, Tiwari S, Murari V, Singh AK, Banerjee T (2015) Wintertime characteristics of aerosols at middle Indo-Gangetic Plain: impacts of regional meteorology and long range transport. Atmos Environ 104:162–175

Kumar M, Singh RK, Murari V, Singh AK, Singh RS, Banerjee T (2016) Fireworks induced particle pollution: a spatio-temporal analysis. Atmos Res 180:78–91

Kumar M, Raju MP, Singh RK, Singh AK, Singh RS, Banerjee T (2017a) Wintertime characteristics of aerosols over middle Indo-Gangetic Plain: vertical profile, transport and radiative forcing. Atmos Res 183:268–282

Kumar M, Raju MP, Singh RS, Banerjee T (2017b) Impact of drought and normal monsoon scenarios on aerosol induced radiative forcing and atmospheric heating rate in Varanasi over middle Indo-Gangetic Plain. J Aerosol Sci 113:95–107

Kumar M, Parmar KS, Kumar DB, Mhawish A, Broday DM, Mall RK, Banerjee T (2018) Long-term aerosol climatology over Indo-Gangetic Plain: trend, prediction and potential source fields. Atmos Environ 180:37–50

Mehta M, Singh N (2018) Global trends of columnar and vertically distributed properties of aerosols with emphasis on dust, polluted dust and smoke-inferences from 10-year long CALIOP observations. Remote Sens Environ 208:120–132

Mhawish A, Banerjee T, Broday DM, Misra A, Tripathi SN (2017) Evaluation of MODIS collection 6 aerosol retrieval algorithms over Indo-Gangetic Plain: implications of aerosols types and mass loading. Remote Sens Environ 201:297–313

Mhawish A, Kumar M, Mishra AK, Srivastava PK, Banerjee T (2018) Remote sensing of aerosols from space: retrieval of properties and applications. In: Remote Sensing of Aerosols, Clouds, and Precipitation. Elsevier Inc, pp 1–38. https://doi.org/10.1016/B978-0-12-810437-8.00003-7

Mhawish A, Banerjee T, Sorek-Hamer M, Lyapustin AI, Broday DM, Chatfield R (2019) Comparison and evaluation of MODIS Multi-Angle Implementation of Atmospheric Correction (MAIAC) aerosol product over South Asia. Remote Sens Environ 224:12–28. https://doi.org/10.1016/j.rse.2019.01.033

Pal DK, Bhattacharyya T, Srivastava P, Chandran P, Ray SK, (2009) Soils of the Indo-Gangetic plains: their historical perspective and management. Curr Sci 1193–1202

Rajput P, Sarin MM, Sharma D, Singh D (2014) Characteristics and emission budget of carbonaceous species from post-harvest agricultural-waste burning in source region of the Indo-Gangetic Plain. Tellus B 66:21026. https://doi.org/10.3402/tellusb.v66.21026

Schulz M, Textor C, Kinne S, Balkanski Y, Bauer S, Berntsen T. et al (2006) Radiative forcing by aerosols as derived from the AeroCom present-day and pre-industrial simulations. Atmos Chem Phys 6(12):5225–5246

Sharma AR, Kharol SK, Badarinath KVS, Singh D (2010) Impact of agriculture crop residue burning on atmospheric aerosol loading–a study over Punjab State, India. Ann Geophys 28(2) (09927689)

Singh N, Murari V, Kumar M, Barman SC, Banerjee T (2017a) Fine particulates over South Asia: Review and meta-analysis of $PM_{2.5}$ source apportionment through receptor model. Environ Pollut 223:121–136

Singh N, Mhawish A, Deboudt K, Singh RS, Banerjee T (2017b) Organic aerosols over Indo-Gangetic plain: sources, distributions and climatic implications. Atmos Environ

Singh N, Banerjee T, Raju MP, Deboudt K, Sorek-Hamer M, Singh RS, Mall RK (2018) Aerosol chemistry, transport, and climatic implications during extreme biomass burning emissions over the Indo-Gangetic Plain. Atmos Chem & Phys 18(19)

Singh N, Mhawish A, Ghosh S, Banerjee T and Mall RK (2019) Attributing mortality from temperature extremes: a time series analysis in Varanasi, India. Sci Total Environ 665:453-464

Soni K, Parmar KS, Kapoor S (2015) Time series model prediction and trend variability of aerosol optical depth over coal mines in India. Environ Sci Pollut Res 22(5):3652–3671

Sorek-Hamer M, Cohen A, Levy RC, Ziv B, Broday DM (2013a) Classification of dust days by satellite remotely sensed aerosol products. Int J Remote Sens 34(8):2672–2688

Sorek-Hamer M, Strawa AW, Chatfield RB, Esswein R, Cohen A, Broday DM (2013b) Improved retrieval of $PM_{2.5}$ from satellite data products using non-linear methods. Environ Pollut 182:417–423

Vadrevu KP, Lasko K, Giglio L, Justice C (2015) Vegetation fires, absorbing aerosols and smoke plume characteristics in diverse biomass burning regions of Asia. Environ Res Lett 10(10):105003

Vaishya A, Babu SNS, Jayachandran V, Gogoi MM, Lakshmi NB, Moorthy KK, Satheesh SK (2018) Large contrast in the vertical distribution of aerosol optical properties and radiative effects across the Indo-Gangetic Plain during the SWAAMI–RAWEX campaign. Atmos Chem Phys 18(23):17669–17685

Van Donkelaar A, Martin RV, Brauer M, Kahn R, Levy R, Verduzco C, Villeneuve PJ (2010) Global estimates of ambient fine particulate matter concentrations from satellite-based aerosol optical depth: development and application. Environ Health Perspect 118(6):847

Vinjamuri KS, Mhawish A, Banerjee T, Sorek-Hamer M, Broday DM, Mall RK, Latif MT (2019) Vertical distribution of smoke aerosols over upper Indo-Gangetic Plain. Under review in Environmental Pollution

Wan X, Kang S, Li Q, Rupakheti D, Zhang Q, Guo J, Chen P, Tripathee L, Rupakheti M, Panday AK, Wang W (2017) Organic molecular tracers in the atmospheric aerosols from Lumbini, Nepal, in the northern Indo-Gangetic Plain: influence of biomass burning. Atmos Chem Phys 17:8867–8885. https://doi.org/10.5194/acp-17-8867-2017

Winker DM, Vaughan MA, Omar A, Hu Y, Powell KA, Liu Z, Hunt WH, Young SA (2009) Overview of the CALIPSO mission and CALIOP data processing algorithms. J Atmos Ocean Tech 26:2310–2323

Young A, Vaughan MA (2009) The retrieval of profiles of particulate extinction from Cloud-Aerosol Lidar Infrared Pathfinder Satellite Observations (CALIPSO) Data: algorithm description. J Atmos Ocean Tech 26:1105–1119

Chapter 9
A Study of Optical and Microphysical Properties of Atmospheric Brown Clouds Over the Indo-Gangetic Plains

Manish Jangid, Saurabh Chaubey and Amit Kumar Mishra

Abstract Atmospheric brown clouds (ABCs), a dense and widespread layer of air, have significant implications to human health, air quality and regional climate. Being a habitat of more than 350 million human population, the Indo-Gangetic Plains (IGP) is also one of the ABCs hotspots of the world. Considering the significant socio-economic impact of ABCs on lives of IGP's populations, understanding of spatio-temporal characteristics of ABCs is necessary over the region. This study focuses on the frequency of ABCs occurrences and associated optical and microphysical properties using data from seven ground-based remote sensors situated across the IGP. We have used total ~5253 days of Level 2 aerosol measurements from seven AERosol Robotic NETwork (AERONET) sites [Karachi, Lahore, Jaipur, New Delhi, Kanpur, Gandhi college and Dhaka University] for three seasons (Winter, Pre-monsoon and Post-monsoon). A well-defined algorithm based on extinction and absorbing properties of particles is used to characterize extreme pollution days (ABCs days) for each sites. Further, we have used spectral dependency of aerosol optical depth (AOD) and absorption aerosol optical depth (AAOD) to characterize aerosol types during ABCs days over IGP. The results indicate a unique feature of ABCs occurrences over the IGP. In general, the maximum frequency of ABCs days is found during pre-monsoon seasons over all the sites. However, other seasons have specific features over specific locations, for example, the maximum occurrence of ABCs days was found over Dhaka in post-monsoon (65%), followed by Delhi during pre-monsoon (61.29%), and Kanpur in pre-monsoon (44.29%). The results show that while extreme pollution days in pre-monsoon is dominated by dust polluted aerosols, biomass burning is the main source during winter and post-monsoon. This study provides a comprehensive climatology of spectral nature of AOD, AAOD, single scattering albedo (SSA), asymmetry parameter and size distribution of ABCs across the region.

M. Jangid · S. Chaubey · A. K. Mishra (✉)
School of Environmental Sciences, Jawaharlal Nehru University, New Delhi 110067, India
e-mail: amitmishra@mail.jnu.ac.in

S. Chaubey
School of Environmental Sciences and Sustainable Development, Central University of Gujrat, Gandhi Nagar, Gujrat 382030, India

T. Gupta et al. (eds.), *Measurement, Analysis and Remediation of Environmental Pollutants*, Energy, Environment, and Sustainability,
https://doi.org/10.1007/978-981-15-0540-9_9

Keywords Aerosols · Atmospheric brown cloud · AOD · AAOD · SSA

9.1 Introduction

Atmospheric brown clouds (ABCs) are a dense and extensive layer of pollutants comprised of both gases and aerosols pollutants (Ramanathan et al. 2005), frequently regarded as extreme pollution events. The principal contributors to these extreme pollution events are soil dust, biomass burning, transportation, urbanization and infrastructure development (Alam et al. 2011). As per recent report of World Health Statistics (2018), about 90% population of the world breath polluted air. The report reveals that about 7 million people have lost their lives due to outdoor (~4.2 million) and indoor (~3.8 million) air pollution in 2016. The relation between the current level of air pollution and associated increased rates of mortality continues to be an issue of global concern (Chowdhury et al. 2018). As per the world health organization (World Health Statistics 2018) report, 14 out of 15 world's most polluted cities are in India. The overall death caused by air pollution in India was about 1.24 million (Watts et al. 2015). Numerous epidemiological studies carried over the Indian sub-continent have shown the effects of a high concentration of particulate matter on human health. The air pollution is known to be associated with various illnesses like heart disease, lung cancer, chronic obstructive pulmonary diseases, pneumonia, eye irritation, etc., (Ostro 1995; Balakrishnan et al. 2019). Health impact analysis over Delhi estimates 7,350–16,200 premature deaths and 6.0 million asthma attacks per year (Guttikunda and Goel 2013).

Besides the human health impact, aerosol components of ABCs also play a dominant role in modulation of earth's radiation budget through aerosol-radiation interaction and aerosol-cloud interaction (Boucher et al. 2013). ABCs influence regional and global climate (direct and indirect effects) and thereby, also modulate the hydrological cycle. Direct radiative effect can be attributed to their ability to scatter and absorb solar radiations (Haywood and Boucher 2000). The presence of absorbing aerosols can significantly reduce the surface solar radiation (-73 W/m^2) for unit increase in aerosol optical depth (Ramanathan 2007). The indirect effects of aerosols include their ability to modify cloud properties (aerosol-cloud interaction), circulation patterns and precipitation (Gautam et al. 2010; Lau and Kim 2006).

UNEP (2008) has reported five ABCs hotspots across the world, one among them that lies in the Indian subcontinent is the Indo-Gangetic plains (IGP) (Ramanathan et al. 2005). IGP is considered to be the "food basket of India". The amount of production of agriculture in this region is mainly dependent on rainfall pattern and surface temperature. Changes in weather and climate in this area significantly affect the food security of the country and the livelihood of many people. There has been a significant increase in the emission of carbonaceous aerosols, volatile organic compounds, oxides of sulphur and nitrogen, and other pollutants due to rapid industrialization and urbanization. Black carbon (BC) and dust constitute the major absorbing particles of aerosol over the IGP (Lau and Kim 2006; Gautam et al. 2010; Mishra et al. 2014).

Many studies have reported the impacts of ABCs on the radiation budget, monsoon pattern and surface dimming. Radiative forcing may be positive or negative resulting in a warming or cooling, respectively (Ramanathan 2007; Menon et al. 2014). ABCs over India have shown dimming effect and reduced level of evaporation which further leads to low rainfall (Ramanathan et al. 2005). They also suggest that apart from a 5% reduction in rainfall from 1930 to 2000, there will be a reduction in rainfall of about 18% by 2020 and a further 4% by 2030.

Due to increasing aerosol loading and associated impacts on various spheres of society, aerosol research has paid considerable attention over the IGP for the last two decades. Various studies have been done to study the aerosol properties over the IGP (Dey et al. 2004; Singh et al. 2005; Ganguly et al. 2005; Gautam et al. 2010; Srivastava et al. 2011; Srivastava and Ramachandran 2012; Mishra and Shibata 2012a). Srivastava et al. (2011) has reported a positive increase in the AOD as one moves from eastern IGP towards the west after the onset of pre-monsoon. Kaskaoutis et al. (2012) reported a dramatic increase in soil-dust load over the IGP (Kanpur and Delhi) during dry pre-monsoon season. The categorization of aerosols into their constituents has been studied extensively using various different methodologies over the IGP. The classification of aerosols shows that Kanpur has a higher level of polluted dust, while continental and carbonaceous aerosols dominate the eastern part of IGP during pre-monsoon (Srivastava et al. 2012). Sinha et al. (2012) has reported a classification based upon the relationship between Angstrom exponent (AE) and the difference in AE [$\Delta AE = AE_{440\text{–}675} - AE_{675\text{–}870}$] including the fine mode fraction over Hyderabad. Mishra et al. (2014) studied the intra-seasonal changes of aerosol properties in pre-monsoon and classified aerosols into dust, biomass burning, urban/Industrial and mixed type aerosols based on AE and absorption Angstrom exponent (AAE).

Studies on changes of aerosol optical and microphysical properties during extreme pollution episodes over the entire IGP are limited. A comparative study of aerosol optical and micro-physical properties along with determining the frequency of extreme pollution days (ABCs days) over the IGP will enable us to understand the factors which contribute to the different types of aerosol over a region on a seasonal basis. Also, this study classifies the aerosols into different types based upon their source of origin during ABCs days over the IGP.

9.2 Data and Methods

9.2.1 Study Area

The Indo-Gangetic Plains (IGP), one of the largest basins in the world, is also one of the five ABCs hotspots of the world. Six core cities/urban agglomerations of the IGP have been studied for extreme pollution days and associated optical and microphysical properties using Aerosol Robotic Network (AERONET) measurements. The study sites are as follows: Karachi (24.94° N, 67.13° E), Lahore

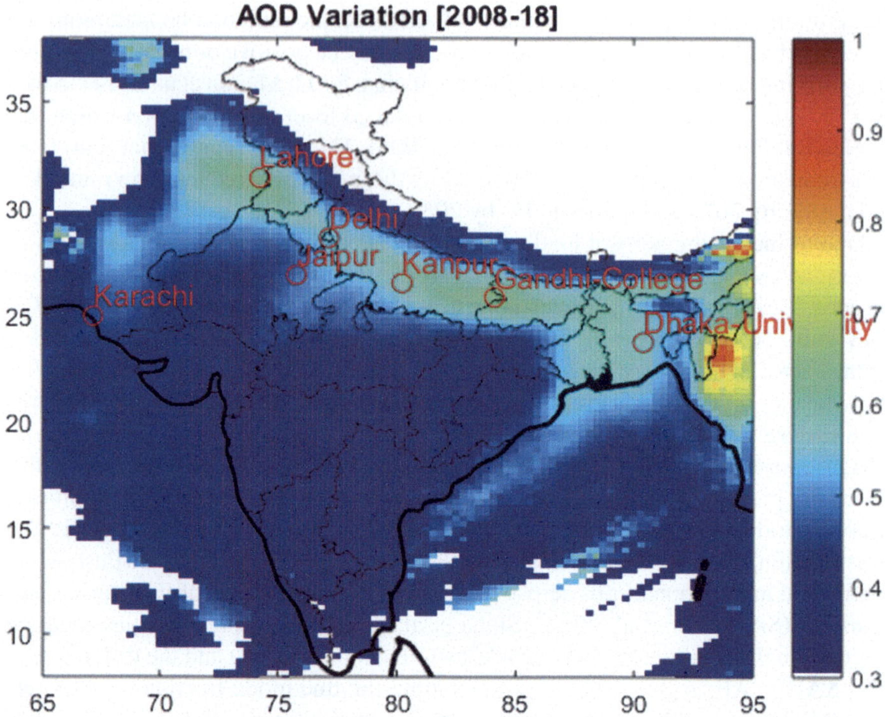

Fig. 9.1 Study region showing the seven AERONET stations over the IGP used in this study. The different color shading represents the 11 years mean (2008–2018) aerosol optical depth (AOD) during extreme pollution events over entire region. AOD values are marked in color-bar. The AOD data used here are taken from MERRA-2, which assimilates various satellite retrieved AOD such that MODIS, MISR etc

(31.47° N, 74.26° E), Delhi (28.70° N, 77.10° E), Kanpur (26.44° N, 80.33° E), Gandhi-college (25.87° N, 84.12° E) and Dhaka-University (23.72° N, 90.39° E). Apart from the six core IGP cities, Jaipur (26.90° N, 75.8° E) has also studied to see the changes of aerosol properties. Figure 9.1 shows the study sites (AERONET location) over the IGP.

9.2.2 Data and Products

AERONET is a global network of programmed sun/sky scanning radiometers maintened by NASA. It provides optical and micro-physical properties of aerosols like aerosol optical depth (AOD), absorbing aerosol optical depth (AAOD), single scattering albedo (SSA), asymmetry parameter (ASY) and volume size distribution. The automatic sky/sun scanning radiometers measure the properties at several wavelengths (380–1020 nm). For the sky radiance, four wavelengths (440, 675, 870,

1020 nm) are utilized. The inversion products such as SSA, AAOD, ASY and volume size distribution over the stations are provided at four wavelengths i.e., 440, 675, 870, 1020 nm (Holben et al. 1998).

AOD, AAOD, and SSA at four wavelengths (440, 675, 870, 1020 nm) are obtained from AERONET website https://aeronet.gsfc.nasa.gov. The daily level 2.0 (quality assured) data products are retrieved at all 7 AERONET stations over the study area. The data for each station is available for the following time period: Karachi (01-01-2007 to 31-12-2018), Lahore (01-01-2006 to 31-12-2018), Kanpur (01-01-2001 to 31-12-2018), Delhi (01-01-2008 to 31-12-2018), Jaipur (01-01-2009 to 31-12-2018), Gandhi-college (01-01-2006 to 31-01-2018) and Dhaka-University (01-01-2012 to 31-12-2018). The details of data products are as follows:

AOD basically gives the measurement of columnar extinction of incoming solar radiation by aerosols, which is used to determine the quantity of aerosol loading in the atmosphere (Dey et al. 2004). AAOD represents columnar aerosol absorbance and is an important parameter in estimation of aerosol radiative forcing. AAOD is retrieved using Eq. (1) given below.

$$\text{AAOD} = (1 - \text{SSA}) \times \text{AOD} \tag{1}$$

SSA can be defined as the ratio of scattering AOD to total AOD (sum of absorption and scattering AOD). SSA is affected by absorption or scattering of aerosol particles.

The Angstrom exponent (AE) shows the wavelength dependence of the aerosol optical depth. The spectral dependence of AOD is given by

$$\tau_{\lambda}/\tau_{\lambda 0} = (\lambda/\lambda_0)^{-\text{AE}}. \tag{2}$$

where, τ_{λ} is the AOD at wavelength λ, and $\tau_{\lambda 0}$ is the AOD at the reference wavelength. The AOD measurements at two different wavelengths is normally used to calculate AE as follows

$$\text{AE} = -\log(\tau_{\lambda}/\tau_{\lambda 0})/\log(\lambda/\lambda_0) \tag{3}$$

AE has a inverse relationship with average size of aerosol i.e., fine particles have higher AE values and coarser particles have lower AE values. AE helps to determine the size distribution of aerosols in a vertical column. Lower AE values, usually less than 1 mean aerosols are dominated by larger size particles and higher AE values (>1) signify the presence of small size particles (Dubovik and King 2000).

Absorption Angstrom exponent (AAE) is another important variable which provides information regarding the wavelength dependence of absorption by aerosol particles and eventually the absorbing nature of aerosols (black carbon, organic matter, and dust). Absorbing aerosols produced from biomass burning/biofuel burning gives high AAE values while black carbon from fossil fuel burning gives AAE value close to unity.

9.2.3 Classification of Extreme Pollution Events (ABCs Days) and Aerosol Types

For the analysis, the AOD and AAOD obtained at 440 and 870 nm is used to estimate AOD and AAOD at 500 nm using $\tau_{500} = \tau_{440} \times (0.50/0.44)^{(-AE)}$, where, τ_{500} is AOD at 500 nm, τ_{440} is AOD at 440 nm, and AE is Angstrom exponent which is derived as $AE = -(\log \tau_{440} - \log \tau_{870})/(\log(0.44) - \log(0.87))$. The aerosol products obtained for each station is categorized into three seasons to study the seasonal variation of each parameter. The three seasons are winter (December, January, February), pre-monsoon (March, April, May), and post-monsoon (October and November). Monsoon season was not included in the study because of unavailability of continuous daily data for most of the stations due to technical issues.

To determine the frequency of extreme pollution days (hereafter ABC days) during a particular season, the definition of ABCs formation (Ramanathan et al. 2008) was used which characterize the region as ABCs dominance when annual AOD $\geq$ 0.3 and AAOD > 10% of AOD. Here, we have used this criteria for daily AOD and AAOD values for extreme pollution events. All those days which follow the mentioned criteria are regarded as ABCs days or say extreme pollution events.

A most common method is to distinguish aerosols is based upon spectral dependence of AOD and AAOD (AE and AAE). The basics of the used methodology are derived from experimental facts such as, fine spherical BC particles (radius ~0.01 μm) can have AAE of 1.0 (Bergstrom et al. 2002). Larger black carbon particles may have AAE values below 1.0. Russel et al. (2010) did similar classification and showed that AAE for dust vary between ~1.2 and 3.0, for U/I aerosols, AAE was found to be in the range 0.75–1.3, and ~1.2–2.0 for biomass burning aerosols. Eck et al. (1999) classified aerosols based on AERONET data and showed that aerosols from biomass burning, urban-industrial pollution and soil-dust had AAE values between ~1.2 and 1.8 for mixed size particles. Following the Giles et al. (2012) and Mishra et al. (2014), we have classified the IGP aerosols into four types based upon AE and AAE values as: dust dominated (AAE $\geq$ 1.8 and AE $\leq$ 0.8), biomass burning (AAE $\geq$ 1.8 and AE $\geq$ 1.2), Urban/Industrial dominated (AAE < 1.8 and AE $\geq$ 1.2), and mixed aerosols which do not fall in abovementioned categories (Giles et al. 2012; Mishra et al. 2014).

9.3 Results

9.3.1 Extreme Pollution Days and Associated Aerosol Parameters Over IGP

The analysis of the occurrence of extreme pollution days for each season over IGP is carried out by separating out the days when daily mean AOD $\geq$ 0.3 and AAOD > 10%

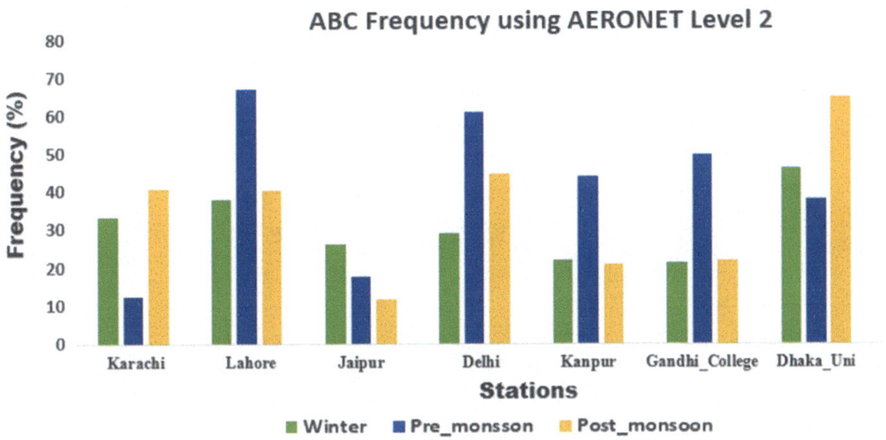

Fig. 9.2 ABCs days' frequency over IGP sites during winter, pre-monsoon and post-monsoon seasons

of AOD (Ramanathan et al. 2008). The frequency map of occurrence of ABC days over IGP for each season is presented in Fig. 9.2 and Table 9.1.

During winter, highest ABCs days' frequency is observed over Dhaka-University (46.17%), followed by Lahore (37.97%) and Karachi (33.61%). Similarly, during post-monsoon, Dhaka-University (65.28%) has the highest ABC days followed by Delhi (44.68%), Karachi (41.02%) and Lahore (40.65%). For Karachi (12.5%) and Jaipur (17.64%), frequency in pre-monsoon is low while the frequency over other locations in IGP -Lahore (67.13%), Delhi (61.29%), Kanpur (44.29%) and Gandhi-college (49.88%) is high. During pre-monsoon, the region is dominated by dust which increases the aerosol loading. High aerosol loading combined with absorbing aerosol from vehicular and industrial emissions gives the high frequency of ABCs days for pre-monsoon season. Kanpur although being one the core region of IGP with high pollution levels, gives the comparatively low frequency of ABCs days during winter and post-monsoon. This can be attributed to the presence of continuous long-term data from 2001 to 2018 while for the rest stations, continuous data before 2006 is not available. The number of observations taken in account for ABCs days' frequency is presented in Table 9.1. In general, results show dominance of ABCs days Dhaka-University and Delhi in almost all the seasons.

The mean AOD, AAOD, and SSA values over the IGP for three seasons during extreme pollution events are calculated and the results obtained are analyzed to determine the dependence of aerosol variables on the season as well as geographical location (Table 9.2). In winter, Karachi has lowest AOD value (0.45 $\pm$ 0.12) which increases as one move from west to east over the IGP: Karachi (0.45 $\pm$ 0.12) towards Delhi (0.74 $\pm$ 0.37), Gandhi-college (0.78 $\pm$ 0.35) and Dhaka-University (0.97 $\pm$ 0.45). Jaipur (0.54 $\pm$ 0.29) has a comparatively low AOD value, suggesting less aerosol loading over the region during winter. A similar results is observed for pre-monsoon and post-monsoon over the IGP. Karachi has the lowest AOD value

Table 9.1 Frequency of ABCs days (extreme pollution events) in percentage. Here, N_0 is the total number of days of data analyzed and N is the total number of ABCs days

Station	Winter			Pre-monsoon			Post-monsoon		
	N_0	N	%	N_0	N	%	N_0	N	%
Karachi	119	40	33.6	168	21	12.62	117	48	41.02
Lahore	158	60	37.97	283	190	67.1	214	87	40.6
Jaipur	220	58	26.3	340	60	17.64	127	15	11.81
Delhi	99	29	29.29	93	57	61.29	47	21	44.68
Kanpur	728	161	22.11	754	334	44.29	557	118	21.18
Gandhi-college	150	32	21.33	425	212	49.88	85	24	28.23
Dhaka-University	301	139	46.17	147	56	38.09	121	79	65.28

Table 9.2 Mean and standard deviation of AOD, AAOD, and SSA over IGP during winter, pre-monsoon and post-monsoon seasons

Station	Season	AOD	AAOD	SSA
Karachi	Winter	0.45 ± 0.125	0.04 ± 0.018	0.87 ± 0.017
	Pre-monsoon	0.51 ± 0.158	0.03 ± 0.012	0.87 ± 0.029
	Post-monsoon	0.46 ± 0.168	0.04 ± 0.016	0.88 ± 0.018
Lahore	Winter	0.64 ± 0.33	0.05 ± 0.027	0.87 ± 0.015
	Pre-monsoon	0.62 ± 0.258	0.06 ± 0.031	0.87 ± 0.029
	Post-monsoon	0.77 ± 0.413	0.06 ± 0.025	0.88 ± 0.015
Jaipur	Winter	0.54 ± 0.295	0.04 ± 0.018	0.87 ± 0.035
	Pre-monsoon	0.50 ± 0.141	0.03 ± 0.014	0.88 ± 0.016
	Post-monsoon	0.66 ± 0.318	0.04 ± 0.021	0.88 ± 0.016
Delhi	Winter	0.74 ± 0.372	0.06 ± 0.034	0.88 ± 0.018
	Pre-monsoon	0.66 ± 0.220	0.06 ± 0.023	0.88 ± 0.013
	Post-monsoon	0.70 ± 0.307	0.06 ± 0.025	0.88 ± 0.018
Kanpur	Winter	0.70 ± 0.314	0.05 ± 0.025	0.87 ± 0.046
	Pre-monsoon	0.60 ± 0.221	0.05 ± 0.023	0.88 ± 0.02
	Post-monsoon	0.79 ± 0.316	0.06 ± 0.023	0.87 ± 0.021
Gandhi-college	Winter	0.78 ± 0.354	0.06 ± 0.027	0.88 ± 0.011
	Pre-monsoon	0.66 ± 0.232	0.07 ± 0.037	0.86 ± 0.037
	Post-monsoon	0.72 ± 0.238	0.06 ± 0.022	0.89 ± 0.007
Dhaka-University	Winter	0.97 ± 0.453	0.09 ± 0.040	0.87 ± 0.016
	Pre-monsoon	0.80 ± 0.268	0.06 ± 0.027	0.87 ± 0.015
	Post-monsoon	0.78 ± 0.35	0.08 ± 0.036	0.86 ± 0.026

during pre-monsoon (0.51 ± 0.16) and post-monsoon (0.46 ± 0.17) while Dhaka-University has the highest value (0.80 ± 0.27 and 0.78 ± 0.35) for these seasons, respectively. Similar seasonal trends in AOD were observed by Mishra et al. (2014).

AAOD also follows a similar trend as AOD. Lowest AAOD value during winter is found over Karachi (0.04 ± 0.018) while the highest value is observed for Dhaka-University (0.09 ± 0.04). In the central IGP, AAOD value ranges from 0.05 to 0.06. High AAOD values towards eastern IGP is due to the presence of absorbing aerosols (carbonaceous particles) which are released in the atmosphere as a result of biomass burning during winter. Srivastava et al. (2012) has reported a similar observation that dust polluted aerosols dominate western IGP while carbonaceous aerosols dominate eastern IGP. A considerably lower value of AAOD is observed during Pre-monsoon as compared to winter for all stations. This can be attributed to the fact that the IGP region is dominated by dust during Pre-monsoon (Mishra et al. 2014). Higher AAOD values during post-monsoon are the result of agricultural crop-residue burning in Punjab and Haryana which results in high absorption AOD loading over IGP. Satellite data have also reported the regional transport of particles from agriculture crop residue burning in IGP (Badrinath et al. 2008; Mishra and Shibata 2012b).

The results obtained for SSA reveals that value ranges from 0.86 to 0.89 over the IGP. The mean SSA value over IGP is 0.88 which indicates that aerosols are relatively absorbing in nature (Ramanathan and Ramana 2005). Winter and pre-monsoon are generally dry over IGP which restricts the growth of aerosol particles, which enhances their life and decrease the SSA (Ramanathan and Ramana 2005). The low SSA values in winter and post-monsoon further verify the presence of absorbing aerosols during these seasons.

9.3.2 Spectral Dependence of Aerosol Parameters Over IGP

Spectral analysis of AOD, AAOD, SSA and ASY was carried out at 5 wavelengths (440, 500, 675, 870, and 1020 nm). Figures 9.3, 9.4, 9.5 and 9.6 indicate the depen-

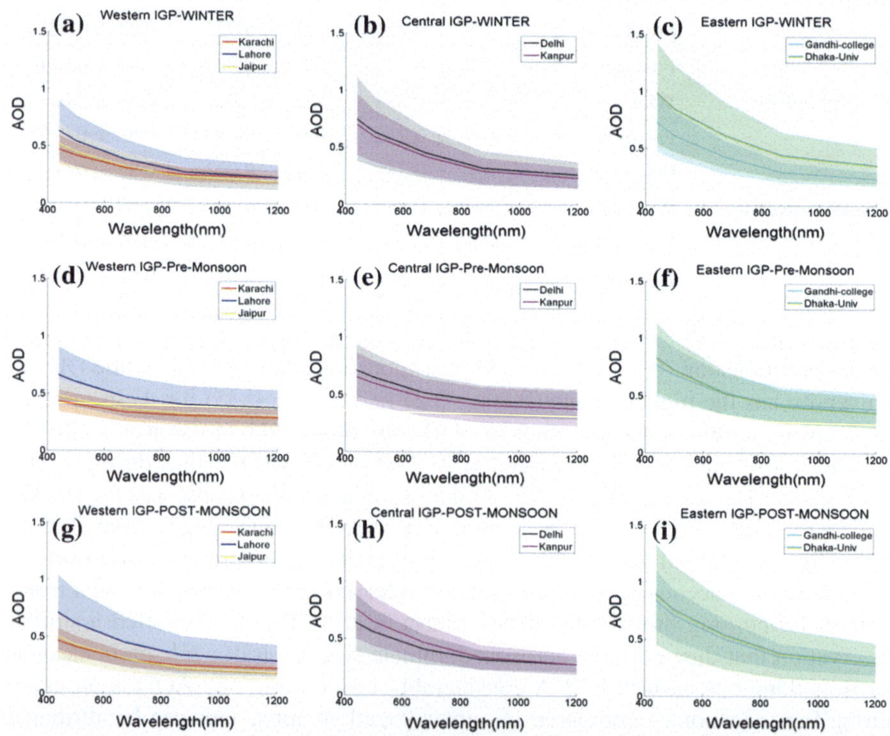

Fig. 9.3 Mean spectral dependence of AOD during extreme pollution events (ABCs days) over western (Karachi, Lahore, Jaipur), central (Delhi and Kanpur) and eastern (Gandhi-college and Dhaka-University) IGP for winter (**a, b, c**), pre-monsoon (**d, e, f**) and post-monsoon (**g, h, i**) seasons, respectively. Shaded regions show the respective standard deviation

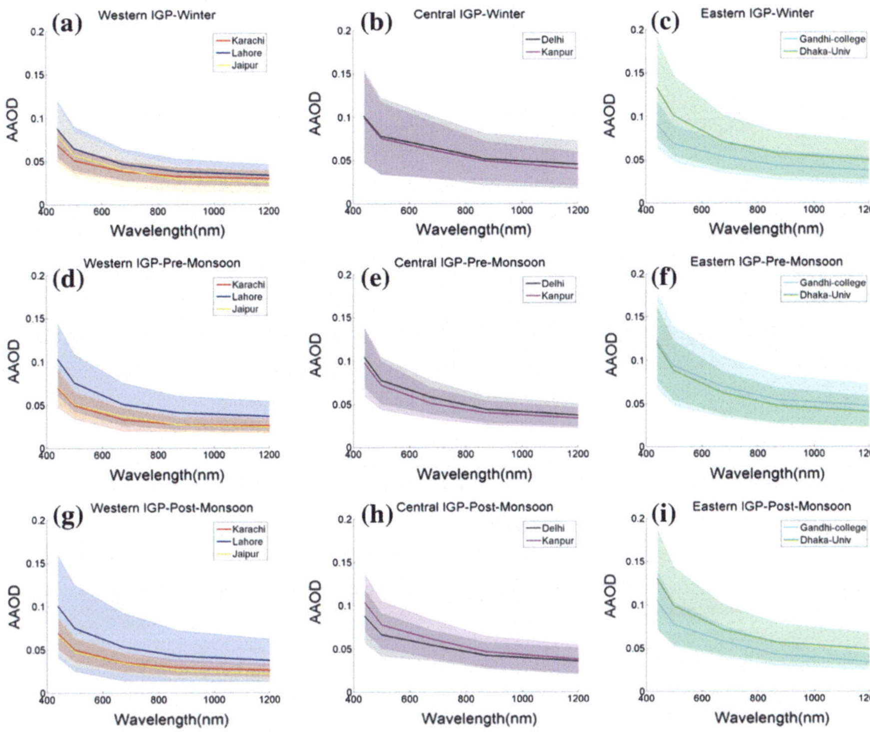

Fig. 9.4 Mean spectral dependence of AAOD during extreme pollution events (ABCs days) over western (Karachi, Lahore, Jaipur), central (Delhi and Kanpur) and Eastern (Gandhi-college and Dhaka-University) IGP for winter (**a**, **b**, **c**), pre-monsoon (**d**, **e**, **f**) and post-monsoon (**g**, **h**, **i**) seasons, respectively. Shaded regions show the respective standard deviation

dency of these parameters on wavelengths. The results clearly indicate a systematic spectral dependence of columnar AOD, AAOD and SSA over the IGP for winter, pre-monsoon, and post-monsoon. In order to present the results, IGP has been divided into western (Karachi, Lahore, Jaipur), central (Delhi, Kanpur) and eastern (Gandhi collage and Dhaka University) IGP.

The spectral variation of AOD exhibits high AOD value at a smaller wavelength (Fig. 9.3) and vice versa as expected from Mie theory. It is clear that the range of AOD is highest for Eastern IGP for all season. The characterization of the AOD's spectral dependence is an important aspect in the modeling and analysis of aerosol parameters retrieval from satellite data. The wavelength dependence of AOD can provide an idea about their types- biomass burning, urban and desert dust aerosols. Biomass burning and urban aerosols exhibit pronounced curvature in the AOD versus wavelength plot (Eck et al. 1999). When large particles dominate in the atmosphere, curvature of the AOD versus wavelength is not significant. Thus, the plot gives information about the aerosol type, which will be further discussed.

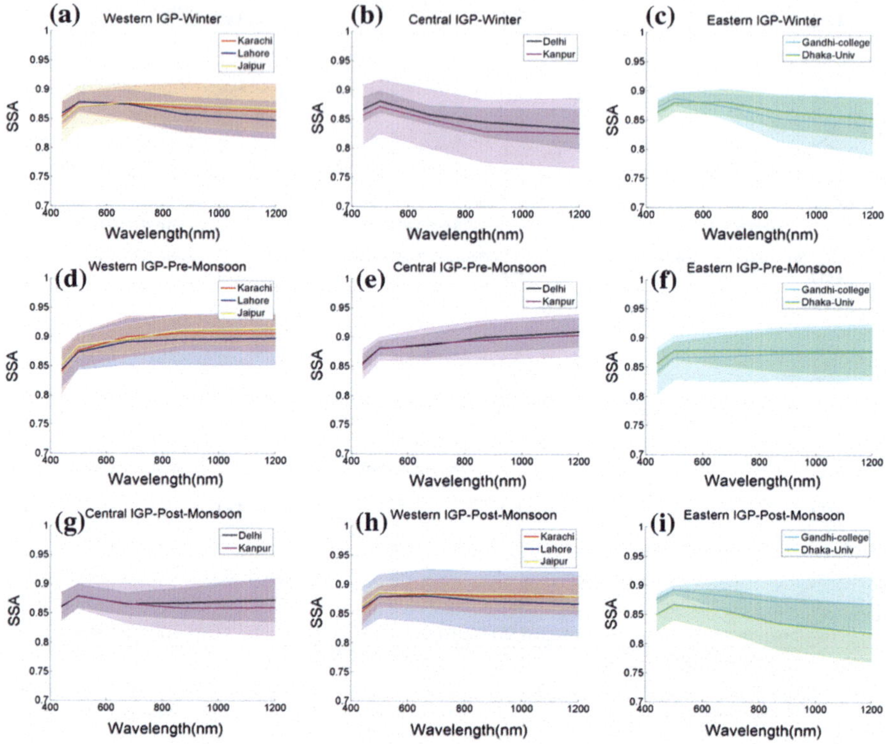

Fig. 9.5 Mean spectral dependence of SSA during extreme pollution events (ABCs days) over western (Karachi, Lahore, Jaipur), central (Delhi and Kanpur) and Eastern (Gandhi-college and Dhaka-University) IGP for winter (**a**, **b**, **c**), pre-monsoon (**d**, **e**, **f**) and post-monsoon (**g**, **h**, **i**) seasons, respectively. Shaded regions show the respective standard deviation

Figure 9.4 gives a clear idea of the dependence of AAOD on wavelength. The dependence follows a similar trend as AOD where AAOD value is highest at a smaller wavelength (440 nm) and decreases with an increase in wavelength. AAOD value increases as one moves from western IGP towards eastern IGP. Also, the highest AAOD value is observed during winter over entire IGP which indicates the impact of biomass burning and prevailing meteorology during winter over IGP. AAOD following a similar trend as AOD indicates that aerosols are more absorbing than scattering.

While AOD and AAOD show a decreasing trend with wavelength, SSA (Fig. 9.5) shows a quite irregular increase with the increase in wavelength. During winter, central IGP has low SSA value suggesting the presence of absorbing type aerosols over Delhi and Kanpur. This can be attributed to the biomass burning and vehicular emission over the region. The variation in the size distribution of aerosols, mostly urban and industrial is due to the particle growth at high humidity and to aerosol interaction with clouds. The variations in the size distribution of aerosols thus influence the radiative properties of aerosols such as AOD, AAOD, and SSA (Kaskaoutis et al.

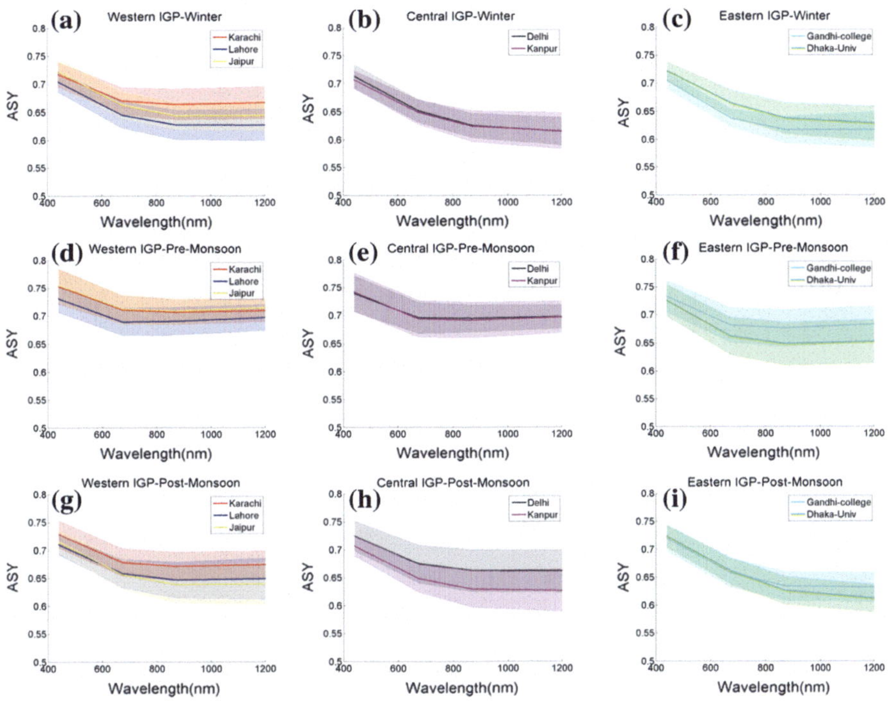

Fig. 9.6 Mean spectral dependence of asymmetry parameter (ASY) during extreme pollution events (ABCs days) over western (Karachi, Lahore, Jaipur), central (Delhi and Kanpur) and Eastern (Gandhi-college and Dhaka-University) IGP for winter (**a**, **b**, **c**), pre-monsoon (**d**, **e**, **f**) and post-monsoon (**g**, **h**, **i**) seasons, respectively. Shaded regions show the respective standard deviation

2006). A monotonically increasing SSA with respect to wavelength indicates the presence of dust particle while monotonically decreasing SSA indicates that aerosol is dominated by black carbon (Si et al. 2017).

Figure 9.6 indicates that for all seasons, the asymmetry parameter (ASY) decreases with increasing wavelength. The ASY represents the nature of scattering (relative dominance of forward and backward scattering). The ASY ranges between 0.7 and 0.75 during winter and post-monsoon while the range becomes close to 0.8. It indicates that during winter and post-monsoon, the particle size distribution is similar and mainly dominated by fine mode, while during pre-monsoon it is dominated by large particles because of high dust loading over the IGP.

9.3.3 Size Distribution of Aerosol During Extreme Pollution Events

Figure 9.7 displays the size distribution of aerosols over IGP during winter, pre-

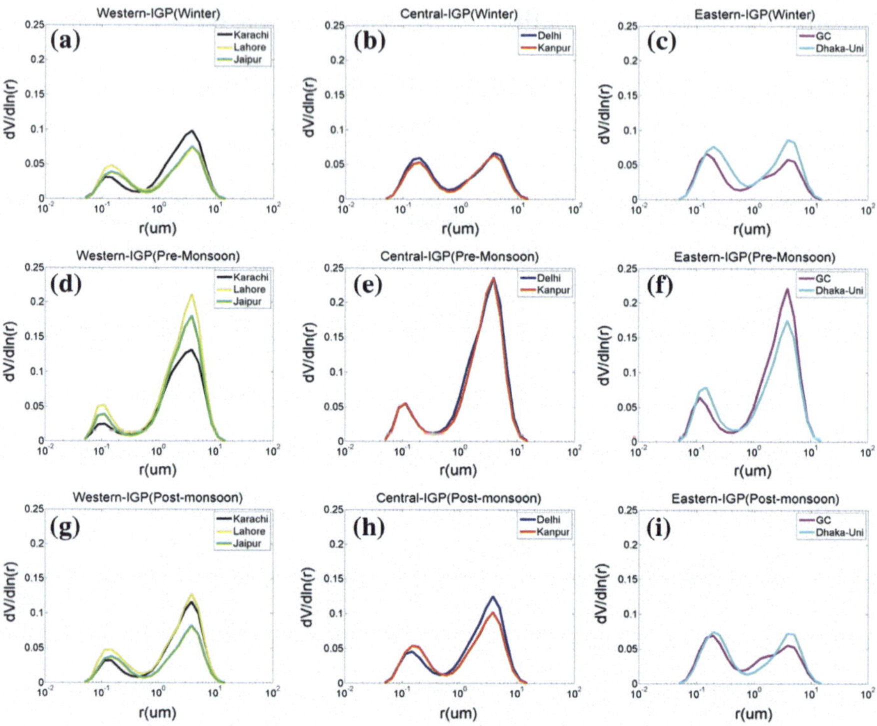

Fig. 9.7 Mean volume size distribution of aerosols during extreme pollution events (ABCs days) over western (Karachi, Lahore, Jaipur), central (Delhi and Kanpur) and eastern (Gandhi-college and Dhaka-University) IGP for winter (**a**, **b**, **c**), pre-monsoon (**d**, **e**, **f**) and post-monsoon (**g**, **h**, **i**) seasons, respectively

monsoon and post-monsoon seasons. The results clearly indicate that fine particles dominate during winter and post-monsoon while coarse particles dominate during pre-monsoon. During winters, volume size distribution of in the range 0.1–1 μm is significantly low as that of the particles ranging from 1 to 10 μm. However, in central and eastern IGP, the distribution is quite similar for both ranged particles. The highest peak is observed for Karachi in the range of radius ~0.5 μm. During pre-monsoon, volume size distribution of the particle ranging from 1 to 10 μm in radius is significantly higher than that of the particles ranging between 0.1 and 1.0 μm. The volume size distribution shows maximum concentration in central IGP. Similar results are obtained during post-monsoon compared to winter where eastern IGP has low difference in the volume size distribution of particle radius between 0.1 and 10 μm. However, western and central IGP experiences a higher volume size distribution of particles ranging between 1 and 10 μm than particles with radius 0.01–1 μm.

9.3.4 Classification of Aerosol During Extreme Pollution Events

Classification of aerosols into their constituent components is important in order to determine their source and contribution to the total aerosol loading over a particular region. Aerosol-type partitioning based upon AE and AAE in an effective method to classify aerosols types (Russell et al. 2010; Giles et al. 2012). Aerosols over the IGP have been classified into 4 types based on the AE and AAE values. Dust: AAE $\geq$ 1.8 & AE $\leq$ 0.8; Biomass burning aerosols: AAE $\geq$ 1.8 & AE $\geq$ 1.2 Urban/Industrial aerosols: AAE < 1.8 & AE $\geq$ 1.20; Mixed aerosols: AE > 0.8 & AE < 1.2 & all AAE values. Figure 9.8 shows the scatter plot between AAE and AE over the IGP sites, depicting different aerosol types over the region.

The percentage composition of aerosol types during each season is shown in Fig. 9.9. During winter, the western IGP (Karachi, Lahore) and Jaipur show a high concentration of mixed type aerosols (>60%). As we move towards eastern IGP, the contribution of carbonaceous aerosols through biomass burning becomes significant. In Kanpur, Patna, and Dhaka, fuel woods are still used to cook food and warming purpose during winters which result in significant biomass contribution over these regions in winter. Pre-monsoon as already mentioned, is dominated by dust coming from the desert regions. The western IGP (Karachi, Lahore, Delhi, Kanpur) shows a high concentration of dust during pre-monsoon. However, the dominance decreases over Gandhi-college and Dhaka-University where mixed type aerosols dominate during Pre-monsoon (57% and 40% respectively).

A similar trend is observed during post-monsoon as observed in winter, where biomass burning sources dominate aerosol type over the eastern IGP while Karachi and Lahore are dominated by mixed type aerosols. The trend can be explained on the basis of agricultural crop residue burning during this season as mentioned earlier. For

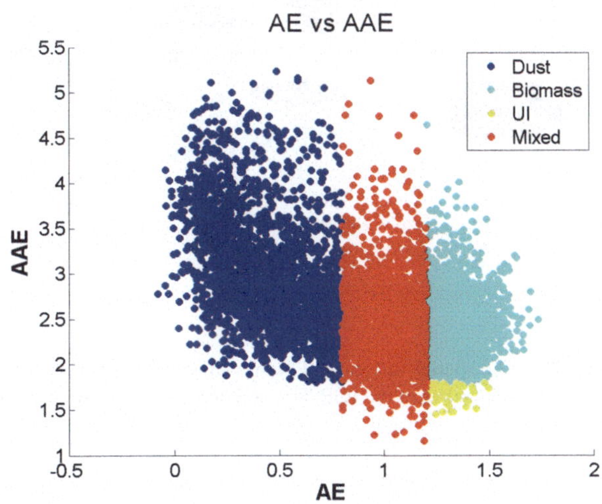

Fig. 9.8 Scatter plot depicting aerosol types based upon AE and AAE values over all IGP sites including all seasons

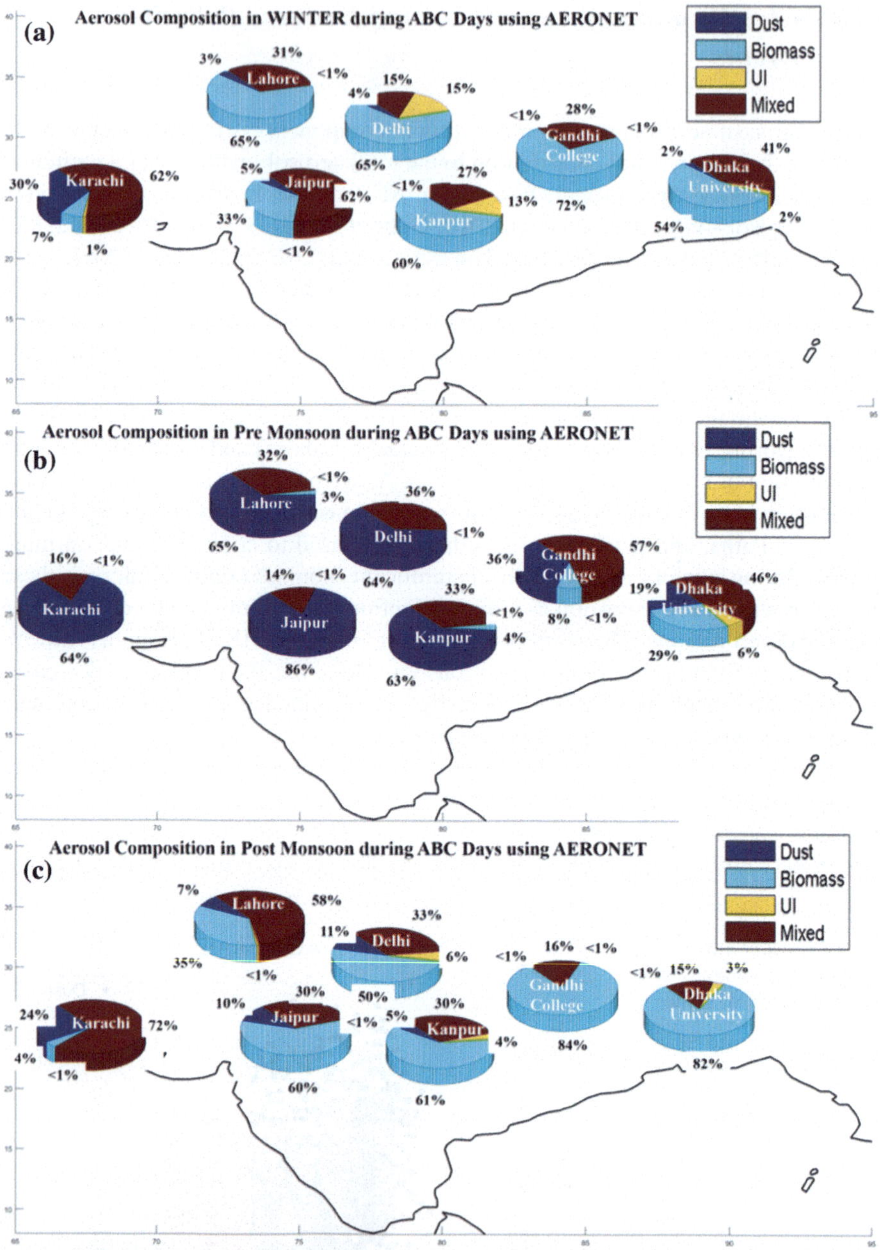

Fig. 9.9 The spatial variation of composition of aerosols during **a** winter, **b** pre-monsoon, **c** post-monsoon seasons over the IGP. The percentage composition has been shown using legends in each figure

all the seasons, the contribution of urban/Industrial sector is found to be very minimal. Result gives us an idea about the fact that the classification algorithm for U/I is still not sufficient in providing accurate results as most of the U/I type aerosol fall under the mixed type category over IGP. Classification of aerosol into their constituents and the trends and differences in the observations could be considered as an area for further investigation. The classification of aerosols based upon similar algorithm over Kanpur and Delhi region has been reported by various studies (Mishra et al. 2014; Srivastava et al. 2011; Kaskaoutis et al. 2012).

9.4 Summary

ABCs phenomena pose a serious threat to the world population and demand an extensive study. Ever since the concept of ABCs has come into the light, a lot of projects have been devoted to understand every possible details about the phenomena and associated impacts. The IGP region is one of the five ABCs hotspots of the world and home of over 500 million South Asians. ABCs affects regional as well as global climate by inducing radiative forcing, altering the precipitation patterns and meteorology of a region. Not only it affects the climate, but it has also led to the reduction in crop productivity which leads to a serious threat of food security. Considering all the aspects, the intensive study is required to understand the concepts, impacts, and challenges of ABCs and possible ways to prevent and save the environment from the same.

A multi-objectives study was carried out over Indo-Gangetic plain to understand the basic trends and dependency of aerosol optical and microphysical properties on geographical locations and seasons. Extreme pollution episodes (ABCs events) over the IGP were determined using pre-requisite algorithm. The results indicate that the highest occurrence of pollution days was observed over Dhaka-university which lies in eastern IGP. This suggests that eastern IGP is more prone to pollution events. Almost 65% days during post-monsoon were found to be polluted in Dhaka followed by Delhi (61%) and Kanpur (44.2%). Results also suggest that eastern IGP receives higher aerosol loading than western IGP which indicates local contribution over the region.

Over the IGP, coarse-sized particles dominate during pre-monsoon while fine particles (biomass burning aerosols) dominate during winter and post-monsoon. Classification of aerosols into dust, biomass burning, urban/industrial and mixed type provides information about the dominance of each type during a particular season. It is observed that pre-monsoon is dominated by dust aerosols while winter and post-monsoon receive a high number of aerosols as a result of biomass burning.

The findings of this study can help us better understand the causes of ABCs over IGP and relate it to the changes in aerosol properties over a particular region and would be helpful in estimating the significance of optical and micro-physical data retrieved from ground-based remote sensing data by comparing it with the results from satellite data. Finally, the study could be crucial for the policymakers and stakeholders to better understand the impacts of human activities like-urbanization, industrialization, ever expansion of vehicular emissions and household activities in rural areas like burning of biomass to cook food, stay warm and agricultural practices like burning the crop residues on the local and regional climate. Further, the mean aerosol parameters (AOD, AAOD, SSA, ASY) in this study could be used to estimate aerosol radiative forcing during extreme pollution conditions.

Acknowledgements The authors would like to acknowledge the Department of Science and Technology (DST) for providing research fund under DST Inspire Faculty scheme (DST/INSPIRE/04/2015/003253). We also thank the PI's of all AERONET sites used in this study for maintaining and providing quality assured data. We also acknowledge the DST-purse grant. Authors would like to thank two anonymous reviewer for their valuable comments and suggestions.

References

Alam K, Trautmann T, Blaschke T (2011) Aerosol optical properties and radiative forcing over mega-city Karachi. Atmos Res 101(3):773–782

Balakrishnan K, Dey S, Gupta T, Dhaliwal RS, Brauer M, Cohen AJ, Stanaway JD, Beig G, Joshi TK, Aggarwal AN, Sabde Y (2019) The impact of air pollution on deaths, disease burden, and life expectancy across the states of India: the Global Burden of Disease Study 2017. Lancet Planet Health 3(1):e26–e39

Bergstrom RW, Russell PB, Hignett P (2002) Wavelength dependence of the absorption of black carbon particles: predictions and results from the TARFOX experiment and implications for the aerosol single scattering albedo. J Atmos Sci 59(3):567–577

Boucher O, Randall D, Artaxo P, Bretherton C, Feingold G, Forster P, Kerminen VM, Kondo Y, Liao H, Lohmann U, Rasch P (2013) Clouds and aerosols. In: Climate change 2013: the physical science basis. Contribution of working group I to the fifth assessment report of the Intergovernmental Panel on Climate Change, pp 571–657. Cambridge University Press

Chowdhury S, Dey S and Smith KR (2018) Ambient PM2.5 exposure and expected premature mortality to 2100 in India under climate change scenarios, Nat Commun 9:318

Dey S, Tripathi SN, Singh RP, Holben BN (2004) Influence of dust storms on the aerosol optical properties over the Indo-Gangetic basin. J Geophys Res Atmos 109(D20)

Dubovik O, King MD (2000) A flexible inversion algorithm for retrieval of aerosol optical properties from sun and sky radiance measurements. J Geophys Res 15:20673–20696

Eck TF, Holben BN, Reid JS, Dubovik O, Smirnov A, O'Neill NT, Slutsker I, Kinne S (1999) Wavelength dependence of the optical depth of biomass burning, urban, and desert dust aerosols. J Geophys Res 104:31333–31349. https://doi.org/10.1029/1999JD900923

Ganguly D, Gadhavi H, Jayaraman A, Rajesh TA, Misra A (2005) Single scattering albedo of aerosols over the central India: implications for the regional aerosol radiative forcing. Geophysical research letters 32(18)

Gautam R, Hsu NC, Lau K-M (2010) Premonsoon aerosol characterization and radiative effects over the Indo-Gangetic Plains: implications for regional climate warming. J Geophys Res 115:D17208. https://doi.org/10.1029/2010JD013819

Giles DM, Holben BN, Eck TF, Sinyuk A, Smirnov A, Slutsker I, Dickerson RR, Thompson AM, Schafer JS (2012) An analysis of AERONET aerosol absorption properties and classifications representative of aerosol source regions. J Geophys Res Atmos 117(D17)

Guttikunda SK, Goel R (2013) Health impacts of particulate pollution in a megacity—Delhi, India. Environ Dev 6:8–20

Haywood J, Boucher O (2000) Estimates of the direct and indirect radiative forcing due to tropospheric aerosols: a review. Rev Geophys 38(4):513–543

Holben BN, Eck TF, Slutsker I, Tanré D, Buis JP, Setzer A, Vermote E, Reagan JA, Kaufman Y, Nakajima T, Lavenu F, Jankowiak I, Smirnov A (1998) AERONET-A federated instrument network and data archive for aerosol characterization. Remote Sens Environ 66:1–16

Kaskaoutis DG, Kambezidis HD, Adamopoulos AD, Kassomenos PA (2006) On the characterization of aerosols using the Ångström exponent in the Athens area. J Atmos Solar Terr Phy 68(18):2147–2163

Kaskaoutis DG, Gautam R, Singh R, Houssos E, Goto D, Singh S, Bartzokas A, Kosmopoulos P, Sharma M, Hsu N, Holben B, Takemura T (2012) Influence of anomalous dry conditions on aerosols over India: transport, distribution and properties. J Geophys Res 117:D09106. https://doi.org/10.1029/2011JD017314

Lau KM, Kim KM (2006) Observational relationships between aerosol and Asian monsoon rainfall, and circulation. Geophys Res Lett 33:L21810. https://doi.org/10.1029/2006GL027546

Menon HB, Shirodkar S, Kedia S, Ramachandran S, Babu S, Moorthy KK (2014) Temporal variation of aerosol optical depth and associated shortwave radiative forcing over a coastal site along the west coast of India. Sci Total Environ 468:83–92

Mishra AK, Shibata T (2012a) Climatological aspects of seasonal variation of aerosol vertical distribution over central Indo-Gangetic belt (IGB) inferred by the space-borne lidar CALIOP. Atmos Environ 46:365–375

Mishra AK, Shibata T (2012b) Synergistic analyses of optical and microphysical properties of agricultural crop residue burning aerosols over the Indo-Gangetic Basin (IGB). Atmos Environ 57:205–218

Mishra AK, Shibata T, Srivastava A (2014) Synergistic approach for the aerosol monitoring and identification of types over Indo-Gangetic Basin in pre-monsoon season. Aerosol Air Qual Res 14:767–782

Ostro B (1995) Fine particulate air pollution and mortality in two Southern California counties. Environ Res 70(2):98–104

Ramanathan V (2007) Global dimming by air pollution and global warming by greenhouse gases: global and regional perspectives. In: Nucleation and atmospheric aerosols, pp 473–483. Springer, Dordrecht

Ramanathan V, Ramana MV (2005) Persistent, widespread, and strongly absorbing haze over the Himalayan foothills and the Indo-Gangetic Plains. Pure Appl Geophys 162(8–9):1609–1626

Ramanathan V, Chung C, Kim D, Bettge T, Buja L, Kiehl JT, Washington WM, Fu Q, Sikka DR, Wild M (2005) Atmospheric brown clouds: impacts on South Asian climate and hydrological cycle. Proc Natl Acad Sci 102(15):5326–5333

Ramanathan V, Agrawal M, Akimoto H, Aufhammer M, Devotta S, Emberson L, Hasnain SI, Iyngararasan M, Jayaraman A, Lawrance M, Nakajima T (2008) Atmospheric brown clouds: regional assessment report with focus on Asia. Published by the United Nations Environment Programme, Nairobi

Russell PB, Bergstrom RW, Shinozuka Y, Clarke AD, DeCarlo PF, Jimenez JL, Livingston JM, Redemann J, Dubovik O, Strawa A (2010) Absorption Ångström exponent in AERONET and related data as an indicator of aerosol composition. Atmos Chem Phys 10:1155–1169

Si Y, Li S, Chen L, Shang H, Wang L, Letu H (2017) Assessment and improvement of MISR Angstrom exponent and single-scattering albedo products using AERONET data in China. Remote Sens 9(7):693

Singh S, Nath S, Kohli R, Singh R (2005) Aerosols over Delhi during pre-monsoon months: characteristics and effects on surface radiation forcing. Geophys Res Lett 32(13)

Sinha PR, Kaskaoutis DG, Manchanda RK, Sreenivasan S (2012) Characteristics of aerosols over Hyderabad in Southern Peninsular India: synergy in the classification techniques. Ann Geophys 30:1393–1410. https://doi.org/10.5194/angeo-30-1393-2012

Srivastava R, Ramachandran S (2012) The mixing state of aerosols over the Indo-Gangetic Plain and Its impact on radiative forcing. Quart J R Meteorolog Soc 139:137–151. https://doi.org/10.1002/qj.1958

Srivastava AK, Tiwari S, Devara PCS, Bisht DS, Srivastava MK, Tripathi SN, Goloub P, Holben BN (2011) premonsoonal aerosol characteristics over the Indo-Gangetic Basin: implications to climate impact. Ann Geophys 29:789–804. https://doi.org/10.5194/angeo-29-789-2011

Srivastava AK, Tripathi SN, Dey S, Kanawade VP, Tiwari S (2012) Inferring aerosol types over the Indo-Gangetic Basin from ground based sunphotometer measurements. Atmos Res 109–110:64–75

UNEP (2008) Atmospheric brown cloud, regional assessment report with focus on Asia. United Nations Environment Programme, Nairobi

Watts N, Adger WN, Agnolucci P et al (2015) Health and climate change: policy responses to protect public health. Lancet 386:1861–914. https://www.thelancet.com/climate-and-health

World Health Statistics (2018) Monitoring health for the SDGs, sustainable development goals. Geneva: World Health Organization. Licence: CC BY-NC-SA 3.0 IGO

Chapter 10
Spatial Variation of Airborne Allergenic Fungal Spores in the Ambient $PM_{2.5}$—A Study in Rajkot City, Western Part of India

Charmi Humbal, Sneha Gautam, Suneel Kumar Joshi and Mahendrapal Singh Rajput

Abstract Fungal spores in the fine particle is an emerging pollutant of the technological age, which can create adversely effect on human health and their surrounding environment. Probably the first time in the western part of India, an investigation was organized to assess the spatial distribution of $PM_{2.5}$ associated fungal spore concentration levels in an urban city. Five urban locations selected to cover probably all major areas of a city to conduct the study by using fine particulate sampler with 24 hours' interval. Highest (101.79 ± 8.09 μg m^{-3}) concentrations of $PM_{2.5}$ have been observed in the industrial area only. The highest (8.0×10^{13} Colony-forming unit (CFU) m^{-3}) in industrial area and lowest (2.0×10^{8} CFU m^{-3}) fungal concentrations were found in the residential area. Spores of seven fungal species (i.e., *Aspergillus, Candida, Fusarium, Penicillium, Alternaria, Cephalosporium and Mucor*) were significantly predominant in all selected locations in the urban area. In these views, *Aspergillus, Candida and Penicillium, and Fusarium* species were the dominant fungi in Industrial, slaughter house and dump site, respectively. The highest concentration of fungal spores was reported in industrial area and poultry farm as compared to other locations. Outcomes of the current work suggested that fungal spores were observed in the respirable fraction (<2.5 μm) and so had the potential to penetrate the deeper part of the lungs. In addition, the meteorological parameters i.e., temperature and relative humidity, were recorded to understand the relationship between meteorology and enhanced viability of fungal spores.

C. Humbal
Department of Environmental Science and Engineering, Marwadi University, Rajkot, Gujarat 360003, India

S. Gautam (✉)
Department of Civil Engineering, Karunya Institute of Technology and Sciences, Karunya Nagar, Coimbatore 641114, India
e-mail: snehagautam@karunya.edu

S. K. Joshi
National Institute of Hydrology, Roorkee 247667, India

M. S. Rajput
Department of Microbiology, Marwadi University, Rajkot, Gujarat 360003, India

T. Gupta et al. (eds.), *Measurement, Analysis and Remediation of Environmental Pollutants*, Energy, Environment, and Sustainability,
https://doi.org/10.1007/978-981-15-0540-9_10

Keywords $PM_{2.5}$ · Bioaerosols · Fungal spores · Health issues

10.1 Introduction

The number of airborne diseases has been reporting significantly in urban cities (Gupta et al. 2019). Significantly well-known diseases (i.e., infectious diseases, respiratory diseases, pre-term birth, abortion, immunotoxic and hormone changes) due to exposure to bioaerosol (Humbal et al. 2018; Douwes et al. 2003). Reddy et al. (2005) reported the statistics of potentially affected people (400 million people) with chronic respiratory diseases globally and 7% of reported deaths in India. Many researchers (Humbal et al. 2018; Schachter et al. 2016; Rohr et al. 2014) suggesting about "Asthma" is one of the common respiratory diseases due to the presence of airborne allergies (Gautam et al. 2018). In Anderson addition, Chakrabarty et al. (2012) reported that a significant number of asthma patients in national average (2468 per 100,000). According to a few researchers (Humbal et al. 2018), airborne fungi and incremental concentration of fungal spores in the surrounding environment might be responsible for asthma diseases.

Temperature (T), and relative humidity (RH) (Després et al. 2012) are significantly responsible for incremental number of fungal spores in the environment. Many researchers reported that increasing the number of fungal spores due to decreasing the RH (Després et al. 2012; Timmer et al. 1998). In addition, wind speed, wind direction, air T, and precipitation generally play an essential role to dispersion of fungal spores from one location to another location (Humbal et al. 2018; Kumar and Attri 2016; Jones and Harrison 2004). A few studies (Elbert et al. 2007; Portnoy et al. 2005) indicated that only 80 numbers (individually) genera out of reported 1.5 million fungi species are identified, which produce allergenic effect and associated with respiratory diseases. Few specific fungi types (i.e., *Cladosporium, Alternaria, Penicillium, Aspergillus, Epicoccum* and other Basidiomycetes) are generally widespread in an ambient air, however, severe health impacts due to direct exposure (respiratory diseases) to Cladosporium (dry spores) are reported due to its longtime existence in the environment (Kallawicha et al. 2015). Few additional studies suggested specific diseases may arise from direct exposure to fungus (i.e., aspergillosis, blastomycosis, coccidioidomycosis, and adiaspiromycosis). The actual impact including environment and human health, of fungi presence with the particle, is poorly identified. Past research (Nasir et al. 2012; Adhikari et al. 2004; Kakde et al. 2001) related to characterizing fungi was updated, based on indoor air (environment) (i.e., shops, academic building, market, etc.). However, limited literature (Kumar and Attri 2016; Chakrabarti et al. 2012) has been reported on incremental levels (monitoring based) of fungi in developing countries, with scanty literature done in the ambient air of western part of India.

The assessment of exposure to biotic matter (i.e., bacteria, fungi, virus, etc.) on human health is very concerned in the western part of India, as reporting high T and RH which plays an important role the in the growth of microbial constituents.

And so, the current study carried out to quantification of fungus and fine particle ($PM_{2.5}$) concentration in different locations in the urban area of western India. The present study highlights few points; (i) The influence of meteorological parameters on fungus and $PM_{2.5}$ concentrations; (ii) Concentration variation of different types of fungi at different locations and (iii) The possible impact of fungi available in the ambient air at different locations in an urban area of western part of India. As per limited studies have been done on the same in India, presented study is possibly first such study carried out in particular region of India.

10.2 Materials and Methods

Five sampling sites (i.e., residential area, industrial area, poultry farm, dump site and slaughter house) have been selected in an urban area, located in Gujarat state, western India, lies between 22° 10′ N and 22° 25′ N, and 70° 40′ E and 70° 55′ E and covers an area of 350 km^2 (Fig. 10.1).

The elevation in the study area varies from 150 to 110 m above sea level (masl). All samplers were installed in and around the source and collected all samples at a height of 1.5 m (general human head height) from the ground level. Fifty samples for 24 h were collected from sampling locations. In addition, T and RH were also recorded simultaneously during bioaerosol sampling by using portable meteorological station. All the samples (fine particles; <2.5 μm = aerodynamic diameter range) were collected using fine particle sampler at sampling locations (Humbal et al. 2019). The sampling methods and related information could be available in published article (Humbal et al. 2019).

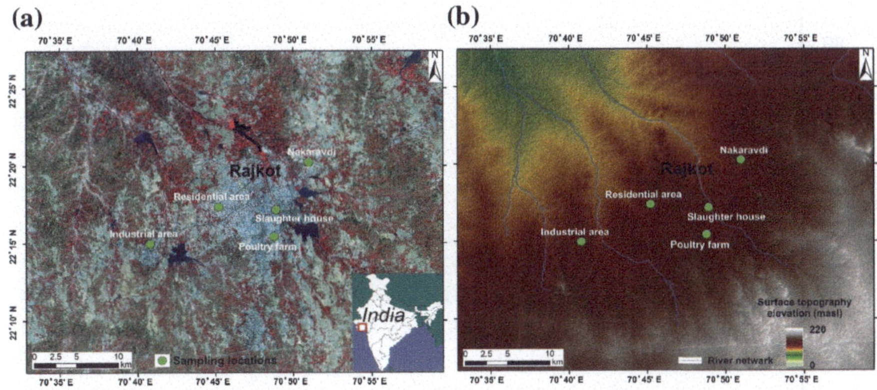

Fig. 10.1 Selected sampling locations in Gujarat state, western India (Humbal et al. 2019)

10.3 Results and Discussion

10.3.1 Fine Particle ($PM_{2.5}$) Concentration at Selected Study Locations

The spatial variation of 24 h average concentration of $PM_{2.5}$ is shown in Fig. 10.2. The average 24 h concentration of fine particle during sampling were (26.26 ± 3.76 μg m^{-3}), (25.93 ± 2.55 μg m^{-3}), (101.79 ± 8.09 μg m^{-3}), (57.26 ± 9.86 μg m^{-3}), and (43.89 ± 3.20 μg m^{-3}) in the residential area, industrial area, slaughter house, poultry farm and dump site respectively (Fig. 10.2).

Most of the times, higher concentration was reported in industrial area than other selected locations, which possible due to the higher discharge of waste material from different sources. The reported $PM_{2.5}$ concentration levels exceeded the Indian National Ambient Air Quality Standard (NAAQS/http://cpcb.nic.in); 60 μg m^{-3}) and WHO (World Health Organization; 25 μg m^{-3}) air quality standards specially in the industrial area and dump site than other selected locations on 40% of the sampling duration/days. Therefore, the real data measurement on field are more important to assess the actual exposure to bioaerosol on living things at different locations.

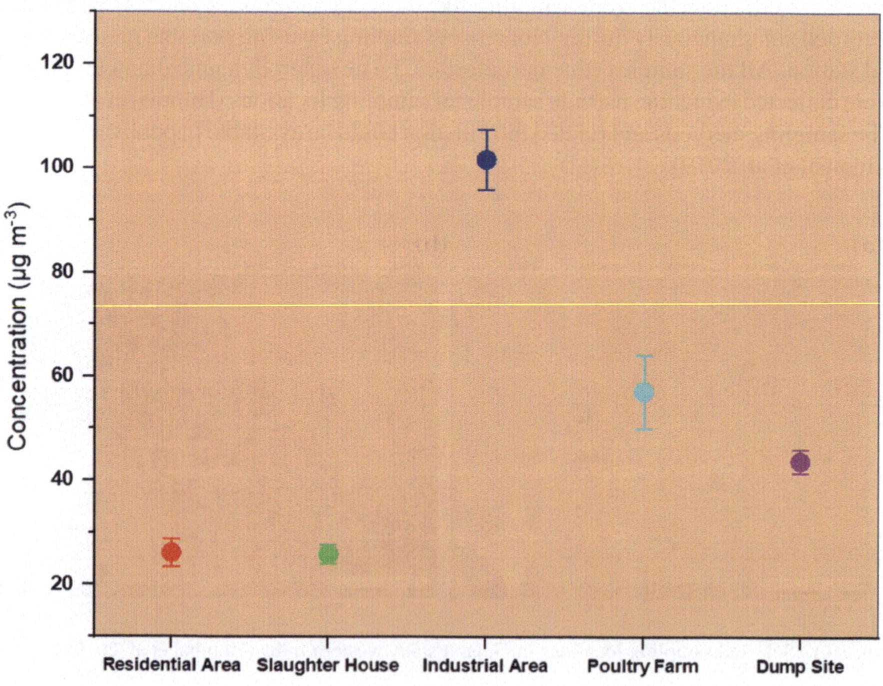

Fig. 10.2 $PM_{2.5}$ concentrations variation at selected locations (Humbal et al. 2019)

10.3.2 Fungal Aerosol Concentrations at Various Locations

The fungal concentrations of selected locations are presented in Fig. 10.3, where concentration (CFU; Colony-forming unit) are higher in industrial area (8×10^{12} CFU m^{-3}) than other sampling sites like residential area (2×10^{8}), poultry farm (2×10^{12} CFU m^{-3}), slaughter house (2×10^{11} CFU m^{-3}) and dump site (2×10^{10} CFU m^{-3}).

The observed results indicating that industrial area are the potential source of biotic matter as compare to other location in urban area, which are also suggested by previous literature (Pasanen et al. 2000; Kulmala et al. 1999). The massive amount of discharge, deposition of waste material and high vegetation are the extreme conditions to support fungal growth in industrial and dump sites (Ritchkoff et al. 2000; Kulmala et al. 1999). The reporting concentration of biotic matter are much higher than the influence limit (1000 CFU m^{-3}) (Goyer et al. 2001), indicating the importance of biotic monitoring to investigate the impact on human health.

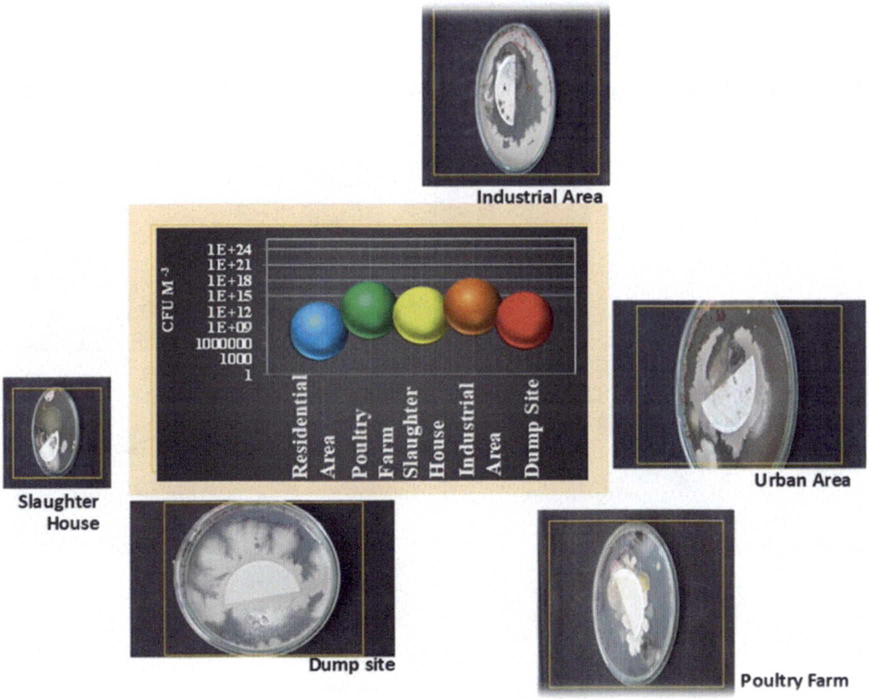

Fig. 10.3 The concentration of fungal aerosol in the study locations

10.3.3 Fungi Diversity at Study Locations

There seven individual fungal spores (i.e., *Aspergillus*, *Candida*, *Fusarium*, *Alterneria*, *Cephalosporium*, *Mucor* and *Penicillium*) have been identified from the total sampling sites (Fig. 10.4). In Fig. 10.4, the microscopic images (i.e., *Alterneria, Mucor, Aspergillus, Penicillium, Candida, and Fusarium species*) have been identified fungal spores in $PM_{2.5}$ samples from selected sampling sites. *Aspergillus* and *Candida* species observed in higher amount in all sampling chosen locations, which are generally well known for their allergic properties (Kumar and Attri 2016).

Significant variability of fungal spores has been observed in the different selected sampling locations (Fig. 10.5). The higher fungal spore concentration was observed in industrial area than other selected locations, indicating the importance of the combination of T, humidity, and food (organic content), especially for the proper growth of fungal spores. On the other hand, wet scavenging of spores might be a reason for the low concentration levels in the residential area. During sampling periods, variations in terms of different concentration and types in slaughter house, implying aggravated fungal growth due to meteorological parameters (T & RH). *Aspergillus*, *Mucor* and *Candida* were most dominant fungal species in industrial, slaughter house and poultry farm. Many studies have been reported that dominant fungal species change with spatial variation. In this regard, Kumar and Attri (2016) reported the higher variation

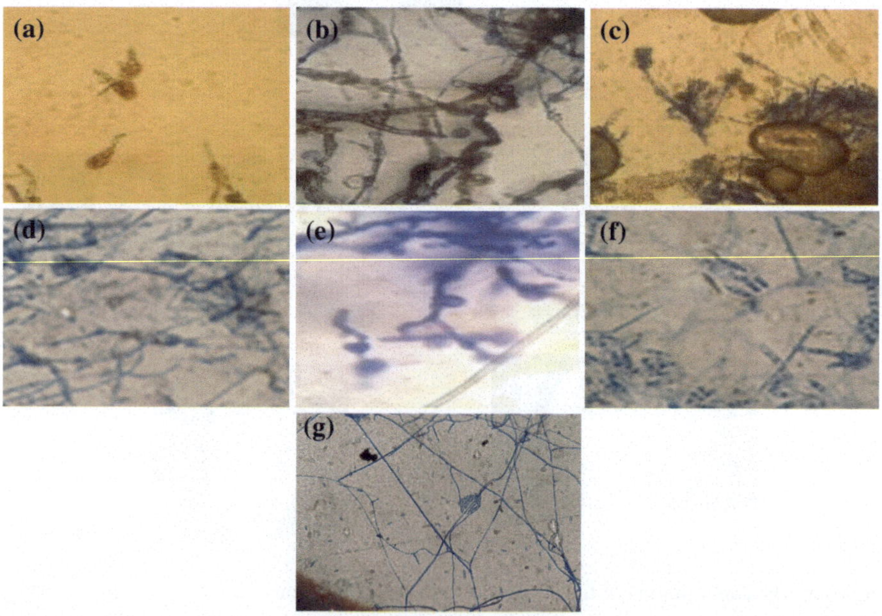

Fig. 10.4 Microscopic image (optimal magnification of 1000 × in the $PM_{2.5}$ samples) of individual fungal spores, sporangia and conidia from selected sampling sites: **a** *Alterneria* sp., **b** *Mucor* sp., **c** *Aspergillus* sp., **d** *Penicillium* sp., **e** *Candida* sp., **f** *Fusarium* sp. and **g** *Cephalosporium* sp.

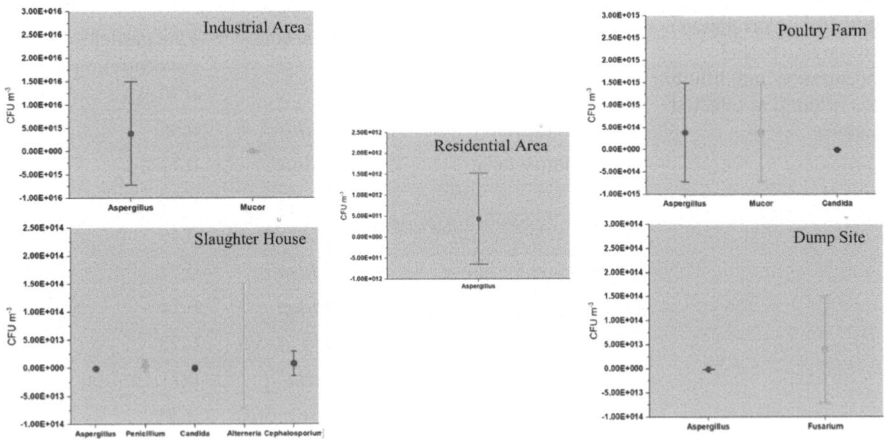

Fig. 10.5 Variations in fungal spores at different selected study locations

in concentration of fungal spores, especially for *Aspergillus* sp. Similarly, Aira et al. (2013) observed the spatial and temporal variation in *Alterneria* spores at different locations.

10.3.4 Correlation Between Fine Particle Concentration and Different Types of Fungi at Different Locations

Fine particles and fungal aerosol are found to be significantly correlated with each other at the significance level of 0.05. The data collected from the urban area with measurement in different locations are used to develop the following empirical relationships between particle concentration and fungal concentration through regression analysis, which are presented in Table 10.1.

Table 10.1 shows that the relationship between fine particle concentration and individual fungi type at selected study locations of urban area. Most of the fungi had moderate to strong correlation with fine particle concentration at different selected locations (Table 10.1). In this regard, *Aspergillus* showed highest correlation with particle concentration at poultry farm (0.72), slaughter house (0.72), and Industrial area (0.99). On other hand, significant association were observed between different types of fungi and selected study area (i.e., poultry farm and slaughter house), where $R^2 = 0.72$, 0.74, 0.71, and 0.72 for *Aspergillus, Penicillium, Candida*, and Alterneria species, respectively in slaughter house and $R^2 = 0.72$, and 0.97, for *Aspergillus*, and *Mucor* species, respectively in poultry farm. In the case of dump site, *Fusarium* had good correlation (0.80) with particle concentration. The incremental concentration of fine particle at different locations have substantial effects on fungal aerosol concentration. In the present study, the higher fungal aerosol concentration was reported

Table 10.1 Correlation between fine particle concentration and different types of fungi at selected locations

Sampling locations	Fungal species	Fine particle concentration ($PM_{2.5}$)
Residential	*Aspergillus*	0.55
Poultry farm	*Aspergillus*	0.72
	Mucor	0.97
	Candida	0.45
Slaughter house	*Aspergillus*	0.72
	Penicillium	0.74
	Candida	0.72
	Alterneria	0.71
	Cephalosporium	0.20
Industrial area	*Aspergillus*	0.99
	Mucor	0.34
Dump site	*Fusarium*	0.80

peaked at an industrial site, slaughter house, and dump site, as the site has already reported higher concentration (Sect. 10.3.1). As per reported relationship are concerned, the higher concentration and lower concentration of particle is may be the reason of higher concentration of fungal aerosol, as they are generally bound to particle.

10.3.5 Correlation Between Meteorological Parameters (T & RH) and Different Types of Fungi at Different Locations

Table 10.2 showing the relationship between meteorological parameters (T & RH) and fungal aerosol. Many studies (Humbal et al. 2019) have been reported the significant effect of meteorological parameters on fungal spores concentration at different variability (time/space). Relationship between T and individual types of fungal spores found was statistically significantly (positive) for all selected study location in urban area.

In the present study, all observed fungi types at different location in the urban area had good to strong correlation with T & RH. *Aspergillus* and *penicillium* showed highest correlation of 0.89 (Industrial area) and—0.69 (slaughter house) with T and RH, respectively.

Table 10.2 Correlation between meteorological parameters (T & RH) and different types of fungi at selected locations

Sampling locations	Fungal species	T	RH
Residential area	*Aspergillus*	0.72	−0.55
Poultry farm	*Aspergillus*	0.80	−0.53
	Mucor	0.74	−0.57
	Candida	0.88	−0.59
Slaughter house	*Aspergillus*	0.84	−0.60
	Penicillium	0.75	−0.69
	Candida	0.60	−0.67
	Alterneria	0.84	−0.61
	Cephalosporium	0.71	−0.60
Industrial area	*Aspergillus*	0.89	−0.62
	Mucor	0.77	−0.69
Dump site	*Fusarium*	0.84	−0.65

10.4 Conclusions

The study showed the variation of particle and fungal spores concentration at the different places (i.e., residential area, industrial area, slaughter house, dump site and poultry farm) in the urban area, western India. Reported concentration of $PM_{2.5}$ at few sampling sites (i.e., industrial area; 101.79 ± 8.09 µg m^{-3}, and slaughter house; 57.26 ± 9.86 µg m^{-3}) are exceeded from the Indian national standards (60 µg m^{-3} CPCB 2009). Higher fungal concentration was reported in an industrial area (8×1012 CFU m^{-3}) and a poultry farm (2×1012 CFU m^{-3}). Fine particle and spores are found to be significantly correlated with each other at the level 0.05. Most of the fungi had good to strong correlation with fine particle concentration at different selected study locations. In this regard, Aspergillus showed highest correlation with particle concentration at poultry farm (0.72), slaughter house (0.72), and Industrial area (0.99). The statistically significant relationship had been observed between T and fungal spores in all selected study locations in an urban area. However, the correlation between T and fungal spores was reported higher in the industrial area as compare to all other locations. Similarly, the negative (−0.69) significant correlation relationship had been observed between RH and fungal spores, i.e., the lower fungal spores concentration, at higher RH and vice versa. The lacuna of present policies/approaches have been highlighted by the outcomes of the present study. Similarly, potential source of bioaerosol in urban area also presented. Incremental concentration of presented microbes is responsible for different types of diseases (i.e., skin, respiratory etc.). The outcomes of the result strongly suggested the comparative monitoring in developing countries.

References

Adhikari A, Sen MM, Gupta-Bhattacharya S, Chanda S (2004) Airborne viable, non-viable, and allergenic fungi in a rural agricultural area of India: a 2-year study at five outdoor sampling stations. Sci Total Environ 326:123–141

Aira M-J, Rodríguez-Rajo FJ, Fernández-González M, Seijo C, ElviraRendueles B, Abreu I, Gutiérrez-Bustillo M, Pérez-Sánchez E, Oliveira M, Recio M, Tormo R, Morales J (2013) Spatial and temporal distribution of *Alternaria* spores in the Iberian Peninsula atmosphere, and meteorological relationships: 1993–2009. Int J Biometeorol 57:265–274. https://doi.org/10.1007/s00484-0120550-x

Chakrabarti HS, Das S, Gupta-Bhattacharya S (2012) Outdoor airborne fungal spora load in a suburb of Kolkata, India: its variation, meteorological determinants and health impact. Int J Environ Health Res 22:37–50

Chakrabarty RK, Garro MA, Wilcox EM, Moosmülle H (2012) India: Guwahati one of the most polluted city in the world. Available. http://chimalaya.org/2012/06/28/indiaguwahati-one-of-the-most-polluted-city-in-the-world/. Accessed 30 Nov 2017

CPCB (Central Pollution Control Board) (2009) National Ambient Air Quality Standards (NAAQS), Gazette Notification, New Delhi (http://www.cpcb.nic.in).

Després V, Huffman JA, Burrows SM, Hoose C, Safatov AS, Buryak G, Fröhlich-NowoiskyJ Elbert W, Andreae MO, Pöschl U, Jaenicke R (2012) Primary biological aerosol particles in the atmosphere: a review. Tellus B: Chemical Physical Meteorology 64:15598

Douwes J, Thorne P, Heederik D (2003) Bioaerosol health effects and exposure assessment: progress and prospects. Annal Occup Hyg 479(3):187–200

Elbert W, Taylor P, Andreae M, Pöschl U (2007) Contribution of fungi to primary biogenic aerosols in the atmosphere: wet and dry discharged spores, carbohydrates, and inorganic ions. Atmos Chem Phys 7:4569–4588

Gautam S, Talatiya A, Patel M, Chabhadiya K, Pathak P (2018) Personal exposure to air pollutants from winter season bonfires in rural areas of Gujarat. Expo Health, India. https://doi.org/10.1007/s12403-018-0287-9

Goyer N, Lavoie J, Lazure L, Marchand G (2001) Bioaerosols in the workplace: evaluations, control and prevention guide. IRSST, Occupational Health and Safety Research Institute Robert Sauvé, Montreal, QC, Canada

Gupta A, Gautam S, Mehta N, Mirang Patel, Teratiya A (2019) Association between changes in air quality and hospital admissions during the holi festival. SN Appl Sci 1:163. https://doi.org/10.1007/s42452-019-0165-5

Humbal C, Gautam S, Trivedi UK (2018) A review on recent progress in observations, and health effects of bioaerosols. Environmental International 118:189–193

Humbal C, Joshi S, Trivedi UK, Gautam S (2019) Evaluating the colonization and distribution of fungal and bacterial bioaerosol in Rajkot, western India using multi-proxy approach. Air Quality Atmosphere and Health (Accepted)

Jones AM, Harrison RM (2004) The effects of meteorological factors on atmospheric bioaeroso concentrations—a review. Sci Total Environ 326:151–180

Kakde UB, Kakde HU, Saoji AA (2001) Seasonal variation of fungal propagules in a fruit market environment, Nagpur (India). Aerobiologia 17:177–182

Kallawicha K, Tsai YJ, Chuang YC, Lung SCC, Wu CD, Chen TH, Chen PC, Chompuchan C, Chao HJ (2015) The spatiotemporal distributions and determinants of ambient fungal spores in the greater Taipei area. Environ Pollut 204:173–180

Kulmala M, Asmi A, Pirjola L (1999) Indoor air aerosol model: the effect of outdoor air, filtration and ventilation on indoor concentrations. Atmos Environ 33:2133–2144

Kumar A, Attri AK (2016) Characterization of fungal spores in ambient particulate matter: a study from the Himalayan region. Atmos Environ 142:182–193

Nasir ZA, Colbeck I, Ahmed S, Sultan S (2012) Bioaerosols in residential micro-environments in low income countries: a case study from Pakistan. Environ Pollut 168:15–22

Pasanen AL, Rautiala S, Kasanen JP, Raunio P, Rantamäki J, Kalliokoski P (2000) The relationship between measured moisture conditions and fungal concentrations in water-damaged building materials. Indoor Air 10:111–120

Portnoy JM, Kwak K, Dowling P, VanOsdol T, Barnes C (2005) Health effects of indoor fungi. Ann Allergy Asthma Immunol 94:313–320

Reddy KS, Shah B, Varghese C, Ramadoss A (2005) Responding to the threat of chronic diseases in India. Lancet 366:1744–1749

Ritchkoff A, Viitanen H, Koskela K (2000) The response of building materials to the mould exposure at different humidity and T conditions. Proc Healthy Build 3:317–322

Rohr AC, Habre R, Godbold J, Moshier E, Schachter N, Kattan M, Grunin A, Nath A, Coull B, Koutrakis P (2014) Asthma exacerbation is associated with particulate matter source factors in children in New York city. Air Qual Atmos Health 7:239–250

Schachter EN, Moshier E, Habre R, Rohr A, Godbold J, Nath A, Grunin A, Coull B, Koutrakis P, Kattan M (2016) Outdoor air pollution and health effects in urban children with moderate to severe asthma. Air Qual Atmos Health 9:251–263. https://doi.org/10.1007/s11869-015-0335-6

Timmer LW, Solel Z, Gottwald TR, Ibañez AM, Zitko SE (1998) Environmental factors affecting production, release, and field populations of conidia of *Alternaria alternata*, the cause of brown spot of citrus. Phytopathology 88:1218–1223

Chapter 11
Measurement, Analysis, and Remediation of Biological Pollutants in Water

Uthradevi Kannan, S. Krishna Prashanth and Shihabudheen M. Maliyekkal

Abstract Clean water is vital for supporting human life and the ecosystem. However, the laxity and mismanagement of water resources have endangered the availability of fresh water significantly. Water pollution and associated diseases claim around 2.1 million human lives every year. The outbreak of water-related microbial infections such as diarrhoea, typhoid, and cholera are the primary cause of the loss of lives. Though there has been remarkable progress in the control and prevention of infectious diseases, microbial risks remain a leading cause of human mortality in India, and the rest of the world and children are the worst affected. In this context, a comprehensive analysis of the source, occurrence, fate, and control of biological contaminants in dirking water is of utmost relevance. The rapid and early detection of the pathogenic organism is also of importance in mitigating the menace. This chapter elucidates the growing significance to address the issue of microbial contamination in drinking water and its associated health implications from the past to the present, recent developments in the technologies for the detection, analysis and the remediation of pathogens in the water.

Keywords Analysis · Biological pollutants · Disinfection · Measurement · Safe water

11.1 Introduction

11.1.1 Microbial Hazards: Growing Concern

Freshwater is a complex resource and is linked to almost everything in the world. Its adequate availability at the point of use is a precondition for the existence of humankind and the sustainability of the planet. Quality and quantity are the two significant attributes of this indispensable resource. Though the earth is covered with

U. Kannan · S. Krishna Prashanth · S. M. Maliyekkal (✉)
Department of Civil and Environmental Engineering, Indian Institute of Technology Tirupati (IITT), Tirupati 517506, AP, India
e-mail: shihab@iittp.ac.in

T. Gupta et al. (eds.), *Measurement, Analysis and Remediation of Environmental Pollutants*, Energy, Environment, and Sustainability,
https://doi.org/10.1007/978-981-15-0540-9_11

70% of the water, only 2.53% of the water is freshwater and in which <0.1% is available for human consumption (USBR 2017).

According to a United Nations (UN) report, the water consumption rate has increased twice than the population growth (Un, n.d.). The Global monitoring body, World Health Organization (WHO) and United Nations International Children's Emergency Fund (UNICEF) joint monitoring programme for water supply, sanitation and Hygiene (JMP) has established reports on current scenario of issues in water supply, sanitation and hygiene, since 1990. The projections to the next few decades in context to water stress and scarcity across the globe is assessed and evaluated by these organizations (Fig. 11.1).

The 2017 update of WHO/UNICEF Joint Monitoring Programme for Water Supply, Sanitation and Hygiene (JMP) reports that in 2015, 2.1 billion individuals lacked "safely managed drinking water" as shown in Fig. 11.2 (UNICEF 2017). The figure also shows that 11% of the global population (844 million people) lacked even a basic drinking water service. Globally, there are about 423 million people collect water from unprotected groundwater, and 159 million people use surface water directly. Cities and towns pose a special and unique water challenge, as they are expected to be home for 66 percent of the world's population by 2050 (UN 2018). The impact of poor water quality on people relying on these sources not only limits the access to safe water but also increases the threat to human health.

Globally, 80% of wastewater flows back into the ecosystem without being treated, and 1.8 billion people use untreated water supply as the source of drinking water, putting them at risk of contracting waterborne diseases (UNESCO 2017).

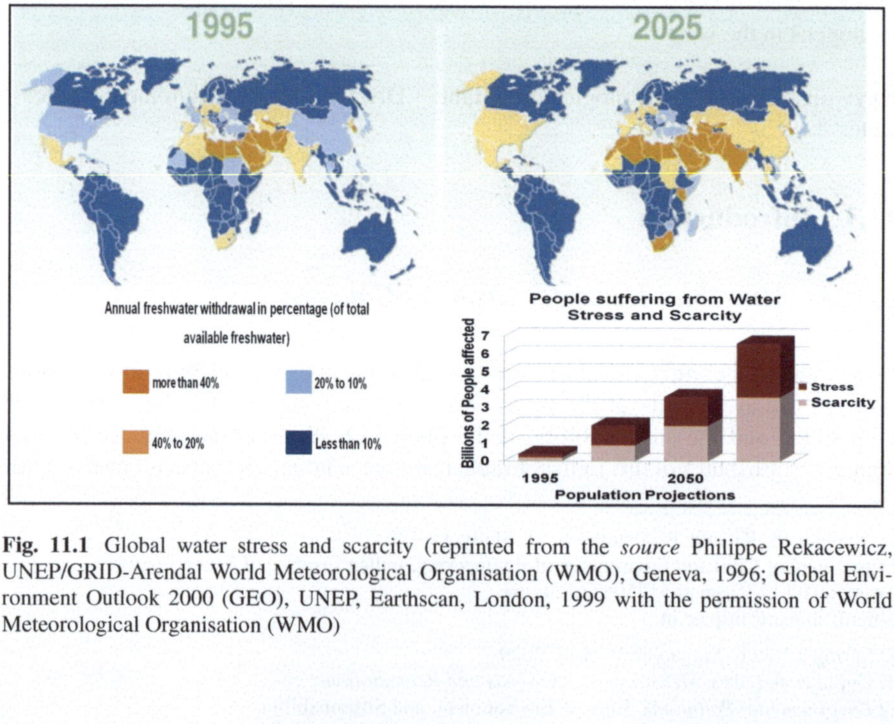

Fig. 11.1 Global water stress and scarcity (reprinted from the *source* Philippe Rekacewicz, UNEP/GRID-Arendal World Meteorological Organisation (WMO), Geneva, 1996; Global Environment Outlook 2000 (GEO), UNEP, Earthscan, London, 1999 with the permission of World Meteorological Organisation (WMO)

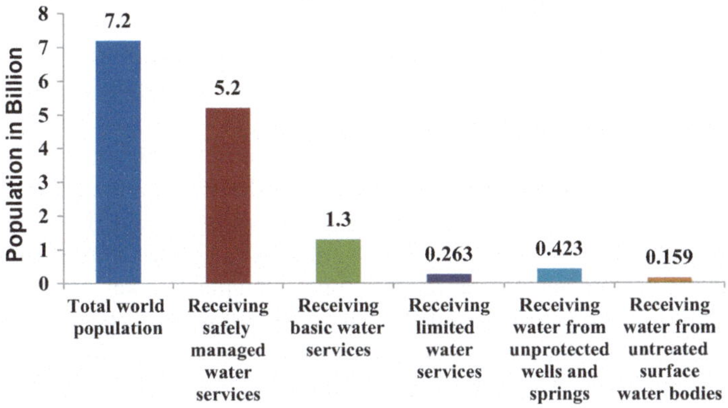

Fig. 11.2 Global drinking water coverage (adapted from the *Source* WHO/UNICEF joint monitoring programme report on water supply, sanitation and hygiene (JMP), 2017 with the permission of WHO)

Water-related illnesses including, dysentery, typhoid, cholera, and schistosomiasis, are prevalent across developing countries, and cholera contributes top of the list outbreaks in 132 countries (WHO 2016). According to a report published in 2015, the waterborne disease remains as an increasing threat among vulnerable and disadvantaged groups across the globe, especially in low-income nations, where 1 in 25 persons is affected due to diarrhoeal diseases. The estimates shows 60% of children under the age of five are affected (UNICEF 2016). Now, waterborne diseases stand as the leading cause of disease and death and accounting 3.4 million loss of lives worldwide (WHO 2001). WHO (2016) reports that diarrhoea is one of the top ten global reasons for death and the second leading cause of death in low-income countries. Around 1.8 million deaths occurred due to diarrhoeal diseases worldwide, and the crude death rate is 58 per 100,000 populations in low-income countries (WHO 2004). Every year around 1.3 to 4 million cholera cases and 21,000 to 143,000 associated deaths are reported worldwide (WHO 2016). According to Global Estimates, there are 20 million reported cases of hepatitis E virus (HEV) infections, which also includes 3.3 million symptomatic cases of hepatitis E infection. The HEV caused approximately 44,000 mortalities in the year 2015, out of which 3.3% were due to viral hepatitis (Rein et al. 2012). As per the Environment and Health Information System, around 13,548 children (0–14 years old) dies in Europe every year due to waterborne diseases (WHO 2007). These statistical facts demonstrate that presence of biological pollutants in water, among others, is a growing concern in both developing and developed countries and it requires adequate attention to meet the safe drinking water needs of the population.

Since the inception of identifying the reasons for human health deterioration due to microbial contamination in water, there have been various theories and subsequent experimental findings to understand the occurrence, health impact, and fate of pathogens in water. For the past few decades, several efforts have been taken towards the development of robust, efficient, and affordable detection and remediation technology in curtailing the waterborne diseases caused by pathogens. Technology to mitigate the issue can be categorised into detection specific and remediation specific.

This chapter is composed primarily to enhance the knowledge on the importance of microbial contamination in drinking water by understanding the occurrences and sources of pathogens. A detailed review of literature is presented on the growing trends in the field of pathogenic detection techniques and remediation technologies from the past to the recent highlighting their principle, mechanism, applications, and limitations through illustrations and discussions thereof. The effort has also been taken to discern the real challenges in implementing these technologies, which may eventually be utilized in bridging the gap between the lab and the field.

11.1.2 Occurrences and Sources of Microbial Contaminants in Water

Presence of microbial pollutants such as bacteria, virus, protozoa, and helminths pose a severe threat to the quality of freshwater. The typical characteristics, source, and impact on human health of these organisms are summarised in Table 11.1. Their occurrences in water bodies vary depending on several factors. These include various chemical and physical characteristics of the catchment area, the intensity and extent of anthropogenic activities, and the domestic animal discharge. However, the human activities such as discharge of untreated or partially treated municipal wastewater, poor sanitation and hygiene, open defecation, industrial and agricultural wastes, and solid/semisolid refuse are the major sources of concern (Planning Commission 2002). A schematic illustration of routes of microbial contamination in water is shown in Fig. 11.3.

As per the 2002 Planning Commission report, there is a higher threat of waterborne diseases in rural areas caused by water contamination due to poorly maintained water and sewer networks, unscientific disposal of solid wastes, poor healthy sanitation and personal hygienic practices. The discharge of urban sewage is identified as a major source of contamination of Indian surface waters (Murty and Kumar 2011). The estimate shows 80% of surface water bodies in the country is polluted by domestic sewerage (Dey 2015). The Arya et al. (2019) report reveals that India produces 61,948 million litres per day (MLD) of urban sewage. The data also shows that more than 70% of the sewage is let out into the environment untreated (Arya et al. 2019).

Groundwater is a prominent source of drinking water to at least 50% of the population worldwide that also accounts for 43% of the water utilized for irrigation (Faures et al. 2001). Unlike surface water, the sub-surface water is considered less vulnerable to microbial pollution due to the barrier effects provided by the covering soil. There are higher chances of subsurface water contamination when these overlaying barriers are breached, allowing exposure to underground pollution sources,

Table 11.1 Causes, health symptoms, and transmission characteristics of some waterborne diseases of concern

S. No.	Disease caused	Causative agent	Predominant symptoms	Latency	Persistence	Infectivity	Ability to multiply
1.	Amoebiasis	Protozoan (*Entamoeba histolytica*)	Abdominal discomfort, fatigue, diarrhoea, weight loss, and flatulence	No	Low to medium	High	No
2.	Cholera	Bacterium (*Vibrio cholerae*)	Vomiting, occasional muscle cramps, and watery diarrhoea	No	Medium to high	Medium to low	Yes
3.	Cryptosporidiosis	Protozoan (*Cryptosporidium parvum*)	Abdominal discomfort and diarrhoea	No	Low to medium	High	No
4.	Hepatitis	Virus (Hepatitis A)	Abdominal discomfort, fever, chills, pain, dark urine jaundice	No	Low to medium	High	No
5.	Typhoid fever	Bacterium (*Salmonella typhi*)	Headache, fever, appetite loss, nausea, constipation, diarrhoea, vomiting, and appearance of abdominal rash	No	Medium to high	Medium to low	Yes
6.	Giardiasis	Protozoan (*Giardia lamblia*)	Diarrhoea, abdominal discomfort	No	Low to medium	High	No

Adapted from WHO (2011)

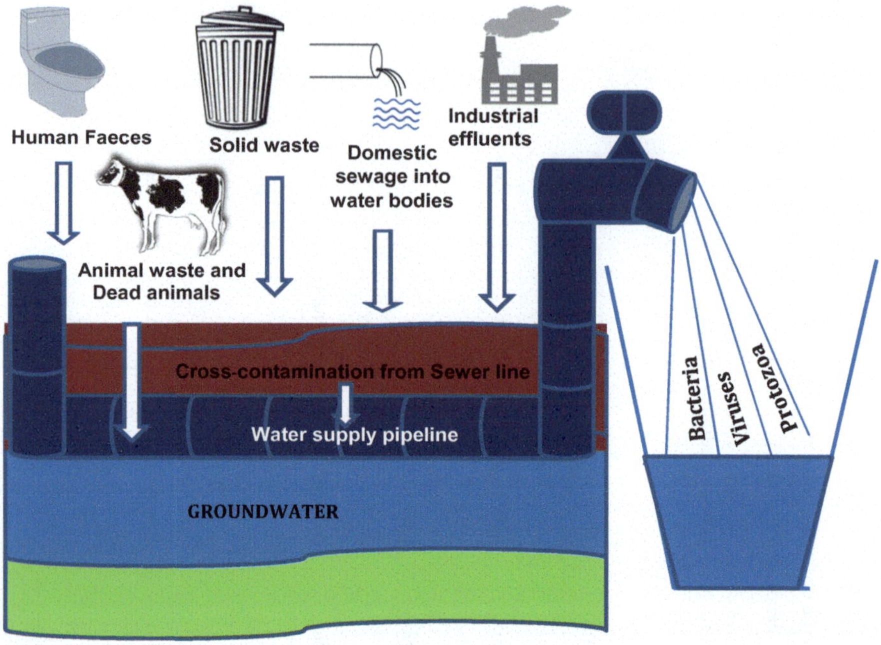

Fig. 11.3 Major sources of microbial contamination of drinking water

such as soak pits, toilets, and sewer lines containing municipal or commercial or industrial wastes. Though the typical presence of human enteric organism is less in groundwater, the pathogens of concern are faecal viruses, which has the potential to enter groundwater system through the porous soil matrices due to their relatively small size. The correlation of groundwater contamination to the global occurrences of the waterborne disease cannot be made typically, as there are various transmission routes. So the exposure-risk relationships are often unclear. However, it is established that several groundwater sources are contaminated with pathogenic organisms and is also responsible for waterborne diseases (Rivera-Jaimes et al. 2018). A study shows that water samples collected from the wells in proximity to the sources of untreated wastewater had higher counts of coliform and faecal coliform, making it unsuitable for both drinking and irrigation purposes (Blumenthal et al. 2000). Saha et al. (2018) and Dey et al. (2017) reported that in northwest Bangladesh, the shallow aquifers are microbiologically contaminated than deep aquifer (Saha et al. 2018; Dey et al. 2017). A study conducted in Kanpur, India documented waterborne disease at an incidence rate of 80.1 per 1000 population (Trivedi et al. 1971). Amongst the shallow wells used by the residents as a source of drinking water, 70% were found to be contaminated by pathogens.

Rural areas of the developing countries using groundwater as a source for drinking are more vulnerable to waterborne diseases than the ones using piped water supplies. The bacteriological quality analysis including Total Coliform, Faecal Coliform, and Faecal *streptococci* showed that the collected groundwater samples from Triffa aquifer basin, Eastern Morocco were contaminated due to unprotected septic tanks and the wastewater dumped in the upstream end of river Cheraa wadi (Yahya et al. 2017). An experimental investigation made by Venkatesan et al. (2014) to study the impact of flooding on microbial contamination in groundwater at Chennai, India after a major flood event revealed higher counts of coliform in subsurface water sources at most affected areas. The rapid escalation in the microbial growth was attributed to the contaminated storm water runoff entering into Adyar River, Tamil Nadu, India (Gowrisankar et al. 2017). The residences with on-site septic systems were likely more affected due to the infiltration of the contaminated river into the groundwater sources (Jamieson et al. 2003).

The increase in levels of pathogenic contamination in estuaries and marine environment is also poses a threat to public health. In context to marine water, the prominent reasons for microbial contamination are failures in septic systems, discharges of sewage from shoreline outfalls, farm animal wastes, and runoff from naturally vegetated areas. The storm water runoffs from urban, commercial, and industrial lands, the practice of open defecation near coastal areas are also responsible for contamination of marine waters (Pandey et al. 2014). World Ocean Network reports that around 90% of wastewater and 70% of industrial waste is being discharged into oceans by the developing countries (Vandeweerd et al. 2002). Estuaries located adjacent to residential areas, when used as a mode of transportation and for recreational activities can cause a significant impact on pathogen levels (Schriewer et al. 2010). The pathogens, including *Salmonella, Vibrio cholerae*, *Cryptosporidium, Giardia,* and *Campylobacter spp.* are reported in estuaries (Rhodes and Kator 1990).

The other potential source of pollution is due to the regrowth of microbes in the water distribution network (Shaheed et al. 2014). In the United States, around 10% of outbreaks are caused by contaminated water due to the improper water distribution network (Craun et al. 2010). The corrosion and poor surface finish in the water supply pipelines enables enhanced colonization of microorganisms and the formation of biofilms (Rakić 2018). These biofilms can act as a short or long-term habitats for pathogenic organisms, such as *Pseudomonas aeruginosa*, *Escherichia coli*, *Legionella* spp., *Campylobacter* spp., noroviruses, adenoviruses, rotaviruses, and parasitic protozoa (*Cryptosporidium parvum*) (Wingender and Flemming 2011). The emergence of new pathogens, mutants of the existing pathogens, and the presence of multi-drug resistance species are also reasons for concern.

11.1.3 Transmission of Waterborne Diseases

It is a fact that microbial hazard is a principal cause of human mortality in the developing world. There are various groups of pathogenic microorganisms, and they have different modes of transmission, as is shown in Fig. 11.4. Drinking water is

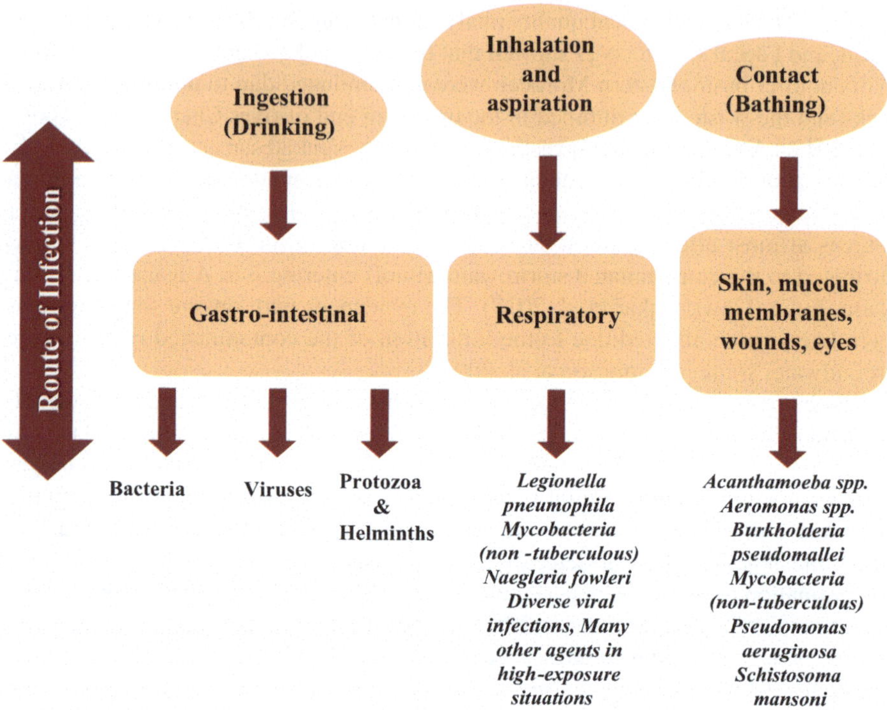

Fig. 11.4 Schematic representation of transmission pathways of waterborne microorganism with examples. *Adapted from the source* Guidelines for drinking-water quality, 4th edition, with the permission of WHO

observed to be the only carrier of the faecal-oral route of pathogenic transmission (WHO 2011).

The transmission characteristics of pathogens can be categorized based on latency, persistence, infectivity, and the ability to multiply. Latency is the lag time between excretion of a pathogen and the stage at which it becomes infective to a new host. Typically, protozoa, bacteria, and enteric viruses have no latent period (Feachem et al. 1983). The most of helminths require a different latent period either for eggs to progress to the transmittable stage or to pass through an intermediate form to complete their life cycles (Cotruvo et al. 2004). Persistence is determined by the span of time that a pathogenic organism exists in the environment outside a human host in viable condition. Persistent microbes can travel through a prolonged route, viz., through a sewage treatment system and can still be infectious to human living away from the original host. In general, persistence increases in the order: bacteria > protozoa > viruses > helminths, whose persistence is measured in months. The infective dose refers to the microbial concentration that can cause infection upon ingestion. Usually, the minimum infective doses for viruses and protozoa are less than that of bacteria.

A summary of various waterborne diseases linked to the protozoa and helminths, its relative infectivity, and persistence in water are discussed in Table 11.1. The pathogens that are not listed in Table 11.1 can also transmit by water, and the list is not complete.

11.1.4 Drinking Water Safety Guidelines

The purpose of the disseminated WHO standard guidelines is to enable countries and regions to develop their own standards conforming to the regulation. It suggests that immediate action must be taken if *E. coli* is detected in drinking water. Monitoring the levels of *E. coli* and faecal coliforms is a common method in the quantification of the pathogen loads in water bodies (Feachem et al. 1983). For decades, public health experts and scientists have assessed water quality in rivers, estuaries, and coastal waters in terms of faecal coliforms and *E. coli* (Pandey and Soupir 2013). However, *E. coli* cannot predict the existence of all pathogenic organisms. For example, *Cryptosporidium* oocysts may survive chlorine disinfection and may be present in the absence of *E. coli*, showing the limitation of using *E. coli* as a potential indicator for faecal contamination. However, *E. coli* is the designated WHO indicator for reliable diagnosis of microbial quality of the water (WHO 2011). The guideline values for assessing the microbial quality are given in Table 11.2.

Despite of establishing definite standard regulations for safe supply of water, the concern of detecting and monitoring these pathogens in water samples collected from various sources have been still a challenging task for both developed and developing countries. The following discussions will describe the existing methods of monitoring and removing pathogenic organisms in water.

Table 11.2 Guideline values for verification of microbial quality

Organisms	Guideline value
All water directly intended for drinking *E. coli* or thermo-tolerant coliform bacteria	Must not be detectable in any 100-ml sample
Treated water entering the distribution system *E. coli* or thermo-tolerant coliform bacteria	Must not be detectable in any 100-ml sample
Treated water in the distribution system	Must not be detectable in any 100-ml sample

Adapted from the *source* Guidelines of Microbial quality, WHO (2011)

11.2 Detection and Analysis of Pathogenic Organisms in Water

Efficient testing and fast detection of pathogenic organisms are vital in the management of water-borne illness. It is the main checkpoint in eliminating the pathogens in drinking water, food, and other biological samples. It also pays a significant role in diagnosing and preventing diseases (Vidic et al. 2017). A typical detection technique should be sensitive, rapid, and affordable. There are several routes, such as culture/growth, optical, molecular, and bio-sensing based are used to detect the pathogenic organism in these samples. The major methods under each detection technique are briefly discussed below. Table 11.3 presents a summary of the testing methods.

11.2.1 Culture/Growth Based Method

Culture or growth based technique is a traditional method employed for the detection of pathogenic organisms in the water. The majority of the testing for bacteria detection is done through this conventional approach. It involves growing and isolation of organisms on Petri-plates containing growth-media, followed by biochemical tests to confirm the presence of pathogenic microorganisms. It is a time-consuming technique, and typically take 5–7 days to obtain the results (Rajapaksha et al. 2019). It is not suitable to detect organisms which are viable but present in the non-culturable state. However, the traditional culture-dependent method is regarded as the standard method for the detection of pathogens, and it is still being used as a regulatory requirement by water treatment companies and laboratories to monitor the microbial quality of drinking water (American Public Health Association, American Water Works Association 1989). The estimation of the most probable number (MPN) or multiple fermentation is a commonly practiced growth-based method to find the concentration of viable microorganisms in the water sample. It is a statistical method that relies on the principle of extinction dilution for testing the quality of water and assesses its suitability for human consumption. The technique typically identifies the presence of an indicator organism of faecal origin to establish the existence of pathogenic microorganisms (Munoz and Silverman 1979). It works based on the principle of fermentation of lactose to produce the acid as well as gas. The presence of coliform is showed by the colour change of the medium, by a change in pH, or by the collection of gas in inverted Durham's kept in the test medium. The total coliforms can be determined by counting the number of tubes showing both colour change and production of gas (Fung and Miller 1970). The MPN analysis is usually performed in 3 steps, including presumptive, confirmatory, and completed test. The presumptive test is the first step and is carried out to identify the presence or abscess of the coliform organism in the water sample. If this screening test is negative, the water sample is considered free from pathogens. If the test is positive, further confirmatory analysis is required to ratify the faecal origin of the coliform organisms. The completed test is performed to check and eliminate the false positive test. These steps are illustrated in Fig. 11.5.

The method is time-consuming and takes normally up to 72 h for obtaining the results. The development of Membrane Filtration (MF) technique shortens the process and reduces the completion time to 24 h. In this technique, the sample is allowed to pass through the membrane filter (pore size of 0.45 μm), and the membrane containing the trapped bacteria is transferred on to a Petri-plate containing the nutrient agar medium. The results were obtained by counting the bacterial colonies, which is grown on the incubated Petri-plates. The number of bacterial colonies grown on the agar medium is counted and is represented as CFU/ml.

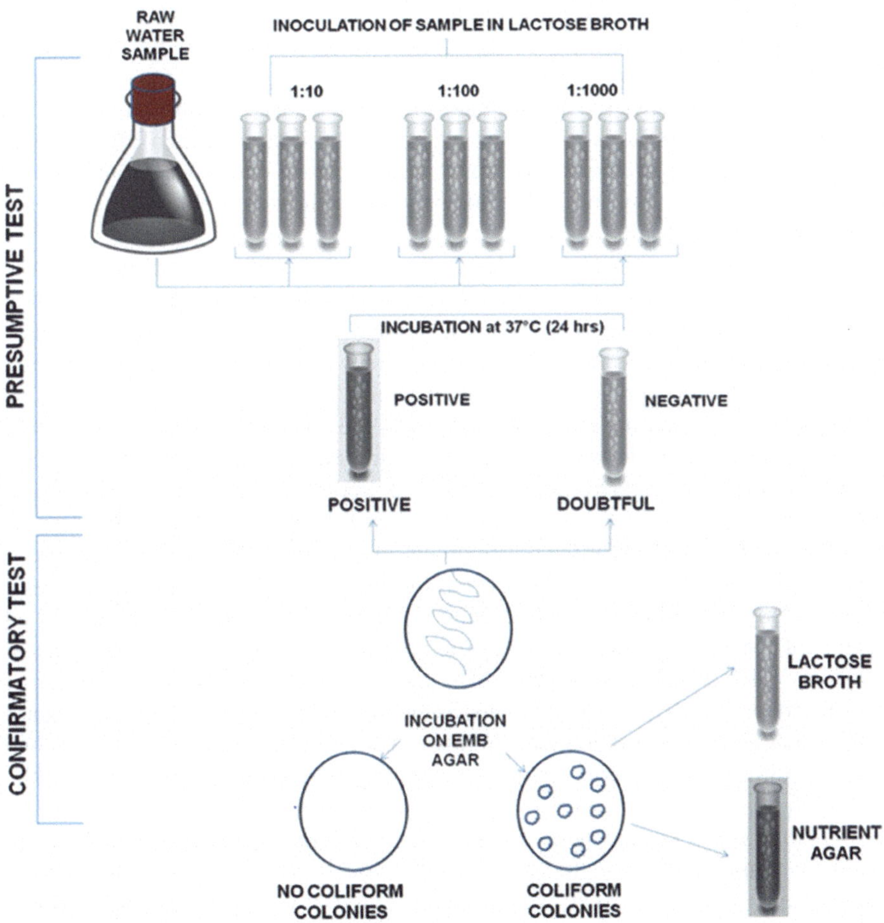

Fig. 11.5 Schematic representation of MPN test

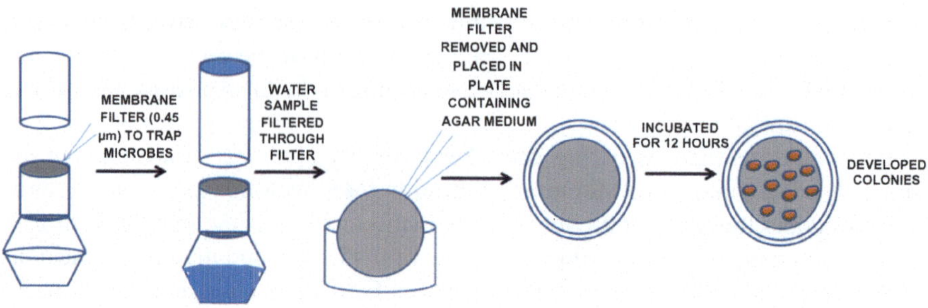

Fig. 11.6 Schematic representation of membrane filtration method for the detection of pathogens

The technique is highly effective for assessing the performance of chlorination as it removes the bactericidal agents through filters (Tankeshwar 2010). The process flow diagram is presented in Fig. 11.6. Apart from standard nutrient agar medium, chromogenic agar medium such as Eosin Methylene blue agar (Leininger et al. 2001) and Macconkey agar (March and Ratnam 1986) is also used for differentiating coliforms from faecal coliforms and isolation of members of family Enterobacteriaceae, respectively. Furthermore, agar medium is available with chromogenic and fluorogenic substrates that helps in the fast and real-time detection of total coliforms and *E. coli* (Manafi and Kneifel 1989).

Pathogenic viruses can also be detected using the culture-based method. The method involves inoculation of virus stock aliquots onto the medium containing susceptible cell monolayers followed by incubation. The inoculated virus gets attach to the cells, and these infected cells release progeny of virus that forms the circular zone of infected cells over the medium called plaque. The result is expressed in plaque forming units (PFU) per ml (Dulbecco and Vogt 1953). Though culture based method are widely used, the technique is time-consuming and requires a lot of resources, including various laboratory equipment and a skilled workforce. The safety concern and less sensitivity of the test may limit the use of the said method in some cases. Therefore, there is a need for the development of rapid and easy to use techniques.

In drinking water treatment and distribution systems, the bio-stability of the water is assessed by assimilable organic carbon (AOC), which represents the dissolved organic carbon that is assimilated by the microbes present in the drinking water (Kooij 1992). In water distribution system, AOC can be correlated with the presence of biofilm and regrowth of microorganisms. Typically, AOC is analysed by the standard method, and it is explained in brief as follows. The AOC determination involves two steps: (i) culturing of microbes and (ii) enumeration of microbes. In the first step, the water sample is inoculated with test microorganisms such as *Pseudomonas sp.* (P-17) and *Spirillum sp* (NOX) and incubated at 15 °C for 9 days. Once the microbial growth attains the stationary phase, the cells are enumerated using the plate count on the agar medium (Tang et al. 2018; Hammes and Egli 2005). The net microbial growth is related to the growth of test organism on acetate (P-17) and oxalate (NOX). The results are represented as acetate-C equivalents. The bio-stable water should have an

AOC concentration of 10 μg/L acetate-C equivalents, which depends on the available chlorine in the water (Kooij 1992).

11.2.2 Optical-Based Methods

The culture-based technique is not suitable for non-culturable organisms. The method is time intensive and laborious. On the contrary, the optical-based methods are nonculture-based technique and is simple, fast, and less costly. The optical-based microscopic techniques are used to visualize the size and morphology of the bacterial cells. However, most of the microbes lack colour and contrast which makes its visualization more difficult. In such cases, the incorporation of fluorescent dyes and stains can help in overcoming the above-said limitations (Claus 1992). The stains are made of salts containing positive or negative ion depends on the chromophore. Typically, the negatively charged bacterial cell wall sticks to the positively charged chromophores, which makes the visualization of microbes under light microscopy easier. The commonly used dyes are safranin, methylene blue, malachite green, eosin, eosin, fuchsin, rose bengal, and crystal violet (Microbiology L.l 2019). The traditional light microscope uses light (400–700 nm) for the illumination to magnify the bacterial cells in the sample. On the other hand, the fluorescence microscope uses much higher light intensity to excite the sample of interest that contains fluorescent dyes. Here, the fluorescent microscope contains the filter cube set that allows the radiation of a wavelength which matches with the fluorescing compounds (Bradburry 1996). The fluorescent microscope enables the real-time detection of dead and live bacterial cells using DEAD/LIVE bacterial viability kit, which contains the fluorescent stains (Boulos et al. 1999). The refinement in the field of microscopes paved the way for the development of differential interference and phase contrast microscope (Keevil 2003), confocal laser scanning microscopes (CLSM), and total internal reflection fluorescence microscopes (TIRF) for the visualization of microorganisms. The phase contrast microscopes are used to visualize the microbes without staining (Keevil 2003). The CLSM (Sheppard et al. 1997) and TIRF (Axelrod 2001) are mainly used to image the structural components of cells, genetic material present inside the cells, and the specific cells within major.

Recently, adenosine triphosphate (ATP), an indicator of the presence of microbial growth, based optical detection method has been developed a potential (Selan et al. 1992). The bioluminescence, a light emission due to chemical reactions in the organism, forms the basis of this detection technique. In this method, a buffering agent that lyses the cell wall of the bacteria is added to the water sample, which is concentrated by membrane filtration. This buffering agent releases the ATP, and the concentration of ATP is measured by light emission intensity (580 nm) produced via luciferin-luciferase assay. The activity of the assay is standardized against known concentrations of ATP, and the results are represented as Relative Light Unit/ml (RLU/ml) (Turner et al. 2010). This ATP based measurement was also used to check the efficiency of the treatment processes, such as ozonation, UV treatment, and chlorination with respect to bio-stability of treated water. This method also acts as the best surrogate for measuring the growth of biomass and for determining the biomass production potential (van der Wielen and van der Kooij 2010). Though the culture-based methods are convenient, simple to perform, low-cost, and rapid, the limited representation of microbial communities is a limitation.

Recently, flow cytometry (FCM), an optical based detection method, is used to identify the individual microbial cells presents in a complex microbial community (Basiji et al. 2007). The working mechanism of FCM is given in the following steps: (i) the microbes present in the suspension is allowed to pass through a laser beam, (ii) the cells present in the suspension are scattered by light, and the fluorogenic substrates are excited to produce emission. The scattered light is captured at a low angle (forward scattering) and high angle (sideward scattering). The fluorescence is detected by using the selective wavelength filters. The method can be used to find the size, shape, and the number of microorganisms present in the sample. The schematic representation is shown in Fig. 11.7. Unlike fluorescence microscopy, FCM does not produce the images, rather the characteristic feature of each cell are presented as histograms or dot plots. The FCM is highly sensitive (<100 cells/ml) and rapid technique (<3 min per sample) (Hammes and Egli 2010). Along with the bacteria, yeast cells (Díaz et al. 2010), algae (Dubelaar and Gerritzen 2000), viruses (Brussaard

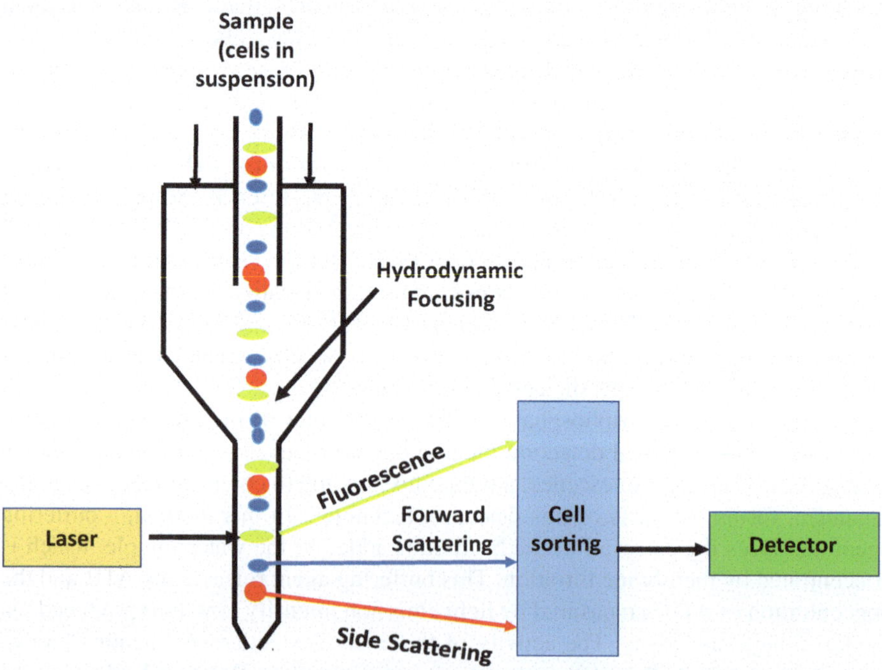

Fig. 11.7 Schematic representation of flow cytometry for the detection of pathogens

et al. 2000), and protozoa (Vesey et al. 1994) present in the water samples can also be detected by FCM (Ambriz-Aviña et al. 2014). The FCM is a single cell technique, and it cannot be used for analysing biofilms present in the drinking water sample. Though there are improvements in the field of optical based detection techniques for the identification and enumeration of the pathogens, the disparity in the order of magnitude of the total number of cells between microscopic counting and plate counting methods needs further attention.

11.2.3 Molecular Based Methods

The molecular-based methods are more sensitive, reliable, robust, and yield conclusive results (Derveaux et al. 2010). The technique is suitable for the detection of a broad spectrum of microorganisms, including emerging pathogens. Unlike conventional techniques, this process functions by detecting specific ribonucleic acid (RNA) or deoxyribonucleic acid (DNA) sequences in the target organism (Garibyan and Avashia 2013; Law et al. 2015). The method allows for simultaneous detection and identification of pathogenic microorganisms.

Polymerase chain reaction (PCR) is one of the popular molecular-based detection ways in practice. It was invented in the late 20th century by Kary Mullis, and he was awarded Nobel Prize in chemistry for the invention in 1993 (McPherson and Møller 2000). It is very sensitive and can detect single bacterial pathogen (Velusamy et al. 2010). It involves amplification of a primer mediated enzymatic DNA and creation of specific DNA fragments (Valones et al. 2009). The amplification typically occurs in a cyclic three-step process (Law et al. 2015). These include DNA melting, annealing, and extension. In DNA melting or DNA denaturation, a double-stranded DNA is physically separated to two-pieces of single-stranded DNA at elevated temperatures (90–97 °C). In the next step, the oligonucleotides or specific primers anneal (50–60 °C) and bind to the complementary sequences of DNA. The two DNA strands then form a template for DNA polymerase to synthesise a new DNA strand. In the finals step, the new DNA strand is used as a template to create the duplicate copies, and the original DNA template is amplified exponentially through a chain reaction.

The developments in the PCR based detection techniques include cold PCR (Milbury et al. 2011), heat pulse extension-PCR (HPE-PCR) (Orpana et al. 2013), and nanoparticle-PCR (Ma et al. 2013). But the limitations with PCR is that it cannot differentiate between live or dead cells, and it will produce false results if there is any contamination in the sample. So, the PCR technique may not be useful for the detection of pathogens present in the wastewater sample. To eliminate the lack of differentiation between live and dead cells, the reverse transcriptase PCR (RT-PCR) was developed (Cangelosi and Meschke 2014). The pathogenic viruses can also be determined by RT-PCR and real-time PCR (Mattison and Bidawid 2009). Recently, multiplex PCR assay has been developed by the researchers to detect 10 viruses in a single tube (Pham et al. 2010; Wolf et al. 2008). Furthermore, advanced molecular methods such as usage of DNA based fluorescent probes and Enzyme-Linked

Immunosorbent Assay (ELISA) could be used for the identifying pathogenic species (Kittigul et al. 2001).

11.2.4 Bio-sensing Based Methods

There is an growing demand for a versatile and sensitive technique that detects pathogens in a rapid manner. Biosensors are devices that work based on the detection of signals produced by the interaction between the bio-recognition elements, such as enzymes, antibodies, aptamers, oligonucleotides probes, nucleic acids, and cell-surface molecules (Rider et al. 2003), and the target analyte species (Zourob et al. 2008).

Here, the biological response is converted to electrical signals, and it is recorded by a detector (Brindha et al. 2018). The schematic diagram showing the working principle of bio-sensors is given in Fig. 11.8. A comparative analysis of various types of biosensors is presented in Table 11.3. In addition to the existing bio-recognition elements, functional nanomaterials are also being used for the detection of pathogens (Krishnan et al. 2019). The method has a few limitations. The credibility and sensitivity of bio-sensors are affected due to the interference caused by organic/inorganic molecules and other contaminants, such as humic substances with the microbes. The natural receptors such as antibodies and enzymes that are immobilized at the transducer surface are prone to the degeneration, which results in loss of selectivity and

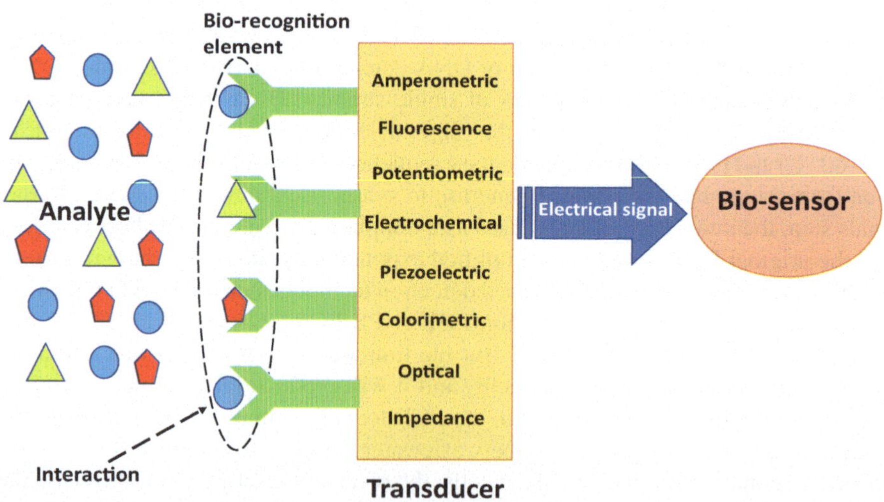

Fig. 11.8 Schematic representation of working mechanism of bio-sensor (courtesy of: http://agscientific.com/blog/2016/01/biosensors-and-diabetes-trending-applications)

Table 11.3 Summary of various detection technologies

Technology/methodologies to detect pathogens	Advantages	Disadvantages
Optical based methods		
Microscopic methods	Can differentiate the pathogens based on the structure; Faster technique	Requires sophisticated instruments
Flow cytometry	Simultaneous analysis of physical and characteristics of cells up to 1000 particles/se, rapid analysis, provides the finger print of microbial community in water sample	Expensive
ATP measurements	Fast: low-cost kits are available	Requires sophisticated instruments
Culture/growth based methods		
Heterotrophic plate count	Reliable results	Time consuming; require well equipped laboratories; expensive; viable but non culturable bacteria can yield false negatives
Assimilable organic carbon	Widely used in drinking water research; can assess the potential growth of biofilms	Quantify the nutrients instead of bacteria, it assumes the growth of bacteria is limited by organic carbon, not applicable for all types of bacteria
Enzyme catalysed multiple tube fermentation or most probable number (MPN) technique	Easy interpretation; effective method for analysing samples with high turbidity	Time consuming; not very accurate; requires a of resources
Molecular method		
Quantitative PCR (Q-PCR) or Real Time PCR (RT-PCR)	Rapid; high sensitive and quantitative; accurate gene quantification; target specific	Difficult to obtain good quality RNA in RT-PCR
Bio sensing method		
Bio sensing based Method	Rapid, sensitive	Costly, antibodies are so sensitive, limited stability of fluorophores

Table 11.4 Comparison of bio-sensing technologies

S. No	Biosensing techniques	Pathogens	Detection limit	Detection time	Reference
1.	Electrochemical biosensor	*E. coli* O157:H7	10 to 10^6 CFU/mL	45 min	Wang and Alocilja (2015)
2.	Electrochemical genosensor	*S. Typhi*	50 pM	Not given	Das et al. (2014)
3.	Potentiometric biosensor	*(i) C. Parvum*	5×10^2 oocysts/mL	60 min	Laczka et al. (2013)
		(ii) V. Cholera	10×10^{-9} to 10×10^{-6} M		
4.	Amperometric biosensor	*E.coli O157:H7*	10^2 CFU/mL	–	Xu et al. (2016)
5.	Impedance based biosensors	*L. Monocytogens*	30 CFU/mL	–	Chen et al. (2015)
6.	IR and Raman spectroscopy based biosensor	*Pseudomonas aeruginosa and E. coli*	10^1 to 10^6 PFU/ml		Al-Qadiri et al. (2008)
		Polio virus, Staphylococca, andl enterotoxin B	1.3 pg/ml	6 h post infection (hpi)	Lee-Montiel et al. (2011)
		E. coli and S. aureus	10^3 cells/ml	–	Wang et al. (2016a, b)
7.	Fluorescence based biosensor	*E. coli, E. aerogens, E. dissolvens, S. aureus*	30 CFU/ml	< 120 min	Dogan et al. (2016)
		Salmonella enteritidis			
8.	Colorimetric biosensor	*E. coli and S. aureus*	30 CFU/ml for *E. coli* & 200 CFU/ml for *S. aureus*	–	Thakur et al. (2015), Safavieh et al. (2014)
9.	Piezo-electric biosensors	*E.coli 0157: H7*	2×10^3 CFU/ml 10^7 to 10^9 CFU/ml	–	Wang et al. (2008)
		S. typhimurium	10^6 to 10^{10} CFU/ml	<30 min	Babacan et al. (2000)

sensitivity. These challenges may be addressed by replacing the natural bio recognition elements with bio-mimetic elements, such as aptamers, peptides, and molecular imprinted polymer that can enhance the sensitivity of the detection (Kumar et al. 2018).

11.3 Remediation of Microbial Contaminated Water

11.3.1 A Brief Review of Disinfection Technologies

Access to clean water is a fundamental human right and is essential for a healthy life. The relationship between the quality of water and health are well documented. One of the primary reasons for the loss of human lives in most of the countries is the consumption of contaminated water. However, providing safe water to every person is a challenging task due to the increasing presence of pollutants and the growing gap between demands and supply. There are several classes of pollutants reported in drinking water and are presented in Fig. 11.9. Among these microbial pollutants requires special attention due to their widespread occurrence and potential to cause adverse ill effects on human health. A brief review of various aspects of the treatment of pathogens is presented below.

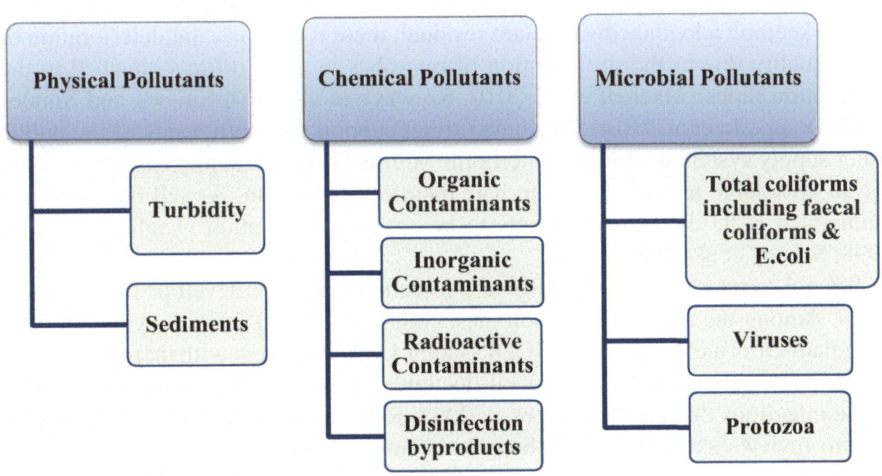

Fig. 11.9 Classification of various pollutants in drinking water

11.3.2 Centralised Versus Point-of-Use Technologies

The goal of disinfection is to alleviate the pathogens responsible for waterborne diseases. Several treatment technologies have been employed to achieve the goal. However, several factors govern the selection of disinfection technologies. These include the ability of the disinfectant to kill a broad spectrum of pathogenic organism, the capacity to provide residual disinfection activity, affordability, the formation of disinfection by-products, and the aesthetic quality of the treated water (National Research Council (US) Safe Drinking Water Committee 1980). The scale of operation is another critical parameter that decides the success of the treatment process. Treatment may be done at a small scale in decentralized plants or can be done at community level at a centralized facility. However, establishing large community water treatment systems is challenging for developing nations. The challenges include capital investment, skilled labour, and governance, access to appropriate technologies, piped water supply networks, water scarcity, maintenance, and recontamination. The poor success in overcoming these challenges has increased the popularity of decentralized or point-of-use (POU) water treatment systems. In developing countries where only limited households have piped water (Kanungo et al. 2010), POU interventions seem to be a sustainable way of providing safe drinking water.

Though technologically advanced countries can afford to use a complex system to meet the stringent regulations, their public water system is also not completely free from pathogenic organisms. Several incidences of the pathogenic microorganism in piped water are reported from developed countries (EPA 1996). Studies show that despite keeping adequate disinfectant residual, there is a significant deterioration of water quality due to the proliferation of microbes in the bio-films attached to the distribution pipes (Machell et al. 2010; Szewzyk et al. 2000; Simoes and Simões 2013; Douterelo et al. 2014). The data further support the inefficiency of the public water supply system to contain waterborne outbreaks. In this context, there is a need for an affordable and efficient alternate disinfection system. A well designed and maintained POU disinfection seems to be an attractive option (Vagliasindi et al. 1998; Sobsey et al. 2008).

Several point-of-use water purification technologies are developed to disinfect water. Among the available technologies, chlorination with safe storage, combined coagulant-chlorine disinfection, SODIS (solar UV radiation + with thermal effects), ceramic filter, bio-sand filter are well documented and capable of reducing waterborne infectious disease (Sobsey et al. 2008; Rose et al. 2006). State-of-the-art literature reviews show that POU household interventions contribute to a 30–40% reduction in diarrheal diseases (Clasen et al. 2007; Fewtrell et al. 2005). According to a recent review, over 18 million people use POU water treatment systems, with 12.8 million using chlorination, 2.1 million using SODIS, 0.934 million using flocculation/chlorination, 0.7 million using bio-sand filtration, and 0.35 million using ceramic filtration (Sobsey et al. 2008; Clasen 2008) have compared these widely promoted and used POU systems for performance and sustainability and identified ceramic and bio-sand based systems are most effective. However, Lantagne et al.

(2008) pointed out the flaws in the comparison and strongly commented that the comparison is biased. The role of participant motivation in reducing dysentery and non-dysentery diarrhoea by disinfection using SODIS among children (0.5 to 6 years) living in peri-urban communities in South Africa was studied (Du Preez et al. 2010). After comparing 383 children in 297 houses using SODIS with 335 children in 267 families with no intervention, the authors concluded that the motivation of participants is also an essential factor for measurable health gain.

Despite of obtaining promising results with the technologies validated in the laboratory and field, except boiling, the large-scale deployment of the technologies has hindered. Some of the reasons highlighted for the failure are (Sobsey et al. 2008): (i) Inability to provide adequate safe water, (ii) difficulty in operation and maintenance, (iii) large user time to treat water, (iv) affordability, (v) weak supply chain for needed replacement of units or parts, (vi) objectionable taste and odour, and (vii) bio fouling. Therefore, addressing these issues is vital for the successful implementation of POU treatment systems. Moreover, public participation, socio-economic considerations, local water quality, and consumer preference also need to be considered as sustainability criteria for developing POU water treatment systems.

11.3.3 Convectional Disinfection Technologies

The microbial quality of water can be improved through physical, chemical, or biological methods. There are several options available under each method. However, commonly practiced techniques involve chlorination, ozonation, filtration, UV irradiation, boiling, and SODIS process. A brief description of each technology is discussed below.

Disinfection with Chlorine: Chlorination is a popular disinfection processes and is achieved by introducing chlorine gas or its derivatives such as sodium hypochlorite (NaOCl) or calcium hypochlorite ($Ca(OCl)_2$) into water. The chlorine gas was first discovered in 1774 by Karl W. Scheele. Later, Humphrey Davy recognized it as a disinfectant in 1810 (Pradeep 2009). The continuous chlorination of public water process was first introduced in 1904 by Sir Alexander Houston of the London. In 1908, the application of calcium hypochlorite to Bubbly Creek water supply of the city of Chicago was initiated in the US (Logan and Savell 1940).

Upon addition of chlorine or its derivative to water, the chlorine agent undergoes hydrolysis and results in the formation of free-chorine species (HOCl and OCl^-). These species are responsible for disinfection of water. The reactions involved in the chlorination process are given below.

$$Cl_2 + H_2O \leftrightarrow HOCl + H^+ + Cl^- \tag{R1}$$

$$NaOCl + H_2O \leftrightarrow HOCl = NaOH \tag{R2}$$

$$Ca(OCl)_2 + 2H_2O \leftrightarrow 2HOCl + Ca(OH)_2 \quad (R3)$$

$$HOCl \leftrightarrow OCl^- + H^+ \quad (R4)$$

Among the free chlorine species, HOCl is the more powerful oxidizing agent (Metcalf 2003). The relative concentration of these species will vary according to the pH of the water. The chlorine can also combine with ammonia present in water and form chloramines. The critical reactions involved in the formations of chloramines are given below.

$$NH_3 + HOCl \leftrightarrow NH_2Cl + H_2O \quad (R5)$$

$$NH_2Cl + HOCl \leftrightarrow NHCl_2 + H_2O \quad (R6)$$

$$NHCl_2 + HOCl \leftrightarrow NCl_3 + H_2O \quad (R7)$$

The formation of chloramines is dependent on the relative concentration of ammonia and chlorine, pH, contact time, and temperature (Metcalf 2003). Though chloramines are less effective compared to free-chlorine species, they are unlikely to produce disinfecting by products (DBPs).

Disinfection with Ozone: Ozonation is the second most widely used disinfection process after chlorination. In 1906, France reported the first use of ozone as a disinfectant (Pradeep 2009). Ozone is generated on-site and is typically produced through electrical discharge method. A schematic diagram showing the generation of ozone is shown in Fig. 11.10. The freed radicals (HO and HO_2) formed as a result of the decomposition are probably responsible for the defection (Metcalf 2003). Unlike chlorine and chloramines, ozone is effective against a broad spectrum of the organisms, including *Giardia lamblia* and *Cryptosporidium parvam*. It is also found to be effective against spores and cysts (Budu-Amoako et al. 2011). However,

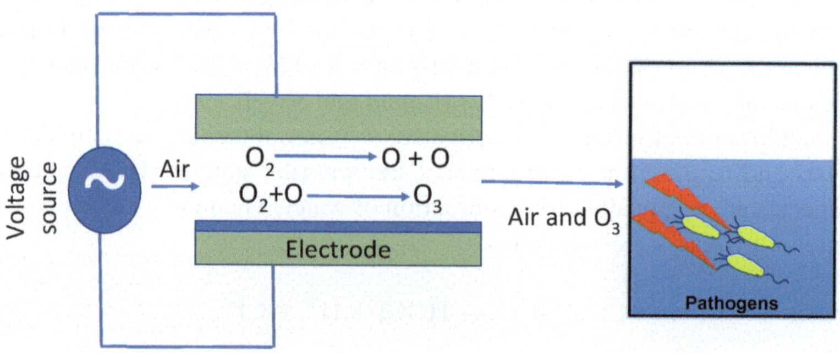

Fig. 11.10 Schematic representation of ozonation (courtesy of: chemistry of ozone disinfection in wastewater, environmental protection agency EPA 2016)

it does not maintain residual ozone concentration and less effective in preventing recontamination of water (EPA 2011).

Disinfection with UV: The first use of UV treatment in municipal water supplies was reported in 1916. Presently, the use of this technology is used in several applications. Unlike ozone and chlorine, UV light is a physical disinfecting agent, and hence, it is free from taste, odour, and harmful by-products even at high dose (EPA 2011). UV radiation at the right wavelength (255–265 nm) has shown active bactericidal and virucidal properties. The schematic diagram of the UV disinfection process is presented in Fig. 11.11b. A comprehensive review of UV based disinfection system is given elsewhere (Nyangaresi et al. 2018; Li et al. 2018).

Disinfection with Solar Radiation: Solar disinfection, popularly known as SODIS process, works based on the germicidal property of UV radiation and thermal heating.

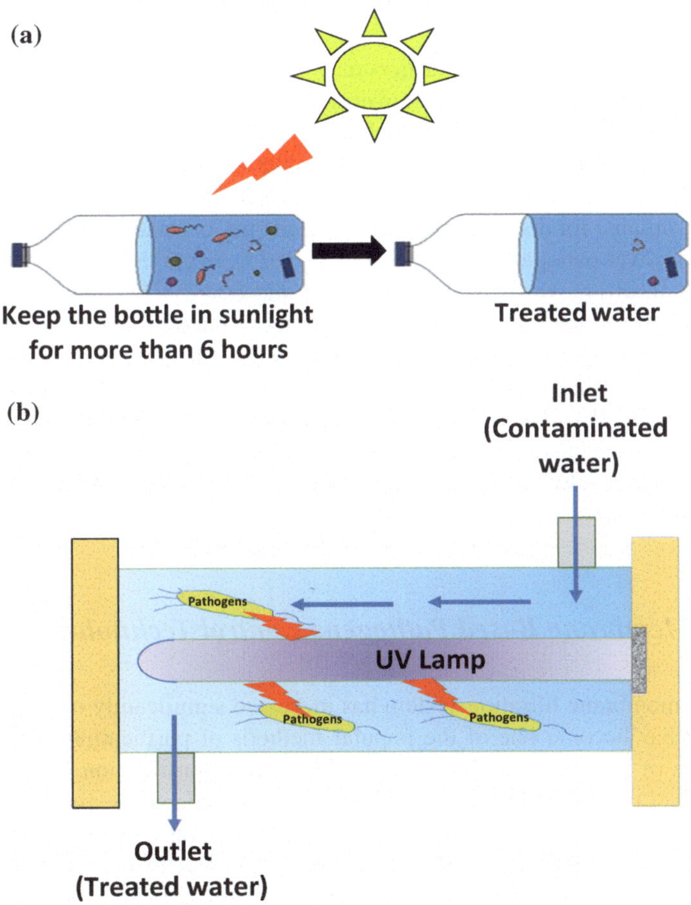

Fig. 11.11 Schematic representation of **a** solar disinfection (SODIS) (courtesy of www.billionbottleproject.org). **b** UV irradiation

In the SODIS process, the water is exposed to natural sunlight instead of light from a UV lamp. The conventional UV system uses UV-C (200–280 nm) radiations, whereas the SODIS process uses UV-A (320–400 nm) radiations. The interaction of UV-A with water generates reactive oxygen species (ROS) and ROS damage the DNA and deactivate the germs in the water. However, the SODIS process is not as effective as conventional UV treatment because UV light constitutes only <5% of the total solar spectrum (McGuigan et al. 2012). The schematic diagram of SODIS process is presented in Fig. 11.11a. Some of the viruses, protozoan species exhibits resistance to chlorination. So, we are in necessity to find the alternative technology that can kill various kinds of pathogenic species.

11.3.4 Nanotechnology Enabled Disinfection Process

Nanotechnology-enabled water purifiers may hold the key for safe drinking water in the future (Shannon et al. 2010; Hossain et al. 2014). The technology uses engineered nanoscale material to improve microbial quality of the water. In comparison to conventional chemical disinfectants, nanoscale materials are not strong oxidants and hence unlikely to produce harmful DBPs. Several natural and engineered nanomaterials are available for disinfection. These include photo catalytic TiO_2 (Dimitroula et al. 2012), silver nanoparticles (Sankar et al. 2013), MgO (Stoimenov et al. 2002; Ganguly et al. 2011), zero-valent iron (Crane and Scott 2012) and so on. Among these disinfectants, the silver-based system is more matured and used in the field level application. The silver nanoparticles are active disinfectant and work for a broad spectrum of bacteria and viruses (Karumuri et al. 2013; De Gusseme et al. 2010; Loo et al. 2013). Though the research and development in this area is not fully matured, the current advancement in nanotechnology may prove to be of significant interest to both developed and developing countries in addressing the problem of safe drinking water.

11.3.5 Membrane Based Pathogens Control Technologies

The use of membrane filtration system has increased significantly over the last two decades. It has become one of the popular methods of purification of water now. The process of membrane includes micro-filtration, ultra-filtration, nano-filtration, and Reverse Osmosis (RO). The classification is based on the pore sizes of the membrane. Among the membranes, RO can efficiently remove bacteria, viruses, and other suspended solids present in the water. It can control the disinfection by-products as well (Van der Bruggen et al. 2003). However, the large amount of rejects, clogging of the membrane, and high energy consumption needs further attention.

Recent years, the use of biomimetic membranes is gaining interest in water treatment. In this technology, aquaporin, a bio-inspired membrane is used. Aquaporin

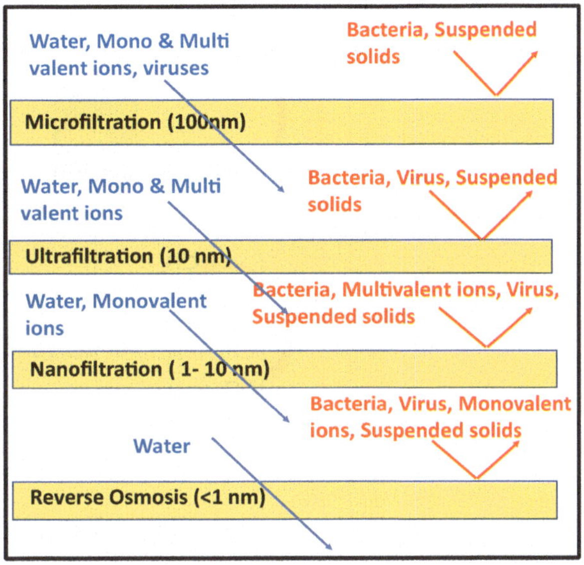

Fig. 11.12 Schematic representation of membrane filtration (courtesy of: https://www.logisticon.com/en/technologies/membrane-filtration)

acts as a water channel and allows only the passage of water through it. This method may reduce the cost of filtration by 30% of the conventional membranes (Wah 2016). The schematic representation of the membrane filtration technique is given in the Fig. 11.12.

11.3.6 Disinfection by Advanced Oxidation Process

The advanced oxidation process (AOP) is a promising technology in the field of water purification. The process typically uses ozone (O_3), UV light, hydrogen peroxide (H_2O_2), or combination thereof. The OH radicals formed during the processes are mainly responsible for the destruction of pathogens. The use of UV/TiO_2 (Matsunaga et al. 1985), UV/H_2O_2 (Gassie et al. 2016), ozone-UV (Crittenden et al. 2012), and photo-Fenton (Rossi et al. 2009) are also studied as next-generation disinfectants. In photo-catalysis, a light source of a specific wavelength is used to excite the electron from valence band to conduction band. The ability of the catalyst to produce the electron-hole pairs decides the efficiency of the process. The free radicals formed during the reaction destroy the pathogens present in the water (Giannakis et al. 2018).

$$TiO_2 \rightarrow e^- + p^+ \qquad \text{(R8)}$$

$$e^- + O_2 \rightarrow {O_2}^- \qquad \text{(R9)}$$

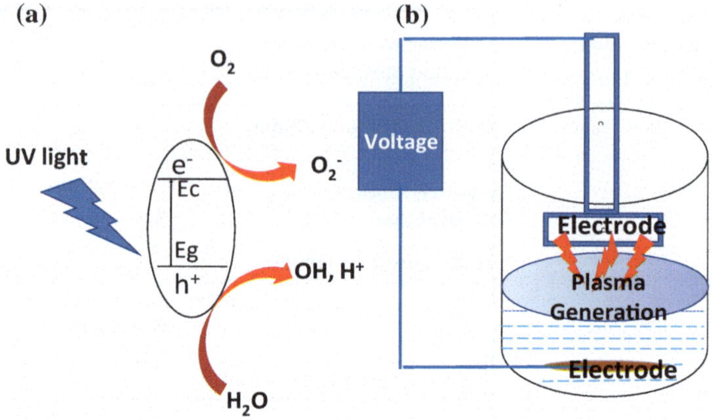

Fig. 11.13 Schematic representation of **a** photo-catalysis, **b** plasma treatment

$$P^{+} + H_2O \rightarrow OH^{\cdot} \quad \text{(R10)}$$

$$H_2O_2 \underset{\rightarrow}{h\vartheta} 2OH^{\cdot} \quad \text{(R11)}$$

$$O_3 + H_2O \underset{\rightarrow}{h\vartheta} O_2 + H_2O_2 \quad \text{(R12)}$$

Equations R8–R10 explain the radical formation in the UV/TiO_2 process, whereas Eqs. R11 and R12 show the radical formation in UV/H_2O_2 and ozone-UV process, respectively. Though AOPs are highly effective against the pathogens, the presence of natural organic matter such as fulvic acid, humic acid, and other ions affects its performance (Keane et al. 2014).

Plasma-based treatment technology is an emerging field for the removal of pathogens present in the drinking water (Rossi et al. 2009; Roth et al. 2010). In plasma technology, the Pulsed power technique (PPP) is found to be an effective method to disinfect the pathogen in short span of time (<6 min) (Singh et al. 2017). The schematic representation of UV photo-catalysis, and plasma techniques are given in Fig. 11.13a, b respectively.

11.4 Conclusion

Potable water is an absolute necessity for humankind as long as life exists. The pursuit of safe drinking water has been the highest priority for humans over centuries. The massive rise in the global population, poorly-managed water systems, and pollution have made the search more challenging. A large number of freshwater bodies across the world are contaminated and has become a major risk to human health and the ecosystem. Realizations on the link that exist between water and health, proving safely managed drinking water has become a criteria agenda in every framework of the developmental organization.

Among the contaminants identified in drinking water, the pathogen causes significant threat due to their widespread occurrence and potential to cause diseases. Though several technologies have been developed to detect and control the pathogens, the lack of reliable and affordable detection and control technologies hindered access to such interventions for a large population, especially from developing world. The emergence of disinfectant and antibiotic-resistant microorganisms also have become a cause of concern for the process of control and detection. There is a need for safe, affordable, and reliable point-of-use disinfection and detections system. There is also scope for developing hybrid systems, which facilitates both detection and removal of pathogens in drinking water. As to move forward, the technologies with abovementioned features should be made available to the public at affordable cost, thereby promoting sustainable, healthy, and productive living condition.

Acknowledgements Authors gratefully acknowledge the Government of India, Ministry of Science and Technology, Department of Science and Technology (DST), Technology Mission Division [Grant no: DST/TM/WTI/WIC/2K17/82(C)] for supporting the work. The author Uthradevi Kannan acknowledges DST—Innovation in Science Pursuit for Inspire Research (INSPIRE), [IF 160288] for the fellowship support to pursue PhD at IIT Tirupati. The authors also thank IIT Tirupati for the support.

References

Al-Qadiri HM, Al-Alami NI, Al-Holy MA, Rasco BA (2008) Using Fourier transform infrared (FT-IR) absorbance spectroscopy and multivariate analysis to study the effect of chlorine-induced bacterial injury in water. J Agric Food Chem 56(19):8992–8997

Ambriz-Aviña V, Contreras-Garduño JA, Pedraza-Reyes M (2014) Applications of flow cytometry to characterize bacterial physiological responses. BioMed Res Int

American Public Health Association, American Water Works Association (1989) Standard methods for the examination of water and wastewater. American Public Health Association

Araya V, Maliyekkal SM, Philip L (2019) Water pollution and treatment technologies – Indian perspective. In: Mjumdar PP, Tiwari VM (eds) Water futures of India: Status of science and technology. Indian National Science Academy, pp 215–252.

Axelrod D (2001) Total internal reflection fluorescence microscopy in cell biology. Traffic 2(11):764–774

Babacan S, Pivarnik P, Letcher S, Rand A (2000) Evaluation of antibody immobilization methods for piezoelectric biosensor application. Biosens Bioelectron 15(11–12):615–621

Basiji DA, Ortyn WE, Liang L, Venkatachalam V, Morrissey P (2007) Cellular image analysis and imaging by flow cytometry. Clinics Lab Med 27(3):653–670

Blumenthal UJ, Mara DD, Peasey A, Ruiz-Palacios G, Stott R (2000) Guidelines for the microbiological quality of treated wastewater used in agriculture: recommendations for revising WHO guidelines. Bull World Health Organ 78:1104–1116

Boulos L, Prevost M, Barbeau B, Coallier J, Desjardins R (1999) LIVE/DEAD® BacLight™: application of a new rapid staining method for direct enumeration of viable and total bacteria in drinking water. J Microbiol Methods 37(1):77–86

Bradburry S (1996) Fluorescence microscopy, contrast techniques in light microscopy
Brindha J, Chanda K, Balamurali M (2018) Physical, chemical and biochemical biosensors to detect pathogens. In: Nanotechnology food security and water treatment. Springer, Berlin, pp 53–86
Brussaard CP, Marie D, Bratbak G (2000) Flow cytometric detection of viruses. J Virol Methods 85(1–2):175–182
Budu-Amoako E, Greenwood SJ, Dixon BR, Barkema HW, McClure J (2011) Foodborne illness associated with cryptosporidium and Giardia from livestock. J Food Prot 74(11):1944–1955
Cangelosi GA, Meschke JS (2014) Dead or alive: molecular assessment of microbial viability. Appl Environ Microbiol 80(19):5884–5891
Chen Q, Lin J, Gan C, Wang Y, Wang D, Xiong Y, Lai W, Li Y, Wang M (2015) A sensitive impedance biosensor based on immunomagnetic separation and urease catalysis for rapid detection of Listeria monocytogenes using an immobilization-free interdigitated array microelectrode. Biosens Bioelectron 74:504–511
Clasen TF (2008) Scaling up household water treatment: looking back, seeing forward. In: Public health and the environment, World Health Organization, Geneva
Clasen T, Schmidt W-P, Rabie T, Roberts I, Cairncross S (2007) Interventions to improve water quality for preventing diarrhoea: systematic review and meta-analysis. BMJ 334(7597):782
Claus D (1992) A standardized Gram staining procedure. World J Microbiol Biotechnol 8(4):451–452
Cotruvo J, Dufour A, Rees G, Bartram J, Carr R, Cliver DO, Craun GF, Fayer R, Gannon VP, World Health Organization (2004) What are the criteria for determining whether a disease is zoonotic and water related? In: Water borne zoonoses: identification, causes and control. IWA
Crane R, Scott T (2012) Nanoscale zero-valent iron: future prospects for an emerging water treatment technology. J Hazard Mater 211:112–125
Craun GF, Brunkard JM, Yoder JS, Roberts VA, Carpenter J, Wade T, Calderon RL, Roberts JM, Beach MJ, Roy SL (2010) Causes of outbreaks associated with drinking water in the United States from 1971 to 2006. Clin Microbiol Rev 23(3):507–528
Crittenden JC, Trussell DW, Hand KJ, Howe KJ, Tchobanoglous G (2012) MWH's water treatment: principles and design. Wiley, New York
Das R, Sharma MK, Rao VK, Bhattacharya B, Garg I, Venkatesh V, Upadhyay S (2014) An electrochemical genosensor for *Salmonella typhi* on gold nanoparticles-mercaptosilane modified screen printed electrode. J Biotechnol 188:9–16
De Gusseme B, Sintubin L, Baert L, Thibo E, Hennebel T, Vermeulen G, Uyttendaele M, Verstraete W, Boon N (2010) Biogenic silver for disinfection of water contaminated with viruses. Appl Environ Microbiol 76(4):1082–1087
Derveaux S, Vandesompele J, Hellemans J (2010) How to do successful gene expression analysis using real-time PCR. Methods 50(4):227–230
Dey S (2015) 80% of India's surface water may be polluted, report by international says. Times of India
Dey NC, Saha R, Parvez M, Bala SK, Islam AS, Paul JK, Hossain M (2017) Sustainability of groundwater use for irrigation of dry-season crops in northwest Bangladesh. Groundwater Sustain Develop 4:66–77
Díaz M, Herrero M, García LA, Quirós C (2010) Application of flow cytometry to industrial microbial bioprocesses. Biochem Eng J 48(3):385–407
Dimitroula H, Daskalaki VM, Frontistis Z, Kondarides DI, Panagiotopoulou P, Xekoukoulotakis NP, Mantzavinos D (2012) Solar photocatalysis for the abatement of emerging micro-contaminants in wastewater: synthesis, characterization and testing of various TiO_2 samples. Appl Catal B 117:283–291
Dogan Ü, Kasap E, Cetin D, Suludere Z, Boyaci IH, Türkyılmaz C, Ertas N, Tamer U (2016) Rapid detection of bacteria based on homogenous immunoassay using chitosan modified quantum dots. Sens Actuators B Chem 233:369–378
Douterelo I, Husband S, Boxall J (2014) The bacteriological composition of biomass recovered by flushing an operational drinking water distribution system. Water Res 54:100–114

Du Preez M, McGuigan KG, Conroy RM (2010) Solar disinfection of drinking water in the prevention of dysentery in South African children aged under 5 years: the role of participant motivation. Environ Sci Technol 44(22):8744–8749
Dubelaar G, Gerritzen P (2000) CytoBuoy: a step forward towards using flow cytometry in operational oceanography. Scientia Marina 64(2):255–265
Dulbecco R, Vogt M (1953) Some problems of animal virology as studied by the plaque technique. In: Cold Spring Harbor symposia on quantitative biology. Cold Spring Harbor Laboratory Press
EPA (1996) National primary drinking water regulations, Monitoring requirements for public drinking water supplies. Final rule. Federal Register 61(94)
EPA (2011) Water treatment manual: disinfection, 2011
Faures J, Eliasson A, Hoogeveen J, Vallee D (2001) AQUASTAT-FAO's information system on water and agriculture. GRID-Magazine of the IPTRID Network (FAO/United Kingdom)
Feachem R, Mara DD, Bradley DJ (1983) Sanitation and disease. Wiley, Washington DC
Fewtrell L, Kaufmann RB, Kay D, Enanoria W, Haller L, Colford JM Jr (2005) Water, sanitation, and hygiene interventions to reduce diarrhoea in less developed countries: a systematic review and meta-analysis. Lancet Infect Dis 5(1):42–52
Fung DY, Miller RD (1970) Rapid procedure for the detection of acid and gas production by bacterial cultures. Appl Microbiol 20(3):527
Ganguly A, Trinh P, Ramanujachary K, Ahmad T, Mugweru A, Ganguli AK (2011) Reverse micellar based synthesis of ultrafine MgO nanoparticles (8–10 nm): characterization and catalytic properties. J Colloid Interface Sci 353(1):137–142
Garibyan L, Avashia N (2013) Research techniques made simple: polymerase chain reaction (PCR). J Invest Dermatol 133(3):e6
Gassie LW, Englehardt JD, Wang J, Brinkman N, Garland J, Gardinali P, Guo T (2016) Mineralizing urban net-zero water treatment: Phase II field results and design recommendations. Water Res 105:496–506
Giannakis S, Voumard M, Rtimi S, Pulgarin C (2018) Bacterial disinfection by the photo-Fenton process: extracellular oxidation or intracellular photo-catalysis? Appl Catal B 227:285–295
Gowrisankar G, Chelliah R, Ramakrishnan SR, Elumalai V, Dhanamadhavan S, Brindha K, Antony U, Elango L (2017) Chemical, microbial and antibiotic susceptibility analyses of groundwater after a major flood event in Chennai. Scientific data 4:170135
Hammes FA, Egli T (2005) New method for assimilable organic carbon determination using flow-cytometric enumeration and a natural microbial consortium as inoculum. Environ Sci Technol 39(9):3289–3294
Hammes F, Egli T (2010) Cytometric methods for measuring bacteria in water: advantages, pitfalls and applications. Anal Bioanal Chem 397(3):1083–1095
Hossain F, Perales-Perez OJ, Hwang S, Roman F (2014) Antimicrobial nanomaterials as water disinfectant: applications, limitations and future perspectives. Sci Total Environ 466:1047–1059
Jamieson RC, Gordon RJ, Tattrie SC, Stratton GW (2003) Sources and persistence of fecal coliform bacteria in a rural watershed. Water Qual Res J 38(1):33–47
Kanungo S, Sah B, Lopez A, Sung J, Paisley A, Sur D, Clemens J, Nair GB (2010) Cholera in India: an analysis of reports, 1997–2006. Bull World Health Organ 88:185–191
Karumuri AK, Oswal DP, Hostetler HA, Mukhopadhyay SM (2013) Silver nanoparticles attached to porous carbon substrates: robust materials for chemical-free water disinfection. Mater Lett 109:83–87
Keane DA, McGuigan KG, Ibáñez PF, Polo-López MI, Byrne JA, Dunlop PS, O'Shea K, Dionysiou DD, Pillai SC (2014) Solar photocatalysis for water disinfection: materials and reactor design. Catal Sci Technol 4(5):1211–1226
Keevil C (2003) Rapid detection of biofilms and adherent pathogens using scanning confocal laser microscopy and episcopic differential interference contrast microscopy. Water Sci Technol 47(5):105–116

Kittigul L, Khamoun P, Sujirarat D, Utrarachkij F, Chitpirom K, Chaichantanakit N, Vathanophas K (2001) An improved method for concentrating rotavirus from water samples. Mem Inst Oswaldo Cruz 96(6):815–821

Kooij D (1992) Assimilable organic carbon as an indicator of bacterial regrowth. J Am Water Works Assoc 84(2): 57–65

Krishnan SK, Singh E, Singh P, Meyyappan M, Nalwa HS (2019) A review on graphene-based nanocomposites for electrochemical and fluorescent biosensors. RSC Adv 9(16):8778–8881

Kumar N, Hu Y, Singh S, Mizaikoff B (2018) Emerging biosensor platforms for the assessment of water-borne pathogens. Analyst 143(2):359–373

Laczka O, Skillman L, Ditcham WG, Hamdorf B, Wong DK, Bergquist P, Sunna A (2013) Application of an ELISA-type screen printed electrode-based potentiometric assay to the detection of *Cryptosporidium parvum* oocysts. J Microbiol Methods 95(2):182–185

Lantagne D, Meierhofer R, Allgood G, McGuigan K, Quick R (2008) Comment on "Point of use household drinking water filtration: A practical, effective solution for providing sustained access to safe drinking water in the developing world". Environ Sci Technol 43(3):968–969

Law JWF, Ab Mutalib NS, Chan K-G, Lee L-H (2015) Rapid methods for the detection of foodborne bacterial pathogens: principles, applications, advantages and limitations. Front Microbiol 5:770

Lee-Montiel FT, Reynolds KA, Riley MR (2011) Detection and quantification of poliovirus infection using FTIR spectroscopy and cell culture. Journal of biological engineering 5(1):16

Leininger DJ, Roberson JR, Elvinger F (2001) Use of eosin methylene blue agar to differentiate Escherichia coli from other gram-negative mastitis pathogens. J Vet Diagn Invest 13(3):273–275

Li X, Cai M, Wang L, Niu F. Yang D, Zhang G (2018) Evaluation survey of microbial disinfection methods in UV-LED water treatment systems. Sci Total Environ

Logan JO, Savell WL (1940) Calcium hypochlorite in water purification. J (Am Water Works Assoc) 32(9):1517–1527

Loo S-L, Fane AG, Lim T-T, Krantz WB, Liang Y-N, Liu X, Hu X (2013) Superabsorbent cryogels decorated with silver nanoparticles as a novel water technology for point-of-use disinfection. Environ Sci Technol 47(16):9363–9371

Ma X, Cui Y, Qiu Z, Zhang B, Cui S (2013) A nanoparticle-assisted PCR assay to improve the sensitivity for rapid detection and differentiation of wild-type pseudorabies virus and gene-deleted vaccine strains. J Virol Methods 193(2):374–378

Machell J, Mounce S, Boxall J (2010) Online modelling of water distribution systems: a UK case study. Drinking Water Eng Sci 3:21–27

Manafi M, Kneifel W (1989) A combined chromogenic-fluorogenic medium for the simultaneous detection of coliform groups and *E. coli* in water. Int J Hygiene Environ Med 189(3): 225–234

March SB, Ratnam S (1986) Sorbitol-MacConkey medium for detection of *Escherichia coli* O157: H7 associated with hemorrhagic colitis. J Clin Microbiol 23(5):869–872

Matsunaga T, Tomoda R, Nakajima T, Wake H (1985) Photoelectrochemical sterilization of microbial cells by semiconductor powders. FEMS Microbiol Lett 29(1–2):211–214

Mattison K, Bidawid S (2009) Analytical methods for food and environmental viruses. Food Environ Virol 1(3–4):107–122

McGuigan KG, Conroy RM, Mosler H-J, du Preez M, Ubomba-Jaswa E, Fernandez-Ibanez P (2012) Solar water disinfection (SODIS): a review from bench-top to roof-top. J Hazard Mater 235:29–46

McPherson M, Møller S (2000) Pcr. Taylor & Francis, New York

Metcalf, L., *Wastewater engineering: treatment and reuse. Metcalf & Eddy Inc*. 2003, McGraw-Hill Inc., New York

Microbiology L.l (2019) Staining microscopic specimens. cited 29 May 2019. Available from: https://courses.lumenlearning.com/microbiology/chapter/staining-microscopic-specimens/

Milbury CA, Li J, Liu P, Makrigiorgos GM (2011) COLD-PCR: improving the sensitivity of molecular diagnostics assays. Expert Rev Mol Diagn 11(2):159–169

Munoz EF, Silverman MP (1979) Rapid, single-step most-probable-number method for enumerating fecal coliforms in effluents from sewage treatment plants. Appl Environ Microbiol 37(3):527–530

Murty M, Kumar S (2011) Water pollution in India: an economic appraisal. India infrastructure report. Water: policy and performance for sustainable development

National Research Council (US) Safe Drinking Water Committee (1980) Drinking water and health: the disinfection of drinking water. National Academies Press (US), New York

Nyangaresi PO, Qin Y, Chen G, Zhang B, Lu Y, Shen L (2018) Comparison of UV-LED photolytic and UV-LED/TiO_2 photocatalytic disinfection for *Escherichia coli* in water. Catalysis Today

Orpana AK, Ho TH, Alagrund K, Ridanpää M, Aittomäki K, Stenman J (2013) Novel Heat pulse extension-PCR–based method for detection of large CTG-repeat expansions in myotonic dystrophy type 1. J Mol Diagn 15(1):110–115

Pandey PK, Soupir ML (2013) Assessing the impacts of E. coli laden streambed sediment on E. coli loads over a range of flows and sediment characteristics. JAWRA J Am Water Resour Assoc 49(6):1261–1269

Pandey PK, Kass PH, Soupir ML, Biswas S, Singh VP (2014) Contamination of water resources by pathogenic bacteria. Amb Express 4(1):51

Pham NTK, Trinh QD, Chan-It W, Khamrin P, Shimizu H, Okitsu S, Mizuguchi M, Ushijima H (2010) A novel RT-multiplex PCR for detection of Aichi virus, human parechovirus, enteroviruses, and human bocavirus among infants and children with acute gastroenteritis. J Virol Methods 169(1):193–197

Planning Commission (2002) Report of the screening commitee on drinking water supply and sanitation (Rural and Urban) for tenth five year plan. P. Commision, New Delhi, India

Pradeep T (2009) Noble metal nanoparticles for water purification: a critical review. Thin Solid Films 517(24):6441–6478

Rajapaksha P, Elbourne A, Gangadoo S, Brown R, Cozzolino D, Chapman J (2019) A review of methods for the detection of pathogenic microorganisms. Analyst 144(2):396–411

Rakić A (2018) Water quality control in the water supply system for the purpose of preventing legionnaires' disease. In: Water challenges of an urbanizing world. IntechOpen

Rein DB, Stevens GA, Theaker J, Wittenborn JS, Wiersma ST (2012) The global burden of hepatitis E virus genotypes 1 and 2 in 2005. Hepatology 55(4):988–997

Rhodes MW, Kator HI (1990) Effects of sunlight and autochthonous microbiota *on Escherichia coli* survival in an estuarine environment. Curr Microbiol 21(1):65–73

Rider TH, Petrovick MS, Nargi FE, Harper JD, Schwoebel ED, Mathews RH, Blanchard DJ, Bortolin LT, Young AM, Chen J (2003) AB cell-based sensor for rapid identification of pathogens. Science 301(5630):213–215

Rivera-Jaimes JA, Postigo C, Melgoza-Alemán RM, Aceña J, Barceló D, de Alda ML (2018) Study of pharmaceuticals in surface and wastewater from Cuernavaca, Morelos, Mexico: occurrence and environmental risk assessment. Sci Total Environ 613:1263–1274

Rose A, Roy S, Abraham V, Holmgren G, George K, Balraj V, Abraham S, Muliyil J, Joseph A, Kang G (2006) Solar disinfection of water for diarrhoeal prevention in southern India. Arch Dis Child 91(2):139–141

Rossi F, Kylián O, Rauscher H, Hasiwa M, Gilliland D (2009) Low pressure plasma discharges for the sterilization and decontamination of surfaces. New J Phys 11(11):115017

Roth S, Feichtinger J, Hertel C (2010) Characterization of *Bacillus subtilis* spore inactivation in low-pressure, low-temperature gas plasma sterilization processes. J Appl Microbiol 108(2):521–531

Safavieh M, Ahmed MU, Sokullu E, Ng A, Braescu L, Zourob M (2014) A simple cassette as point-of-care diagnostic device for naked-eye colorimetric bacteria detection. Analyst 139(2):482–487

Saha R, Dey NC, Rahman S, Galagedara L, Bhattacharya P (2018) Exploring suitable sites for installing safe drinking water wells in coastal Bangladesh. Groundwater for Sustainable Development 7:91–100

Sankar MU, Aigal S, Maliyekkal SM, Chaudhary A, Kumar AA, Chaudhari K, Pradeep T (2013) Biopolymer-reinforced synthetic granular nanocomposites for affordable point-of-use water purification. Proc Natl Acad Sci 110(21):8459–8464

Schriewer A, Miller WA, Byrne BA, Miller MA, Oates S, Conrad PA, Hardin D, Yang H-H, Chouicha N, Melli A (2010) Presence of bacteroidales as a predictor of pathogens in surface waters of the central California coast. Appl Environ Microbiol 76(17):5802–5814

Selan L, Berlutti F, Passariello C, Thaller M, Renzini G (1992) Reliability of a bioluminescence ATP assay for detection of bacteria. J Clin Microbiol 30(7):1739–1742

Shaheed A, Orgill J, Montgomery MA, Jeuland MA, Brown J (2014) Why? improved? water sources are not always safe. Bull World Health Organ 92:283–289

Shannon MA, Bohn PW, Elimelech M, Georgiadis JG, Marinas BJ, Mayes AM (2010) Science and technology for water purification in the coming decades. In: Nanoscience and technology: a collection of reviews from nature Journals. World Scientific, pp 337–346

Sheppard C, Shotton D, Sheppard C (1997) Confocal laser scanning microscopy. BIOS Scientific Publishers, Oxford

Simoes LC, Simões M (2013) Biofilms in drinking water: problems and solutions. Rsc Adv 3(8):2520–2533

Singh RK, Babu V, Philip L, Ramanujam S (2017) Disinfection of water using pulsed power technique: effect of system parameters and kinetic study. In: Sustainability issues in civil engineering. Springe, Berlin, pp 307–336

Sobsey MD, Stauber CE, Casanova LM, Brown JM, Elliott MA (2008) Point of use household drinking water filtration: a practical, effective solution for providing sustained access to safe drinking water in the developing world. Environ Sci Technol 42(12):4261–4267

Stoimenov PK, Klinger RL, Marchin GL, Klabunde KJ (2002) Metal oxide nanoparticles as bactericidal agents. Langmuir 18(17):6679–6686

Szewzyk U, Szewzyk R, Manz W, Schleifer K-H (2000) Microbiological safety of drinking water. Ann Rev Microbiol 54(1):81–127

Tang P, Wu J, Liu H, Liu Y, Zhou X (2018) Assimilable organic carbon (AOC) determination using GFP-tagged *Pseudomonas fluorescens* P-17 in water by flow cytometry. PLoS ONE 13(6):e0199193

Tankeshwar (2010) Membrane filter technique for bacteriological examination of water. Cited 26 May 2019. Available from: https://microbeonline.com/membrane-filter-technique/

Thakur B, Amarnath CA, Mangoli S, Sawant SN (2015) Polyaniline nanoparticle based colorimetric sensor for monitoring bacterial growth. Sens Actuators B Chem 207:262–268

Trivedi BK, Gandhi HS, Shukla NK (1971) Bacteriological water quality and incidence of water borne diseases in a rural population. Indian J Med Sci 25(11):795–801

Turner DE, Daugherity EK, Altier C, Maurer KJ (2010) Efficacy and limitations of an ATP-based monitoring system. J Am Assoc Lab Anim Sci 49(2):190–195

UN (n.d.) Water scarcity. Available at: http://www.unwater.org/water-facts/scarcity/

UN DESA (2018) Revision of World Urbanization Prospects. Cited 27 May 2019. Available at: https://www.un.org/development/desa/publications/2018-revision-of-world-urbanization-prospects.html

UNESCO (2017) United Nations world water development report. 23 May 2019; at: http://www.unesco.org/new/en/natural-sciences/environment/water/wwap/wwdr/2017-wastewater-the-untapped-resource/

UNICEF (2016) One is too many: Ending child deaths from pneumonia and diarrhoea

UNICEF (2017) Progress on drinking water, sanitation and hygiene: 2017 update and SDG guidelines

USBR (2017) Water facts—worldwide water supply. Available at: https://www.usbr.gov/mp/arwec/water-facts-ww-water-sup.html

Vagliasindi FG, Belgiorno V, Napoli RM (1998) Water treatment in remote and rural areas: a conceptual screening protocol for appropriate POU/POE technologies. In: Environmental engineering and renewable energy. Elsevier, London, pp 329–336

Valones MAA, Guimarães RL, Brandão LAC, Souza PRE, Carvalho AAT, Crovela S (2009) Principles and applications of polymerase chain reaction in medical diagnostic fields: a review. Braz J Microbiol 40(1):1–11

Van der Bruggen B, Vandecasteele C, Van Gestel T, Doyen W, Leysen R (2003) A review of pressure-driven membrane processes in wastewater treatment and drinking water production. Environ Prog 22(1):46–56

van der Wielen PW, van der Kooij D (2010) Effect of water composition, distance and season on the adenosine triphosphate concentration in unchlorinated drinking water in the Netherlands. Water Res 44(17):4860–4867

Vandeweerd V, Bernal P, Belfiore S, Goldstein K, Cicin-Sain B (2002) A guide to oceans, coasts, and islands at the World Summit on sustainable development

Velusamy V, Arshak K, Korostynska O, Oliwa K, Adley C (2010) An overview of foodborne pathogen detection: in the perspective of biosensors. Biotechnol Adv 28(2):232–254

Venkatesan KD, Balaji M, Victor K (2014) Microbiological analysis of packaged drinking water sold in Chennai. Int J Med Sci Public Health 3(4):472–477

Vesey G, Hutton P, Champion A, Ashbolt N, Williams KL, Warton A, Veal D (1994) Application of flow cytometric methods for the routine detection of Cryptosporidium and Giardia in water. Cytometry J Int Soc Anal Cytol 16(1):1–6

Vidic J, Manzano M, Chang C-M, Jaffrezic-Renault N (2017) Advanced biosensors for detection of pathogens related to livestock and poultry. Vet Res 48(1):11

Wah TY (2016) Nature-inspired membrane set to reduce purification costs. Membr Technol 2016(5):7

Wang Y, Alocilja EC (2015) Gold nanoparticle-labeled biosensor for rapid and sensitive detection of bacterial pathogens. J Biol Eng 9(1):16

Wang L, Wei Q, Wu C, Hu Z, Ji J, Wang P (2008) The Escherichia coli O157: H7 DNA detection on a gold nanoparticle-enhanced piezoelectric biosensor. Chin Sci Bull 53(8):1175–1184

Wang W, Wang W, Liu L, Xu L, Kuang H, Zhu J, Xu C (2016a) Nanoshell-enhanced Raman spectroscopy on a microplate for staphylococcal enterotoxin B sensing. ACS Appl Mater Interfaces 8:15591–15597

Wang C, Wang J, Li M, Qu X, Zhang K, Rong Z, Xiao R, Wang S (2016b) A rapid SERS method for label-free bacteria detection using polyethylenimine-modified Au-coated magnetic microspheres and Au@ Ag nanoparticles. Analyst 141(22):6226–6238

Wingender J, Flemming H-C (2011) Biofilms in drinking water and their role as reservoir for pathogens. Int J Hyg Environ Health 214(6):417–423

Wolf S, Hewitt J, Rivera-Aban M, Greening GE (2008) Detection and characterization of F + RNA bacteriophages in water and shellfish: application of a multiplex real-time reverse transcription PCR. J Virol Methods 149(1):123–128

World Health Organization (WHO) (2001) Water for health: taking charge

World Health Organization (WHO) (2004) Water, sanitation and hygiene links to health: facts and figures

World Health Organization (WHO) (2007) Outbreak of water borne diseases. Cited 27 May 2019. Available from: http://www.euro.who.int/__data/assets/pdf_file/0006/97359/1.1.pdf?ua=1

World Health Organization (WHO) (2011) Guidelines for drinking-water quality. WHO Chron 38(4):104–108

World Health Organization (WHO) (2016) Number of reported cholera cases. Cited 23 May 2019. Available at: https://www.who.int/gho/epidemic_diseases/cholera/cases_text/en/

Xu M, Wang R, Li Y (2016) An electrochemical biosensor for rapid detection of E. coli O157: H7 with highly efficient bi-functional glucose oxidase-polydopamine nanocomposites and Prussian blue modified screen-printed interdigitated electrodes. Analyst 141(18):5441–5449

Yahya HSA, Jilali A, Mostareh MMM, Chafik Z, Chafi A (2017) Microbiological, physicochemical, and heavy metals assessment of groundwater quality in the Triffa plain (eastern Morocco). Applied Water Science 7(8):4497–4512

Zourob M, Elwary S, Turner AP (2008) Principles of bacterial detection: biosensors, recognition receptors and microsystems. Springer Science & Business Media, Berlin

Chapter 12
Occurrence, Contamination, Speciation and Analysis of Selenium in the Environment

M. S. V. Naga Jyothi, B. J. Ramaiah and Shihabudheen M. Maliyekkal

Abstract Selenium (Se) is an essential micronutrient for animals and human. It serves as an antioxidant and anti-carcinogenic agent in humans. However, there is only a minuscule gap in intake concentration between deficiency and toxicity limits. Natural source of Se in the environment is due to weathering of selenium-containing parent bedrocks and atmospheric deposition. However, the magnitude of the release of Se into the environment has intensified by anthropogenic activities. The bioavailability and toxicity of Se depend on its chemical forms, exposure time, and concentration. Several studies have documented both the beneficial and adverse effects of Se on human and other biotic forms. In this context, understanding the fundamentals of selenium chemistry, occurrence, sources, toxicity, and exposure mechanism are important to control its intake and threshold uptake limit. The chapter reviews the literature relevant to the release of Se into the natural habitat, speciation, and its effects on humans and animals. The manuscript also renders outline on analytical techniques based on speciation and detection of selenium compounds.

Keywords Analysis · Environment · Occurrence · Selenium · Source · Speciation

12.1 Introduction

Selenium (Se) is a trace metal with 0.000005% abundance in the earth's crust. In the year 1817, Jöns Jacob Berzelius and Johan Gottlieb Gahn, Chemists from Sweden discovered Se while inspecting the quality of sulfuric acid produced at a plant in Falun, Sweden. The presence of Se as an impurity in the sulfuric acid sample was traced back to sulfide ore from the mines of Falun (Trofast 2011). The word selenium originates from the Greek word *Selene*, which means moon. The name was chosen to match the sister element tellurium (*Tellus*), which means earth in Greek (Tan et al. 2016).

M. S. V. Naga Jyothi · B. J. Ramaiah · S. M. Maliyekkal (✉)
Department of Civil and Environmental Engineering, Indian Institute of Technology Tirupati, Tirupati 517 619, AP, India
e-mail: shihab@iittp.ac.in

T. Gupta et al. (eds.), *Measurement, Analysis and Remediation of Environmental Pollutants*, Energy, Environment, and Sustainability,
https://doi.org/10.1007/978-981-15-0540-9_12

Se is an essential micronutrient to support various metabolic activities in humans and animals. Typically, a human cell contains more than a million atoms of Se. On the earth's crust, there is no known predominant ore with pure selenium, and it typically occurs as a concomitant in the ores of sulfide-containing heavy metals such as lead, zinc, and copper. The seleniferous soils which are produced from these ores can cause significant impact on surrounding biotic life. The effect would vary based on its chemical forms, concentration present in the soil, and bioavailability, which intern function of soil reduction potential (Eh), pH, clay and soil organic carbon (SOC) content, and climate and soil interaction (Jones et al. 2017).

The Se can release into the environment through both natural and human activities. However, anthropogenic activities are responsible for the higher release of Se into the environment. Both organic and inorganic forms of Se exist in nature (soil and plants). The organic forms present in plants are precursors of selenoproteins in human and animals. Some of these proteins aid the synthesis of thyroid hormone and, destroy cancer cells (Bellinger et al. 2009). Ingesting drinking-water and certain types of food are the primary sources of Se in human. Se is majorly found in Brazil nuts, fish, meat, chicken, and eggs. However, the concentration of Se in the food may vary due to variation in geological distribution of Se and soil conditions (Navarro-Alarcon and Cabrera-Vique 2008). But intake of Se beyond requirement causes hepatotoxicity, gastrointestinal problems and dermatologic effects in humans, and emaciation, hoof malformation, and loss of hair in animals (Vinceti et al. 2001). Due to the minuscule gap between the toxicity and the deficiency, the upper tolerable daily intake of Se in human is limited to 400 μg by the Food and Nutrition Board, Institute of Medicine of the National Academies, US (Institute of Medicine 2000).

In this context, understanding the fundamentals of selenium chemistry, occurrence, sources, toxicity, use, and exposure mechanism is essential to control its intake and threshold uptake limit. The chapter reviews the literature relevant to the release of Se in the natural environment in general and aquatic environment in particular. The Se speciation and its beneficial and adverse health effects on human are discussed. The chapter also presents the outline of the major analytical techniques employed for quantifying Se in the aqueous medium.

12.2 Industrial Use of Selenium and Its Derivatives

Commercially, selenium is obtained as a byproduct of the electrolytic refining of copper. In the process, it collects in the anode and is recovered from the anode residues (USGS 2000). This route is responsible for >80% of its global production. According to estimates, the world refinery production of Se has increased from 1260 ton in 2000 to 2,800 ton in 2018 (USGS 2000, 2019) due to its growing demand in diverse applications. Figure 12.1 shows the estimated global use of Se by 2015 (USGS 2015). One of the significant use of the element is in metallurgy. It is employed as an additive to steel, copper and lead alloys for faster production and provide better machinability (Indian Bureau of Mines 2017) and is also used as a chemical additive

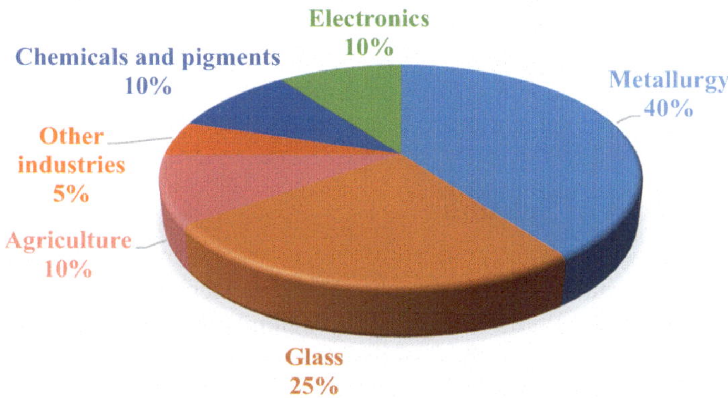

Fig. 12.1 Estimated global use of selenium in the year 2015 (USGS 2015)

for electrodeposition of manganese to improve current efficiency and hinder the evolution of harmful hydrogen gas (Lu et al. 2014). It is also used commonly as a glass decolorizing agent. Glass manufacturing industry requires nearly 6.5–8.0 kg of Se to produce 1 ton of glass, and the primary purpose is either to remove the green color due to the presence of iron impurities or to provide pink and red color anti-glare pot/flat glass (Kirkpatrick and Roberts 1919). The other uses of the element include pigments in paints, ceramics, and plastics, semiconductor in electrical rectifiers and photocells, xeroradiography for medical diagnosis, and photoreceptor in xerographic copiers (USGS 2004; Udoye and Jafarzadeh 2010). In photocopiers, Se is applied as semiconductor onto the photoreceptor drum due to its photoconductive property. However, in current photocopiers, Se is being replaced by organic solvents. Some of the selenium-based pigments, such as CdSSe and ZnSCdSSe-$BaSO_4$, are also phased-out due to the high cost and toxicity of cadmium in the environment (USGS 2004).

The potential modern uses of Se includes thin-film photovoltaic cells and energy-efficient windows. The production of copper-indium-gallium-selenide (CIGS) based thin-film photovoltaic cells would be an important commercial use of the element in the 21st century. Though silicon-based photovoltaic cells still dominate the market, eventually CIGS based technology would replace the relatively inefficient silicon-based technology due to higher efficiency and lower production cost (Bleiwas 2010). It also finds application in medicine and personal care products. Due to the antioxidant property, Se compounds can be used to treat cancer cells (Fernandes and Gandin 2015). For Alzheimer's disease, modified Se nanoparticles are used to inhibit Aβ aggregation into amyloid fibrils in the brain (Gupta et al. 2019). Selenium disulfide is popularly used as an antidandruff substance in shampoos (Cummins and Kimura 1971). In the commercial market, Se is available in four grades including: (i) commercial refined Se with a purity of >99.5 wt%, (ii) pigment grade with a purity

>99.7 wt%, (iii) high grade with a minimum of 99.999 wt% Se, and (iv) ultrahigh grade having purity in the range of 99.999–99.9999 wt% Se (Hoffmann and King 2001).

12.3 Properties of Selenium

Selenium is a non-metal with atomic number 34 and molecular weight 78.971 g/mole. In the periodic table, it belongs to the chalcogen family (Group XVI) with chemical properties analogous to sulfur. Most of the known compounds of Se are similar to that of sulfur compounds. It is a p-semiconductor and displays photoconductivity and photovoltaic action. Physical properties of Se are given below (Periodic table 2019):

Electronic configuration	$(Ar)3d^{10}4s^24p^4$
Atomic radius, Å	1.90
Covalent radius, Å	1.18
Ionic radius, Å	1.98
Melting point	220.8 °C
Boiling point	685 °C
Common oxidative states	−2, 0, +4, +6
Electron affinity, eV	−4.21
Electronegativity	2.55
State at 20 °C	Solid
Specific heat, J kg^{-1} K^{-1}	321

Numerous allotropes of Se exist, and a few are shown in Table 12.1. Depending on the rate of cooling from its melting point, it can occur in amorphous or crystalline forms. Liquid selenium on cooling turns into viscous at 60 °C and becomes hard and brittle at 30–40 °C giving grey vitreous form (Saunders 1900). This vitreous form is similar to its amorphous form where reduction of selenous acid produces a red brick powder. Rapid melting of amorphous form gives black selenium beads. The crystalline structures of Se are monoclinic (α, β) and hexagonal/trigonal (Murphy et al. 1977). Monoclinic forms consist of Se_8 rings with a Se–Se bond length of

Table 12.1 Physical properties of allotrope of selenium

Allotrope	Melting temperature (°C)	Density (g/cm^3)
Amorphous	50	4.28
Monoclinic	170–180	4.44 at 0 °C
Metallic	217	4.79 at 0 °C

233.5 pm, and Se–Se–S angle is 105.7°. Monoclinic forms convert into a hexagonal structure at a temperature more than 110 °C. The α-monoclinic consists of flat hexagonal and polygonal crystals, whereas β-monoclinic contains needle-like crystals (National Research Council 1983). The most stable grey hexagonal crystal or metallic selenium is produced by the heating of black Se in quinolone, pyridine, aniline or other basic organic solvent followed by slow cooling to condense below its melting point at 217 °C. Metallic Se consists of helical polymeric chains with the Se–Se distance of 237.3 pm and Se–Se–Se angle of 130.1°.

The naturally occurring isotopes of Se are ^{74}Se (0.89%), ^{82}Se (8.73%), ^{76}Se (9.32%), ^{77}Se (9.63%), ^{78}Se (23.77%) and ^{80}Se (49.61%) (Sarquis and Mickey 1880). Major isotopes like ^{77}Se, ^{78}Se, ^{79}Se and ^{80}Se are produced by fission reaction and are stable. ^{79}Se is also stable and typically occurs in uranium ores in trace quantity. Unlike other radioactive isotopes, ^{82}Se is considered as stable due to long half-life period of 1.08×10^{20} years. Till now, 32 radioactive isotopes of Se has been discovered with a decay period ranging from >180 ns to 295,000 years (Stolz et al. 2005).

12.4 Occurrence, Source, and Distribution

12.4.1 Occurrence of Selenium

Se is the 66th most abundant element in the earth crust, and it exists in both organic and inorganic forms. The inorganic forms include selenide minerals, selenite salts, and selenite. It is found in sulfide minerals, wherein it is partially replaced by sulfur. Examples are pyrite (FeSe), tiemannite (HgSe), klockmannite (CuSe), clausthalite (PbSe), chrisstanleyite (AgSe), and stilleite (ZnSe) (Khamkhash et al. 2017). These sulfide minerals are the main source of Se produced for the commercial market. Se can also combine with organic molecules and forms organoselenium compounds. They exist in the forms of selenols (R-SeH), selenides (R-Se-R) and diselenides (R-Se-Se-R) analogous to that of sulfur compounds such as thiol (R-SH) and sulfides (R-S-R). Human proteins like other compounds are selenoxides (R-Se(O)-R), selenones (R-Se(O)$_2$R), Se alkyl halides (R-Se-Cl; R-Se-Br), Se acids (R-Se-(O)OH; RSe-OH; RSe(O)H), and selenoheterocyclic (Perrone et al. 2015). In biotic life, organic selenium exists in the form of seleno-aminoacids called selenocysteine (SeCys) and selenomethionine (SeMet). In plants, water-soluble selenium competes with sulfur to enter into roots through sulfate transporters and assimilates to form selenium amino acids (Gupta and Gupta 2017). It is taken up by animals and humans and produces selenoproteins through metabolic activity of cells.

12.4.2 Sources and Distribution of Selenium

Selenium is distributed in nature widely. However, the distribution is not uniform. It can release into the environment (water, soil, and air) either by natural ways or anthropogenic activities. The major natural events governing the release are volcanic eruptions, marls, biogeochemical processes, and weathering of Se containing rocks and soils (He et al. 2018). The Se compounds such as SeO_2 and H_2Se are volatile, and their presence is reported in volcanic releases along with sulfur gases. Estimates show that over geologic time, nearly 1 kg/m^2 of Se has been deposited on the surface of the earth from volcanic activities (USGS 1973). The processes such as adsorption, groundwater transport, chemical and biochemical actions, dry and wet deposition, the isomorphic substitution of selenium for sulfur, and metabolic activities by plants and animals also affect the relative distribution of the element. Once it is released to the atmosphere, it undergoes various transformations during its life cycle, as shown in Fig. 12.2.

The anthropogenic activities such as burning of coal in thermal power plants, metal extraction from ores by froth flotation and chemical treatment, oil refining, agricultural practices in seleniferous soils, and municipal landfills can accelerate the release of Se into nature (Lemly 2004). Estimates show that human activities are responsible for 37–40% of Se release into the environment and coal burning being the primary contributor (Wen and Carignan 2007). The world average of Se content in coal for hard coals and brown coals are 1.6 ± 0.1 and 1.0 ± 0.15 ppm, respectively (Yudovich and Ketris 2006). In coal beneficiation, conventional coal cleaning process removes 11–25% of Se (Tillman et al. 2012). Smelting of ores containing metals like Cu, Ni, and Zn can also release volatile Se compounds into the air (Germani et al. 1981). No sooner it gets deposit onto particles, settles or precipitates and discharges into the stream. Agriculture in seleniferous soils leaches Se in dissolved form and release

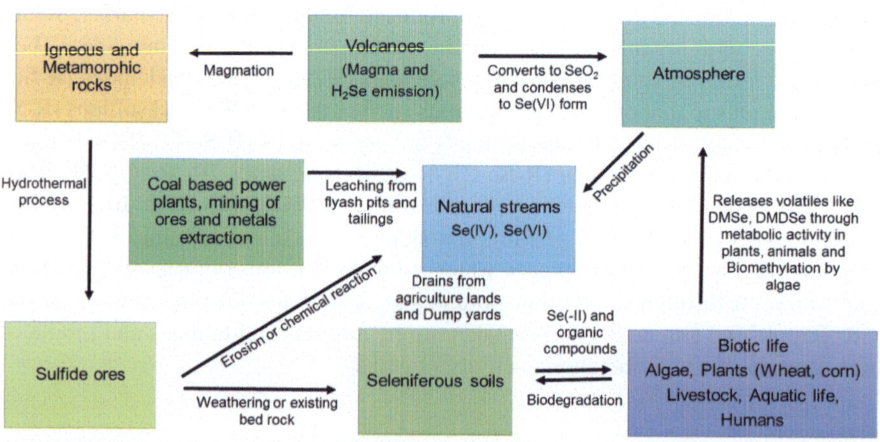

Fig. 12.2 Flow diagram showing biogeochemical processes that cycle selenium in the environment

into water bodies. The biogeochemical process, like methylation, can release volatile Se organic compounds such as dimethyl selenide (DMSe) and dimethyl diselenide (DMDSe) into the atmosphere (Wallschlager and Feldmann 2010). Freshwater algae such as *Chlorella sp.* methylate ~90% of selenite (20 μM) present in the tailings from the coal-based power plant to DMSe (Wallschlager and Feldmann 2010). Microorganisms like *Corynebacterium* produces volatiles from dissolved and undissolved forms of Se present in the soil.

12.5 Speciation in Environment

Selenium is a redox-sensitive element and it can exists in various oxidation states including 0, +1, +2, +3, +4, +5, +6, and −2 However, the principal oxidation states are +6, +4, −2, and 0 (Mehdi et al. 2013). Each oxidation state shows distinct chemical behavior. Table 12.2 shows the common selenium species found in natural systems. The speciation of Se may vary according to the origin of Se. For instance, tailings from coal-fired power plants contain Se in inorganic forms (chemical action), whereas drainage from agricultural fields may contain both organic and inorganic forms (bacterial action). The mobility and bioavailability of Se species can vary based on the processes such as precipitation, adsorption, and biotransformation which are influenced by the pH, amount of competing ions, organic matter, and redox conditions (Fernández-Martínez and Charlet 2009). The chemicals speciation not only affects the mobility and bioavailability; it also matters in the point of environmental remediation and toxicity. Pourbaix diagram in Fig. 12.3 shows the stability of Se compounds in water and soil with respect to the pH, activity of chemical species (pE) and redox potential (Eh) (Brookins 1988).

12.6 Contamination of Selenium in the Environment

Irrigation discharges, discharge from coal-based power plants, leaching from the mining industry or existing geological are the principal causes of Se contamination in the environment. The first episode of the contamination was documented in 1950 at Kesterson Reservoir, California. The pollution was attributed to the discharge of sub-surface irrigation drainage from the agriculture fields of San Joaquin Valley to Kesterson reservoir, a pond system within Kesterson National Wildlife Refuge (KNWR). The twelve interconnected evaporation ponds were contaminated with nearly 350 μg/L of Se (Presser and Ohlendorf 1988). Belews Lake in North Carolina is contaminated due to discharge from flyash settling basin containing 150–200 μg/L of Se (Lemly 1999). A similar case was reported in Martin reservoir, Texas, where discharges from 2 flyash basins generate 1–34 μg/L of Se in the water body. It created a severe impact on fish species like green sunfish and redear sunfish with damage of gills, heart, liver, and kidneys (Lemly 1999). Contamination of Se in

Table 12.2 The list of common selenium species found in natural systems (Rayman 2000, 2012; Papp et al. 2007)

Species/enzyme	Chemical formula	Description
Inorganic species		
Se (VI)	SeO_4^{-2}, H_2SeO_4 and $HSeO_4^-$	Highly mobile; present in aerated soils and alkaline waters with Eh 100–400 mV (pK_a—3.0)
Se (IV)	SeO_3^{-2} and $HSeO_3^-$	Present in moderately reduced soils and slight acidic waters with Eh −400 to 100 mV (pK_a—2.6)
Elemental selenium	Se(0)	Precipitated after microbial reduction and to less extent by inorganic process
Se (-II)	H_2Se	Present with metals in ores; forms volatile compounds by biological or chemical processes (pK_a—3.8)
Organic species		
Dimethyl selenide	C_2H_6Se	Produced by biomethylation of dead matter in aqueous systems; emitted from non-accumulator plants like alfalfa species
Dimethyl Diselenide	$C_2H_6Se_2$	Produced by biomethylation of dead matter in aqueous systems; produced by accumulator plants like astragalus species
Selenocysteine	$C_3H_7NO_2Se$	Selenoamines produced in plants and it is building blocks of selenoproteins in animals and humans
Selenomethionine	$C_5H_{11}NO_2Se$	Produced in plants and a precursor of dimethyl selenide in plants

(continued)

Table 12.2 (continued)

Species/enzyme	Chemical formula	Description
Selenoproteins		
Glutathione peroxidase (GPx)		Acts as antioxidant enzyme and help to remove hydrogen peroxides, lipid, phospholipid and cholesterol hydroperoxides at cytosol, gastro intestine, plasma, etc.
Iodothyronine deiodinase (Dio)		Regulates the production of active thyroid hormone T3 (Triiodothyronine) from Thyroxin, T4 in specified places—thyroid, liver, kidney (Dio_1), brain pituitary, muscle (Dio_2), cerebral cortex, skin, placenta, pregnant uterus (Dio3)
Selenoprotein P		Contains 10 SeC residues per protein subunit; through plasma, it transports Se from liver to brain, kidney; and testis; protects endothelial cell against damage from peroxynitrite
Thioredoxin reductases [TexR]		Meant for regeneration of antioxidant systems and reduction of nucleotides for DNA synthesis
Selenoprotein (15 kDa)		Located in endoplasmic reticulum (ER) and controls glycoprotein folding within ER

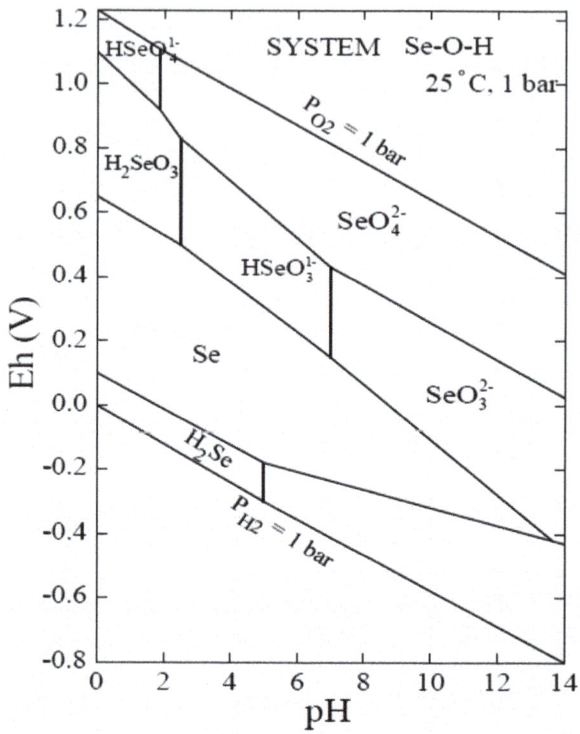

Fig. 12.3 Eh-pH diagram for Se–O–H system. Reprinted from Eh-pH diagrams for geochemistry with permission of Springer

water bodies is significant in North America due to discharge from power plants and mining industries. The presence of Se in the environment is also reported from other parts of the world as well. The major occurrences are compiled in Fig. 12.4.

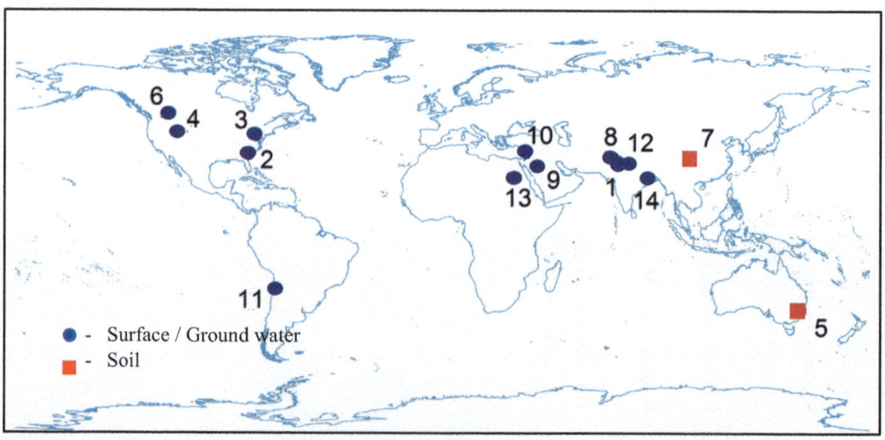

Fig. 12.4 Representation of Se contaminated places in world

Point	Location	Concentration (μg/L)	Description	Reference
1	Jainpur and Barwa, Punjab, India	45–241	Irrigation with Se enriched water and high content in soil due to deposition by flood water from Shivalik range, are the cause of selenosis in that area	(Bajaj et al. 2011)
2	Savannah river, South Carolina	100	Direct release of ash slurry from coal based powerplant affected aquatic life	(Cherry and Guthrie 1977)
3	Appalachian mountain	42	Major rivers and its tributaries near to mountain range are polluted due to surface coal mining	(Tan et al. 2016)
4	Black foot river, Idaho	>400	Intense phosphate mining elevated harmful levels in aquatic food web by polluting river	(Young et al. 2010)
5	Lake Macquarie, Australia	2 mg/kg	Ash dams leachings and stack emissions from lead and zinc smelters, coal based power plants	(Schneider et al. 2017)
6	Elk River Valley	>300	Leaching from waste rock deposits into black foot river effected aquatic life	(Tan et al. 2016)
7	Naore Valley, China	10–30 mg/kg	High Se intake from crops growing in seleniferous soils derived from pyrite bed rock and black shale effecting biotic life	(Kunli et al. 2004)
8	Soan Sakesar Valley, Pakistan	2103	Main source of Se in ground water is from limestone formation, rich in shales and fossils	(Afzal et al. 2000)

(continued)

(continued)

Point	Location	Concentration (μg/L)	Description	Reference
9	Amman Zarqa Basin, Jordan	0.09–742	Phosphate rock is the main source increasing ground water levels. Use of fertilizers made from phosphate rock is also a reason	(Al and Abdel-Fattah 2010)
10	Yarmouk Basin, North Jordan	10.6–53.6 (dry) 7.3–47.2 (wet)	Se mobilisation into soils by use of phosphate fertilizers containing Se. Through irrigation it percolated into ground	(Al-Taani et al. 2012)
11	Atacama Desert, Chile	<800	Groundwater contains Se near to copper deposits	(Leybourne and Cameron 2008)
12	Sirmaur district, Himachal Pradesh	2150–4475	Se source is from geological strata and industries	(Kashyap et al. 2018)
13	Lower Olifant river catchment	207.82	Tribal people relied on that source are effected by draining from open coal mining into system	(Genthe et al. 2018)
14	East Bokaro coalfield, Jharkand	1.3–6.9	Open coal mining	(Mahato et al. 2016)

12.7 Effects of Selenium on Humans and Animals Health

Selenium is a necessary supplement for animals and human if it is taken in the required amounts. The bioavailability and toxicity of Se can vary based on its chemical forms, exposure time, and concentration. In humans, low intake is associated with Keshan and Kashin beck diseases. Keshan disease is multifocal myocarditis linked with cardiac enlargement and heart failure. In Kashin beck disease, lesions appear in the cartilage of the whole skeleton system. The typical symptoms are joint pains, stiffness in extremities, and shortening of hands and feet (Kolsteren 1992). Less intake of Se can cause white muscle disease, a myopathy of heart, and skeletal muscles in animals (Delesalle et al. 2017). It can also affect the reproductive system with deformed offspring or reduced fertility. However, excess intake can results in alkali disease with symptoms like emaciation, deformation and shedding of hooves, loss of hair, erosion of the joints in long bones, and loss of vitality (Moxon 1937). Nevertheless, uptake of Se in measured quantity benefits in cancer reduction, improve the immune system and decrease particular cardiovascular problems due to oxidation effects (Hatfield et al. 2014). On the contrary, high intake may revert to peroxidation damage to DNA by the generation of oxygen radicals (Lee and Jeong 2011).

Aquatic species are more sensitive to bioaccumulation of water-soluble Se. For fish, the dietary threshold should be within 3 μg/g, and body Se residue should be close to 4 μg/g (Hamilton 2004). But it is worth to note that the adverse effects would vary based on factors such as age, gender, the chemical form present in the diet, transport capacity through cellular cytoplasmic membranes, the efficiency of bioconversion from the inorganic to organic forms as well as exposure duration, frequency, and exposure route (Lee and Jeong 2011).

12.8 Water Quality Criteria and Standards

Drinking water and food, especially animal protein, are the predominant sources of Se intake in human. Hence, ingestion of Se from these sources should be taken into account while fixing the permissible limits for drinking water. Considering this, each country has set their drinking water standards based on the average total daily intake (TDI) in mg/kg or μg/day and percentage of Se contribution from drinking water. Separate guidelines are given for surface waters, as the permissible limit for aquatic animals is 10 times lesser than that of in drinking water. World Health Organization (WHO) 2011 has set provisional drinking water limit to 40 μg/L based on 20% of the upper tolerable intake of 400 μg/day with daily consumption of 2 L of water (WHO 2011). The United States Environment Protection Agency (USEPA) set a value of 50 μg/L (USEPA 2019). Depending on the factors of availability, countries can fix their standards. The standards fixed for drinking water and aquatic animals in various countries are given in Table 12.3 (Kumkrong et al. 2018).

Table 12.3 Water quality standards of Se for various matrix

Function	Place	Limit (μg/L)
Drinking water	Canada:	
	i. British Columbia, Ontario, Quebec	10
	ii. Alberta, Nova Scotia, Prince Edward Island	50
	Germany, Ireland, Japan, India, Malaysia, Russia	10
	Abu Dhabi	40
	United States:	
	i. California, Colorado, Texas, Wyoming	50
	ii. Idaho	170
	South Africa	20
Fresh water for aquatic life	United States:	
	i. Colorado, Utah, Pennsylvania	4.6
	ii. North Carolina, South Carolina, Washington	5
	Australia and New Zealand	10
	Germany	3

12.9 Sampling Techniques

Sampling procedure involves the selection of sites and storage container and preservation of samples for the analysis. In the case of river water, leachate from landfills, and wastewater effluents, a clear understanding of factors affecting the distribution of Se is required to select the sampling sites where spatial distribution is not homogeneous. The samples can be a grab or integrated sample. Grab samples can be taken at a required depth. In the case of an integrated sample, two or more samples are mixed based on depth, area, time of collection, and discharge. For wastewater effluents, samples are taken near to discharge outlet where the effluent is well mixed with a water source. Groundwater samples are taken from existing bore wells at various depths or screening or from the test wells after pumping out the water for some time. The EPA, WHO, and IS follow a similar method of sample preservation. According to the guidelines, sample bottles are to be rinsed with water followed by HNO_3 and distilled water. Other conditions are as follows (USEPA 2016; UNEP/WHO 1996):

Container	Polyethylene or PTFE
Preservative	Sample adjusted to pH < 2 by (1:1) HNO_3
Holding time	For acidified sample 6 months if not 14 days

The procedure for total recoverable Se of the aqueous sample containing >1% (w/v) suspended solids is given in USEPA, method—200.2 (USEPA 1994a). The sample prepared by this method can be analyzed by Inductively Coupled Plasma Optical Emission Spectrometer (ICP-OES), Inductively Coupled Plasma Mass Spectrometer (ICP-MS) or Stabilized Temperature Graphite Furnace Atomic Absorbance Spectrometer (ST-GFAAS). Conventional flame and graphite furnace technique is not recommended due to chloride in the sample, which may reduce analyte signals. The typical sample preparation procedure is as follows: Step1: To 100 mL sample in a Griffin beaker, add 2 mL (1:1) HNO_3 and 1 mL (1:1) HCl and heat at 85 °C on a hot plate. Step 2: Reduce the volume to 50 mL and reflux for 30 min more. Step 3: Cool it, remove undissolved solids, and dilute it to 50 mL for analysis (USEPA 1994a).

12.10 Analytical Techniques for Measurement of Selenium

Selenium is a trace metal with a minuscule gap between beneficial and toxicity limit. Hence, sensitive analytical techniques are essential to understand its presence in biological and environmental samples. Over the years, a variety of analytical techniques are developed to determine the trace quantities of Se in environmental and biological samples. These include Atomic Absorbance Spectrometry (AAS), ICP-OES, ICP-MS, Gas Chromatography-Mass Spectrometry (GC-MS), and UV-VIS spectrophotometry. Several other techniques, including X-ray fluorescence analysis, neutron activation analysis (NAA), fluorimetry, etc. are also developed. A few of the established and commonly used techniques are discussed in this section.

12.10.1 Spectrophotometric Method

A typical spectrophotometer comprises a spectrometer, which produces electromagnetic radiation of any wavelength, and a photometer to measure the intensity of the electromagnetic radiation. The equipment works based on the absorption of light in relation to the molecular structure of the analyte. Based on the wavelength and the intensity of radiation absorbed, the analyte is quantified—equipment such as UV-VIS spectrophotometer and AAS work based on this principle.

UV-VIS Spectrophotometric method: Diaminonaphthalene method (IS 3025 (Part-56) 2003) is an established spectrophotometric method. In this method, the analyte Se(IV) is made to react with 2,3-diaminonaphthalene (DAN) to form fluorescent piazselenol with an absorbance peak at 480 nm. Cyclohexane is used to extract piazselenol from the final compound (BIS 2003). To analyze total Se [Se(IV) and Se(VI)] with no organic selenium compounds, 10 ml of the sample is digested with 10 ml conc. HCl for 60 min at 95 °C in a water bath. The minimum detection limit is 2 μg/L, and the working range is 10–2000 μg/L. The possible interference due to colored organics extractable by cyclohexane is removed by ion exchange resin such as Amberlite XAD-8. The negative interference from Fe^{+2} (up to 2.5 mg) can be eliminated by adding ethylenediaminetetraacetic acid (EDTA). In another approach, Ulusoy (2015) has demonstrated a cloud point extraction technique to separate Se(IV) by complexation with 4,5-diamino-6-hydroxy-2-mercapto pyrimidine and extracted into the micelle phase of surfactants like Triton X-114 and sodium dodecyl sulfate (Ulusoy 2015). The surfactant-rich phase is separated by centrifugation and diluted with 20% ethanol prior to analysis by UV-VIS spectrophotometer. The absorbance peak is at 458 nm. The minimum detection limit is 6.06 μg/L, and the working range is 20–1500 μg/L. The recovery rate of Se(IV) of 250 μg/L from water in the presence of cations and anions is 95–98%.

Atomic absorption or emission spectrometric methods: This is a single element analysis technique. The technique uses the absorption or emission spectrum of free atoms present in a gaseous sample to measure the target analyte present in it. Here, the sample is atomized with the help of a flame or electro-thermal method. The methods, including hydride generation, glow-discharge, and cold-vapor atomization are also employed. In the conventional flame technique (FAAS), air-acetylene is the source for the flame to produce the required temperature for atomization, and its working principle is graphically represented in Fig. 12.5. The sample passes through a nebulizer to form aerosols and emit into the flame. The radiation source is generated from a hollow cathode lamp (HCL) or electrodeless discharge lamp (EDL).

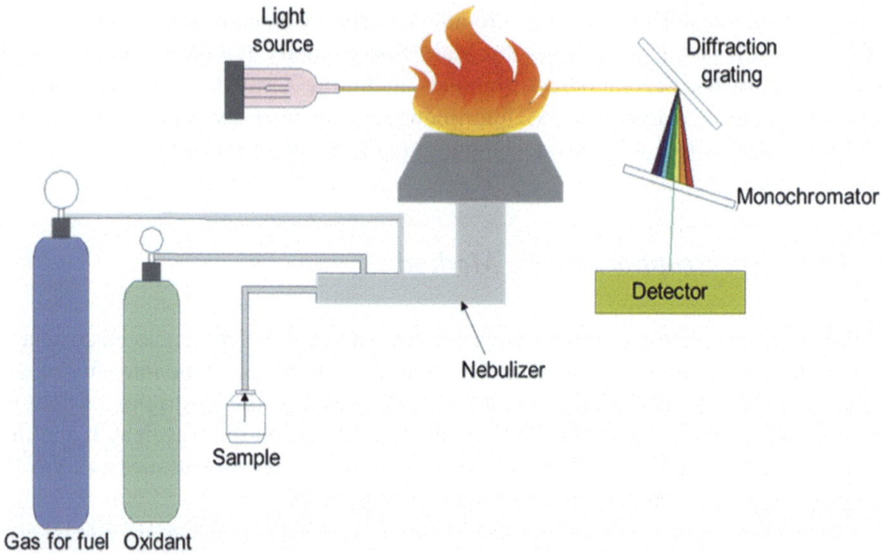

Fig. 12.5 Schematic diagram showing the working principle of FAAS

However, this method is relatively less sensitive and not have sufficient lower detection limit for measuring selenium concentration in water samples. Another limitation of FAAS is the background absorption of incident radiation by the flame, especially in the UV region (absorption peak for Se is 196.1 nm) (Rann and Hambly 1965). Almost 70 and 50% of the radiation is absorbed by air-C_2H_2 and hydrogen-air flame, and the limit of detection is 0.2–3 and 1 mg/L of Se, respectively (Verlinden et al. 1981). The use of NO-C_2H_2 atomized the refractory oxide compounds, but the detection limits to 1.8–2.5 mg/L (Verlinden et al. 1981).

Unlike FAAS, flameless, or graphite furnace ASS (GF-AAS) uses a lined graphite furnace to atomize the sample, which is shown in Fig. 12.6. This technique is also known as Electrothermal Atomic Absorption Spectroscopy (ETAAS). Here, a graphite tube receives the sample and which is then heated to automize the analyte.

The method is more sensitive. However, light scattering and atomic absorption by the sample matrix can limit its application. Moreover, the acidic sample causes intercalation of graphite at high temperatures and eventually releases at the time of analyte atomization (Bulska 2009). High matrix sample containing chloride, sodium, potassium, and sulfate gives positive interference to the analyte due to molecular absorption. Proper control of temperature is required for complete pyrolysis of the matrix. Matrix modifiers are used to achieve matrix pyrolysis at low temperature by keeping the analyte stable.

It enhances absorbance signal of analyte either by ionizing the salts before the target analyte or making thermally stable compounds with metal modifiers. Platinum molybdenum and Nickel are commonly used as metal modifiers. These modifiers can

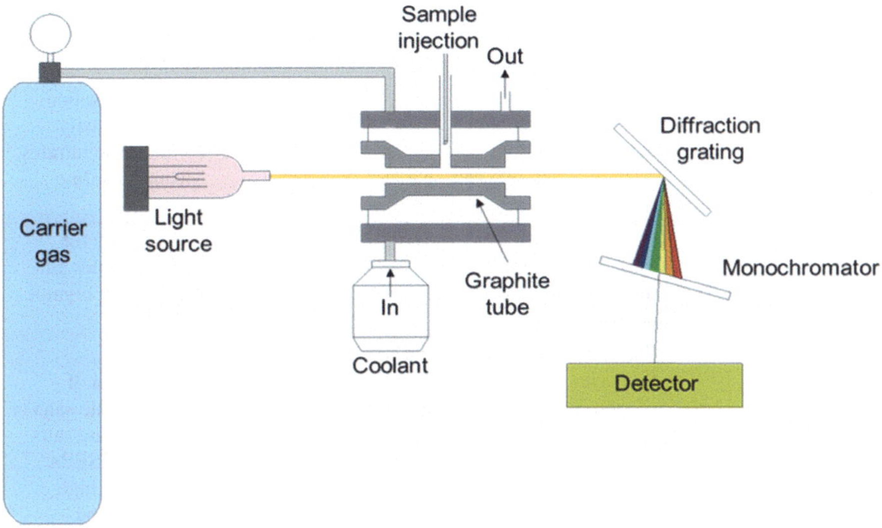

Fig. 12.6 Schematic diagram showing the working principle of GF-AAS

react with Se and give thermally stable metal selenides. Different types of modifiers and their functions are summarized in Table 12.4 (Bulska 2009).

As a substitute to matrix modifiers, hydride generation technique (HGAAS) came into use where the analyte is separated from the sample in the form of volatile hydride, which is free from the sample matrix. However, the method demands a large sample volume compared to GF-AAS technique. In this method, the dissolved Se(IV) present in the sample is reduced to Se(-II) by reducing agents such as $NaBH_4$, zinc, and stannous chloride in acidic medium. Depending on the dose of chemicals, sample matrix, order of adding reducing agent and acid, and reaction time for hydride formation create interference in the signals from Se present in the sample (Pierce and Brown 1976). Trace elements such as Ni, Co, and Cu can reduce the analyte signal. Ni and Co can form their respective hydrides by oxidizing Se hydrides, and Cu forms insoluble copper selenide (Bax et al. 1988). Hydride generation by $NaBH_4$ is high if 4 N HCl is used to acidify the sample (McDaniel et al. 1976). Addition of 0.5% $NaBH_4$ prior to 1:1 HCl is not advisable and can will reduce Se signal 100%. This is due to absorption by Cd^{2+}, Co^{2+}, Ni^{2+}, Cu^{2+}, V^{2+}, Pb^{2+}, Sn^{2+}, Mo^{3+}, Ag^{+}, Zn^{2+}, MnO_4^{-}, VO_3^{-}, $S_2O_8^{2-}$, and NO_3^{-} with Se(IV) concentration greater than 16.7 mg/L (Pierce and Brown 1976). HCl addition before $NaBH_4$ can eliminate the above-mentioned cations interference. The concentration of Cu, Pb, and Ni greater than 1 mg/L, and hydride forming elements like Bi, Sn, Te, and Sn in the range of 0.1–1 mg/L can inhibit the Se signals (APHA 1992). The elements such as Ca^{2+}, Mg^{2+}, K^{+}, Na^{+}, SO_4^{2-}, and Cl^{-} concentration up to 333 mg/L has no effect on the analyte signals above 1 μg/L of Se (Pierce and Brown 1976). Presence of CrO_7^{2-}, VO_3^{-}, and MoO_4^{-} above 6.7 mg/L can reduce the signals to 40, 60 and 40% respectively, but do not have a major impact in the real scenario due to its rare presence in drinking water

Table 12.4 Modifiers typically used in GFAAS

Modifier	Function	Compound	Comment
Matrix modifier	Increases the volatility of sample matrix	NH_4NO_3	Converts alkaline halides to easily combustible nitrates. Meant for saline water
Matrix/analyte modifier	Similar to matrix modifier and reduces the volatization of analyte	H_3PO_4	Can be used for saline samples and degradation organic matter
Metal modifier	Stabilizes Se, As and Sb during pyrolysis of matrix	$Ni(NO_3)_2$, $Mg(NO_3)_2$, $Pd(NO_3)_2$	Typically, combination of Ni-Mg nitrate salts or Pd-Mg nitrate salts are used (USEPA 200.9). But high usage of these modifiers may create memory interference which cannot precisely analyze that metals further
Gaseous modifier	Additional gas is purged into Ar medium during matrix pyrolysis	H_2 and O_2	It prevents the e-oxidation of analyte hydrides by forming respective acids like HCl, H_2SO_4. But O_2 is corrosive above 500 °C
Long term performance modifier	By electroplating inside graphite tube	Palladium and iridium coating	Can withstand up to 2400 °C with well intercalated in graphite layers

(Pierce and Brown 1976). The method is suitable for analyzing inorganic Se only. To analyze organoselenium compounds as total Se, the sample should be pretreated, exposing to UV radiation to convert Se(-II) and Se (0) to Se VI). The detection limit of Se in drinking water as per US EPA 200.9 is <0.6 μg/L (USEPA 1994b).

12.10.2 Plasma Spectrometric Methods

ICP-MS is one of the sophisticated techniques employed for analyzing Se in an aqueous medium. The instrument comprises of ICP and mass spectrometer (MS), and its working principle is shown in Fig. 12.7. The RF coils present at one end of the

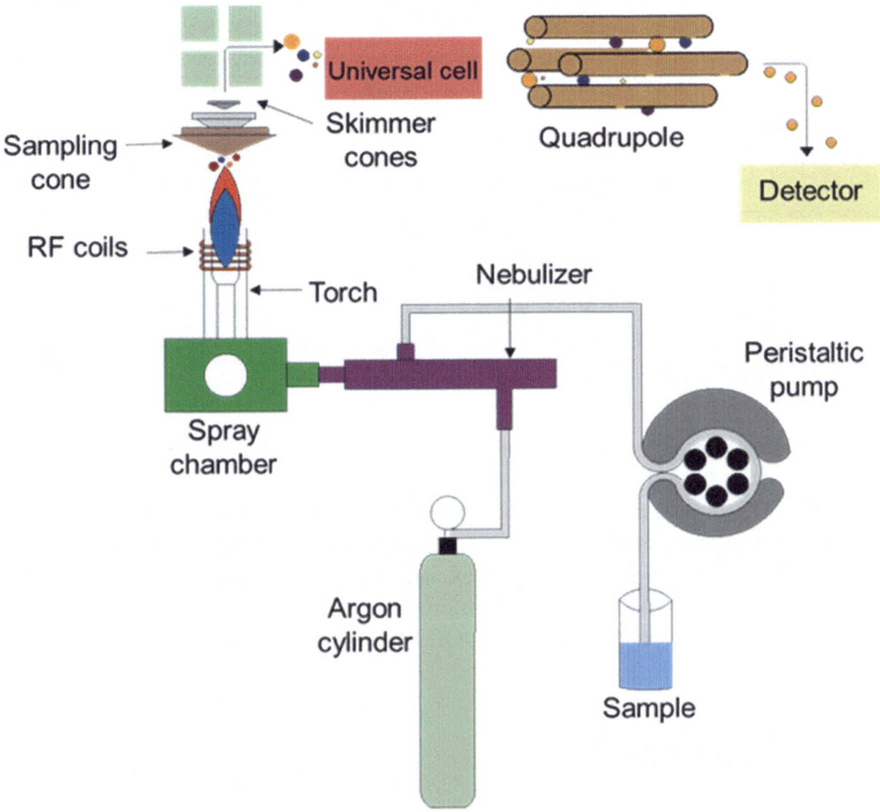

Fig. 12.7 Schematic diagram showing the working principle of ICP-MS

ICP torch ionizes the Argon gas and the emitted electrons in the magnetic field creates high-temperature plasma source at 5500–6500 K temperature. At this temperature, sample aerosol will undergo drying, dissociation, and ionization. The charged ions are then directed to the MS wherein they are separated depending on mass to charge ratio (m/z). An electron multiplier tube detector identifies and quantifies each ion.

The technique is more robust and suitable for removing interference due to anions. However, results can be deviated due to polyatomic and isobaric interference and elemental carryover. The polyatomic ions ($^{40}Ar_2^+$, $^{38}Ar_2^+$, $^{36}Ar_2^+$) can create interference in the analysis of Se by reacting with chlorine, calcium, sulfur, and nitrogen (Kumkrong et al. 2018). Isobaric interference occurs when other element isotopes resemble Se. Table 12.5 summarizes possible Isobaric and polyatomic interference of Se by ICP-MS. These can be avoided by using different cell modes—collision and reaction cell mode. Collision cell mode breaks the polyatomic ions by colliding with other inert gas atoms. In the reaction cell, interference is eliminated by shifting Se to different mass spectrum. In the gas phase of Ar and NH_3, new polyatomic species is

Table 12.5 Interferences for Se isotope in ICP-MS

Isotope	Abundance (%)	Isobaric interference	Polyatomic interference
^{74}Se	0.87	^{74}Ge	$^{37}Cl_2^+$, $^{36}Ar^{38}Ar^+$, $^{38}Ar^{36}S^+$, $^{40}Ar^{34}S^+$
^{76}Se	9.02	^{76}Ge	$^{40}Ar^{36}Ar^+$, $^{40}Ar^{36}S$
^{77}Se	7.58	$^{154}Gd^{2+}$	$^{36}Ar^{40}Ar^{1}H^+$, $^{40}Ar^{37}Cl^+$, $^{38}Ar_2^1H^+$, $^{12}C^{19}F^{14}N^{16}O_2^+$
^{78}Se	23.52	$^{78}Kr^+$, $^{156}Gd^{2+}$	$^{40}Ar^{38}Ar^+$, $^{38}Ar^{40}Ca^+$
^{80}Se	49.82	$^{80}Kr^+$, $^{160}Gd^{2+}$	$^{40}Ar_2^+$, $^{32}S^{16}O_3^+$
^{82}Se	9.19	$^{82}Kr^+$	$^{12}C^{35}Cl_2^+$, $^{40}Ar_2^1H_2^+$, $^{34}S^{16}O_3^+$

Reprinted from selenium analysis in waters. Part 1: Regulations and standard methods with permission of Elsevier

formed under the mechanism of atom transfer, charge exchange, formation of new compounds, condensation, and analyte association/condensation (D'llio et al. 2011).

Besides MS, atomic emission or optical emission spectroscopic (AES/OES) technique is also employed to quantify the analyte. However, in ICP-OES, the quantification is done based on the analysis of excited atoms and ions at the characteristics wavelength of the analyte. ICP-MS is more sensitive with parts per trillion as the lowest limit of detection. However, ICP-OES can detect only up to parts per billion.

12.11 Conclusion

Selenium is a trace nutrient with proven health benefits. However, the ingestion of Se above the required limit can be detrimental to human and animals. Though its average abundance in earth crust is only 0.5 mg/g, the industrial and agricultural activities can accelerate its release into the environment. The coal and phosphate mining, metal extraction, and large-scale pumping of groundwater are major causes of enhanced release of Se into nature. The volcanic eruptions, biogeochemical processes, and weathering of rocks and soils can also release Se into the environment. However, the chemical forms, exposure time, and its concentration dictate the bioavailability and toxicity. Se intake in human is mostly through drinking water and consumption of specific food. Hence, human exposure to Se may vary from one region to another region. Due to the pros and cons of Se intake and variability in exposure, it is difficult to fix a universal Se standard, and the standard should be region specific. Developing detection technologies with high sensitivity is also essential to achieve the overall goal. Spectrophotometric methods can detect the element in μg/L in low matrix samples. However, the technique such as ICP-MS and ICP-OES are proven effective for a wide range of sample matrix.

Acknowledgements Authors gratefully acknowledge the Department of Science and Technology (DST), Technology Mission Division, Government of India [Grant no: DST/TM/WTI/WIC/2K17/82(C)]. The authors would like to express gratitude towards Springer publisher for rendering license to reuse Eh-pH diagram of selenium from the book chapter 'Eh-pH diagrams for geochemistry'. We are grateful for Elsevier publisher for allowing to use table of isobaric and polyatomic interferences for selenium analysis by ICP-MS from 'Selenium analysis in waters. Part 1: Regulations and standard methods'. The authors also thank IIT Tirupati for the support.

References

Afzal S, Younas M, Ali K (2000) Selenium speciation studies from Soan-Sakesar Valley, Salt Range, Pakistan. Water Int 25:425–436. https://doi.org/10.1080/02508060008686850

Al Kuisi M, Abdel-Fattah A (2010) Groundwater vulnerability to selenium in semi-arid environments: Amman Zarqa Basin, Jordan. Environ Geochem Health 32:107–128. https://doi.org/10.1007/s10653-009-9269-y

Al-Taani AA, Batayneh A, El-Radaideh N, Al-Momani I, Rawabdeh A (2012) Monitoring of selenium concentrations in major springs of Yarmouk Basin, North Jordan. World Appl Sci J 18:704–714. https://doi.org/10.5829/idosi.wasj.2012.18.05.3181

APHA (1992) Method 3114: standard methods for the examination of water and wastewater

Bajaj M, Eiche E, Neumann T, Winter J, Gallert C (2011) Hazardous concentrations of selenium in soil and groundwater in north-west India. J Hazard Mater 189:640–646. https://doi.org/10.1016/j.jhazmat.2011.01.086

Bax D, Agterdenbos J, Worrell E, Kolmer JB (1988) The mechanism of transition metal interference in hydride generation atomic absorption spectrometry. Spectrochim Acta Part B: Atom Spectrosc 43:1349–1354. https://doi.org/10.1016/0584-8547(88)80174-5

Bellinger FP, Raman AV, Reeves MA, Berry MJ (2009) Regulation and function of selenoproteins in human disease. Biochem J 422:11–22. https://doi.org/10.1042/BJ20090219

BIS (2003) Methods of sampling and test (physical and chemical) for water and wastewater, Part 56: selenium. BIS, New Delhi

Bleiwas DI (2010) Byproduct mineral commodities used for the production of photovoltaic cells. USGS Circular 1365, 10p. https://pubs.usgs.gov/circ/1365/

Brookins DG (1988) Selenium. In: Eh-pH diagrams for geochemistry. Springer, Berlin, Heidelberg. https://doi.org/10.1007/978-3-642-73093-1_4

Bulska E (2009) Modifiers in graphite furnace atomic absorption spectrometry – mechanisms and applications. Encyclopedia of analytical chemistry: applications, theory and instrumentation. https://doi.org/10.1002/9780470027318.a9125

Cherry DS, Guthrie RK (1977) Toxic metals in surface waters from coal ash. J Am Water Resour Assoc 13:1227–1236. https://doi.org/10.1111/j.1752-1688.1977.tb02093.x

Cummins LM, Kimura ET (1971) Safety evaluation of selenium sulfide antidandruff shampoos. Toxicol Appl Pharmacol 20:89–96. https://doi.org/10.1016/0041-008X(71)90092-5

D'llio, Violante N, Majorani C, Petrucci F (2011) Dynamic reaction cell ICP-MS for determination of total As, Cr, Se and V in complex matrices: still a challenge? A review. Anal Chim Acta 698:6–13. https://doi.org/10.1016/j.aca.2011.04.052

Delesalle C, Brujin MD, Wilmink S, Vandendriessche H, Mol G, Boshuizen B, Plancke L, Grinwis G (2017) White muscle disease in foals: focus on selenium soil content. A case series. BMC Vet Res 13:121–130. https://doi.org/10.1186/s12917-017-1040-5

Fernandes AP, Gandin V (2015) Selenium compounds as therapeutic agents in cancer. Biochim Biophys Acta: Gen Subj 1850:1642–1660. https://doi.org/10.1016/j.bbagen.2014.10.008

Fernández-Martínez A, Charlet L (2009) Selenium environmental cycling and bioavailability: a structural chemist point of view. Rev Environ Sci Biotechnol 8:81–110. https://doi.org/10.1007/s11157-009-9145-3

Genthe B, Kapwata T, Roux WL, Chamier J, Wright CY (2018) The reach of human health risks associated with metals/metalloids in water and vegetables along a contaminated river catchment: South Africa and Mozambique. Chemosphere 199:1–9. https://doi.org/10.1016/j.chemosphere.2018.01.160

Germani MS, Small M, Zoller WH, Moyers JL (1981) Fractionation of elements during copper smelting. Environ Sci Technol 15:299–305. https://doi.org/10.1021/es00085a005

Gupta M, Gupta S (2017) An overview of selenium uptake, metabolism, and toxicity in plants. Front Plant Sci 7:1–14. https://doi.org/10.3389/fpls.2016.02074

Gupta J, Fatima MT, Islam Z, Khan RH, Uversky VN, Salahuddin P (2019) Nanoparticle formulations in the diagnosis and therapy of Alzheimer's disease. Int J Biol Macromol 130:515–526. https://doi.org/10.1016/j.ijbiomac.2019.02.156

Hamilton SJ (2004) Review of selenium toxicity in the aquatic food chain. Sci Total Environ 326:1–31. https://doi.org/10.1016/j.scitotenv.2004.01.019

Hatfield DL, Tsuji PA, Carlson BA, Gladyshev VN (2014) Selenium and selenocysteine: roles in cancer, health and development. Trends Biochem Sci 39:112–120. https://doi.org/10.1016/j.tibs.2013.12.007

He Y, Xiang Y, Zhou Y, Yang Y, Zhang J, Huang H, Shang C, Luo L, Gao J, Tang L (2018) Selenium contamination, consequences and remediation techniques in water and soils: a review. Environ Res 164:288–301. https://doi.org/10.1016/j.envres.2018.02.037

Hoffmann JE, King MG (2001) Selenium and selenium compounds. Kirk-Othmer Encycl Chem Technol. https://doi.org/10.1002/0471238961.1905120508150606.a01.pub2

Indian Bureau of Mines (2017) Indian minerals yearbook 2017 (Part-II: metals & alloys) selenium. Government of India, Ministry of Mines, Nagpur, 7p

Institute of Medicine (2000) Dietary reference intakes for vitamin C, vitamin E, selenium, and carotenoids. Institute of Medicine (US) Panel on Dietary Antioxidants and Related Compounds, National Academies Press (US), Washington, DC, 309p. https://doi.org/10.17226/9810

Jones GD, Droz B, Greve P, Gottschalk P, Poffet D, McGranth SP, Seneviratne SI, Smith P, Winkel LHE (2017) Selenium deficiency risk predicted to increase under future climate change. Proc Natl Acad Sci 114:2848–2853. https://doi.org/10.1073/pnas.1611576114

Kashyap R, Verma KS, Uniyal SK, Sk Bhardwaj (2018) Geospatial distribution of metal(loid)s and human health risk assessment due to intake of contaminated groundwater around an industrial hub of northern India. Environ Monitor Assess 190:1–18. https://doi.org/10.1007/s10661-018-6525-6

Khamkhash A, Srivastava V, Ghosh T, Akdogan G, Ganguli R, Aggarwal S (2017) Mining-related selenium contamination in Alaska, and the state of current knowledge. Minerals 7:1–13. https://doi.org/10.3390/min7030046

Kirkpatrick FA, Roberts GG (1919) Production of selenium red glass. J Am Ceram Soc 2:895–904. https://doi.org/10.1111/j.1151-2916.1919.tb18751.x

Kolsteren P (1992) Kashin-beck disease. Ann Soc Belg Med Trop 72:81–91. http://193.190.239.98/bitstream/handle/10390/4908/1992asbm0081.pdf?sequence=1

Kumkrong P, Leblanc KL, Mercier PHJ, Mester Z (2018) Selenium analysis in waters. Part 1: regulations and standard methods. Sci Total Environ 640–641:1611–1634. https://doi.org/10.1016/j.scitotenv.2018.05.392

Kunli L, Lirong X, Jian'an T, Douhu W, Lianhua X (2004) Selenium source in the selenosis area of the Daba region, South Qinling Mountain, China. Environ Geol 45:426–432. https://doi.org/10.1007/s00254-003-0893-z

Lee KH, Jeong D (2011) Bimodal actions of selenium essential for antioxidant and toxic pro-oxidant activities: the selenium paradox (review). Mol Med Rep 5:299–304. https://doi.org/10.3892/mmr.2011.651

Lemly AD (1999) Selenium impacts on fish: an insidious time bomb. Hum Ecol Risk Assess: Int J 5:1139–1151. https://doi.org/10.1080/10807039.1999.10518883
Lemly AD (2004) Aquatic selenium pollution is a global environmental safety issue. Ecotoxicol Environ Saf 59:44–56. https://doi.org/10.1016/S0147-6513(03)00095-2
Leybourne MI, Cameron EM (2008) Source, transport, and fate of rhenium, selenium, molybdenum, arsenic, and copper in groundwater associated with porphyry-Cu deposits, Atacama Desert, Chile. Chem Geol 247:208–228. https://doi.org/10.1016/j.chemgeo.2007.10.017
Lu J, Dreisinger D, Glück T (2014) Manganese electrodeposition – a literature review. Hydrometallurgy 141:105–116. https://doi.org/10.1016/j.hydromet.2013.11.002
Mahato MK, Singh PK, Tiwri AK, Singh AK (2016) Risk assessment due to intake of metals in groundwater of East Bokaro coal field. Expos Health 8:265–275. https://doi.org/10.1007/s12403-016-0201-2
McDaniel M, Shendrikar AD, Reiszner KD, West PW (1976) Concentration and determination of selenium from environmental samples. Anal Chem 48:2240–2243. https://doi.org/10.1021/ac50008a047
Mehdi Y, Hornick JL, Istasse L, Dufrasne I (2013) Selenium in the environment, metabolism and involvement in body functions. Molecules 18:3292–3311. https://doi.org/10.3390/molecules18033292
Moxon AL (1937) Alkali disease or selenium poisoning. Bulletins, 331p
Murphy KE, Altman MB, Wunderlich B (1977) The monoclinic-to-trigonal transformation in selenium. J Appl Phys 48:4122–4131. https://doi.org/10.1063/1.323439
National Research Council (1983) Selenium in nutrition, revised edition. The National Academics Press, Washington, DC. https://doi.org/10.17226/40
Navarro-Alarcon M, Cabrera-Vique C (2008) Selenium in food and the human body: a review. Sci Total Environ 400:115–141. https://doi.org/10.1016/j.scitotenv.2008.06.024
Papp LV, Lu JUN, Holmgren A, Khanna KK (2007) From selenium to selenoproteins: synthesis, identity, and their role in human health. Antioxid Redox Signal 9:775–806. https://doi.org/10.1089/ars.2007.1528
Periodic table. http://www.rsc.org/periodic-table/element/34/selenium. Accessed 11 Apr 2019
Perrone D, Monteiro M, Nunes JC (2015) The chemistry of selenium. In: Selenium: chemistry, analysis, functions and effects. The Royal Society of Chemistry, pp 3–15. https://doi.org/10.1039/9781782622215-00003
Pierce FD, Brown HR (1976) Inorganic interference study of automated arsenic and selenium determination with atomic absorption spectrometry. Anal Chem 48:693–695. https://doi.org/10.1021/ac60368a019
Presser TS, Ohlendorf HM (1988) Biogeochemical cycling of selenium in the San Joaquin Valley, California, USA. Environ Manag 11:805–821. https://doi.org/10.1007/BF01867247
Rann CS, Hambly AN (1965) Distribution of atoms in an atomic absorption flame. Anal Chem 37:879–884. https://doi.org/10.1021/ac60226a024
Rayman MP (2000) The importance of selenium to human health. Lancet 356:233–241. https://doi.org/10.1016/S0140-6736(00)02490-9
Rayman MP (2012) Selenium and human health role of selenium: selenoproteins. Lancet 379:1256–1268. https://doi.org/10.1016/S0140-6736(11)61452-9
Sarquis M, Mickey CD (1880) Selenium. Part 1: its chemistry and occurrence. J Chem Educ 57:886–889. https://doi.org/10.1021/ed057p886
Saunders AP (1900) The allotropic forms of selenium. J Phys Chem 6:423–513. https://doi.org/10.1021/j150024a001
Schneider L, Maher W, Potts J, Gruber B, Batley G, Taylor A, Chariton A, Krikowa F, Zawadzki A (2017) Recent history of sediment metal contamination in Lake Macquarie, Australia, and an assessment of ash handling procedure effectiveness in mitigating metal contamination from coal-fired power stations. Sci Total Environ 490:659–670. https://doi.org/10.1016/j.scitotenv.2014.04.055

Stolz A, Baumann T, Frank NH, Ginter TN, Hitt GW, Kwan E, Mocko M, Peters W, Schiller A, Sumithrarachchi CS, Thoennessen M (2005) First observation of ^{60}Ge and ^{64}Se. Phys Lett 627:32–37. https://doi.org/10.1016/j.physletb.2005.08.130

Tan LC, Nancharaiah YV, Hullebusch EDV, Lens PNL (2016) Selenium: environmental significance, pollution, and biological treatment technologies. Biotechnol Adv 34:886–907. https://doi.org/10.1016/j.biotechadv.2016.05.005

Tillman DA, Duong DNB, Harding NS (2012) Environmental aspects of fuel blending. Solid fuel blending: principles, practices and problems, Elsevier, 21p. https://doi.org/10.1016/b978-0-12-380932-2.00006-4

Trofast J (2011) Berzelius' discovery of selenium. Chem Int 33:16–19. https://doi.org/10.1515/ci.2011.33.5.16

Udoye CI, Jafarzadeh H (2010) Xeroradiography: stagnated after a promising beginning? A historical review. Eur J Dent 4:95–99

Ulusoy HI (2015) Simple and useful method for determination of inorganic selenium species in real samples based on UV-VIS spectroscopy in a micellar medium. Anal Methods 7:953–960. https://doi.org/10.1039/c4ay02691h

UNEP/WHO (1996) Field work and sampling. Water Quality Monitoring – a practical guide to the design and implementation of freshwater quality studies and monitoring programmes, 348p (chapter 1)

USEPA (1994a) Sample preparation procedure for spectrochemical determination of total recoverable elements, Method 200.2, Revision 2.8

USEPA (1994b) Determination of trace elements by stabilized temperature graphite furnace atomic absorption. Method 200.9

USEPA (2016) Quick guide to drinking water sample collection

USEPA. National primary drinking water regulations. https://www.epa.gov/ground-water-and-drinking-water/national-primary-drinking-water-regulations. Accessed 28 Feb 2019

USGS (1973) United States mineral resources. U.S. Government Printing Office, Washington, DC, 722p. https://doi.org/10.3133/pp820

USGS (2000) Mineral commodity summaries 2000. USGS, Reston, VA, 196p. https://doi.org/10.3133/70140094

USGS (2004) Mineral commodity summaries 2004. USGS, Reston, VA, 200p. https://doi.org/10.3133/mineral2004

USGS (2015) Mineral commodity summaries 2015. USGS, Reston, VA, 196p. https://doi.org/10.3133/70140094

USGS (2019) Mineral commodity summaries 2019. USGS, Reston, VA, 204p. https://doi.org/10.3133/70202434

Verlinden M, Deelstra H, Adriaenssens E (1981) The determination of selenium by atomic-absorption spectrometry: a review. Talanta 28:637–646. https://doi.org/10.1016/0039-9140(81)80091-4

Vinceti M, Wei ET, Malagoli C, Bergomi M, Vivoli G (2001) Adverse health effects of selenium in humans. Rev Environ Health 16:233–252. https://doi.org/10.1515/REVEH.2001.16.4.233

Wallschlager D, Feldmann J (2010) Formation, occurrence, significance, and analysis of organoselenium and organotellurium compounds in the environment. In: Organometallics in environment and toxicology: metal ions in life sciences. The Royal Society of Chemistry, 6p. https://doi.org/10.1039/9781849730822-00319

Wen H, Carignan J (2007) Reviews on atmospheric selenium: emissions, speciation and fate. Atmos Environ 41:7151–7165. https://doi.org/10.1016/j.atmosenv.2007.07.035

WHO (2011) Guidelines for drinking water quality, 564p

Young T, Finley K, Adams WJ, Besser J, Hopkins WA, Jolley D, McNaughton E, Presser TS, Shaw DP, Unrine J (2010) Appendix A: selected case studies of ecosystem contamination by selenium. Ecological assessment of selenium in the aquatic environment, pp 257–292. https://doi.org/10.1201/ebk1439826775-a1

Yudovich YE, Ketris MP (2006) Selenium in coal: a review. Int J Coal Geol 67:112–126. https://doi.org/10.1016/j.coal.2005.09.003

Chapter 13
Bioleaching of Selected Metals from E-Waste Using Pure and Mixed Cultures of *Aspergillus* Species

Amber Trivedi and Subrata Hait

Abstract Printed circuit board (PCB) is an essential part present in electronic waste (e-waste). Rich metallic content including base, precious, and toxic metals makes PCB a secondary metal reservoir. Recycling of PCB is necessary to conserve natural resources and to protect the environment. Bioleaching process is preferred over the conventional metallurgical techniques for metal recycling from e-waste due to its better efficiency and environmental compatibility. Metal bioleaching from e-waste employing heterotrophic microorganisms like fungi is the comparatively less explored area. In this context, bioleaching of selected metals such as Cu, Ni, and Zn from e-waste in the form of desktop PCB using pure and mixed cultures of *Aspergillus* species was attempted. *Aspergillus niger* was chosen for its organic acid production ability to presumably help in bioleaching of metals. As the metals are usually embedded in the polymer matrix in the PCB, *Aspergillus tubingensis* was selected for its polymer-degrading ability. The bioleaching experiments were performed using pulverized waste PCB (WPCB) in the particle size range of 0.038–1 mm for a period of 33 days at 1 g/L of pulp density. Results showed that the pure culture of *Aspergillus niger* was able to leach a maximum of 71% Cu, 32% Ni, and 79% Zn. Similarly, the corresponding maximum metal leaching efficiency employing a pure culture of *Aspergillus tubingensis* was 54% Cu, 41% Zn, and 14% Ni. Using mixed cultures of *Aspergillus* species, there was a marginal increase in metal bioleaching for Cu and Ni with extraction efficiency of 76 and 36%, respectively. Extraction efficiency of 63% for Zn was observed using the mixed culture. Results indicated the practical feasibility of fungal bioleaching using pure and mixed cultures of *Aspergillus* species for metal recycling from e-waste for prospective beneficiation.

Keywords Electronic waste · Printed circuit board · Metals · Fungal bioleaching · Pure and mixed cultures · *Aspergillus* spp

A. Trivedi · S. Hait (✉)
Department of Civil and Environmental Engineering, Indian Institute of Technology, Patna, Bihar 801106, India
e-mail: shait@iitp.ac.in

T. Gupta et al. (eds.), *Measurement, Analysis and Remediation of Environmental Pollutants*, Energy, Environment, and Sustainability,
https://doi.org/10.1007/978-981-15-0540-9_13

13.1 Introduction

Sustained technological developments in electrical and electronic equipment (EEE) have increased the demand for EEE like computers, mobile phones, televisions (Awasthi et al. 2017; Iannicelli-Zubiani et al. 2017). The increased use of EEE along with their shorter shelf life has led to a significant increase in the generation of waste EEE (WEEE) or electronic waste (e-waste) globally. It has been reported that around 44.7 Mt of e-waste was generated in the year 2016 and is expected to grow to 52.2 Mt by 2021 (Baldé et al. 2017). In India, 1.95 Mt of WEEE is generated in the year 2016 (Baldé et al. 2017). Printed circuit board (PCB) or printed wiring board is an essential part of any EEE comprising around 6% (wt/wt) of EEE (Das et al. 2009). Metallic content of around 27–30% and non-metallic content with 70–73% makes PCB matrix complex and heterogeneous in nature (Zhou and Qiu 2010). Rich metallic content of PCBs up to 24% Cu, 6.3% Pb, and 3% Ni, which is considerably high in comparison with natural reservoirs, makes e-waste to be categorized as a metal-rich secondary reserve (Zhou and Qiu 2010). However, e-waste containing toxic metals such as Cr, Hg, Ni, and Pb beyond permissible limit turns it hazardous (Li et al. 2004; Priya and Hait 2017). Thus, e-waste management for metal recycling is a necessity not only for resource circulation but also for the environmental point of view (Niu and Li 2007). However, conventional metal recycling techniques such as pyro- and hydro-metallurgy techniques are associated with secondary environmental pollution and high-energy requirements (Cui and Zhang 2008; Ilyas et al. 2010; Pant et al. 2012).

Now-a-days, bioleaching technique or bioextraction of metals from e-waste using diverse microorganisms are gaining attention worldwide due to its eco-friendly nature. Hydrometallurgy coupled with biotechnology by employing microorganisms is known as biometallurgy. Bioleaching is an eco-friendly, effective, and low-cost metal recycling technique (Beolchini et al. 2012; Brandl et al. 2001; Priya and Hait 2018). Numerous microorganisms consisting of mostly autotrophic bacteria have been reported to be used for bioleaching of metals from electronic waste (Brandl et al. 2001; Priya and Hait 2017, 2018). Microorganisms produce a diverse range of metabolites and biomass during bioleaching, which aids the metal extraction from the solid matrix (Lundgren et al. 1986). However, metal bioleaching employing heterotrophic fungi is the comparatively less explored area. Fungi have a shorter lag phase and faster leaching rate by showing more tolerance to toxic constituents. Further, fungi can grow in acidic as well as in alkaline conditions. Furthermore, the fungi excreted metabolites like organic acids assist to leach the metals (Wu and Ting 2006). Fungal species like *Aspergillus spp.*, *Penicillium spp.*, and yeast like *Y. lipolytica* are well known for the production of organic acids (Magnuson and Lasure 2004; Sauer et al. 2008; Singleton 2001). Fungal bioleaching of metals from e-waste is mostly limited to the pure culture of *Aspergillus* spp. and *Penicillum* spp. The specific attribute of fungal species like the polymer-degrading ability of *Aspergillus tubingensis* is yet to be explored for fungal bioleaching of metals from e-waste as metals are embedded in the polymer matrix of PCB. Moreover, a combination or consortia of fungal cultures with different attributes are yet to be explored for metal bioleaching from e-waste. *Aspergillus tubingensis*, a polymer degrading fungal species in combination with *Aspergillus niger*, known for producing organic acids can be explored for bioleaching of metals from PCB. In this context, the bioleaching of selected metals, viz. Cu, Ni, and Zn from the waste PCB (WPCB) of obsolete computer was comparatively assessed employing pure and mixed fungal cultures of *Aspergillus tubingensis* and *Aspergillus niger*.

13.2 Materials and Methods

13.2.1 WPCB Collection and Sample Preparation

WPCBs (Fig. 13.1) of obsolete computer were collected and brought to the laboratory from local EEE repairing shops in Bihta, Patna. After removing the mounted parts manually, WPCBs were cut into smaller parts. Smaller WPCB parts were then pulverized using a cutting mill (SM200, Retsch, Germany) and separated into particle size fraction of 0.038–1 mm using a sieve shaker (AS200, Retsch, Germany).

Fig. 13.1 Typical WPCB of obsolete computer used in the present study

Particle size fraction ranging from 0.038–1 mm was subsequently used for metallic content quantification, morphological characterization, and bioleaching experiment.

13.2.2 Fungal Cultures

In bioleaching study, pure and mixed fungal cultures of *Aspergillus* spp. i.e., *Aspergillus niger* (MTCC-281) and *Aspergillus tubingensis* (ATCC-76608) were employed. *Aspergillus niger* was chosen for its organic acid production ability to presumably help in bioleaching of metals. As the metals are usually embedded in the polymer matrix in the PCB, *Aspergillus tubingensis* (ATCC-76608) was selected for its polymer-degrading ability. The pure culture of the *Aspergillus niger* and *Aspergillus tubingensis* was obtained from the Microbial Type Culture Collection and Gene Bank (MTCC), India and American Type Culture Collection (ATCC), USA, respectively. Fungal cultures were inoculated in separate 1000 mL conical flasks containing 500 mL of potato dextrose broth (PDB) medium in each upon autoclaving. The flasks were then incubated in an incubated shaker (SIF 5000R, Jeio Tech, South Korea) for 48 h at 200 rpm and 30 °C.

13.2.3 Metallic Content and Morphology of Pulverized WPCB and the PCB Residue

Morphology of the pulverized WPCB and the PCB residue after bioleaching was magnified using a scanning electron microscope (SEM) (EVO 50, Carl Zeiss, Germany). To compute the bioleaching efficiency, the content of selected metals i.e. Cu, Ni, and Zn in the pulverized WPCB sample was quantified following the acid digestion as per the USEPA method 3052 (Magnuson and Lasure 2004) and filtration using 0.22 μm filter with appropriate dilution. The diluted samples were then analyzed using inductively coupled plasma mass spectrometer (ICP-MS) (7800, Agilent, USA) for metal quantification.

13.2.4 Bioleaching Experiments

For bioleaching experiments, 100 mL of autoclaved PDB medium with adjusted pH of 5 in 250 mL conical flasks were added with the pulverized WPCB sample in the size fraction of 0.038–1 mm at an e-waste pulp density of 1 g/L. In order to comparatively investigate the metal leaching efficiency of two *Aspergillus* spp., bioleaching experiments were conducted in three sets. In the first and second sets, 1 mL spore suspension each of pure culture of *Aspergillus niger* and *Aspergillus tubingensis* was inoculated in the autoclaved flasks separately. In the third set, 1 mL each of spore suspension of both the fungal spp. was inoculated in the autoclaved flasks. All the inoculated flasks were then incubated in an incubated shaker (SIF 5000R, Jeio Tech, South Korea) at 170 rpm and 30 °C. Control reactors without the addition of fungal culture were also maintained under similar experimental conditions. The dissolution of the selected metals such as Cu, Zn, and Ni from the pulverized WPCB matrix in the bioreactors and control reactors was analyzed at every 3 days by analyzing metal concentration in the leachate using ICP-MS (7800, Agilent, USA). All bioleaching experiments were conducted in triplicate and average values along with standard deviations were reported.

13.2.5 Analytical Measurements

The microscopic structure of the pulverized WPCB particles and the PCB residue upon bioleaching was magnified using SEM (EVO 50, Carl Zeiss, Germany). Elemental analysis of the WPCBs of obsolete computer for the targeted elements i.e. Cu, Zn, and Ni was performed using ICP-MS (7800, Agilent, USA) in the digestate prepared using following the microwave-assisted acid digestion as per the USEPA 3052 method (USEPA 1995). Bioleachate samples collected from all three sets of bioleaching reactors employing pure and mixed fungal cultures along with control reactors were measured for the targeted metals using ICP-MS following filtration through a filter having pore size of 0.22 μm for removal of fungal mycelium and suspended particles followed by dilution with deionized (DI) water. Analyses were done in triplicate.

13.3 Results and Discussion

13.3.1 Metallic Content and Morphology of Pulverized WPCB and the PCB Residue

The metal quantification of the pulverized WPCBs of obsolete computer revealed that Cu, among the targeted metals, is found to be the highest in abundance with a content of 21.5%, followed by 0.10% Zn and 0.08% Ni. Since the presence of the base metals like Cu, Zn, and Ni was found to be in abundance, the pulverized WPCB of obsolete computer can be considered as a secondary metal reservoir.

Morphology of pulverized WPCBs in the particle size fraction of 0.038–1 mm and the PCB residue obtained upon bioleaching was observed under SEM. The SEM image of the pulverized PCBs (shown in Fig. 13.2a) revealed the non-uniformity in particle shapes and sizes. The pulverized WPCB samples were observed to consist of predominantly rod- and lumps-shaped agglomerated fiberglass small particles along

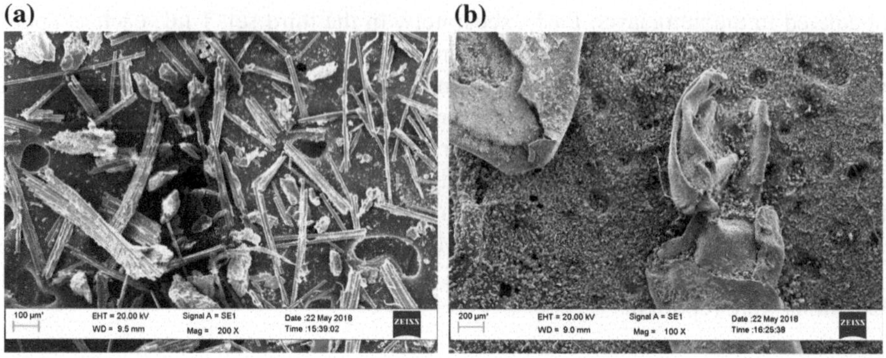

Fig. 13.2 SEM-micrograph of **a** raw pulverized WPCB, and **b** PCB residue upon fungal bioleaching

with metallic particles. Further, the rough surface of the pulverized PCB was featured with several shiny spots indicative of metal abundance. The variations in the surface texture of particles can be ascribed to the response of metal diversity, glass fibers, etc. when passed through the different forces during sample preparation using milling operation (Murugan et al. 2008). Upon pulverization, the adherence of fungus to the pulverized particles is likely to aid the bioleaching of metals. The SEM image of the PCB residue (shown in Fig. 13.2b) obtained from the mixed fungal culture bioreactor showed degraded surface featured with fine particulates in the backdrop suggesting metal dissolution from the pulverized PCB particles.

13.3.2 Bioleaching of Targeted Metals from E-Waste Using Pure and Mixed Fungal Cultures

Bioleaching results revealed that the metal leaching from pulverized PCB in the pure cultures of *Aspergillus tubingensis* and *Aspergillus niger* differed significantly at 1 g/L of e-waste pulp density with better dissolution in case of *Aspergillus niger*. The metal extraction was observed to increase gradually during the initial period and showed an asymptotic trend after 30 days of bioleaching in the pure and mixed cultures of *Aspergillus tubingensis* and *Aspergillus niger* (Fig. 13.3). The maximum metal leaching employing a pure culture of *Aspergillus tubingensis* was observed to be 54% Cu, 41% Zn, and 14% Ni. Similarly, the corresponding maximum metal leaching efficiency employing a pure culture of *Aspergillus niger* was 71% Cu, 32% Ni, and 79% Zn at the pulp density of 1 g/L.

Using the mixed cultures of *Aspergillus tubingensis* and *Aspergillus niger*, there was a marginal increase in metal bioleaching for Cu and Ni with extraction efficiency of 76 and 36%, respectively. For Zn, the maximum extraction efficiency of 63% was observed using the mixed culture. However, the maximum Zn extraction efficiency in mixed culture was observed to be more than that in the pure culture of

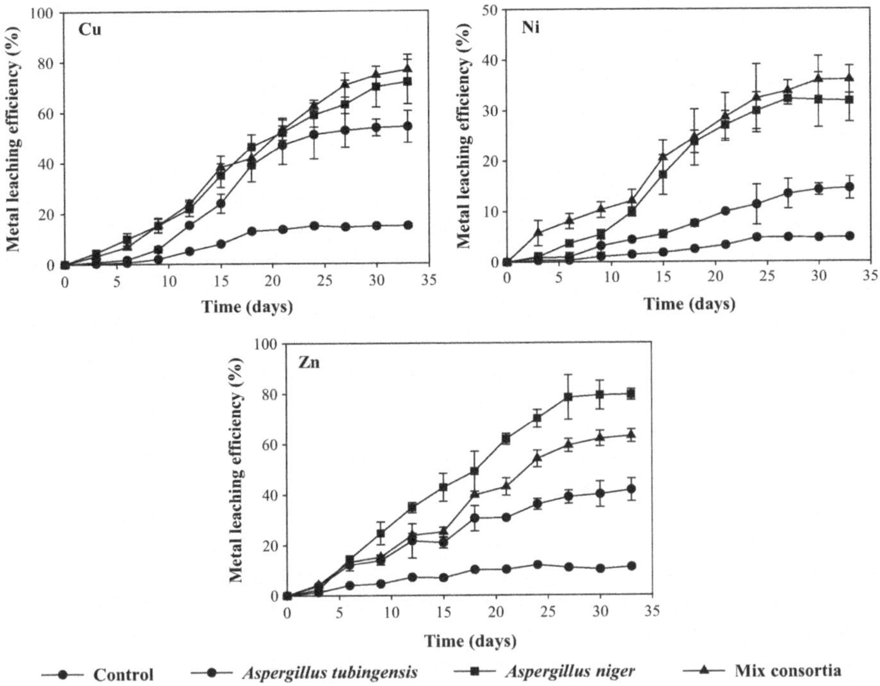

Fig. 13.3 Bioleaching of selected metals from WPCB of obsolete computer using pure and mixed cultures of *Aspergillus niger and Aspergillus tubingensis*

Aspergillus tubingensis and less than that in the pure culture of *Aspergillus niger*. A compilation of literature on the bioleaching of metals from e-waste using different fungal species along with the maximum metal extraction efficiency and bioleaching period is provided in Table 13.1. It is evident from Table 13.1 that the maximum metal extraction efficiency achieved for the selected metals in the present study is comparable with the results reported in the literature. Owing to the organic acid production ability of *Aspergillus* spp., the inoculated fungi were thriving on the PDB medium for the production of various organic acid like citric acid and oxalic acid which further kept pH of the solution low (Singleton 2001). Some other studies have also shown that citric and oxalic acids are the predominant organic acid produced by *Aspergillus* spp. facilitating metal bioleaching (Kim et al. 2016; Horeh et al. 2016; Santhiya and Ting 2005).

13.3.3 Future Implications

The present study has shown that the pure and mixed fungal culture of *Aspergillus tubingensis*, a polymer-degrading species and *Aspergillus niger*, an organic acid

Table 13.1 Bioleaching of metals from e-waste using various fungal species

E-waste	Fungal species	Duration (day)	Maximum metal extraction efficiency (%)	References
E-waste scrap	*A. niger and P. simplicissimum*	21	Al: 95, Cu: 65, Ni: 95, Pb: 95, Zn: 95	Brandl et al. (2001)
Lithium ion batteries	*A. niger*	30	Al: 65, Cu: 100, Li: 95, Ni: 38	Kim et al. (2016)
Catalyst	*A. niger*	70	Al: 55, Ni: 65	Santhiya and Ting (2005)
Converter	*P. simplicissimum*	13	Zn: 93	Schinner and Burgstaller (1989)
Hydrocracking catalyst	*P. simplicissimum*	30	Al: 25, Mo: 92.7, Ni: 66.43	Amiri et al. (2011)
Fluid catalytic cracking catalyst	*A. niger*	14	Al: 30, Ni: 9, Sb: 64, V: 36	Aung and Ting (2005)
Electronic scrap	*A. niger*	42	Cu: 68.3, Pb: 27.9	Kolenčík et al. (2013)
Electronic scrap	*P. chrysogenum*	24	Al: 96, Cu: 48, Ni: 73, Zn: 25	Ilyas et al. (2013)
PCB	*A. niger*	3	Cu: 80	Jadhav and Hocheng (2015)
PCB	Mixed culture	27	Al:15, Pb: 20, Sn: 8, Zn: 49	Xia et al. (2018)
Computer PCB	*A. niger*	21	Cu: 85%, Ni: 80, Zn: 100	Faraji et al. (2018)
Computer PCB	*A. tubingensis and A. niger*	33	Cu: 76%, Ni: 36, Zn: 63	This study

producing species can efficiently bioleach metals from waste PCB at a pulp density of 1 g/L in 33 days. This highlights the practical feasibility of fungal bioleaching of metals from e-waste using the pure and mixed cultures of *Aspergillus tubingensis* and *Aspergillus niger*. Therefore, it is essential to improvise the bioleaching efficiency of metals from WPCB by employing different fungal species in pure and mixed forms along with the addition of suitable organic ligands to develop a hybrid process. Acclimatized fungal species with prior exposure to the e-waste environment may lead to enhanced metal bioleaching. Moreover, the metal extraction rate can be increased by optimizing the different factors governing the fungal bioleaching process like pH, inoculum size, continuous supply of nutrients, etc.

13.4 Conclusions

The metal extraction from e-waste employing pure and mixed forms of *Aspergillus* fungal strains revealed that *Aspergillus niger* was able to leach out more metals as compared to *Aspergillus tubingensis*. At a pulp density of 1 g/L, the pure culture of *Aspergillus niger* showed maximum metal extraction of 79% Zn, followed by 71% Cu and 32% Ni. At the same pulp density, the corresponding metal extraction efficiency for Cu, Zn, and Ni using *Aspergillus tubingensis* was 54, 41, and 14%, respectively. Except for Zn, the mixed culture of *Aspergillus niger* and *Aspergillus tubingensis* showed a marginal increase in the metal bioleaching reaching extraction efficiency of 76 and 36% for Cu and Ni, respectively at an e-waste pulp density of 1 g/L. In the present study, the metal extraction efficiency for Zn using the mixed fungal culture of *Aspergillus* strains was 63%. However, the metal extraction rate and the time required for bioleaching can be further enhanced by employing a consortium of fungal strains along with the addition of suitable organic ligands to develop a hybrid process.

References

Amiri F, Mousavi SM, Yaghmaei S (2011) Enhancement of bioleaching of a spent Ni/Mo hydroprocessing catalyst by *Penicillium simplicissimum*. Sep Purif Technol 80(3):566–576

Aung KMM, Ting YP (2005) Bioleaching of spent fluid catalytic cracking catalyst using *Aspergillus niger*. J Biotechnol 116(2):159–170

Awasthi AK, Zlamparet GI, Zeng X, Li J (2017) Evaluating waste printed circuit boards recycling opportunities and challenges, a mini review. Waste Manag Res 35(4):346–356

Baldé CP, Forti V, Gray V, Kuehr R, Stegmann P (2017) The global e-waste monitor 2017: quantities, flows and resources. United Nations University, International Telecommunication Union, International Solid Waste Association

Beolchini F, Fonti V, Dell'Anno A, Rocchetti L, Vegliò F (2012) Assessment of biotechnological strategies for the valorization of metal bearing wastes. Waste Manag 32(5):949–956

Brandl H, Bosshard R, Wegmann M (2001) Computer-munching microbes: metal leaching from electronic scrap by bacteria and fungi. Hydrometallurgy 59(2–3):319–326

Cui J, Zhang L (2008) Metallurgical recovery of metals from electronic waste: a review. J Hazard Mater 158(2–3):228–256

Das A, Vidyadhar A, Mehrotra SP (2009) A novel flowsheet for the recovery of metal values from waste printed circuit boards. Resour Conserv Recycl 53(8):464–469

Faraji F, Golmohammadzadeh R, Rashchi F, Alimardani N (2018) Fungal bioleaching of WPCBs using *Aspergillus niger*: observation, optimization and kinetics. J Environ Manage 217:775–787

Horeh NB, Mousavi SM, Shojaosadati SA (2016) Bioleaching of valuable metals from spent lithium-ion mobile phone batteries using *Aspergillus niger*. J Power Sources 320:257–266

Iannicelli-Zubiani EM, Giani MI, Recanati F, Dotelli G, Puricelli S, Cristiani C (2017) Environmental impacts of a hydrometallurgical process for electronic waste treatment: a life cycle assessment case study. J Clean Prod 140:1204–1216

Ilyas S, Ruan C, Bhatti HN, Ghauri MA, Anwar MA (2010) Column bioleaching of metals from electronic scrap. Hydrometallurgy 101(3–4):135–140

Ilyas S, Chi RA, Lee JC (2013) Fungal bioleaching of metals from mine tailing. Miner Process Extr Met Rev 34(3):185–194

Jadhav U, Hocheng H (2015) Hydrometallurgical recovery of metals from large printed circuit board pieces. Sci Rep 5:14574

Kim MJ, Seo JY, Choi YS, Kim GH (2016) Bioleaching of spent Zn–Mn or Ni–Cd batteries by *Aspergillus* species. Waste Manag 51:168–173

Kolenčík M, Urík M, Čerňanský S, Molnárová M, Matúš P (2013) Leaching of zinc, cadmium, lead and copper from electronic scrap using organic acids and the *Aspergillus niger* strain. Fresenius Environ Bull 22(12a):3673–3679

Li J, Shrivastava P, Gao Z, Zhang HC (2004) Printed circuit board recycling: a state-of-the-art survey. IEEE Trans Electron Packag Manuf 27(1):33–42

Lundgren DG, Valcova-Valchanova M, Reed R (1986) Chemical reactions important in bioleaching and bioaccumulations. In: Ehrlich HL, Holmes DS (eds) Biotechnology for the mining, metal-refining, and fossil fuel processing industries. John Wiley & Sons, New York, pp 7–22

Magnuson JK, Lasure LL (2004) Organic acid production by filamentous fungi. Advances in fungal biotechnology for industry, agriculture, and medicine. Springer, Boston, MA, pp 307–340

Murugan RV, Bharat S, Deshpande AP, Varughese S, Haridoss P (2008) Milling and separation of the multi-component printed circuit board materials and the analysis of elutriation based on a single particle model. Powder Technol 183:169–176

Niu X, Li Y (2007) Treatment of waste printed wire boards in electronic waste for safe disposal. J Hazard Mater 145(3):410–416

Pant D, Joshi D, Upreti MK, Kotnala RK (2012) Chemical and biological extraction of metals present in E waste: a hybrid technology. Waste Manag 32(5):979–990

Priya A, Hait S (2017a) Qualitative and quantitative metals liberation assessment for characterization of various waste printed circuit boards for recycling. Environ Sci Pollut Res 24(35):27445–27456

Priya A, Hait S (2017b) Comparative assessment of metallurgical recovery of metals from electronic waste with special emphasis on bioleaching. Environ Sci Pollut Res 24(8):6989–7008

Priya A, Hait S (2018) Extraction of metals from high grade waste printed circuit board by conventional and hybrid bioleaching using *Acidithiobacillus ferrooxidans*. Hydrometallurgy 177:132–139

Santhiya D, Ting YP (2005) Bioleaching of spent refinery processing catalyst using *Aspergillus niger* with high-yield oxalic acid. J Biotechnol 116(2):171–184

Sauer M, Porro D, Mattanovich D, Branduardi P (2008) Microbial production of organic acids: expanding the markets. Trends Microbiol 26(2):100–108

Schinner F, Burgstaller W (1989) Extraction of zinc from industrial waste by a *Penicillium sp*. Appl Environ Microbiol 55(5):1153–1156

Singleton I (2001) Fungal remediation of soils contaminated with persistent organic. Fungi in bioremediation 23:79

USEPA (1995) Microwave assisted acid digestion of siliceous and organically based matrices USEPA method 3052, 3rd edn. United States Environmental Protection Agency, Washington, DC

Wu HY, Ting YP (2006) Metal extraction from municipal solid waste (MSW) incinerator fly ash—chemical leaching and fungal bioleaching. Enzyme Microb Technol 38(6):839–847

Xia M, Bao P, Liu A, Wang M, Shen L, Yu R, Qiu G (2018) Bioleaching of low-grade waste printed circuit boards by mixed fungal culture and its community structure analysis. Resour Conserv Recycl 136:267–275

Zhou Y, Qiu K (2010) A new technology for recycling materials from waste printed circuit boards. J Hazard Mater 175(1–3):823–828

Chapter 14
Recent Advances in Micro-extraction Based Analytical Approaches for Pesticides Analysis in Environmental Samples

Anshuman Srivastava, Minu Singh, Shiv Singh and Sheelendra Pratap Singh

Abstract Pesticides are auxiliary input to induce the quality and production of food. It is also considerate to minimize the postharvest losses of crops and to fulfill the increasing demand. Although pesticides are useful in food production, their versatile use had resulted in the occurrence of contamination and residues in all necessities of our life, i.e., food, water, and air. Nowadays, consumer perception about food safety and quality leads to transparency and traceability in various agricultural practices. The pesticide had a diverse meaning, it encompasses many hundreds of toxic chemicals, exhibiting enormously diverse chemical and physical properties. Thus to determine the level of pesticide residue in different commodities, a simple, effective and robust sample preparation and analytical determination technique is required. Further, technique includes both identification and quantification of analytes in food commodities, which strictly fulfill the international regulation on maximum residue limits. This chapter briefly discusses the modern analytical approaches, including different sample preparation technique followed by chromatography and mass spectrometry techniques.

Anshuman Srivastava and Minu Singh are equally contributed to this work.

A. Srivastava · M. Singh · S. P. Singh (✉)
Pesticide Toxicology Laboratory, Regulatory Toxicology Group, CSIR—Indian Institute of Toxicology Research (CSIR-IITR), MG Marg, Lucknow, Uttar Pradesh 226001, India
e-mail: sheelendra@iitr.res.in

S. Singh (✉)
Light Weight Metallic Materials, Council of Scientific and Industrial Research-Advanced Materials and Processes Research Institute, Hoshangabad Road, Bhopal, Madhya Pradesh 462026, India
e-mail: sshiv.singh@ampri.res.in

Nanomaterial Toxicology Group CSIR-Indian Institute of Toxicology Research (CSIR-IITR), Vishvigyan Bhawan 31, Mahatma Gandhi Marg, Lucknow, Uttar Pradesh 226001, India

S. P. Singh
Analytical Chemistry Laboratory, Regulatory Toxicology Group, CSIR-Indian Institute of Toxicology Research (CSIR-IITR), Vish Vigyan Bhavan 31 Mahatma Gandhi Marg, Lucknow, Uttar Pradesh 226001, India

T. Gupta et al. (eds.), *Measurement, Analysis and Remediation of Environmental Pollutants*, Energy, Environment, and Sustainability,
https://doi.org/10.1007/978-981-15-0540-9_14

Keywords Environmental pollutant · Micro-extraction · QuEChERS · Analytical approach · Sample preparation · Carbon · Pesticides

14.1 Introduction

Since the beginning of the industrial revolution until the current surge in technological advancement, the successive build-up of environmental pollutants has always been a challenging issue. Environmental pollutants directly affect living beings and bring an irreversible deterioration to the earth. With the progression in the technological areas, various strategies were implemented to control the exposure of environmental pollutant. However, the measurement, analysis and the remediation of the emerging environmental pollutants have always been the pragmatic step for the management and implementation of various mitigation strategies. Several types of pollutant have emerged which are very relevant to discuss, such as pesticides. Pesticides are an integral part of the modern agriculture, and their enormous and continual use leads to damage farmland and seriously affects human beings.

In analytical chemistry, sample preparation at trace level is very critical and yet very difficult for residue analysis. Especially in environmental samples, complex nature of it makes it a challenging task for the analysis of analytes. Thus, sample preparation is a prior step required before the instrumental analysis for the development of a procedure for the quantification purpose. Generally, it is the most time-consuming step in the whole experimental analysis, so a suitable technique should be optimized and evaluated for the accuracy and the precision of the result.

Nowadays, one of the most common environmental pollutants is the presence of pesticide residue in food samples on which the whole population is dependent. Though, the pesticide are ubiquitously present but their extraction techniques is a challenging issue in recent times for enhanced separation, clean up, and enrichment of analytes.

The prime function of using pesticide is to curb unwanted pests, weeds, and insects (targeted organisms) and also to boost up the agricultural crop productivity (Rowe et al. 2016). Thus, pesticides represent those compounds or mixtures which are inherently toxic in nature that are deliberately released in the environment to mitigate pests and insects. It facilitates crop yield, but it also exhibits detrimental and undesirable consequences (oncogenic and cytotoxic) on humans, especially workers engaged in its production, transportation, vendors of fruits and vegetables. They also cause nervous disorders, fertility ailments, immunological, and pulmonary diseases (Xu et al. 2017). As a consequence, to fulfill the constant demand for increasing population dependency on pesticides have been raised (Nsibande and Forbes 2016). Therefore, explicit and constant monitoring of pesticide residue, which accumulates in the food chain is required to assure health complications and environment welfare (Han et al. 2016; He et al. 2015; Kolberg et al. 2011). The accurate monitoring of pesticide residue is very important. Thus a highly sensitive yet selective method for sample preparation is discussed in this chapter.

In today's trend, the unanimously followed procedure for the extraction of pesticides in food commodities is the Quick, Easy, Cheap, Effective, Rugged, and Safe (QuEChERS) method. So, for multi pesticide residue analysis QuEChERS is considered as significant development because of its economic, convenient, and user-friendly advantages. It comprises of two different steps (1) solid-liquid partitioning step followed by salting out effect and (2) dispersive solid-phase extraction (d-SPE) for the clean up of the extract. It was the first time reported in 2003 by Anastassiades et al. (2003a) for high water content matrices in various agricultural products. Later, the originally proposed method was constantly modified for detection of pesticides comprising planar, polar, and non-polar types in food and environmental matrices (Lee et al. 2017; González-Curbelo et al. 2015).

In this way, due to its intrinsic benefits and diverse implication, this method has achieved exclusive approval and extended its boundaries beyond its traditional application to extraction of various analytes (polyphenols, antibiotics, neuro-active insecticides, and morphine) from different matrices (e.g., food pollutant Reichert et al. 2018; Kaczyński and Łozowicka 2017; Zheng et al. 2019), environmental pollutant (Zaidon et al. 2019; Pang et al. 2016) and biological fluids (Taliansky-Chamudis et al. 2017; Correia-Sá et al. 2018; Winborn and Kerrigan 2019). The main aim of the extraction method is to facilitate easy and streamlined extraction procedures with minimal cost and hassle-free techniques to simplify the analysis process.

The conventional approaches like, solid phase extraction (SPE) and liquid-liquid extraction (LLE) provides good extraction efficiency and sample cleanup, but are comprehensive in nature. These conventional techniques have limitations such as the use of the abundant amount of hazardous organic solvents and time-consuming extraction phase. Besides these drawbacks, especially in LLE, the formation of emulsion and acquirement of limited preconcentration are important factors affecting the precision of the methodology and the extraction efficiency. Thus, in addition to these limitations, achieving environmental friendly was a major concern which shifts the focus towards microextraction techniques as an alternate substitute to the conventional techniques. These techniques follow the "green chemistry "rules by using less organic toxic solvents and the sample size is also reduced for the accurate and precise determination of various analytes. Some of the typical approaches which are discussed and highlighted in this chapters are solid phase microextraction (SPME) (Arthur and Pawliszyn 1990), Dispersive liquid-liquid microextraction (DLLME) (Rezaee et al. 2006), and single drop microextraction (SDME) (Jeannot and Cantwell 1996). These techniques follow the essential virtues that lead to safe and secure green analytical chemistry (de la Guardia and Garrigues 2012). Currently, a new trend is followed with an involved amalgamation of classical extraction approach with microextraction as well as individually.

This chapter discussed several analytical and advanced techniques used for the detection of pesticides in environmental commodities. Moreover, a brief description of the classification of pesticides with its toxicity on targeted and non-targeted organisms is discussed. A detailed description of classical sample preparation techniques and the microextraction techniques are explained in this chapter.

14.2 Pesticide Classification

Pesticide classification and its toxicity are designed as per origin, chemical composition, and target entities. The broad classification is done into two categories (i) synthetic pesticides and (ii) biopesticides. The former one is produced synthetically having an inorganic chemical composition which directly targets pest or deactivates the pest. These are further classified to (1) insecticides (2) fungicides (3) herbicides (4) rodenticide (5) nematicides (Sharma 2006). The second one is derived from plants, animals, bacteria, and certain minerals, including their metabolites or genes to cast away the pests. Pesticides are further grouped into chemical families as (1) organochlorine (OC) (2) organophosphate (OP) (3) carbamates (C) (4) pyrethroids (P).

OC class of pesticides has high persistent, high microbial and chemical resistants, and high toxicity and lipophilic characteristics. They are also carcinogenic and neurotoxicants, and can bioaccumulate in nature (Genuis et al. 2016). The estimated half-life of OC is around 10–30 years, which is very long enough to accumulate in the body and shows potential toxic effects to human health, thus a joint concern of many countries is to ban the usage of this class of pesticides (Rani et al. 2017). Initially, it had been introduced to control malaria, since it has low cost so later on OC such as aldrin, dieldrin, hexachlorocyclohexane got popular in developing countries, especially in Asia (Choi et al. 2009). OC can accumulate in the food chain, so an effort had been made to withdraw it and replace it with other class of compounds.

Following the subsequent replacement of OC pesticides, a new class of organophosphorus pesticides got introduced, which is easily biodegradable and can act as an inhibitor of acetylcholinesterase (AChE) enzymes. It can be said that OP and C are considered for further uses as they replace the OC class of pesticides. The toxic effect of OP is that it directly affects amphibians, insects, mammals by phosphorylation of AChE. This enzyme is vital for normal functioning of nerve fiber within the central nervous system. Now, various OP combinations had been used in the form of insecticides and fungicides. They have chemically active ester moiety, which easily gets volatile in the environment. They are hydrophobic, lipophilic, and highly soluble in inorganic solvents (Patel et al. 2019).

C are such class of pesticides which are almost morphologically and chemically to similar OP. These are originated from carbamic acid and lead to carbamylation of AChE. Thus, their working structure is identical to the OP class, so they are called cholinesterase inhibitors. They are used on a large scale in farming and to shield crops to inhibit pests (Zhang et al. 2015). When compared to OC class, these two are readily available, cheap, and have low accumulation rate in the environment, less half-life, and a wide range of applications, and can control wide spectrum of pests (Wei et al. 2018).

Pyrethroids are "chrysanthemum derived flower" derivatives, which can be considered as mild toxic compared to other class of pesticides. These are synthetically produced from the chrysanthemum flower extract and chemically modified to increase its degradation period and combine with synergists, enhance potency, and conceding with human's capability to detoxify the pesticide. The mechanism of their action is to inhibit the ionic conductance of the nervous system in mammals or insects by delaying the sodium channel pathway. The degradation of SP is sluggish, which increase its persistence time in the environment than the naturally occurring pyrethrin, which gets the breakdown by sunlight, heat, and moisture.

Although there is an abundant use of pesticides and its various combination for mitigating pests and insects, it has many negative bearings in the environment as well as for those who are using it without having much information. It's hazardous mechanism is also because of its accumulation tendency in the food. The uncontrolled use of pesticides increases the residue concentration in environmental matrices (soil, water, and food) that leads to adverse effects on human health. The hazardous effect of pesticides includes congenital disabilities, fetal deaths, carcinogen, neuro disruptor, hormone disruptor and, allergens (Van Maele-Fabry et al. 2010; Wickerham et al. 2012).

In this context, for the well being of human, monitoring the usage of pesticide in raw food commodities is essential. For the estimation of maximum residue limit (MRLs) value in different food commodities various government agencies are working globally i.e., European Union (EU) and Codex Alimentarius Commission (CAC) (Codex Alimentarius 2012) and for method development and validation SANTE guideline is used. Hence these guideline provides a standard for the optimization of pesticides concentration in food to secure food safety for human.

Pesticide application in crop field to enhance the production is responsible for pest resistant as well as categorize itself a severe environmental pollutant. Significant measures should be taken to reduce the hazardous effects on human health irrespective of its dependency on crop yields/production. So, various techniques and methods have been established for pesticides residue analysis.

14.3 Traditional Sample Preparation Techniques

Analytical approaches in pesticide analysis are broadly bifurcated in two steps (1) sample preparation, which includes suitable homogenization, extraction, and extracts clean up. (2), Data analysis, which includes accurate and precise result. Thus, in multi-residues pesticide analysis, the recent analytical approach relies mainly to cover a maximum number of desired analytes and quantifying it at a very low level. The key point in residue analysis is thorough extraction procedure followed by clean-up of samples to avoid any co extractives interferences.

14.3.1 Liquid-Liquid Extraction (LLE)

In sample preparation, LLE was always believed as conventional preconcentration and matrix segregation technique in analytical chemistry. It is a kind of separation

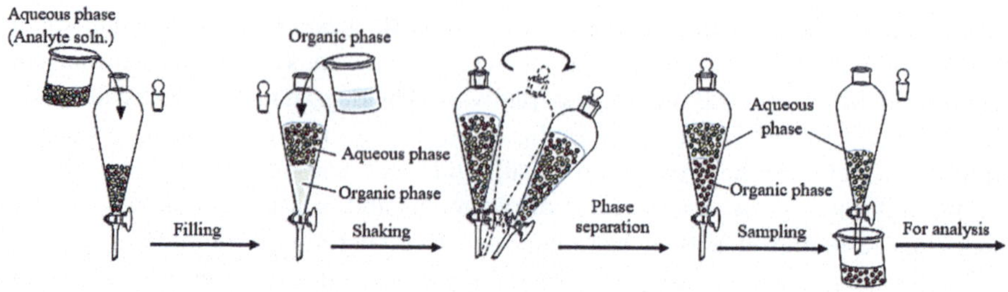

Fig. 14.1 Schematic representation showing the liquid-liquid extraction process

process based on the phenomena of "partitioning", in which there is a transfer of one solute phase from one solvent to the other solvent. Both the solvents are either immiscible or semi/partially miscible within themselves as shown in Fig. 14.1. Mostly, one solvent is an aqueous medium (water) and the other one is a non-polar organic liquid. LLE consists of two steps: the first step is mixing (contacting), and the second step is the separation of phase.

It is essential to consider both the steps in the selection of proper solvent and to regulate the mode of operation. Simultaneously, vigorous mixing is suggested which enhance the transfer of desired analytes from one solvent to the other phase. This also helps in the formation of emulsification which facilitates the phase separation. Practically, a rule of "distribution coefficient" 'K' is followed. We can say that equilibrium is attained when the chemical potential of extractable solute is consistent in both phases.

Some of the organic solvent used for extraction purpose in pesticide determination are hexane (Cabras et al. 2001), ethyl acetate (EtAc)/cyclohexane (Sannino et al. 2004), acetonitrile (MeCN) (Zhang et al. 2010; Farajzadeh et al. 2016), dichloromethane/acetone (Pose Juan et al. 2006; Fatoki and Awofolu 2003; Gonzalez-Rodriguez et al. 2009). Recently, several studies have been done which applies to combine LLE with other extraction methods like the study of pesticides on fruit juice (Farajzadeh et al. 2015a) in this LLE is combined with improved extraction technique liquid-liquid microextraction prior to their analysis on gas-chromatography with electron capture detector (GC-ECD). The limit of detection obtained was 2–12 μg/L as recovery was 87–96%. In this study first time, counter current extraction and salting out phenomena for preconcentration was applied for fruit juices. Similarly, a study conducted on an aqueous sample by the same salting out effect and LLE method on counter current mode with later associated with dispersive liquid-liquid microextraction with a limit of detection (LOD) of 0.5–1 μg/L using gas-chromatography with flame ionization detector (GC-FID) (Farajzadeh et al. 2015b).

Another, LLE study was done on soybean oil for 95 pesticides a multi-residue method in which LLE is combined with centrifugation, freezing and dispersive solid phase extraction as a cleanup step and samples were analyzed on gas-chromatography mass spectrometry (GC-MS/MS). The objective of this study was to minimize the co-extractive hindrance due to the presence of fat in oil, the obtained recoveries were in the acceptable range of 89–114% (Nguyen et al. 2010). A similar study on vegetable oil samples using LLE combined with DLLME extraction method was done for pyrethroids pesticides on GC-FID followed by confirmation on GC/MS/MS. LOD and recovery were in the range 0.02–0.16 mg/kg and 85–109%, respectively (Farajzadeh et al. 2014). For multi-residue analysis in five different juice samples study was conducted using a hollow fiber microporous method extracted by LLE and further analyzed on liquid chromatography mass spectrometry (LC/MS/MS) giving good repeatability and low LOD (Bedendo et al. 2012). Based on the above study LLE was combined with a new extraction technique to overcome the drawback as it consumes many solvents and is time exhaustive process.

14.3.2 Solid Phase Extraction Method (SPE)

This preconcentration technique in sample preparation is a crucial step in analysis, especially in trace or ultra-trace levels of analytes as well as in sophisticated matrices (like environmental, food, biological, and pharmaceuticals). A suitable sample pre-treatment method not only assures selectivity and sensitivity but also increases the analytes compatibility and reduces the disturbance in the long run of the analytical instrument. Conventionally used preconcentration method of sample preparation is solid phase extraction (Andrade-Eiroa et al. 2016; Ruiz-Gil et al. 2008; Duca et al. 2014). The solid phase extraction was first reported 50 years ago (Liška 2000; Poole 2000). The basic principle behind SPE is that it is a partition technique between the liquid phase and solid sorbent phases. It is a sample preconcentration method by which the compounds get dissolved in the liquid mixtures gets separated from the other compounds as per their chemical and physical properties. There are two phases: Mobile phases, SPE uses the affinity for the dissolved solutes in the liquid for the solid through which the samples to be allowed to pass (stationary phase), to isolate the mixture into the desired and non-desired compounds. The analytes containing the sample is allowed to pass through the stationary phase, which is collected or discarded, depending on either it contains our targeted analytes or not. If any portion is retained on the stationary phase, it is again rinsed and collected for an additional step, in this, the stationary phase is rinsed with a suitable eluent. The summarized representation is shown in Fig. 14.2. There are five steps involved in SPE. (1) The first step is to condition the cartridge with the solvent so that the sorbent gets evenly wet. (2) After the conditioning, the wet sorbent is equilibrated with the solvent so that afterward there should be proper loading. (3) In this, the loading is done with the solution containing the desired analytes which are percolated and allowed to pass through the solid phase. (4) In this step, the sorbent is washed out to eliminate all the impurities. (5) In the final step, the desired analytes are eluted and collected.

There are three types of SPE procedures (1) normal SPE, (2) Reversed-phase SPE, (3) Ion exchange SPE.

Extraction in the SPE method is achieved by the use of particular or monolith packed sorbent (50–60 μm) to a short length column (called as cartridges). These

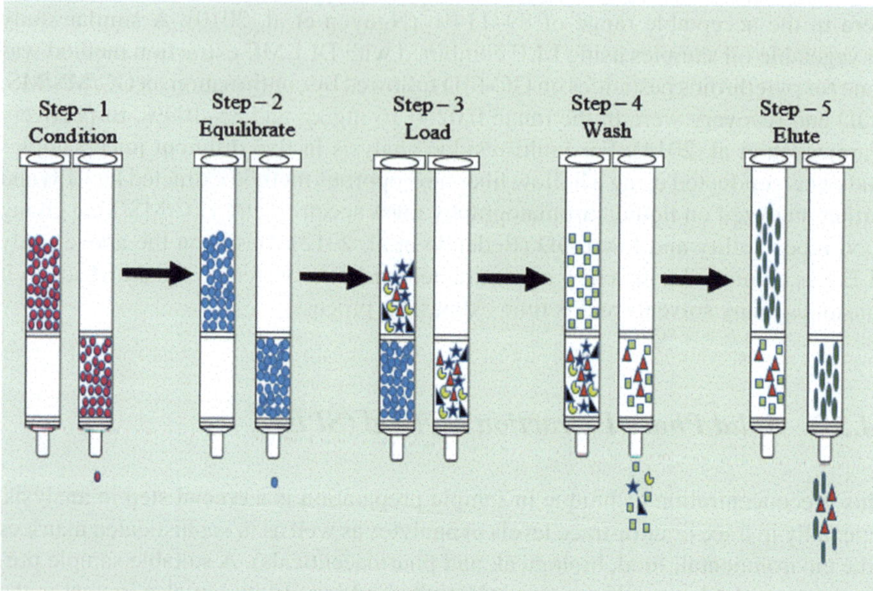

Fig. 14.2 Schematic representation of solid-phase extraction process

packed cartridges are porous in nature, siloxane –bonded silica, which are tightly packed which allow processing of sample by moderate suction. Corresponding to today's trends, the number of sorbents is increased deliberately, even these are class or compound specific.

The major application of SPE is in the determination of organic pollutant in water samples (Piri-Moghadam et al. 2016). Various implications of SPE have been reported in food samples like a multi-residue pesticide method was developed using the above-said technique in red wine. A mixture of acetonitrile and hexane was selected for elution solvent and a mixture of methanol and water for rinsing purpose. Very low LOQ was achieved using this method about 10 μg/L covering most of the pesticide and recovery obtained was between the ranges 70–120%. Real sample analysis was done for almost 32 wine samples (Pelajić et al. 2016). Another study was conducted on tea samples for 131 pesticides. The main aim of the study was to develop a robust method. Using functionalized nanomaterial multi-walled carbon nanotubes (MWCNT) for optimization in tea sample as a sample pre-treatment procedure. In validation parameters, good linear coefficient was achieved at about 0.99 (R^2). Very low LOQ was achieved about 0.5–5.0 μg/kg and recoveries obtained were 87.2–113.9%. The above method was simple and robust and is sensitive to a large number of pesticides (Zhu et al. 2019).

A similar study using MWCNT as sorbent material for magnetic SPE method was done in water, soil and river sediments. In this study a multi-residue enantiomeric study was performed on LC/MS/MS for solid and liquid matrices for 18 chiral pesticides. The functionalized carbon material shows high surface area which had applied in magnetic SPE to detect pesticides. The proposed method showed good linearity with linearity coefficient greater than 0.9912. Method quantification level for solids and liquid matrices were below 0.50 and 2.04 ng/L. Mean recovery was 81.1–106.3% with intraday relative standard deviation (RSD) of 2.1–11.9% and interday RSD of 2.6–12.7% for water matrices and recoveries between 80.3 and 105.9% for solid matrices with intraday RSD 2.3–10.9% and interday RSD of 4.0–13.4.

Another SPE conducted in vegetable matrices for 15 multiclass pesticides using SPE procedure as a clean-up method with elution ratio of acetone toluene (3:1). The whole analysis was conducted on GC-ECD with LOD lower when compared to MRL proposed by the European Union (EU) 0.02–4.5 ng/g. Good linearity was achieved with linearity ranges of 0.990–0.995 (Farina et al. 2017). Another study done using SPE for multi-residue pesticide analysis using GC/MS/MS from food matrices as an effective preconcentration technique. A combination of SPE with DLLME method was applied for conducting this study. All variables for optimization were critically optimized from extraction solvent, elution solvent to the extraction time. The preconcentration enrichment factor obtained was in the range of 2362–10,593 for 100 ml of sample. LOD obtained was 0.5–1 ng/kg RSD $\leq$ 11.8% and a good linearity $\geq$ 0.9915 (Shamsipur et al. 2016).

Magnetic porous carbon material derived from bimetallic metal-organic framework was used as an adsorbent in SPE procedure (Liu et al. 2017). Synthesized new material with functionalized large surface area worked as an effective adsorbent to analyze OCs class of pesticides. In the elution process, dichloromethane was used prior to analysis on GC/MSMS. The optimized method successfully determines analytes of OC class with LOD and LOQ between 0.39 and 0.70 ng/L and 1.45–2.0 ng/L respectively. The metal organic framework (MOF) derived carbon material proves as excellent adsorbent, fast extraction efficiency, and simple operation (Li et al. 2019a).

Recently various modified versions of SPE were adopted like magnetic SPE, d-SPE, matrix SPE because it consumed a large amount of solvent with handling issues so it is not considered as a user-friendly process.

14.3.3 Matrix Solid-Phase Dispersion Extraction (MSPD)

MSPD is a process which is patent protected, which was first time introduced in 1989 by Barker et al. (1989), for simultaneous disrupting and extraction of solid samples (Barker 2000). This process owns several merits which categorize this process a simplistic and flexible process superior over other classical extraction processes (Barker 2007). It is an integrated method in which extraction and clean-up process is in a single step with mechanical blending of the sample with solid support such as C18 (Tsai et al. 2014), silica, florisil, and diatomaceous earth (Chu et al. 2005), which makes it a hassle free, low-cost, less solvent usage thus it is an eco-compatible process (Kristenson et al. 2006) shown in Fig. 14.3. This is particularly applied for the extraction of biological samples which are extremely viscous solid and semi-solids. It is based on the principles of chemistry and physics, involving forces which are

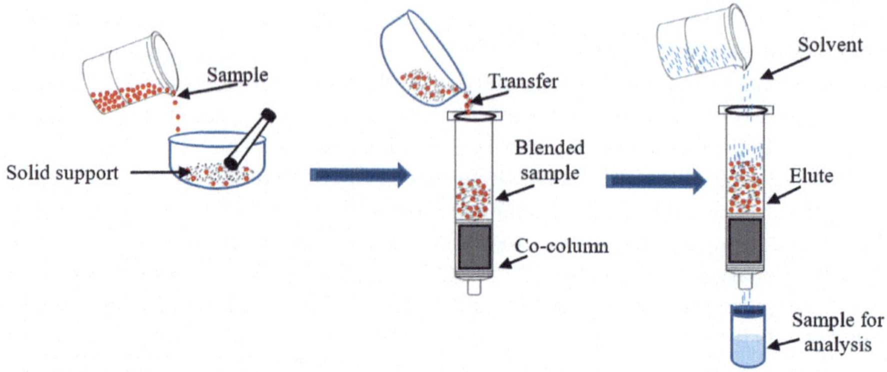

Fig. 14.3 Schematic representation of matrix solid-phase dispersion extraction process

applied to the sample and are mechanically blended for sample disintegration. Thus, contact is developed between the matrix and the solid support bonded phase (SPE). The Matrix solid phase dispersion MSPD extraction, sample/matrix is put in the glass or agate by a mortar, which contains octadecylsilyl (ODS)—derivatized silica (C_{18}). These two components are manually blended using a pestle for 30 s. If anybody wishes to add internal standard or spikes, it may be added previously to this step. Then the blended material is transferred and enclosed in the column, which is suitable for doing systematic elution with the help of solvents. The whole column is filled with the final blended mixture which is properly distributed thus generating a different phase which shows exclusive characteristics for the fractionation of samples. Thus, in this way, a relevant solvent is used to clean the column to isolate and extract the analytes.

A study reported using drinking water treated sludge for 15 pesticides of various chemical composition in which an alternative support material was optimized. The main focus of this study was to develop alternative solid supports in MSPD and it compares it with conventional solid support. According to the study, chitin was considered as a best solid support in recovery study with lowest RSD. The obtained recoveries were 75–13% with RSD values of less than 20% for all the present analytes. This method of extraction uses less solvent (5 ml), the low mass of the sample (1.5 g) and chitin volume (0.5 g) as environmental friendly support (Soares et al. 2017). MSPD procedure was used to detect phenylurea herbicides in yam samples using capillary electrophoresis (CE) with electrochemiluminescence (ECL) (Hu 2015; Li et al. 2017a). Dichloromethane as an extraction solvent to determine metrafenone fungicides in vegetables using MSPD–HPLC method.

A study was done SPE derivatization MSPD on cartridge solid support for the determination of cyclamate in food samples. In this method, the syringe was loaded with the samples, $KMnO_4$, and silica as dispersant which used as MSPD reactor. Better clean up achieved using on cartridge derivatization process rather than conventional derivatization process. LOD and LOQ obtained were very low 0.3 and 1 mg/L with recoveries 91.6–101.3% (Li et al. 2017b). Multiresidue pesticide analysis of 16

OC pesticide and 7 polychlorinated phenyls in oil crops (peanut and soybean) was studied. The focus of this work was to use sulphuric acid impregnated silica as dispersion sorbent using MSPD procedure. The method was optimized for enantiomer of chiral OCs pesticides by GC-ECD. In general, the MSPD is an inexpensive, less solvent and simple method for analysis of OCs pesticides (Zhan et al. 2016).

Another study reported the use of MSPD procedure for detection of dithiocarbamates in the list of vegetables, nuts, cereals and fruits by liquid chromatography (Blasco et al. 2004). For Multiresidue pesticide (monocrotophos, acetate, dimethoate, organochlorine, triazine, benzimidazole, neonicotinoid, carbamate, phenylurea, diacyl hydrazine) in fruit sample MSPD procedure was optimized for the determination of pesticides. The method uses dichloromethane as eluent and diatomaceous earth as a dispersant in MSPD and further detected by liquid chromatography (Radisic et al. 2013). Recently, MSPD was associated with the molecularly imprinted polymer in the synthesis of dummy molecularly imprinted silica nanoparticles (DMISPs). For the detection of acrylamide in a processed sample using a sol-gel synthesis of molecularly imprinted polymer (MIP) which act as dispersant sorbent. The above method was easy and simple to use and compatible with the user as it requires less amount of solvent (Arabi et al. 2016). A study reported in honeybee for multi-residue pesticide analysis based on MSPD method detected by GC/MS/MS was reported. C18 sorbent was optimized with florisil clean-up and acetonitrile: methanol (99:1) was best optimized and evaluated for pesticides, LOD and recoveries were in the range (Balsebre et al. 2018). This preconcentration method is a simple, rapid and solvent-free method for determination of environmental pollutants.

14.3.4 QuEChERS Based Procedures

QuEChERS based methods were first time introduced at European Pesticide Residue Workshop (EPRW) 2002 in Rome, which later published by Anastassiades et al. (2003b). The original QuEChERS method was developed to extract polar and non-polar pesticides from high water content foods mainly vegetables and fruits. Basically, this method involves (1) extraction of analytes with use of organic solvent (2) Partitioning step with the use of salt mixture (3) Clean-up of the extract with the sorbent followed by analysis of the extract directly on the instrument. So, this method is exclusively applied for the multi-pesticide residue to target a wide range of analytes and quantification in various matrices described in Fig. 14.4.

14.3.4.1 Comparison of Different Types of QuEChERS Methods

Figure 14.5 shows the step by step comparison of the three most commonly used QuEChERS method.

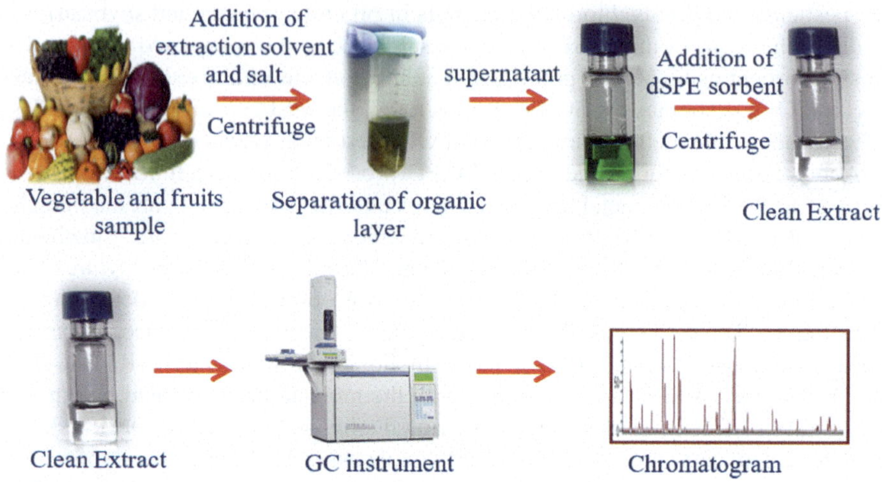

Fig. 14.4 Schematic representation of QuEChERS method

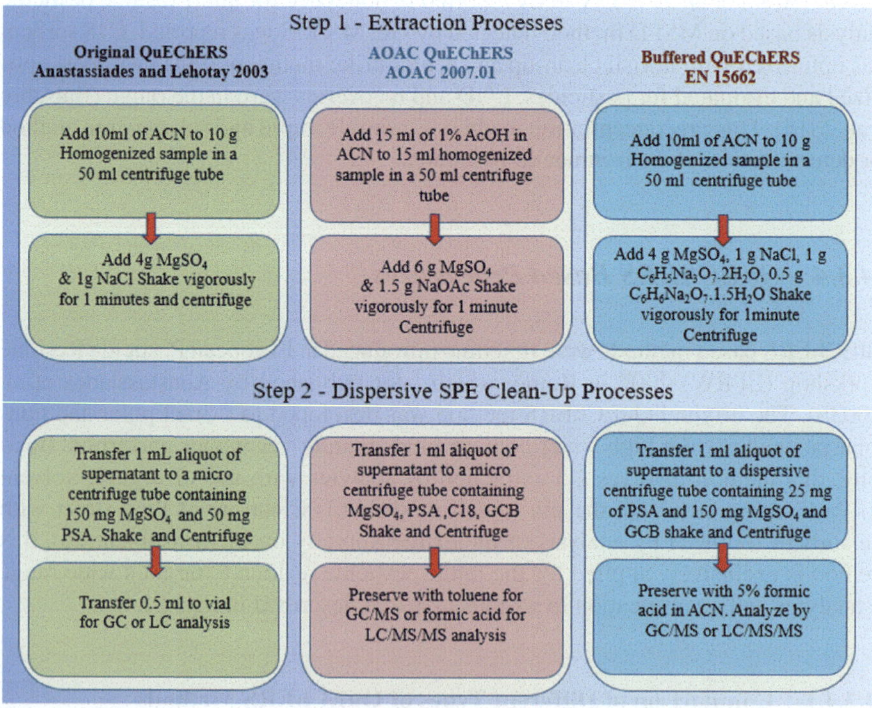

Fig. 14.5 Summary of the original QuEChERS method with most relevant modifications

(1) **Original QuEChERS Method (by Anastassiades, Lehotay, et al.)**

This method was introduced in 2003. In this method, sodium chloride is used to reduce polar interferences. This method provides the cleanest extraction because it uses fewer reagents.

(2) **AOAC 2007.01**

In this method, 1% acetic acid in acetonitrile and sodium acetate buffer is used to protect base sensitive analytes from degradation. United States Department of Agriculture USDA study has demonstrated that this method provides superior recovery for pH-sensitive compounds when compared to the other two QuEChERS methods.

(3) **EN 15662**

This European method includes sodium chloride to limit polar interferences and several buffering reagents to preserve base sensitive analytes. Sodium hydroxide used in the citrus step should be avoided as it can add impurities to the extract as well as damage the sorbent used in the clean-up step.

A recent implication of QuEChERS was studied for simultaneous pesticide and mycotoxins analysis in raw coffee beans using QuEChERS procedure. In this study, $MgSO_4$ was applying in portioning step while moving on the clean-up step. This step was chosen after optimization of QuEChERS method in which extraction solvent MeCN was acidified with either one acetic acid or formic acid, using or not using buffers and similarly with clean–up extract or avoiding it to escape unneeded steps prior to LC/MS/MS detection (Reichert et al. 2018). Another study conducted using the latest analytical approach LC/MS/MS (Ribeiro Begnini Konatu et al. 2017), in this the optimization points were choice of extraction solvents, salts in partitioning step and final clean-up of the raw extract. It reveals analysis in lettuce with accuracy and sensitivity with LOQs 5 μg/kg and recovery 70–120%.

A comparative study of QuEChERS with other extraction procedure was conducted in honey and honey bees. In this study, various parameters were optimized and evaluated for comparisons like the cost value of equipment and solvents, time, sensitivity, specificity, and accuracy. The QuEChERS was compared with SPE in which the former one shows good response in terms of recovery and precision in both matrices whereas the later on does not show good recovery in both matrices (Calatayud-Vernich et al. 2016). One more comparative study of multi-pesticide residue analysis in tamarind matrix in this citrate buffer version of QuEChERS is assessed and optimized on GC/ECD later confirmed on GC/MS/MS. The fine extraction procedure was compared and the final result shows that the one with greater ultrasonic bath time gives good result in terms of recovery (70–115%) (Paz et al. 2015).

A multpesticide residue method of 54 pesticides in fruits and vegetables was evaluated with UPLC/MS/MS. The optimized conditions were changed in pre-treatment, selection of adsorbent, extraction time and solvent depend on different matrices to create a sensitive, accurate and precise method (Xiu-ping et al. 2017). A simple study for detection of multi-pesticide residue in vegetables was optimized for 18

OCs classes of pesticides. It involves the extraction step with MeCN then salting out effect by the addition of salt and finally the clean-up with the use of $MgSO_4$, primary secondary amine (PSA) and C18 (Boes et al. 2015). A method for 200 multi-pesticide residues in cereals which low moisture content ($\leq$75%) as stated in original QuEChERS was established and optimized. Several parameters were optimized like the use of extraction solvent (EtAc and MeCN) with different buffers followed by dispersive solid phase extraction. This established method is simple and rapid and covering a wide range of pesticides for such matrices which have low water content and is quite tedious in extraction (He et al. 2015).

Recently, a number of modified QuEChERS miniaturized methods were also developed as a novel change in QuEChERS methods. Like the use of MWCNT as an effective clean-up sorbent for determination of 171 pesticides in cowpea. The MWCNT proves to be more efficient than PSA and graphitized carbon black (GCB). Low LOD and LOQs were achieved 0.001–0.003 mg/kg and 0.002–0.009 mg/kg respectively and recovery of 74–129% (Han et al. 2016). Another carbon-based study for multpesticide residue analysis using multi-filter clean- up was established in the complex matrix (crops) which have high matrix interferences (Qin et al. 2017). Thus it can be well stated that for residue analysis QuEChERS is well-established extraction method in the complex as well as in simple matrices and now various modification are also being developed in this methods which to proves an effective approach.

14.3.5 Gel Permeation Chromatography (GPC)

It is a type of chromatography based on size exclusion chromatography which separates the compounds of interest on the basis of size particles. GPC works on the basis of size or another term "hydrodynamic volume" of analytes. GPC work phenomena depend on the separation technique which is different from the other separation technique where separation is done on the basis of chemical and physical properties. The main application of this type of chromatography analysis is for polymer study. Traditionally, it was used to separates out the high molecular weight lipids from targeted compounds (Stalling et al. 1972; Specht et al. 1995; Roos et al. 1987).

Application of GPC for the determination of polyaromatic hydrocarbons as a clean-up method using GC/MS/MS is described. GPC was assessed and evaluated as a better clean up as SPE for the final extract of the cigarette smoke. By use of GPC column matrix interference was greatly reduced than traditional SPE. For selectivity and sensitivity of the method, GC/MS/MS was selected as an analytical instrument in which for quantitative "Pseudo multiple reactive monitoring" (PMRM) mode was selected and for qualitative classic multiple reaction monitoring (MRM) mode was selected.

For determination of OCs class pesticides by GPC-SPE extraction procedure a study was done on milk and milk powder. The extraction was done by acetone and hexane (1:1) later cleaned up by GPC-SPE so that lipids and other interfering substance gets cleaned up by this two-step procedure (Zheng et al. 2014). Likewise, detection of pesticide in a complex matrix such as olive oil was determined by GPC to clean up purification mechanism to make sure the best method for determination in olive oil. For removal lipophilic pesticides GPC proves its potential as a better method than others. 83.8–110.3% recovery was obtained with good sensitivity (Sánchez et al. 2006). A modified QuEChERS method accompanies (GCB/magnetite/PSA) as an adsorbent for the detection of ten pesticide residue in tobacco samples. An amalgamation of modified QuEChERS with online gel permeation chromatography was established. For the detection of target analytes following condition was optimized 3.3–5.3 time limit for online GPC, adsorbent amount 80 mg with 1 min of clean up time. The final result showed accuracy with precision for this method with the recovery of 68.8–132.2% (Luo et al. 2015).

14.3.6 Soxhlet Extraction

In 1879, the first time Von Soxhlet introduced a new and unique extraction mechanism (Soxhlet extractor) which was used a long time as commonly used leaching technique (Soxhlet 1879). It is a discontinuous extraction procedure in a boiling flask in which there is an extraction solvent which is boiled and then it is evaporated. Then it is recondensed in distillation unit (column) above. Later, it gets dropped down on the solid sample material which is to be extracted. The process is repeated again (empty and fill) and thus the compound to be extracted is filled in the boiling flask placed down. From starting point merit and demerit of Soxhlet extraction was the topic for discussion followed by modification and later merger with a new technique to lessen the leaching time and automating the Soxhlet apparatus. Conventional Soxhlet set up comes with many disadvantages like wastage of abundant solvent, long time for completion of extraction procedure which leads to thermal decomposition. Conventional Soxhlet extraction was applied to the determination of fat in milk (Soxhlet 1879). Later various modifications of the conventional approach were modified like high-pressure Soxhlet extraction (Ndiomu and Simpson 1988; Bernal et al. 1992), Automated Soxhlet extractor (http://www.foss.co.uk/Solutions/ProductsDirect/SoxtecSystems.aspx), Ultrasonic assisted Soxhlet extractor (Cañizares et al. 2004), Microwave-assisted Soxhlet extraction (Virot et al. 2007).

One of the applications of Soxhlet extraction for the extraction of polycyclic aromatic hydrocarbon (PAHs) and OCs from soil was reported for three methods (Wang et al. 2007). In this experiment Soxhlet extraction, microwave-assisted extraction (MAE) and accelerated solvent extraction (ASE) were compared. Overall it can be said that accelerated solvent extraction better than above two in terms of limit of detection and quantification. Following the solvent consumption and time required for extraction MAE and ASE was better than Soxhlet. One more experiment of microwave-assisted extraction compared to Soxhlet was studied and evaluated for OCs in plant samples. Pesticides were extracted with solvent ratio hexane: ethyl acetate (80:20) with the recovery of 75.5–132.7% for Soxhlet and 81.5–108.4% for MAE. Both this method proved to be good for extraction of OCs from plant samples (Barriada-Pereira et al. 2003). Another study investigated the conventional Soxhlet extraction with the two recent extraction procedure MAE and ASE for detection of polyphenols and polychlorinated biphenyls in fish samples (Wang et al. 2010).

14.4 Microextraction Techniques

Sample preparation technique is a very crucial and key factor in the advancement of analytical approaches. The appropriate selection of sample preparedness set the path for accurate and precise output throughout the experiment. As mentioned above the classical techniques like LLE, SPE and Soxhlet are quite a time consuming and utilizes the abundant amount of organic solvent which is not at all environment as well as user-friendly. So, we should move towards microextraction a technique which lies in accordance with the green chemistry rules. Benefits in adopting these new approaches are less use of solvent, reduce the size of the samples with precise determination of analytes in different matrices of a large number of microextraction techniques below some of the generally used techniques are explained and discussed in this chapter.

14.4.1 Dispersive Liquid-Liquid Microextraction (DLLME)

A liquid phase microextraction technique introduced by Assadi and their colleagues (Spietelun et al. 2014) as DLLME shown in Fig. 14.6 which can be categorized equivalent to LLE. In DLLME procedure, an amalgamation of the solvents used as extraction solvent (mL) and next the disperser solvent (mL) are forcefully injected into the aqueous sample which leads to the formation of a cloudy solution which finally results in the creation of high surface area for the extractor solvent and aqueous phase (Zang et al. 2009). Commonly used extraction solvents in DLLME are chloroform, carbon tetrachloride, chlorobenzene, tetrachloroethylene, and dispersive solvents are acetonitrile, methanol, and acetone. Various advantages have been associated with DLLME like less usage of solvent, reduce extraction time, greater enrichment factor with extraction efficiency.

A lot of work has been done regarding the analysis of environmental pollutant. A miniaturized environmental friendly technique, QuEChERS is associated with low-density solvent dispersive liquid-liquid microextraction (LDS-DLLME) and Ultra-low temperature at −80 °C techniques for detection of PAHs in baby foods. This sensitive GC/MS method was used to detection of almost 12 PAHs in commercial baby foods at a very low level recommended by the European Commission for baby 1 μg/g. The developed method was verified and optimized for all method validation parameters and recovery output of 72–112%. DLLME follow the green chemistry

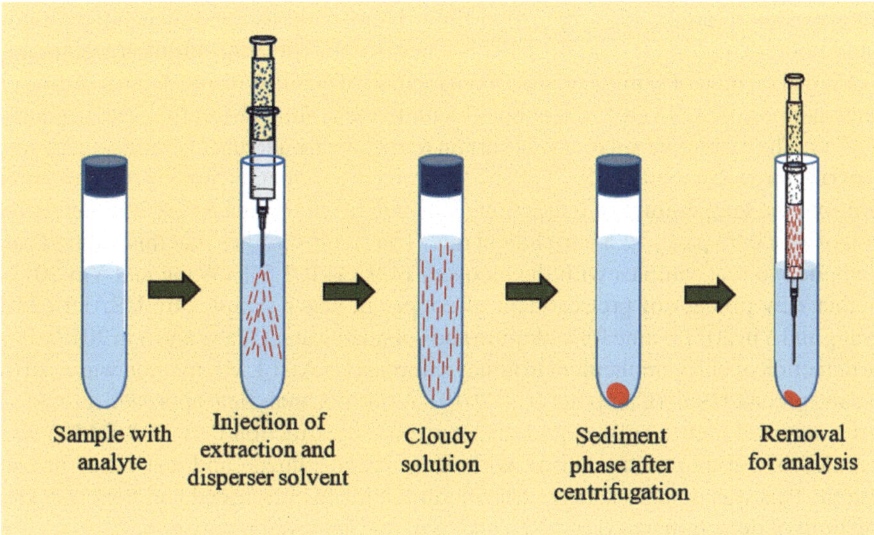

Fig. 14.6 Schematic representation Dispersive liquid-liquid microextraction procedure

principle, in this experiment less amount 75 μL of toluene was used as an extraction solvent in LDS-DLLME and in QuEChERS acetonitrile was used as disperser solvent (Petrarca and Godoy 2018). Recently, a new generation of insecticide was determined by the use of DLLME in soil matrices which has been confirmed by LC/MS/MS in MRM mode. The extraction of analytes was done in acetonitrile using SPE; this MeCN later becomes the dispersant solvent for DLLME. The extraction efficiency in DLLME was optimized like 1.5 ml of disperser solvent, 125 μL of extraction solvent and aqueous solution 10 mL. Final the sample was reconstituted in acetonitrile and then injected in LC/MS. The obtained detection limit was 0.0015–0.0090 ng/g with recovery value of 87–114% (Pastor-Belda et al. 2015).

An inexpensive and rapid method of detection of multi-pesticide residue in nutraceutical (nutrition + pharmaceutical) drop containing 40% alcohol was optimized using the DLLME procedure coupled with GC/MS. Several parameters in DLLME method was explored like the type of dispersant solvent, volume, and type of extraction solvent, centrifugation time, pH, extraction time. The proposed method shows the very low limit of detection 0.001–0.910 μg/L with quantification 0.004–3.003 μg/L, less use of organic phase (80 L), simple, the ease with high extraction efficiency (Szarka et al. 2018).

Advancement in technology leads the way to improve the already existing DLLME techniques with the synergism of other sample preparation techniques like ultrasonic assistant LLME, vortex assistant LLME, and air assistant LLME are discussed. Regarding the presence of trace level of metal analysis in the complicated matrix, the analysis gets very difficult as this is an urge to achieve the sensitivity with accuracy and precision to attain a low detection limit. Ultrasound assisted dispersive liquid-liquid microextraction (USADLLME) is a new sample preconcentration technique.

A novel method of sample pretreatment method of quantifying trace level of zinc in water samples by USADLLME coupled with flame atomic absorption spectrometry (FAAS). The extractant solvent was carbon tetrachloride and the chelating agent was (2-pyridylazo)-2-naphthol (PAN). The optimized parameters were pH, extraction duration, the temperature of extraction. This condition reveals a very low detection limit of zinc 0.95 μgL, 12.5 enrichment rate. The result signifies that this method can be applied to real samples with the recovery of 92.4–101.5% (Wang and Yin 2017) another new process of pretreatment of copper in water samples by USADLLME (Wang and Yin 2017) same for cadmium and cobalt in water (Wang and Yin 2017). For the detection of chloramphenicol in honey samples, USADLLME method was used to fast and precise result (Campone et al. 2019). A one pot analytical approach termed as vortex assisted matrix solid liquid dispersive microextraction (VA-MSLDME) was executed by mixing of the matrix with PSA, ACN, toluene and last water in one system. This was done to analyze triazole fungicide in cottonseed and honeysuckle. The limit of detection was 0.05–20 μg/kg with the achieved recovery of 82.9–97.8% (Xue et al. 2016) another application of this technique is in the detection of lead metal in mussel and seawater samples (Erarpat et al. 2017).

Air assisted liquid-liquid microextraction (AALLME) is a new addition to the LLME technique reported in 2012 (Farajzadeh and Mogaddam 2012). This technique is very user-friendly in which, very less quantity of organic extractant is needed and the whole process can be emulsified without using of dispersive solvent by continuously suction and injection of the mixture of aqueous and extractant organic solvent in narrow neck tube with the help of syringe (You et al. 2015). Till now AALLME was effectively applied for detection of fungicide in juice samples coupled with GC/ECD using the first time the solidification of floating organic droplet. A low-density solvent was used with near equal melting point to room temperature thus at the end of the process the extractant droplet can be collected by solidifying at a temperature less than the melting point. The optimised condition reveals LOD of 0.02–0.25 μgL (You et al. 2015). Similarly, AALLME was reported for OP in fruit juice samples with no use of organic dispersive solvent (You et al. 2013).

14.4.2 Single Drop Microextraction (SDME)

SDME is miniaturized kind of solvent extraction technique, especially for aqueous and gaseous samples. In 1990, the first time use of a single liquid drop for extraction of analytes was proposed by Liu and Dasgupta (1995). Actually, this microextraction method is used as an enrichment technique for analytes. Thus, it's a kind of solvent-free method when compared with conventional LLE. Indirect, SDME immersion process, a syringe needle containing an organic solvent, is allowed to puncture the vial septum nearly touching below the meniscus of the sample. Afterward, when the plunger is pressed a single drop of insoluble in water is suspended in the aqueous sample. The compound to be analyzed must be more soluble in the extraction solvent rather than water solvent (Jeannot et al. 2010; Andruch et al. 2011; Filippou et al. 2017). Most importantly SDME is a simple, less sophisticated, free from the dependency of apparatus only common microsyringe is the only apparatus needed. In spite of it SDME only comes with an advantage that the single drop formed has instability issues.

Application of SDME in food samples was studied for acrylamide detection followed by GC-ECD. This novel method was used to derivatize acrylamide to 2,3-DBPA as SDME optimization product. 3.0 ml of the derivatized product was extracted with the help of 1.0 μL n-octanol droplet suspended from the tip of the needle. After that, the final extract was allowed to run on GC-ECD (Saraji and Javadian 2019).

14.4.3 Hollow Fiber Liquid Phase Microextraction (HFLPM)

Analytes determination and sample extraction, especially in complex matrices, are being a challenging work to accomplish. Comparing the conventional extraction technique with the advanced one in terms of the output result, environment-friendly, simpler the later one always comes in priority. Discussing the new miniaturized technique the membrane-based microextraction technique (HFLPM) was the first time came into the picture in 1999 by Pedersen-Bjergaard in context to the shortcomings with the use of SDME (Pedersen-Bjergaard and Rasmussen 1999). The problem associated with SDME is the single drop instability which gets disappeared when allowed to come in the contact of the sample solution (Pawliszyn and Pedersen-Bjergaard 2006). In this procedure, the solvent used in the organic phase is immobilized and allowed to sustain in the voids of the hydrophobic fiber which is hollow in shape by the help of capillary movement after that by the use of microsyringe the acceptor phase is endorsed in the hollow fiber. Lastly, to enhance the mass transfer phenomena continuous stirring by the use of magnet is done in the container. Thus, the main steps involved in this process is (1) the target analyte which is the donor phase; (2) for the immobilization process porous nature polypropylene fibre hollow shape; (3) solvent that to be immobilized via the pores; (4) lastly, the acceptor phase which could be of acidic, basic or organic which gets filled in the voids of the fibre lumen.

Recent application of HF-LPM was reported for the remediation of the emerging pollutant in water from the sewage treatment plant and the surface water. This microextraction technique used polypropylene hollow porous fiber for the determination of pollutant. The samples were analyzed on LC/MS/MS. Various parameters were optimized for this technique like the selection of organic solvent, pH of the sample, length of the fiber, sample amount, the overall time of the extraction and last salting out effect. This technique provides a high enrichment factor from 6 to 4177 with a lower detection limit of 1.09–98.15 ng/L (Salvatierra-stamp et al. 2018). For the detection of perfluorinated (PFC) compounds in environmental water. This study was done by automated bundle hollow fiber-liquid phase microextraction coupled with ultra-high performance liquid chromatography-tandem mass spectrometry

(UHPLC-MS/MS). Same conditions were optimized like the above study giving an enrichment factor of 9–40 for extraction of the 10 ml sample. The coupling of BHF-LPME with UHPLC/MS/MS was good enough for the real sample analysis of sensitive trace level determination of PFCs. New MIP based HF-LPM approach for the detection of fluoroquinolones in aqueous media and biological samples. A resultant MIP based microextraction techniques with SPME develop a sensitive and selective method for the optimization purpose. Parameters like total time for extraction, rater of stirring, and the media loading were studied as these were the key factors in the whole experimental analysis. The final MIP product made showed selectivity towards the hydrophilic antibiotics in toluene and in polar media such as for those containing water about 20%. The limit for detection is about 0.1–10 μg/L with recovery in the range (Barahona et al. 2019). Another novel work related to MIP for detection of triazine herbicide in water is reported by use of technique HF-LPME. It utilizes polymer coated with hollow fiber with LLME the extraction based and the analytical instrument used is HPLC-UV detector. In a concise way, a very thin film of organic toluene gets immobilized in the cavities of the prepared MIP-HF. The targeted analytes are extracted by LLE from the desires sample in the immobilized toluene which is allowed to gets diffuse to the tailor-made voids of MIP. The obtained detection of limit is 0.05–0.1 μg/L. These automated coextraction techniques benefits in tackling the low-performance capability and low selectivity in the MIP especially in aqueous media (Barahona et al. 2016).

14.5 Merits and Demerits of Extraction Techniques

S. No.	Sample preparation technique	Merits	Demerits
1	Liquid-liquid extraction	Simple operation Simple and low cost apparatus are used	Low selectivity Require large volume of solvent Formation of emulsion that are difficult to break
2	Solid phase extraction	Clean extract Easy to automate High recovery of compound Applicable on the compound which are soluble or suspended in liquid medium	Costly and labour intensive Not directly applicable on solid matrices

(continued)

(continued)

S. No.	Sample preparation technique	Merits	Demerits
3	Matrix solid-phase dispersion	Simple and fast extraction Economical and environmental friendly Applicable for solid, semisolid and viscus environmental and biological samples	Not applicable for aqueous sample
4	QuEChERS	No chlorinated solvent included Non-labour-intensive (only a few steps involve) Robust result	Not applicable for aqueous sample Tedious weighing of salts
5	Gel permeation chromatography	Short analysis time Small amount of mobile phase required Good sensitivity	A limited number of peak can be resolved Filtration requires to prevent instrument from dust
6	Soxhlet extraction	Inexpensive method Temperature can me set during extraction	Require large volume of solvent and excessive extraction time
7	Dispersive liquid-liquid microextraction	Small amount of solvent required Precise and robust result	Use of chlorinated solvent Only applied to aqueous sample
8	Single drop microextraction	Simple and cost effective Little use of solvent	Low sensitivity and precision
9	Hollow fiber liquid phase microextraction	High repeatability Cheap and good cleanup ability	Possibility of blockage of fibers pore Relatively Long extraction time

14.6 Chromatographic Techniques Used for the Determination of Pesticide

From last few decay chromatography techniques are the most extensively utilized technique in the field of analytical chemistry. Chromatography is raised on the principle where analyte molecules are applied on to the surface or into the liquid and solid stationary phase to separate each other while moving with the help of the mobile phase. Chromatography techniques are commonly used for identification, purification, and separation of a mixture of analytes for qualitative and quantitative analysis. Factors which effect on separation process include molecular weight, polarity affinity to mobile and stationary phase, etc.

14.6.1 Gas Chromatography

Gas chromatography (GC) is a versatile analytical technique. This technique is generally used for the identification of organic compounds at precisely controlled conditions. The principal of the GC is based on that greater the affinity of the compounds to the stationary phase, the compound will be retained by the column and longer it will be eluted and detected. The column is the major component in GC where the separation occurs thus column also denote as the heart of GC. For GC analysis compound must be volatile in nature and thermally stable. Since temperature influence, the volatility of compounds, therefore, the column is positioned in a thermostatically controlled oven. To achieve the better separation and enriched chromatography response and reduction of analysis time column parameters can be changed like: composition of column stationary phase, column inner diameter, film thickness or operational parameters. A schematic representation is shown in Fig. 14.7.

GC has been one of the adequate approaches to quantify many of the organic contaminants that transpire at trace level in complex food and environmental samples. A study was conducted by using GC-ECD and GC-flame photometric detection (FPD) in honeybee based on matrix solid-phase dispersion (MSPD) for twelve pesticides. 65% of the sample found contaminated with eight different pesticides (Balsebre et al. 2018). For the pesticides analysis in the aqueous sample, a method was developed by using dumbbell-shaped magnetic stir-bar sorbent was set and analysis conducted on GC-ECD. The detection limit was 0.4–4.5 ng/L and relative standard deviation(RSD) found between 2.4 and 6.3% with acceptable recovery range of 83.2–98.7% (Huang et al. 2018). A chromatography analysis performed in different vegetables, fruits and grain samples by QuEChERS based extraction procedure using phenolic resin based activated carbon fibers as a reversed-dispersive solid-phase extraction solvent

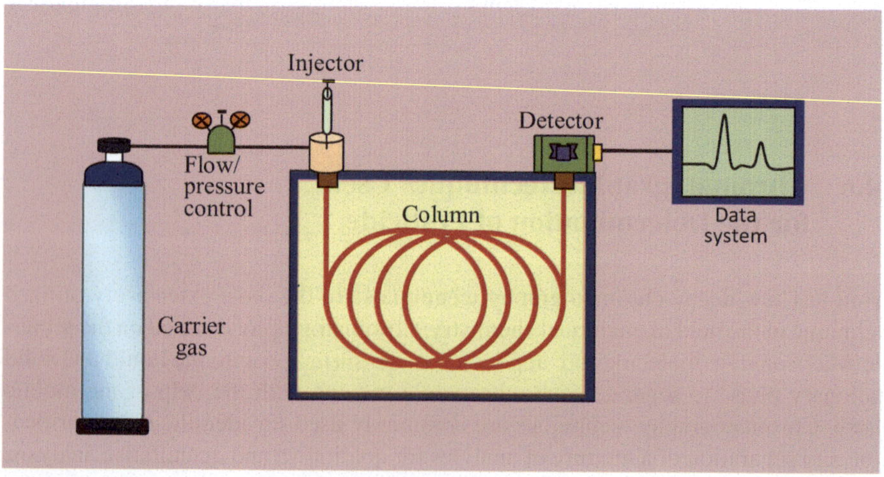

Fig. 14.7 Gas chromatography instrumentation

in the place of primary secondary amine. The analysis was carried out by 26 different class of pesticides with acceptable recovery between 70 and 120% and the limit of detection and limit of quantification were 1.13–5.48 ng/g and 3.42–16.60 ng/g (Singh et al. 2018). A simple and effective method was developed and validated in Alpinia oxyphylla for the analysis of 26 organochlorine pesticides using gas chromatography equipped with ECD. The optimized method was sensitive with detection and quantification limit were 0.1–2.0 μg/kg. The recovery was found in the range between 70 and 110% with RSD below 15% (Li et al. 2014). In another study, a sensitive GC method was developed by using a modified QuEChERS for the risk assessment of fipronil and their metabolites in soil and sugarcane. The recovery was 80.7–98.5% and the limit of detection and quantification were 0.0015–0.002 μg/g and 0.005 μg/g. No dietary risk was found due to fipronil ($RQ_d < 1$).

14.6.2 High-Performance Liquid Chromatography

High-performance liquid chromatography (HPLC) is a powerful analytical technique. It has the ability to quantitate, separate and identify, the compounds that are not thermally stable and dissolved in liquid. HPLC is highly used in a different field such as food samples analysis, industrial chemical, environmental chemical and also in pharmaceutical samples analysis (Moldoveanu and David 2013).

The typically HPLC instrument depicts in Fig. 14.8 includes a degasser, sampler, pumps, and a detector. The mobile phase that is always a pure organic solvent or combination with an aqueous phase or buffer is used. HPLC is an effective separation technique that encloses the small volume of sample injected into a tube packed column where individual components of the samples are passed through with the liquid mobile phase which forced by high pressure delivered by a pump. The components

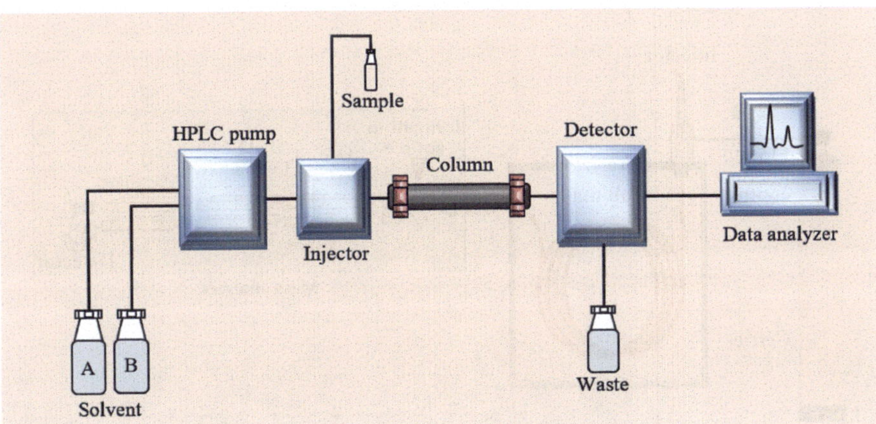

Fig. 14.8 High-performance liquid chromatography instrumentation

are separated in the column by various physical interaction between molecules and the packing material. These separated components are detected through a detector and output from this detector called as a liquid chromatogram. The different study was conducted for the better separation and identification of pesticides in diverse environmental and food samples. A study was performed to analyze the Chlorpyrifos and prophenofos in vegetables and fruits samples a method developed on HPLC using UV-spectrophotometer at 277 and 289 nm respectively. The calibration curve was carried out in range 6–16 μg/ml. The HPLC separation was attained on Hibar C18column (250 mm × 4.6 mm, 5 μ) using acetonitrile: water (90:10, v/v) as mobile phase. Extraction was done by solid phase and liquid-liquid process. Out of this process, solid phase extraction was found more effective in term of % recovery (Harshit et al. 2017). The analysis was conducted for the pesticide residue analysis in the agricultural product using acetonitrile and water (1:1 v/v) as extraction with HPLC-DAD. The average recovery was found in the range of 70–120% for all pesticide (Watanabe et al. 2014). HPLC coupled with diode array detector was used for the determination of neonicotinoid insecticides in bovine milk samples by using solid-phase extraction as clean-up process. The average recovery was found in the range between 85.1 and 99.7% at spiking level 0.01, 0.05 and 0.1 mg/kg. RSD was found below 10% and the limit of detection and quantification ranged from 0.01 to 0.04 mg/kg (Seccia et al. 2008).

14.6.3 Gas Chromatography-Mass Spectroscopy

Gas chromatography-mass spectroscopy (GCMS/MS) is an analytical tool that combines the feature of gas chromatography and mass spectrometry as shown in Fig. 14.9. In this technique, the analysis occurs on the basis of the mass/charge ratio. The separation befalls in the gas chromatographic column where the analyte is carried through

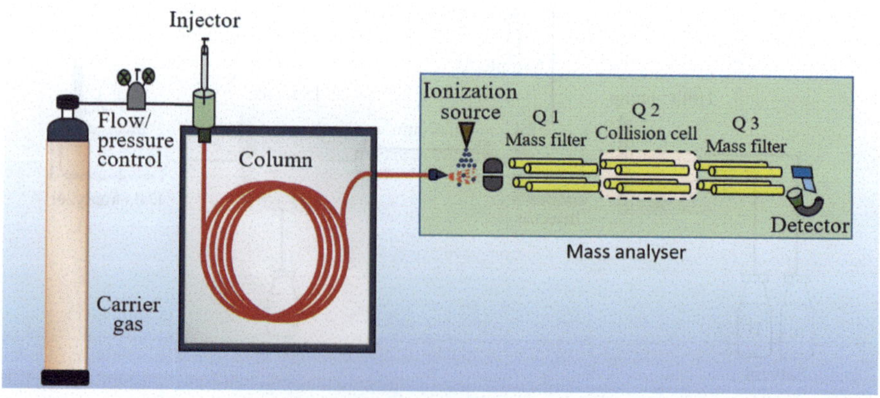

Fig. 14.9 High-performance liquid chromatography instrumentation

the inert carrier gas. The separation occurs on the basis of interaction of analytes to the stationary phase. While the sample travels through the stationary phase the difference in the physical property, boiling point, and their chemical properties help to separate the analyte from a mixture. The time spends by the analyte after the injection to elute from the column is recognized as the retention time. As the separated analyte transpires from the column, they further flow into the MS through the controlled temperature transfer line. After that depends on the ionization mode (electron impact ionization (EI) and chemical ionization (CI), the analytes not only ionize but also break into their ion fragment. All ionized analyte (molecular ions or product-ion) forms a fingerprint mass spectrum, which gives a unique identity to each molecule. Particularly very helpful in the determination of organic volatile and non-volatile compound in vegetables and fruits samples and also in closely related fields. A monitoring study was performed in vegetables and fruits samples for 210 different class of pesticides and validated using GCMS/MS instrument. Ethyl acetate used as an extraction solvent and validated method was applied to 292 samples collected from different countries with acceptable recovery and precision (Seccia et al. 2008). A QuEChERS extraction based method was developed on GCMS/MS for 77 pesticides in wine sample. For extraction procedure 10 ml sample used with acetonitrile for liquid-liquid partition by adding 4 g $MgSO_4$ and 3 g NaCl and clean up with 150 mg $MgSO_4$ and 50 mg PSA the recovery were in range 70–110%, with inter-day and intra-day precision was 22 and 15% respectively. The limit of quantification was in the range of 0.003–0.05 mg/ml (Jiang et al. 2009). A high throughput multi-residue method was validated in cereals using QuEChERS as extraction procedure on GCMS/MS. The calibration linearity range was 2–200 µg/kg and the recovery study was performed at 5, 10, 20, 50, 100, and 200 µg/kg concentration are recovery was found in acceptance criteria between 70 and 120% with associated RSD $< 20\%$ (He et al. 2015). A convenient method was developed on commercially available chenpi, for 133 pesticides on GCMS/MS. The sample was prepared through the QuEChERS extraction method using acetonitrile as an extraction solvent. In clean up procedure octadecyl-modified silica (C18) and graphitized carbon were used and aminopropyl (NH_2) used instead of PSA. Recovery was achieved ranged from 70.0 to 112.2% with RSD of 0.2 to 14.4% (Li et al. 2019b). A study was performed on GCMS/MS for the steroid profile which correlated with subclinical hypercortisolism. The steroid profiling covers 91 steroids from sera of 54 patients through unilateral ($n = 29$) and bilateral ($n = 25$) adrenal incidentalomas were compared (Hana et al. 2018).

14.6.4 Liquid Chromatography-Mass Spectrometry (LCMS/MS)

LCMS/MS is a powerful hyphenated technique that combines the resolving power of Liquid chromatography with mass spectrometry. Coupled chromatography MS

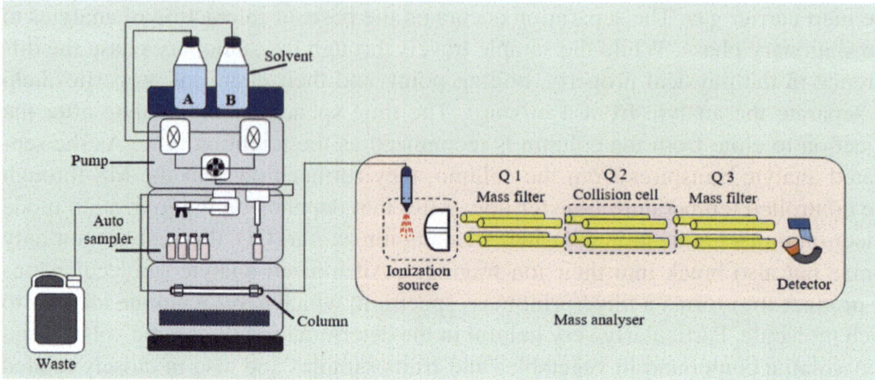

Fig. 14.10 Liquid chromatography-mass spectrometry instrumentation

detection system is recently popular in the analysis of the non-volatile and thermally unstable compound. This joint technique is highly effective to complex sample from environmental and biological origin. The separation process occurs as similar to HPLC. Mass spectrometer instrument not fully compatible with the LC system because it is not possible to directly transfer the eluate from the LC column to MS source. Therefore, the LC-MS/MS system holds an interface that well transfers the separated components from the LC column to MS source. The interface should not hinder over the ionization efficiency and vacuum condition of the MS system. A proper schematic representation of LC-MS/Ms is shown in Fig. 14.10. Nowadays, the utmost widely applied LCMS/MS interfaces are grounded on electrospray ionization (ESI), atmospheric pressure photo-ionization (APPI) and atmospheric pressure chemical ionization (APCI).

For the determination of neonicotinoid insecticide in sugarcane juice by QuEChERS method using electrospray ionization source on LCMS/MS. The optimized method was validated by assessing linearity, precision, accuracy limit of detection and quantification. LOD and LOQ were found in range 0.0007–0.002 μg/kg and 0.002–0.005 μg/kg respectively and RSD between 2.52 and 14.57% (Suganthi et al. 2018). A LCMS/MS method was developed for pesticide determination in sweet pepper using QuEChERS extraction procedure. The method provides good recovery between 70 and 120% and satisfactory precision ≤20% (da Costa Morais et al. 2018). A QuEChERS method coupled with LCMS/MS was validated for the determination of 102 pesticide residues in green tea produced in Jiangxi province, China. The recovery was found in range 62–125% with RSD 18%. The LOD and LOQ ranged from 0.03 to 15 μg/kg and 0.1–5 μg/kg. Besides the level of 11 pesticides in 18 samples were existent to be greater than the values permitted in EC regulation No. 396/2005 (Huang et al. 2019). One more study conducted on the polyphenol-rich agricultural product using QuEChERS extraction procedure by LCMS/MS. The method was successfully applied in tea, apple, broccoli, and shallot samples or 20 pesticides. The clean-up was performed using 150 mg PSA, and GCB 10 mg. The mean recovery was in ranged from 73 to 106% and RSD less than 13% and limit of quantification was found in range from 0.01 to 0.02 mg/kg (Guo et al. 2019).

14.7 Advanced Detection Technique

14.7.1 Molecular Imprinting Polymer (MIP)

MIP is a new analytical approach in the field of separation and detection of several analytes. In this technology, artificial recognition sites are produced on the polymer surface which is corresponding to the template in their spatial arrangement of the functional group (Algieri et al. 2014). The produced synthetic receptor is capable to bind those target molecule, which exactly fits into the binding sites with high affinity and specificity. Between the polymer and the template different types of interaction, forces work like hydrogen bond, hydrophobic interaction, electrostatic interaction, and van der Waals forces. Nowadays MIP technology has been widely used for the creation of biosensors for targeted environmental toxicant separation and analysis.

In MIP formation process different steps involve (1) template, (2) functional monomers, (3) cross-linker, (4) polymerization initiator, (5) porogenic solvent, (6) extraction solvent (Fig. 14.11).

A sensitive MIP capacitive sensor was developed for the detection of methyl-parathion. The detection of methyl-parathion was conceded on a polyquercetin-polyresorcinol-gold nanoparticles electrode using MIP and electro polymerization method. The sensor's linear response was in the range between 7×10^{-8} mol/L and 1×10^{-6} mol/L with a detection limit of 3.4×10^{-10} mol/L (Li et al. 2012). A non-covalent MIP was created for the selective analysis of fenarimol in food samples. The developed MIP was successfully applied are recovery found between 91.16 and 99.52%. the LOD and LOQ in samples were 0.03–0.06 and 0.12–0.21 μg/ml respectively (Khan et al. 2016). MIP were synthesized for phosmet insecticide by the batch polymerization process. The pre-exponential factor K_F was determined for MIP was

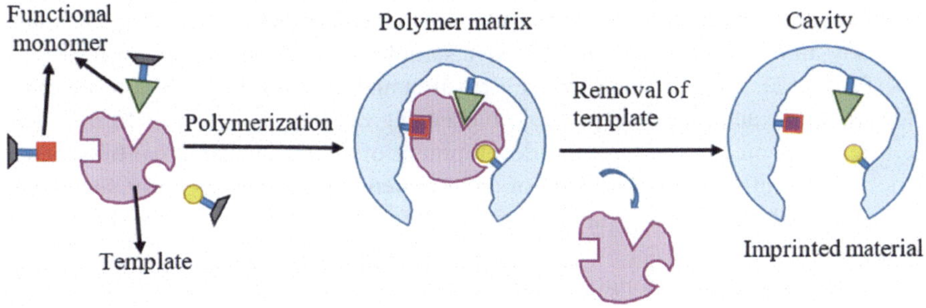

Fig. 14.11 Molecular imprinting polymer procedure

1.439 $mg^{(1-N)}$ g^{-1} L^N, that is more than 10 times higher for non imprinted polymer (NIPs) (0.125 $mg^{(1-N)}$ g^{-1} L^N, that indicate an increase of the binding number and average affinity (Aftim et al. 2017). In this study electrochemical impedimetric sensor based MIP was developed for the detection of methidathion organophosphorus insecticide. Immobilization of synthesized MIP on the printed electrode done through the sol-gel process and the developed method was successfully applied on real samples (tap water) (Bakas et al. 2014).

14.7.2 Biosensor (Electrochemical Sensor, Optical Nano Sensors, Piezoelectric)

A biosensor is an analytical device, generally used for the detection of chemical elements that combines a biological component with a physicochemical detector. In the field of environmental and biological toxicant analysis biosensors are widely used as fast, robust cost-effective and real-time analytical technique (Mehrotra 2016). For the detection of multiplexed pollutant, it is needed a portable, effective and reliable biosensing device. Biosensors can be classified according to their work principle such as electrochemical sensor (Selvolini et al. 2018) (impedance and amperometric, biosensors), optical (including surface plasma resonance and optical fiber) and piezoelectric (with quartz crystal microbalance biosensors) (March et al. 2009).

Pesticides are major toxicant existing in the environment, biosensors including electrochemical, immunosensors and enzymatic biosensors are reported for the detection and monitoring of various persistent environmental pollutant (Justino et al. 2017).

14.7.2.1 Biosensors for Environmental Toxicant Monitoring

The environmental monitoring of organic pollutant, toxic elements, pesticide, and pathogens are necessary for the development of society. Traditionally used analytical methods employed for the environmental monitoring of toxicants includes chromatography techniques (example, GC, HPLC, GCMS/MS, LCMS/MS) but they are costly and need effective and selective sample preparation or pretreatment process (Patel et al. 2010). Thus to make detection and monitoring process more cost-effective, easy, and sensitive portable biosensing devices are needed. In this way the role of nanotechnology on the development of smart and effective biosensing devices is essential to the detection of environmental pollutant, recently developed biosensors include nanomaterial to improves the analytical performances (Holzinger et al. 2014).

For the pesticides monitoring biosensors are developed based on acetylcholinesterase (AChE) and organophosphorus hydrolase (OPH) enzymes (Bucur et al. 2018).

14.8 Future Prospective and Conclusion

Monitoring of persistent environmental pollutants comprising pesticides has received considerable attention due to its adverse effect on human health. Therefore, several analytical techniques have been developed for the detection and quantification of pesticides in the various matrices. Microextraction techniques are now extensively used in many research labs. The increase in the demand for these sample preparation and extraction technique has lead to the development of new approaches and the improvement of existing sampling methods. Moreover, the less consumption of extraction solvent, high extraction efficiency, and selectivity have made these techniques a better alternatives to classical extraction. Nowadays, more focus is on the multi-residue methods so that more analytes can be cover in less time.

Another important factor is that the method which should be opted must be environmental as well a user-friendly and follows the principle of chemistry. As already discussed, the conventional techniques requires more solvent, time consuming, and labour intensive which are not fit for complex matrices so we should adopt advanced microextraction techniques which uses lss organic solvent in the extracting phase, less dependency on sophisticated analytical instruments, better clean up and high enrichment factor like in HFME, SDME and in various modifications of DLLME. Another important conclusion of this book chapter is taking the advantage all miniaturized technique the coupling of the LLME with the other techniques open the future prospectives for analysis of analytes covering almost all matrices.

In the special context of pesticides analysis, QuEChERS had always been the most preferred technique for analysis. But, for the pretreatment of solid samples, the steps should be very simplified so that it allows the proper and efficient precipitation of the concerned analytes. So we should adopt an appropriate mechanical approach for complete homogenization of the samples by adopting methods like microwave, ultrasound, vortex. Proper selection of the solvent should be made so that lipophilic pesticides can be easily extracted and gives good extraction recovery. Another concern is achieving good recovery with less matrix effect to improve the output result. In the field of analytical chemistry, the sample preparation had always been the bottleneck as proper extraction is very critical and crucial mainly when trace level analysis is to be procured in environmental matrices. Lastly, focus and strategies should be made when the sample available is less abundant, to reduce the solvent cost, sorbent and all other components used in QuEChERS and minimum waste discard when compared with the classical extraction approaches.

Acknowledgements The authors would like to express their gratitude to CSIR-Indian Institute of Toxicology Research and also thankful to CSIR-Advanced Material and Processed Research Institute Bhopal for providing necessary infrastructural facilities. The authors gratefully acknowledging to the department of science and technology, New Delhi, India in the form of a research grant (DST/INSPIRE/04/2015/001859) and project GAP-168 to facilitate the research work. The authors supportively acknowledge the Gun Ei Chemical Industry Co. Ltd., for providing the ACFs.

References

Aftim N, Istamboulié G, Piletska E, Piletsky S, Calas-Blanchard C, Noguer T (2017) Biosensor-assisted selection of optimal parameters for designing molecularly imprinted polymers selective to phosmet insecticide. Talanta 174:414–419. https://doi.org/10.1016/j.talanta.2017.06.035

Algieri C, Drioli E, Guzzo L, Donato L (2014) Bio-mimetic sensors based on molecularly imprinted membranes. Sensors (Basel, Switzerland) 14(8):13863–13912. https://doi.org/10.3390/s140813863

Anastassiades M, Lehotay SJ, Štajnbaher D, Schenck FJ (2003a) Fast and easy multiresidue method employing acetonitrile extraction/partitioning and "dispersive solid-phase extraction" for the determination of pesticide residues in produce. J AOAC Int 86(2):412–431

Anastassiades M, Lehotay S, štajnbaher D, Schenck F (2003) Fast and easy multiresidue method employing acetonitrile extraction/partitioning and "dispersive solid-phase extraction" for the determination of pesticide residues in produce, vol 86

Andrade-Eiroa A, Canle M, Leroy-Cancellieri V, Cerdà V (2016) Solid-phase extraction of organic compounds: a critical review (Part I). TrAC Trends Anal Chem 80:641–654. https://doi.org/10.1016/j.trac.2015.08.015

Andruch V, Kocúrová L, Balogh IS, Škrlíková J (2011) Recent advances in coupling single-drop and dispersive liquid–liquid microextraction with UV—vis spectrophotometry and related detection techniques, vol 102. https://doi.org/10.1016/j.microc.2011.10.006

Arabi M, Ghaedi M, Ostovan A (2016) Development of dummy molecularly imprinted based on functionalized silica nanoparticles for determination of acrylamide in processed food by matrix solid phase dispersion. Food Chem 210:78–84. https://doi.org/10.1016/j.foodchem.2016.04.080

Arthur CL, Pawliszyn J (1990) Solid phase microextraction with thermal desorption using fused silica optical fibers. Anal Chem 62(19):2145–2148. https://doi.org/10.1021/ac00218a019

Bakas I, Hayat A, Piletsky S, Piletska E, Chehimi MM, Noguer T, Rouillon R (2014) Electrochemical impedimetric sensor based on molecularly imprinted polymers/sol–gel chemistry for methidathion organophosphorous insecticide recognition. Talanta 130:294–298. https://doi.org/10.1016/j.talanta.2014.07.012

Balsebre A, Báez ME, Martínez J, Fuentes E (2018) Matrix solid-phase dispersion associated to gas chromatography for the assessment in honey bee of a group of pesticides of concern in the apicultural field. J Chromatogr A 1567:47–54. https://doi.org/10.1016/j.chroma.2018.06.062

Barahona F, Díaz-Álvarez M, Turiel E, Martín-Esteban A (2016) Molecularly imprinted polymer-coated hollow fiber membrane for the microextraction of triazines directly from environmental waters. J Chromatogr A 1442:12–18. https://doi.org/10.1016/j.chroma.2016.03.004

Barahona F, Albero B, Tadeo JL, Martín-Esteban A (2019) Molecularly imprinted polymer-hollow fiber microextraction of hydrophilic fluoroquinolone antibiotics in environmental waters and urine samples. J Chromatogr A 1587:42–49. https://doi.org/10.1016/j.chroma.2018.12.015

Barker SA (2000) Matrix solid-phase dispersion. J Chromatogr A 885(1):115–127. https://doi.org/10.1016/S0021-9673(00)00249-1

Barker SA (2007) Matrix solid phase dispersion (MSPD). J Biochem Biophys Methods 70(2):151–162. https://doi.org/10.1016/j.jbbm.2006.06.005

Barker SA, Long AR, Short CR (1989) Isolation of drug residues from tissues by solid phase dispersion. J Chromatogr A 475(2):353–361. https://doi.org/10.1016/S0021-9673(01)89689-8

Barriada-Pereira M, Concha-Graña E, González-Castro MJ, Muniategui-Lorenzo S, López-Mahía P, Prada-Rodríguez D, Fernández-Fernández E (2003) Microwave-assisted extraction versus Soxhlet extraction in the analysis of 21 organochlorine pesticides in plants. J Chromatogr A 1008(1):115–122. https://doi.org/10.1016/S0021-9673(03)01061-6

Bedendo GC, Jardim ICSF, Carasek E (2012) Multiresidue determination of pesticides in industrial and fresh orange juice by hollow fiber microporous membrane liquid–liquid extraction and detection by liquid chromatography–electrospray-tandem mass spectrometry. Talanta 88:573–580. https://doi.org/10.1016/j.talanta.2011.11.037

Bernal JL, del Nozal MJ, Jiménez JJ (1992) Use of a high-pressure Soxhlet extractor for the determination of organochlorine residues by gas chromatography, vol 34. https://doi.org/10.1007/bf02290238

Blasco C, Font G, Picó Y (2004) Determination of dithiocarbamates and metabolites in plants by liquid chromatography–mass spectrometry. J Chromatogr A 1028(2):267–276. https://doi.org/10.1016/j.chroma.2003.12.002

Boes E, Rosmalina RT, Ridwan YS, Nugraha WC, Yusiasih R (2015) Development of validated method using QuEChERS technique for organochlorine pesticide residues in vegetable. Procedia Chem 16:229–236. https://doi.org/10.1016/j.proche.2015.12.052

Bucur B, Munteanu F-D, Marty J-L, Vasilescu A (2018) Advances in enzyme-based biosensors for pesticide detection. Biosensors (Basel) 8(2):27. https://doi.org/10.3390/bios8020027

Cabras P, Angioni A, Garau V, Pirisi FM, Cabitza F, Pala M, Farris GA (2001) Fenhexamid residues in grapes and wine, vol 18. https://doi.org/10.1080/02652030120571

Calatayud-Vernich P, Calatayud F, Simó E, Picó Y (2016) Efficiency of QuEChERS approach for determining 52 pesticide residues in honey and honey bees. MethodsX 3:452–458. https://doi.org/10.1016/j.mex.2016.05.005

Campone L, Celano R, Piccinelli AL, Pagano I, Cicero N, Sanzo RD, Carabetta S, Russo M, Rastrelli L (2019) Ultrasound assisted dispersive liquid-liquid microextraction for fast and accurate analysis of chloramphenicol in honey. Food Res Int 115:572–579. https://doi.org/10.1016/j.foodres.2018.09.006

Cañizares MP, Garcı-Mesa JA, de Luque Castro MD (2004) Fast ultrasound-assisted method for the determination of the oxidative stability of virgin olive oil. Anal Chim Acta 502(2):161–166. https://doi.org/10.1016/j.aca.2003.10.022

Choi MPK, Kang YH, Peng XL, Ng KW, Wong MH (2009) Stockholm Convention organochlorine pesticides and polycyclic aromatic hydrocarbons in Hong Kong air. Chemosphere 77(6):714–719. https://doi.org/10.1016/j.chemosphere.2009.08.039

Chu X-G, Hu X-Z, Yao H-Y (2005) Determination of 266 pesticide residues in apple juice by matrix solid-phase dispersion and gas chromatography–mass selective detection. J Chromatogr A 1063(1):201–210. https://doi.org/10.1016/j.chroma.2004.12.003

Codex Alimentarius. Accessed on July 2012

Correia-Sá L, Norberto S, Delerue-Matos C, Calhau C, Domingues VF (2018) Micro-QuEChERS extraction coupled to GC–MS for a fast determination of Bisphenol A in human urine. J Chromatogr B 1072:9–16. https://doi.org/10.1016/j.jchromb.2017.10.060

da Costa Morais EH, Collins CH, Jardim ICSF (2018) Pesticide determination in sweet peppers using QuEChERS and LC–MS/MS. Food Chem 249:77–83. https://doi.org/10.1016/j.foodchem.2017.12.092

de la Guardia M, Garrigues S (2012) Handbook of green analytical chemistry. Wiley, Chichester

Duca R-C, Salquebre G, Hardy E, Appenzeller BMR (2014) Comparison of solid phase- and liquid/liquid-extraction for the purification of hair extract prior to multi-class pesticides analysis. J Chromatogr B 955–956:98–107. https://doi.org/10.1016/j.jchromb.2014.02.035

Erarpat S, Özzeybek G, Chormey DS, Bakırdere S (2017) Determination of lead at trace levels in mussel and sea water samples using vortex assisted dispersive liquid-liquid microextraction-slotted quartz tube-flame atomic absorption spectrometry. Chemosphere 189:180–185. https://doi.org/10.1016/j.chemosphere.2017.09.072

Farajzadeh MA, Mogaddam MRA (2012) Air-assisted liquid–liquid microextraction method as a novel microextraction technique; application in extraction and preconcentration of phthalate esters in aqueous sample followed by gas chromatography–flame ionization detection. Anal Chim Acta 728:31–38. https://doi.org/10.1016/j.aca.2012.03.031

Farajzadeh MA, Khoshmaram L, Nabil AAA (2014) Determination of pyrethroid pesticides residues in vegetable oils using liquid–liquid extraction and dispersive liquid–liquid microextraction followed by gas chromatography–flame ionization detection. J Food Compos Anal 34(2):128–135. https://doi.org/10.1016/j.jfca.2014.03.004

Farajzadeh M, Feriduni B, Afshar mogaddam M (2015) Development of a new extraction method based on counter current salting-out homogenous liquid–liquid extraction followed by dispersive liquid–liquid microextraction: application for the extraction and preconcentration of widely used pesticides from fruit juices, vol 367. https://doi.org/10.1016/j.talanta.2015.06.024
Farajzadeh MA, Feriduni B, Afshar Mogaddam MR (2015b) Development of counter current salting-out homogenous liquid–liquid extraction for isolation and preconcentration of some pesticides from aqueous samples. Anal Chim Acta 885:122–131. https://doi.org/10.1016/j.aca.2015.05.031
Farajzadeh MA, Feriduni B, Afshar Mogaddam MR (2016) Development of a new extraction method based on counter current salting-out homogenous liquid–liquid extraction followed by dispersive liquid–liquid microextraction: Application for the extraction and preconcentration of widely used pesticides from fruit juices. Talanta 146:772–779. https://doi.org/10.1016/j.talanta.2015.06.024
Farina Y, Abdullah MP, Bibi N, Khalik WMAWM (2017) Determination of pesticide residues in leafy vegetables at parts per billion levels by a chemometric study using GC-ECD in Cameron Highlands, Malaysia. Food Chem 224:55–61. https://doi.org/10.1016/j.foodchem.2016.11.113
Fatoki OS, Awofolu RO (2003) Methods for selective determination of persistent organochlorine pesticide residues in water and sediments by capillary gas chromatography and electron-capture detection. J Chromatogr A 983(1–2):225–236
Filippou O, Bitas D, Samanidou V (2017) Green approaches in sample preparation of bioanalytical samples prior to chromatographic analysis. J Chromatogr B 1043:44–62. https://doi.org/10.1016/j.jchromb.2016.08.040
Genuis SJ, Lane K, Birkholz D (2016) Human elimination of organochlorine pesticides: blood, urine, and sweat study. Biomed Res Int 2016:10. https://doi.org/10.1155/2016/1624643
González-Curbelo M, Socas-Rodríguez B, Herrera-Herrera A, González-Sálamo J, Hernández-Borges J, Rodríguez-Delgado M (2015) Evolution and applications of the QuEChERS method. TrAC Trends Anal Chem 71:169–185
Gonzalez-Rodriguez RM, Cancho-Grande B, Simal-Gandara J (2009) Multiresidue determination of 11 new fungicides in grapes and wines by liquid-liquid extraction/clean-up and programmable temperature vaporization injection with analyte protectants/gas chromatography/ion trap mass spectrometry. J Chromatogr A 1216(32):6033–6042. https://doi.org/10.1016/j.chroma.2009.06.046
Guo J, Tong M, Tang J, Bian H, Wan X, He L, Hou R (2019) Analysis of multiple pesticide residues in polyphenol-rich agricultural products by UPLC-MS/MS using a modified QuEChERS extraction and dilution method. Food Chem 274:452–459. https://doi.org/10.1016/j.foodchem.2018.08.134
Han Y, Song L, Zou N, Chen R, Qin Y, Pan C (2016) Multi-residue determination of 171 pesticides in cowpea using modified QuEChERS method with multi-walled carbon nanotubes as reversed-dispersive solid-phase extraction materials. J Chromatogr B 1031:99–108. https://doi.org/10.1016/j.jchromb.2016.07.043
Hana V Jr, Kosak M, Hána V, Hill M (2018) Steroid profile using gas chromatography tandem mass spectrometry (GC-MS/MS) in search for a steroid which correlates most with subclinical hypercortisolism. https://doi.org/10.1530/endoabs.56.P161
Harshit D, Charmy K, Nrupesh P (2017) Organophosphorus pesticides determination by novel HPLC and spectrophotometric method. Food Chem 230:448–453. https://doi.org/10.1016/j.foodchem.2017.03.083
He Z, Wang L, Peng Y, Luo M, Wang W, Liu X (2015) Multiresidue analysis of over 200 pesticides in cereals using a QuEChERS and gas chromatography–tandem mass spectrometry-based method. Food Chem 169:372–380. https://doi.org/10.1016/j.foodchem.2014.07.102
Holzinger M, Le Goff A, Cosnier S (2014) Nanomaterials for biosensing applications: a review. Front Chem 2:63. https://doi.org/10.3389/fchem.2014.00063
Hu Y (2015) Simultaneous determination of phenylurea herbicides in yam by capillary electrophoresis with electrochemiluminescence detection. J Chromatogr B 986–987:143–148. https://doi.org/10.1016/j.jchromb.2015.02.016

Huang Y-W, Lee HK, Shih H-K, Jen J-F (2018) A sublimate sorbent for stir-bar sorptive extraction of aqueous endocrine disruptor pesticides for gas chromatography-electron capture detection. J Chromatogr A 1564:51–58. https://doi.org/10.1016/j.chroma.2018.06.018

Huang Y, Shi T, Luo X, Xiong H, Min F, Chen Y, Nie S, Xie M (2019) Determination of multi-pesticide residues in green tea with a modified QuEChERS protocol coupled to HPLC-MS/MS. Food Chem 275:255–264. https://doi.org/10.1016/j.foodchem.2018.09.094

http://www.foss.co.uk/Solutions/ProductsDirect/SoxtecSystems.aspx

Jeannot MA, Cantwell FF (1996) Solvent microextraction into a single drop. Anal Chem 68(13):2236–2240. https://doi.org/10.1021/ac960042z

Jeannot MA, Przyjazny A, Kokosa JM (2010) Single drop microextraction—development, applications and future trends. J Chromatogr A 1217(16):2326–2336. https://doi.org/10.1016/j.chroma.2009.10.089

Jiang Y, Li X, Xu J, Pan C, Zhang J, Niu W (2009) Multiresidue method for the determination of 77 pesticides in wine using QuEChERS sample preparation and gas chromatography with mass spectrometry. Food Addit Contam Part A: Chem Anal Control Expo Risk Assess 26(6):859–866. https://doi.org/10.1080/02652030902822794

Justino CIL, Duarte AC, Rocha-Santos TAP (2017) Recent progress in biosensors for environmental monitoring: a review. Sensors (Basel) 17(12):2918. https://doi.org/10.3390/s17122918

Kaczyński P, Łozowicka B (2017) One-step QuEChERS-based approach to extraction and cleanup in multiresidue analysis of sulfonylurea herbicides in cereals by liquid chromatography–tandem mass spectrometry. Food Anal Methods 10(1):147–160

Khan S, Bhatia T, Trivedi P, Satyanarayana GN, Mandrah K, Saxena PN, Mudiam MK, Roy SK (2016) Selective solid-phase extraction using molecularly imprinted polymer as a sorbent for the analysis of fenarimol in food samples. Food Chem 199:870–875. https://doi.org/10.1016/j.foodchem.2015.12.091

Kolberg DI, Prestes OD, Adaime MB, Zanella R (2011) Development of a fast multiresidue method for the determination of pesticides in dry samples (wheat grains, flour and bran) using QuEChERS based method and GC–MS. Food Chem 125(4):1436–1442. https://doi.org/10.1016/j.foodchem.2010.10.041

Kristenson EM, Brinkman UAT, Ramos L (2006) Recent advances in matrix solid-phase dispersion. TrAC Trends Anal Chem 25(2):96–111. https://doi.org/10.1016/j.trac.2005.05.011

Lee J, Kim L, Shin Y, Lee J, Lee J, Kim E, Moon J-K, Kim J-H (2017) Rapid and simultaneous analysis of 360 pesticides in brown rice, spinach, orange, and potato using microbore GC-MS/MS. J Agric Food Chem 65(16):3387–3395

Li H, Wang Z, Wu B, Liu X, Xue Z, Lu X (2012) Rapid and sensitive detection of methyl-parathion pesticide with an electropolymerized, molecularly imprinted polymer capacitive sensor. Electrochim Acta 62:319–326. https://doi.org/10.1016/j.electacta.2011.12.035

Li J, Liu D, Wu T, Zhao W, Zhou Z, Wang Z (2014) A simplified procedure for the determination of organochlorine pesticides and polychlorobiphenyls in edible vegetable oils, vol 151. https://doi.org/10.1016/j.foodchem.2013.11.047

Li J, Li Y, Xu D, Zhang J, Wang Y, Luo C (2017a) Determination of metrafenone in vegetables by matrix solid-phase dispersion and HPLC-UV method. Food Chem 214:77–81. https://doi.org/10.1016/j.foodchem.2016.07.061

Li J, Liu Y, Liu Q, Hui J, Liu Y (2017b) On-cartridge derivatisation using matrix solid phase dispersion for the determination of cyclamate in foods. Anal Chim Acta 972:46–53. https://doi.org/10.1016/j.aca.2017.04.032

Li D, He M, Chen B, Hu B (2019a) Metal organic frameworks-derived magnetic nanoporous carbon for preconcentration of organophosphorus pesticides from fruit samples followed by gas chromatography-flame photometric detection. J Chromatogr A 1583:19–27. https://doi.org/10.1016/j.chroma.2018.11.012

Li S, Yu P, Zhou C, Tong L, Li D, Yu Z, Zhao Y (2019b) Analysis of pesticide residues in commercially available chenpi using a modified QuEChERS method and GC-MS/MS determination. J Pharm Anal. https://doi.org/10.1016/j.jpha.2019.01.005

Liška I (2000) Fifty years of solid-phase extraction in water analysis—Historical development and overview, vol 885. https://doi.org/10.1016/s0021-9673(99)01144-9
Liu S, Dasgupta PK (1995) Liquid droplet. A renewable gas sampling interface. Anal Chem 67(13):2042–2049. https://doi.org/10.1021/ac00109a023
Liu Y, Gao Z, Wu R, Wang Z, Chen X, Chan TWD (2017) Magnetic porous carbon derived from a bimetallic metal–organic framework for magnetic solid-phase extraction of organochlorine pesticides from drinking and environmental water samples. J Chromatogr A 1479:55–61. https://doi.org/10.1016/j.chroma.2016.12.014
Luo Y-B, Zheng H-B, Jiang X-Y, Li X, Zhang H-F, Zhu F-P, Pang Y-Q, Feng Y-Q (2015) Determination of pesticide residues in tobacco using modified QuEChERS procedure coupled to on-line gel permeation chromatography-gas chromatography/tandem mass spectrometry. Chin J Anal Chem 43(10):1538–1544. https://doi.org/10.1016/S1872-2040(15)60870-2
March C, Manclus JJ, Jimenez Y, Arnau A, Montoya A (2009) A piezoelectric immunosensor for the determination of pesticide residues and metabolites in fruit juices. Talanta 78(3):827–833. https://doi.org/10.1016/j.talanta.2008.12.058
Mehrotra P (2016) Biosensors and their applications—A review. J Oral Biol Craniofac Res 6(2):153–159. https://doi.org/10.1016/j.jobcr.2015.12.002
Moldoveanu SC, David V (2013) Chapter 1—Basic information about HPLC. In: Moldoveanu SC, David V (eds) Essentials in modern HPLC separations. Elsevier, pp 1–51. https://doi.org/10.1016/B978-0-12-385013-3.00001-X
Ndiomu DP, Simpson CF (1988) Some applications of supercritical fluid extraction. Anal Chim Acta 213:237–243. https://doi.org/10.1016/S0003-2670(00)81359-8
Nguyen TD, Lee MH, Lee GH (2010) Rapid determination of 95 pesticides in soybean oil using liquid–liquid extraction followed by centrifugation, freezing and dispersive solid phase extraction as cleanup steps and gas chromatography with mass spectrometric detection. Microchem J 95(1):113–119. https://doi.org/10.1016/j.microc.2009.11.009
Nsibande SA, Forbes PBC (2016) Fluorescence detection of pesticides using quantum dot materials—A review. Anal Chim Acta 945:9–22. https://doi.org/10.1016/j.aca.2016.10.002
Pang N, Wang T, Hu J (2016) Method validation and dissipation kinetics of four herbicides in maize and soil using QuEChERS sample preparation and liquid chromatography tandem mass spectrometry. Food Chem 190:793–800. https://doi.org/10.1016/j.foodchem.2015.05.081
Pastor-Belda M, Garrido I, Campillo N, Viñas P, Hellín P, Flores P, Fenoll J (2015) Dispersive liquid–liquid microextraction for the determination of new generation pesticides in soils by liquid chromatography and tandem mass spectrometry. J Chromatogr A 1394:1–8. https://doi.org/10.1016/j.chroma.2015.03.032
Patel KN, Patel JK, Patel MP, Rajput GC, Patel HA (2010) Introduction to hyphenated techniques and their applications in pharmacy. Pharm Methods 1(1):2–13. https://doi.org/10.4103/2229-4708.72222
Patel H, Rawtani D, Agrawal YK (2019) A newly emerging trend of chitosan-based sensing platform for the organophosphate pesticide detection using acetylcholinesterase—a review. Trends Food Sci Technol 85:78–91. https://doi.org/10.1016/j.tifs.2019.01.007
Pawliszyn J, Pedersen-Bjergaard S (2006) Analytical microextraction: current status and future trends. J Chromatogr Sci 44(6):291–307
Paz M, Correia-Sá L, Becker H, Longhinotti E, Domingues VF, Delerue-Matos C (2015) Validation of QuEChERS method for organochlorine pesticides analysis in tamarind (Tamarindus indica) products: peel, fruit and commercial pulp. Food Control 54:374–382. https://doi.org/10.1016/j.foodcont.2015.02.005
Pedersen-Bjergaard S, Rasmussen KE (1999) Liquid–Liquid–Liquid microextraction for sample preparation of biological fluids prior to capillary electrophoresis. Anal Chem 71(14):2650–2656. https://doi.org/10.1021/ac990055n
Pelajić M, Peček G, Mutavdžić Pavlović D, Vitali Čepo D (2016) Novel multiresidue method for determination of pesticides in red wine using gas chromatography–mass spectrometry and solid phase extraction. Food Chem 200:98–106. https://doi.org/10.1016/j.foodchem.2016.01.018

Petrarca MH, Godoy HT (2018) Gas chromatography–mass spectrometry determination of polycyclic aromatic hydrocarbons in baby food using QuEChERS combined with low-density solvent dispersive liquid–liquid microextraction. Food Chem 257:44–52. https://doi.org/10.1016/j.foodchem.2018.02.135

Piri-Moghadam H, Ahmadi F, Pawliszyn J (2016) A critical review of solid phase microextraction for analysis of water samples. TrAC Trends Anal Chem 85:133–143. https://doi.org/10.1016/j.trac.2016.05.029

Poole CF (2000) Solid-phase extraction, encyclopedia of separation science, vol 3. Academic Press, p 1405

Pose Juan E, Cancho-Grande B, Rial-Otero R, Simal-Gandara J (2006) The dissipation rates of cyprodinil, fludioxonil, procymidone and vinclozolin during storage of grape juice, vol 17. https://doi.org/10.1016/j.foodcont.2005.07.009

Qin Y, Zhang J, Li Y, Wang Q, Wu Y, Xu L, Jin X, Pan C (2017) Automated multi-filtration cleanup with nitrogen-enriched activated carbon material as pesticide multi-residue analysis method in representative crop commodities. J Chromatogr A 1515:62–68. https://doi.org/10.1016/j.chroma.2017.08.009

Radisic M, Dujaković NN, Vasiljević TM, Laušević M (2013) Application of matrix solid-phase dispersion and liquid chromatography–ion trap mass spectrometry for the analysis of pesticide residues in fruits, vol 6. https://doi.org/10.1007/s12161-012-9448-9

Rani M, Shanker U, Jassal V (2017) Recent strategies for removal and degradation of persistent & toxic organochlorine pesticides using nanoparticles: a review. J Environ Manage 190:208–222. https://doi.org/10.1016/j.jenvman.2016.12.068

Reichert B, de Kok A, Pizzutti IR, Scholten J, Cardoso CD, Spanjer M (2018) Simultaneous determination of 117 pesticides and 30 mycotoxins in raw coffee, without clean-up, by LC-ESI-MS/MS analysis. Anal Chim Acta 1004:40–50. https://doi.org/10.1016/j.aca.2017.11.077

Rezaee M, Assadi Y, Milani Hosseini M-R, Aghaee E, Ahmadi F, Berijani S (2006) Determination of organic compounds in water using dispersive liquid–liquid microextraction. J Chromatogr A 1116(1):1–9. https://doi.org/10.1016/j.chroma.2006.03.007

Ribeiro Begnini Konatu F, Breitkreitz MC, Sales Fontes Jardim IC (2017) Revisiting quick, easy, cheap, effective, rugged, and safe parameters for sample preparation in pesticide residue analysis of lettuce by liquid chromatography–tandem mass spectrometry. J Chromatogr A 1482:11–22. https://doi.org/10.1016/j.chroma.2016.12.061

Roos AH, Van Munsteren AJ, Nab F, Tuinstra LGMT (1987) Universal extraction/clean-up procedure for screening of pesticides by extraction with ethyl acetate and size exclusion chromatography, vol 196. https://doi.org/10.1016/s0003-2670(00)83074-3

Rowe C, Gunier R, Bradman A, Harley KG, Kogut K, Parra K, Eskenazi B (2016) Residential proximity to organophosphate and carbamate pesticide use during pregnancy, poverty during childhood, and cognitive functioning in 10-year-old children. Environ Res 150:128–137. https://doi.org/10.1016/j.envres.2016.05.048

Ruiz-Gil L, Romero-González R, Garrido Frenich A, Martínez Vidal JL (2008) Determination of pesticides in water samples by solid phase extraction and gas chromatography tandem mass spectrometry. J Sep Sci 31(1):151–161. https://doi.org/10.1002/jssc.200700299

Salvatierra-stamp V, Muñiz-Valencia R, Jurado JM, Ceballos-Magaña SG (2018) Hollow fiber liquid phase microextraction combined with liquid chromatography-tandem mass spectrometry for the analysis of emerging contaminants in water samples. Microchem J 140:87–95. https://doi.org/10.1016/j.microc.2018.04.012

Sánchez AG, Martos NR, Ballesteros E (2006) Multiresidue analysis of pesticides in olive oil by gel permeation chromatography followed by gas chromatography–tandem mass-spectrometric determination. Anal Chim Acta 558(1):53–61. https://doi.org/10.1016/j.aca.2005.11.019

Sannino A, Bolzoni L, Bandini M (2004) Application of liquid chromatography with electrospray tandem mass spectrometry to the determination of a new generation of pesticides in processed fruits and vegetables. J Chromatogr A 1036(2):161–169. https://doi.org/10.1016/j.chroma.2004.02.078

Saraji M, Javadian S (2019) Single-drop microextraction combined with gas chromatography-electron capture detection for the determination of acrylamide in food samples. Food Chem 274:55–60. https://doi.org/10.1016/j.foodchem.2018.08.108

Seccia S, Fidente P, Montesano D, Morrica P (2008) Determination of neonicotinoid insecticides residues in bovine milk samples by solid-phase extraction clean-up and liquid chromatography with diode-array detection. J Chromatogr A 1214(1–2):115–120. https://doi.org/10.1016/j.chroma.2008.10.088

Selvolini G, Băjan I, Hosu O, Cristea C, Săndulescu R, Marrazza G (2018) DNA-based sensor for the detection of an organophosphorus pesticide: profenofos. Sensors (Basel) 18(7):2035. https://doi.org/10.3390/s18072035

Shamsipur M, Yazdanfar N, Ghambarian M (2016) Combination of solid-phase extraction with dispersive liquid–liquid microextraction followed by GC–MS for determination of pesticide residues from water, milk, honey and fruit juice. Food Chem 204:289–297. https://doi.org/10.1016/j.foodchem.2016.02.090

Sharma BK (2006) Environmental chemistry. Goel Publication House, New Delhi

Singh S, Srivastava A, Singh SP (2018) Inexpensive, effective novel activated carbon fibers for sample cleanup: application to multipesticide residue analysis in food commodities using a QuEChERS method. Anal Bioanal Chem 410(8):2241–2251. https://doi.org/10.1007/s00216-018-0894-0

Soares KL, Cerqueira MBR, Caldas SS, Primel EG (2017) Evaluation of alternative environmentally friendly matrix solid phase dispersion solid supports for the simultaneous extraction of 15 pesticides of different chemical classes from drinking water treatment sludge. Chemosphere 182:547–554. https://doi.org/10.1016/j.chemosphere.2017.05.062

Soxhlet F, DP 1879. 232 (461)

Specht W, Pelz S, Gilsbach W (1995) Gas-chromatographic determination of pesticide residues after clean-up by gel-permeation chromatography and mini-silica gel-column chromatography6. Communication(*): Replacement of dichloromethane by ethyl acetate/cyclohexane in liquid-liquid partition and simplified conditions for extraction and liquid-liquid partition, vol 353. https://doi.org/10.1007/s0021653530183

Spietelun A, Marcinkowski Ł, de la Guardia M, Namieśnik J (2014) Green aspects, developments and perspectives of liquid phase microextraction techniques. Talanta 119:34–45. https://doi.org/10.1016/j.talanta.2013.10.050

Stalling D, Tindle RC, Johnson JL (1972) Cleanup of pesticide and polychlorinated biphenyl residues in fish extracts by gel permeation chromatography, vol 55

Suganthi A, Bhuvaneswari K, Ramya M (2018) Determination of neonicotinoid insecticide residues in sugarcane juice using LCMSMS. Food Chem 241:275–280. https://doi.org/10.1016/j.foodchem.2017.08.098

Szarka A, Turková D, Hrouzková S (2018) Dispersive liquid-liquid microextraction followed by gas chromatography–mass spectrometry for the determination of pesticide residues in nutraceutical drops. J Chromatogr A 1570:126–134. https://doi.org/10.1016/j.chroma.2018.07.072

Taliansky-Chamudis A, Gómez-Ramírez P, León-Ortega M, García-Fernández AJ (2017) Validation of a QuEChеRS method for analysis of neonicotinoids in small volumes of blood and assessment of exposure in Eurasian eagle owl (Bubo bubo) nestlings. Sci Total Environ 595:93–100. https://doi.org/10.1016/j.scitotenv.2017.03.246

Tsai D-Y, Chen C-L, Ding W-H (2014) Optimization of matrix solid-phase dispersion for the rapid determination of salicylate and benzophenone-type UV absorbing substances in marketed fish. Food Chem 154:211–216. https://doi.org/10.1016/j.foodchem.2014.01.013

Van Maele-Fabry G, Lantin A-C, Hoet P, Lison D (2010) Childhood leukaemia and parental occupational exposure to pesticides: a systematic review and meta-analysis. Cancer Causes Control 21(6):787–809. https://doi.org/10.1007/s10552-010-9516-7

Virot M, Tomao V, Colnagui G, Visinoni F, Chemat F (2007) New microwave-integrated Soxhlet extraction: an advantageous tool for the extraction of lipids from food products. J Chromatogr A 1174(1):138–144. https://doi.org/10.1016/j.chroma.2007.09.067

Wang Z, Yin C (2017) State-of-the-art on ultrasonic oil production technique for EOR in China. Ultrason Sonochem 38:553–559. https://doi.org/10.1016/j.ultsonch.2017.03.035
Wang W, Meng B, Lu X, Liu Y, Tao S (2007) Extraction of polycyclic aromatic hydrocarbons and organochlorine pesticides from soils: a comparison between Soxhlet extraction, microwave-assisted extraction and accelerated solvent extraction techniques. Anal Chim Acta 602(2):211–222. https://doi.org/10.1016/j.aca.2007.09.023
Wang P, Zhang Q, Wang Y, Wang T, Li X, Ding L, Jiang G (2010) Evaluation of Soxhlet extraction, accelerated solvent extraction and microwave-assisted extraction for the determination of polychlorinated biphenyls and polybrominated diphenyl ethers in soil and fish samples. Anal Chim Acta 663(1):43–48. https://doi.org/10.1016/j.aca.2010.01.035
Watanabe E, Kobara Y, Baba K, Eun H (2014) Aqueous acetonitrile extraction for pesticide residue analysis in agricultural products with HPLC–DAD. Food Chem 154:7–12. https://doi.org/10.1016/j.foodchem.2013.12.075
Wei J-C, Wei B, Yang W, He C-W, Su H-X, Wan J-B, Li P, Wang Y-T (2018) Trace determination of carbamate pesticides in medicinal plants by a fluorescent technique. Food Chem Toxicol 119:430–437. https://doi.org/10.1016/j.fct.2017.12.019
Wickerham EL, Lozoff B, Shao J, Kaciroti N, Xia Y, Meeker JD (2012) Reduced birth weight in relation to pesticide mixtures detected in cord blood of full-term infants. Environ Int 47:80–85. https://doi.org/10.1016/j.envint.2012.06.007
Winborn J, Kerrigan S (2019) Quantitative analysis of desomorphine in blood and urine using solid phase extraction and gas chromatography-mass spectrometry. J Chromatogr B 1106–1107:43–49. https://doi.org/10.1016/j.jchromb.2018.12.022
Xiu-ping Z, Lin M, Lan-qi H, Jian-Bo C, Li Z (2017) The optimization and establishment of QuEChERS-UPLC–MS/MS method for simultaneously detecting various kinds of pesticides residues in fruits and vegetables. J Chromatogr B 1060:281–290. https://doi.org/10.1016/j.jchromb.2017.06.008
Xu M-L, Gao Y, Han XX, Zhao B (2017) Detection of pesticide residues in food using surface-enhanced raman spectroscopy: a review. J Agric Food Chem 65(32):6719–6726. https://doi.org/10.1021/acs.jafc.7b02504
Xue J, Li H, Liu F, Jiang W, Hou F (2016) Vortex-assisted matrix solid–liquid dispersive microextraction for the analysis of triazole fungicides in cotton seed and honeysuckle by gas chromatography. Food Chem 196:867–876. https://doi.org/10.1016/j.foodchem.2015.09.093
You X, Xing Z, Liu F, Jiang N (2013) Air-assisted liquid–liquid microextraction used for the rapid determination of organophosphorus pesticides in juice samples. J Chromatogr A 1311:41–47. https://doi.org/10.1016/j.chroma.2013.08.080
You X, Xing Z, Liu F, Zhang X (2015) Air-assisted liquid–liquid microextraction by solidifying the floating organic droplets for the rapid determination of seven fungicide residues in juice samples. Anal Chim Acta 875:54–60. https://doi.org/10.1016/j.aca.2015.03.033
Zaidon SZ, Ho YB, Hamsan H, Hashim Z, Saari N, Praveena SM (2019) Improved QuEChERS and solid phase extraction for multi-residue analysis of pesticides in paddy soil and water using ultra-high performance liquid chromatography tandem mass spectrometry. Microchem J 145:614–621. https://doi.org/10.1016/j.microc.2018.11.025
Zang X-H, Wu Q-H, Zhang M-Y, Xi G-H, Wang Z (2009) Developments of dispersive liquid-liquid microextraction technique. Chin J Anal Chem 37(2):161–168. https://doi.org/10.1016/S1872-2040(08)60082-1
Zhan J, Li J, Liu D, Liu C, Yang G, Zhou Z, Wang P (2016) A simple method for the determination of organochlorine pollutants and the enantiomers in oil seeds based on matrix solid-phase dispersion. Food Chem 194:319–324. https://doi.org/10.1016/j.foodchem.2015.07.067
Zhang J, Rodila R, Gage E, Hautman M, Fan L, King LL, Wu H, El-Shourbagy TA (2010) High-throughput salting-out assisted liquid/liquid extraction with acetonitrile for the simultaneous determination of simvastatin and simvastatin acid in human plasma with liquid chromatography. Anal Chim Acta 661(2):167–172. https://doi.org/10.1016/j.aca.2009.12.023

Zhang Y, Arugula MA, Wales M, Wild J, Simonian AL (2015) A novel layer-by-layer assembled multi-enzyme/CNT biosensor for discriminative detection between organophosphorus and non-organophosphorus pesticides. Biosens Bioelectron 67:287–295. https://doi.org/10.1016/j.bios.2014.08.036

Zheng G, Han C, Liu Y, Wang J, Zhu M, Wang C, Shen Y (2014) Multiresidue analysis of 30 organochlorine pesticides in milk and milk powder by gel permeation chromatography-solid phase extraction-gas chromatography-tandem mass spectrometry. J Dairy Sci 97(10):6016–6026. https://doi.org/10.3168/jds.2014-8192

Zheng W, Yoo K-H, Choi J-M, Park D-H, Kim S-K, Kang Y-S, El-Aty AA, Hacımüftüoğlu A, Jeong JH, Bekhit AE-D (2019) A modified QuEChERS method coupled with liquid chromatography-tandem mass spectrometry for the simultaneous detection and quantification of scopolamine, L-hyoscyamine, and sparteine residues in animal-derived food products. J Adv Res 15:95–102

Zhu B, Xu X, Luo J, Jin S, Chen W, Liu Z, Tian C (2019) Simultaneous determination of 131 pesticides in tea by on-line GPC-GC–MS/MS using graphitized multi-walled carbon nanotubes as dispersive solid phase extraction sorbent. Food Chem 276:202–208. https://doi.org/10.1016/j.foodchem.2018.09.152

Chapter 15
Role of Microorganisms in Degradation and Removal of Anticonvulsant Drugs: A Review

Neha Alok Sinha and Vipin Kumar

Abstract Over the past few years, large amounts of pharmaceutical compounds along with their active metabolites are emerging as environmental pollutants, which enters into water systems from different sources, such as sewage (human and animal excretion), landfill sites, wastewater from pharmaceutical industries and hospitals. In recent years the wide use of anticonvulsant drugs is one of the environmental concern due to its frequent use and determination in the environment. Anticonvulsant drugs and their metabolites are present in trace concentrations of $\mu g\ L^{-1}$ to mg L^{-1} in water bodies. Biodegradation of these drug depends upon the different factors, such as physicochemical properties of drug, environmental factors (i.e. pH, light, temperature) and type of microbial strain. Different microbes like fungal strains (*Cunninghamella elegans* ATCC 9245, *Umbelopsis ramanniana* R-56) and *Bacillus and Streptomyces* species of bacteria have been reported with the ability to degrade pharmaceutical compounds via metabolic and metabolic pathways. This chapter summarizes the knowledge about biotransformation or biodegradation of anticonvulsant drugs and their metabolic pathways.

Keywords Biodegradation · Anticonvulsant drugs · Microorganisms · Metabolic pathways

15.1 Introduction

Pharmaceuticals compounds are important for the well-being of humans as well as animal health or their life cycle. Recently, due to the contamination of various environmental components, biodegradation of these compounds have become a thrust research area. These "emerging contaminants" have been identified in the environment for years, originates from several sources such as confined animal feeding operation, surface runoff and wastewater treatment plant (WWTP) effluent (Boxall

N. A. Sinha (✉) · V. Kumar
Department of Environmental Science and Engineering, Indian Institute of Technology (Indian School of Mines), Dhanbad 826004, India
e-mail: alok@iitism.ac.in

T. Gupta et al. (eds.), *Measurement, Analysis and Remediation of Environmental Pollutants*, Energy, Environment, and Sustainability,
https://doi.org/10.1007/978-981-15-0540-9_15

2004). Large amount of pharmaceuticals excreted as metabolites through urine and feces end up in the wastewater treatment plant (Monteiro and Boxall 2010). WWTPs effluents are characterized as an important pathway for aquaculture revelation to pharmaceuticals compound if they are not removed through the wastewater treatment process (Zuccato et al. 2005; Ottmar et al. 2010). Hospitals, pharmaceutical industries, industrial wastewater treatment plants (Bottoni et al. 2010), and agriculture (Emke et al. 2014; Muñoz et al. 2010) can also be a major source of aquatic contamination (Jiang and Zhou 2013; Brozinski et al. 2012).

The sewage sludge, used as a crop fertilizer and when irrigated through discharged waste water enters pharmaceuticals and mixed with soil and domestic water (Radke et al. 2010). The contaminant reach groundwater through percolation and are also found in coastal marine waters (Barbosa et al. 2016).

Wastewater treatment plants uses activated sludge process to remove pharmaceutical pollutants which are present at concentrations of mg L^{-1}. These treatment plants are not suitable for degrading high concentration of pharmaceuticals due to complexity of pharmaceutical pollutants present in wastewater (Sutton et al. 2014). Almost 15,000 human and 3000 veterinary pharmaceuticals are present in the worldwide. All drugs consists of an active substance, the miscellaneous quantity of supplementary substances to be used as pharmaceutical doses. Recent concern has been emerged due to impact of pharmaceutical products on the environment and has been reflected in the literature since the 1990s due increase in water pollution. These emerging compounds are used in a very high range, with growing diversity day by day (Kümmerer 2009). Pharmaceutical term covers different compounds with chemically complex structures and several physiochemical properties. These compounds are polar, charged and hydrophilic compounds (Tanna et al. 2013). Many compounds belong to the acidic groups. Polar compounds are commonly dissolved in water and normally spread in aquatic and agriculture environments (Escher et al. 2011). The exposure risk of pharmaceuticals shows the negative eco toxicological effects such as carcinogenicity, genotoxicity, chronic toxicity, acute toxicity and pharmacological effects in the hormones and nervous system (Cordy et al. 2004). Different types of pharmaceutical compounds concentration leads to changes in microbial population, their structure and affects the food chain. Anticonvulsant drugs such as carbamazepine and Sulfamethoxazole, discharged into the water and show their resistance properties (Zimmerman 2005). Humans and animals are successively infected with these drug-resistant bacteria and these drugs are not suitable for treatment. The drugs resistance is one of the main reason 15,000 deaths of people (Atalay and Ersöz 2016). An overview of the removal of anticonvulsant drugs pharmaceutical compounds through biodegradation.

15.2 Source of Pharmaceuticals in Water

The intake of pharmaceutical compounds has been developing extensively in the previous years, due to various reasons such as the inverting age structure, population

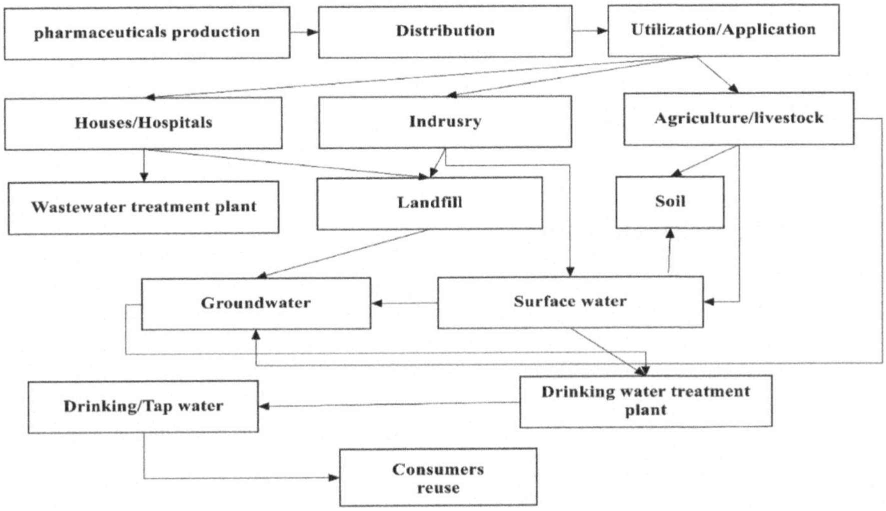

Fig. 15.1 Origin and pathway of pharmaceuticals compounds (Caracciolo et al. 2015)

growth, presence and target of new age groups, and also innovation of new groups of drug (Khetan and Collins 2007). As a result, the threat of water contamination with these chemicals pollutants has been increased gradually. Pharmaceuticals and their metabolites can enter the environment through several routes. Human body modifies the pharmaceutical compound into metabolites before excretion (Fig. 15.1). After excretion, further transformation of the pharmaceutical compounds takes place in wastewater treatment plants and this leads to generation of several pharmaceutical intermediates metabolites. Some metabolites are formed through specific treatment technique and are known to exhibit higher toxicity levels when compared with original compounds (Illés et al. 2014).

15.3 Biodegradation of Pharmaceuticals Through Microbial Species

A natural group of microorganisms in water and soil are very beneficial for monitoring the quality of the ecosystem (Topp et al. 2013). Microbes present in natural environment degrades pharmaceutical pollutants via different (metabolic and cometabolic) pathways, and this mechanism is known as 'biodegradation' (Tiwari et al. 2017). In the process of biodegradation, microbe breakdown the toxic chemical and complex compounds into simpler, less toxic products by secreting enzymes. Biological degradation is helpful in the removal of pharmaceuticals compound in WWTP. The degradation efficiency of pharmaceutical pollutants generally depends on their solubility in wastewater. If pollutants solubility in water is very low or poor (hydrophobic compound) it will be recollected in sewage wastewater and sludge which requires further biodegradation, i.e., micropollutant degraded through catabolic microbial enzymes or different microbial strain, uses these as a sole carbon source (Ozdemir et al. 2015; Loos et al. 2010).

15.4 Anticonvulsant Drugs

Anticonvulsant drugs are mainly used to control an epilepsy disease in people. This is not a single disease, it is shows a set of symptoms that have different type of causes in different persons. The main symptoms epilepsy to inequity in the brain electrical activity. This symptoms are sources of seizuers that may affect all part of the body and causes consciousness. These drugs are divided into several groups such as, hydantoins which includes phenytoin and mephenytoin. Also divided a various number of other drugs which are not linked to larger groups 'of drugs i.e., ethosuximide felbamate, gabapentin, carbamazepine and other several drugs. These pharmaceuticals are recommended only with a physician's instructions and available in forms of capsules, tablets and liquids (Caracciolo et al. 2015) (Table 15.1).

Table 15.1 Classification of anticonvulsant drugs (Goldenberg 2010)

Primary generalized tonic–clonic seizures	Partial seizures	Absence seizures	Myoclonic and atonic seizures
Main Groups of Anticonvulsant Drugs			
Valproic acid Lamotrigine Topiramate	Phenytoin Oxcarbazepine Carbamazepine	Valproic acid Ethosuximide	Topiramate Lamotrigine
Alternative Anticonvulsant Drugs			
Carbamazepine Oxcarbazepine Phenobarbital Primidone Felbamate Sulfamethoxazole Zonisamide Phenytoin	Felbamate Eslicarbazepine Rufinamide Levetiracetam Topiramate Tiagabine Vigabatrin Lacosamide Pregabalin Zonisamide Gabapentin Phenobarbital Primidone	Clonazepam Lamotrigine	Felbamate Clonazepam

15.4.1 Carbamazepine

An anticonvulsant drug carbamazepine (5H-dibenzo, azepine-5-carboxamide, Fig. 15.2), is used for the treatment of seizures and depression in the world (Evgenidou et al. 2015). It is commonly detected pharmaceutical in effluents of wastewater treatment plant (WWTP) (Zhang et al. 2012). It is very persistent and has been normally detected in surface water and rarely present in groundwater. It is present in the ng L^{-1} to g L^{-1} concentration in the environment. The environmental risk of carbamazepine has been increased tremendously due to its common presence in the environment. These exerts toxicity effect on aquatic organisms and have impact on humans (Castiglioni et al. 2006).

15.4.1.1 Carbamazepine Biodegradation/Biotransformation by Microorganisms

Biodegradation of Carbamazepine has been achieved by prokaryotic *Streptomyces violascens* ATCC 31560 and *Streptomyces diastatochromegenes* ATCC 31561 (Fent et al. 2006). It has been shows that fungal strain *Cunninghamella elegans* and *Umbelopsis ramanniana* are suitable microbial strains for biodegradation of carbamazepine Biodegradation efficiency (%) drug sample (D%) defined by the following expression (Jelic et al. 2011)

$$D\% = \frac{C_0 - C}{C_0} \times 100 \tag{1}$$

where, C_0 is the initial concentration of drug sample and C is the concentration of drug sample after biodegradation.

The degradation efficiency of carbamazepine through different microbial species are shown in Table 15.2.

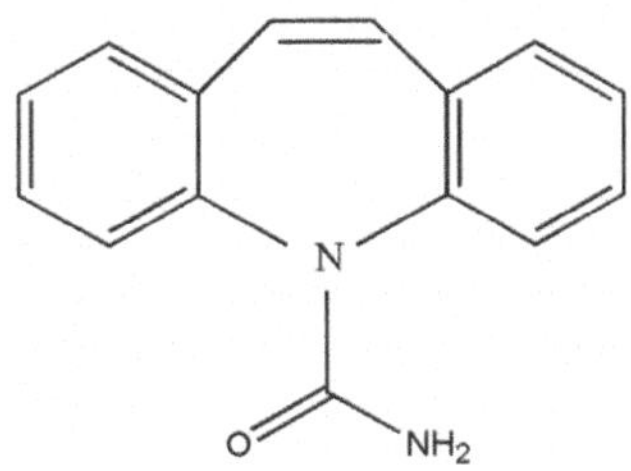

Fig. 15.2 The structure of Carbamazepine

Table 15.2 Carbamazepine degradation through different microbial species

Degrading microbes	Degradation efficiency (%)	References
Trametes versicolor	77	Kittelmann et al. (1993)
Ganoderma lucidum	58	Kang et al. (2008)
Trichoder Maharzianum	72	Rodarte-Morales et al. (2012)
Pleurotus ostreatus	68	Marco-Urrea et al. (2009)
Phanerochaete chrysosporium	90	Buchicchio et al. (2016)
Pseudomonas CBZ-4	46	Li et al. (2016)
Paraburkholderia xenovorans LB400	100	Li et al. (2013)
Pseudomonas putida *Acetinobacter US1* *Micrococcus SBS-8* *Bacillus halodurans*	55–60	Aukema et al. (2016)

15.4.1.2 Identification of Carbamazepine Metabolites

Carbamazepine (CBZ) is mainly absorbed and found in the human liver. All most 30 different metabolites have been identified (Lertratanangkoon and Horning 1982). The pathways of carbamazepine intermediate compounds in humans and rats have been characterized by two fungal strain *Cunninghamella elegans* ATCC 9245 or *Umbelopsis ramanniana* R-56 and three metabolites has been identified through each strain and are represented in Table 15.3. All metabolites are shown in different peaks (nm) and adding one parent compound (O_2).

The liquid chromatography-mass spectrometric analysis identified six intermediates compound of Carbamazepine after biodegradation. 10, 11-dihydro-10, 11-epoxycarbamazepine is an important intimidated metabolites (C3) and (M3) and 3-hydroxycarbamazepine (C2 and M2) has been identified from carbamazepine biodegradation by fungal species. *Cunninghamella elegans* was able to form 2-hydroxycarbamazepine (C1) by-product and *Umbelopsis ramanniana* formed *one*

Table 15.3 Carbamazepine metabolites identified by *Umbelopsis ramanniana* R-56 and *Cunninghamella elegans* ATCC 9245 (Zimmerman 2005)

Metabolites	LC-MS peaks (nm)	Compound
C1	216, 284	2-Hydroxycarbamazepine (CBZ-2OH)
C2	213, 297	3-Hydroxycarbamazepine (CBZ-3OH)
C3	211	10,11-epoxycarbamazepine (CBZ-EP)
M1	216, 309	Unidentified
M2	225, 297	3-Hydroxycarbamazepine (CBZ-3OH)
M3	215	10,11-epoxycarbamazepine (CBZ-EP)

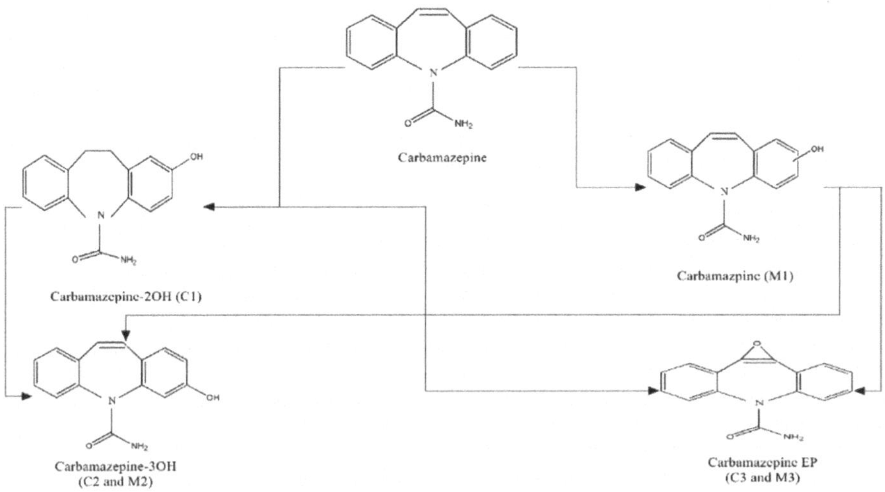

Fig. 15.3 Biodegradation pathway of carbamazepine by *C. elegans* (C1, C2, and C3) ATCC and *U. ramanniana* R-56 (M1, M2 and M3) (Kang et al. 2008)

intermediate compound identified as an unknown hydroxylated ring compound (M1). The possible degradation metabolic pathways of carbamazepine drugs by two microbial strain *U. ramanniana* R-56 and *C. elegans* ATCC 9245 are represented in Fig. 15.3. Both fungal strain *C. elegans* and *U. ramanniana*, degraded the carbamazepine through different mixed mono-oxidation reactions. These fungal strains formed CBZ-EP as a major metabolite and CBZ-3OH as one of the minor metabolites (Fig. 15.3). But, *C. elegans* produced CBZ-2OH while *U. ramanniana* produced the unidentified CBZ-1OH or CBZ-4OH metabolites (Figs. 15.3). Neither of the fungal strain formed the hydration product of CBZ-EP, CBZ-trans DiOH, in the duration of 25 days. Also both fungal strain *U. ramanniana* and *C. elegans* followed an oxidation reaction patterns by producing CBZ-EP metabolite, which was transformed into CBZ-trans-DiOH. These two strain may play an important biological role in degradation of CBZ through oxidation and hydroxylation reactions in the environment (Ha et al. 2016).

15.4.2 *Sulfamethoxazole*

Sulfamethoxazole is a synthetic drug (Fig. 15.4), used as anticonvulsant and herbicidal agents (Bahr et al. 2007). It is generally present in the aquaculture environment

Fig. 15.4 The structure of Sulfamethoxazole

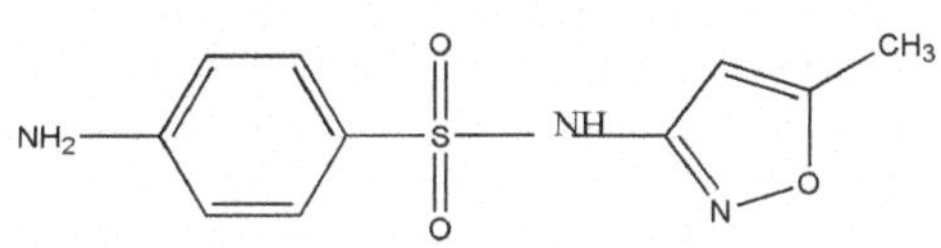

Table 15.4 Sulfamethoxazole degradation through different microbial species

Degrading microbes	Degradation efficiency	References
Shewanella sp. MR-4	100%	Drillia et al. (2005)
Achromobacter denitrificans PR1	99–100%	Höltge and Kreuzig (2007)
Geobacillus thermoleovorans S-07	72.96–85.96%	Mao et al. (2018)
Terrabacter sp. 2APm3	90	Pan et al. (2017)
Kribbella sp. SDZ-3S-SCL47	60.6%	Tappe et al. (2015)
Acinetobacter sp. W1	95–100%	Mulla et al. (2016)
Rhodococcus sp. BR2	24–44%	Wang and Wang (2018)
Acinetobacter Pseudomonas	41.7–100%	Ricken et al. (2013)

(Shoaib Ahmad Shah et al. 2013). It was detected up to 113 ng l^{-1} in drinking water 7.9–1900 ng l^{-1} in surface water and 38–450 ng l^{-1} in groundwater.

N4-acetyl-sulfamethoxazole (Ac-Sulfamethoxazole), 10–15% is sulfamethoxazole-N1-glucuronide and 20–25% of the Sulfamethoxazole as naturally excreted from wastewater (Michael et al. 2013). This compound is one of the most recommended anticonvulsant drugs in worldwide. (Cribb and Spielberg 1992), and used in aquaculture, animals' treatment, and breeding of livestock for handling the growth of dieses (Michael et al. 2013).

15.4.2.1 Sulfamethoxazole Degrading Microbial Species

Sulfamethoxazole was totally degraded after 1 day of inoculated *Microbacterium microbial* Strain *BR1* (Herzog et al. 2013). Another strain *Achromobacter* sp. *BR3*, degraded Sulfamethoxazole with degradation up to 24–44% in 16 days, at 30 °C and pH-7 (Huang et al. 2012; Cribb and Spielberg 1992). *Pseudomonas psychrophila* HA-4 degraded 35% Sulfamethoxazole in 8 days having temperature range from 5.0 to 30 °C, and the maximum degradation was achieved at 10 °C. Furthermore, the biodegradation efficiency of sulfamethoxazole through strain HA-4 was almost 34% after 192 h at 10 °C temperature. The 23.03% unsaturated fatty acid was observed to be at 10 °C. Its shows that strain *Pseudomonas psychrophila* HA-4 was capable to sulfamethoxazole degradation at low temperatures (Nguyen et al. 2017; Jiang et al. 2014). Sulfamethoxazole degradation through different microbial species is shown in Table 15.4.

15.4.2.2 Identification of Sulfamethoxazole Metabolites

Sulfamethoxazole metabolites are detected by Liquid chromatography-mass spectrometric analysis. 4-aminophenol and 3-amino-5-methylisoxazole were the most commonly identified metabolites represented in Fig. 15.5. *Microbacteriumsp.* sp.

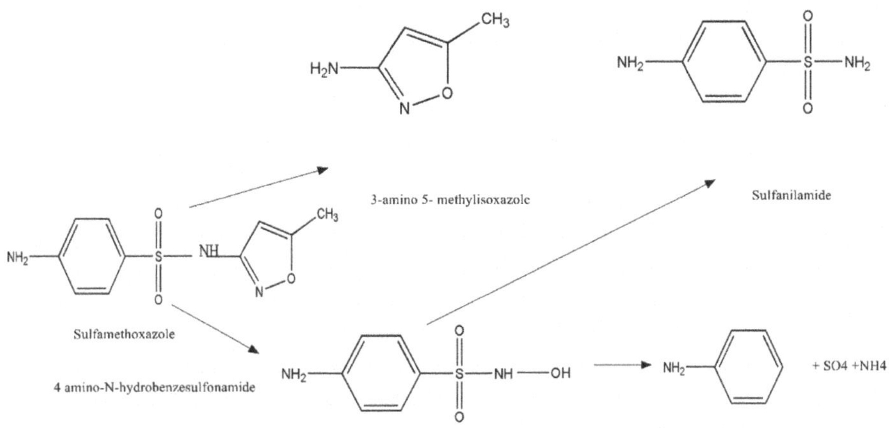

Fig. 15.5 Proposed metabolic pathways of Sulfonamide by *Microbacteriumsp* [66]

and *Achromobacter denitrificans* were identified for the degradation of sulfamethoxazole intermediate compound (Bouju et al. 2012). The activated sludge process was capable to degrade sulfamethoxazole into 3-amino 5-methylisoxazole, Buthydroxyl-*N*-(5-methyl-1, 2-oxazole-3-yl) benzene-1-sulfamethoxazole was recognized as a secondary metabolite when sulfamethoxazole was used a single source of carbon (Yang et al. 2016). Also, Sulfamethoxazole primarily breaks down into 3-amino 5-methylisoxazole and 4-amino benzene sulfonic acid in microbial fuel cell reactors. 3-amino-5-methylisoxazole further degraded into 4-amino-2-butanol due to the opening of isoxazole ring and proposed the Sulfamethoxazole degradation pathway. Sulfamethoxazole was originated through NADH-dependent ipso-hydroxylation and releases 3-amino-5-methylisoxazole and benzoquinoneimine metabolites, and it again breaks down to 4-aminophenol. Also 1, 2, 4-trihydroxybenzene, has been identified as another metabolite which illustrated the biodegradation pathway of Sulfamethoxazole Chen and Xie (2018).

15.5 Conclusion

Anticonvulsant drugs removed from the environment depending upon the existence of different type of microorganism that have the capability to degrade them. Several pharmaceuticals are potentially biodegrade through metabolic and cometabolic pathways. Frequent exposure of water and soil microbial groups in some cases developed enriched biodegradation activity and decreases their persistence in environment. Some factors can significantly affect removal efficiency such as bioavailability for microbial degradation to occur, solubility, temperature, time and pH. The human and veterinary pharmaceuticals most commonly used worldwide, these are constantly added to ecosystems and thus shows their virtual-persistence towards biodegradation. This is because these drugs are designed to be active at low concentrations and therefore may in some instances represent an ecotoxicological concern even at environmentally-relevant concentrations. With respect to environmental microorganisms, impacts of concern would be any that compromise key ecosystem services, for example nutrient biogeochemical cycling. Therefore, anticonvulsant drugs are emerging as a major concern because of its potential enrichment towards growth of resistant microorganisms. Overall, much more awareness about environmental discharges of pharmaceuticals their determination, complete degradation pathways and impacts on human and ecosystem health is desirable.

Acknowledgements The authors thank the financial support from Indian Institute of Technology (Indian School of Mines), Dhanbad under Junior Research Fellowship scheme funded by Ministry of Human Resource Development (MHRD).

References

Atalay S, Ersöz G (2016) Novel catalysts in advanced oxidation of organic pollutants. Springer International Publishing

Aukema KG, Escalante DE, Maltby MM, Bera AK, Aksan A, Wackett LP (2016) In silico identification of bioremediation potential: carbamazepine and other recalcitrant personal care products. Environ Sci Technol 51(2):880–888

Bahr C, Schumacher J, Ernst M, Luck F, Heinzmann B, Jekel M (2007) SUVA as control parameter for the effective ozonation of organic pollutants in secondary effluent. Water Sci Technol 55(12):267–274

Barbosa MO, Moreira NF, Ribeiro AR, Pereira MF, Silva AM (2016) Occurrence and removal of organic micropollutants: an overview of the watch list of EU Decision 2015/495. Water Res 94:257–279

Bottoni P, Caroli S, Caracciolo AB (2010) Pharmaceuticals as priority water contaminants. Toxicol Environ Chem 92(3):549–565

Bouju H, Ricken B, Beffa T, Corvini PFX, Kolvenbach BA (2012) Isolation of bacterial strains capable of sulfamethoxazole mineralization from an acclimated membrane bioreactor. Appl Environ Microbiol 78(1):277–279

Boxall AB (2004) The environmental side effects of medication: how are human and veterinary medicines in soils and water bodies affecting human and environmental health? EMBO Rep 5(12):1110–1116

Brozinski JM, Lahti M, Meierjohann A, Oikari A, Kronberg L (2012) The anti-inflammatory drugs diclofenac, naproxen and ibuprofen are found in the bile of wild fish caught downstream of a wastewater treatment plant. Environ Sci Technol 47(1):342–348

Buchicchio A, Bianco G, Sofo A, Masi S, Caniani D (2016) Biodegradation of carbamazepine and clarithromycin by Trichoderma harzianum and Pleurotus ostreatus investigated by liquid chromatography–high-resolution tandem mass spectrometry (FTICR MS-IRMPD). Sci Total Environ 557:733–739

Caracciolo AB, Topp E, Grenni P (2015) Pharmaceuticals in the environment: biodegradation and effects on natural microbial communities. A review. J Pharm Biomed Anal 106:25–36

Castiglioni S, Zuccato E, Crisci E, Chiabrando C, Fanelli R, Bagnati R (2006) Identification and measurement of illicit drugs and their metabolites in urban wastewater by liquid chromatography—tandem mass spectrometry. Anal Chem 78(24):8421–8429

Chen J, Xie S (2018) Overview of sulfonamide biodegradation and the relevant pathways and microorganisms. Sci Total Environ 640:1465–1477

Cordy GE, Duran NL, Bouwer H, Rice RC, Furlong ET, Zaugg SD, Meyer MT, Barber LB, Kolpin DW (2004) Do pharmaceuticals, pathogens, and other organic waste water compounds persist when waste water is used for recharge? Groundwater Monit Rem 24(2):58–69

Cribb AE, Spielberg SP (1992) Sulfamethoxazole is metabolized to the hydroxylamine in humans. Clin Pharmacol Ther 51(5):522–526

Drillia P, Dokianakis SN, Fountoulakis MS, Kornaros M, Stamatelatou K, Lyberatos G (2005) On the occasional biodegradation of pharmaceuticals in the activated sludge process: the example of the antibiotic sulfamethoxazole. J Hazard Mater 122(3):259–265

Emke E, Evans S, Kasprzyk-Hordern B, de Voogt P (2014) Enantiomer profiling of high loads of amphetamine and MDMA in communal sewage: a Dutch perspective. Sci Total Environ 487:666–672

Escher BI, Baumgartner R, Koller M, Treyer K, Lienert J, McArdell CS (2011) Environmental toxicology and risk assessment of pharmaceuticals from hospital wastewater. Water Res 45(1):75–92

Evgenidou EN, Konstantinou IK, Lambropoulou DA (2015) Occurrence and removal of transformation products of PPCPs and illicit drugs in wastewaters: a review. Sci Total Environ 505:905–926

Fent K, Weston AA, Caminada D (2006) Ecotoxicology of human pharmaceuticals. Aquat Toxicol 76(2):122–159

Goldenberg MM (2010) Overview of drugs used for epilepsy and seizures: etiology, diagnosis, and treatment. Pharm Ther 35(7):392

Ha H, Mahanty B, Yoon S, Kim CG (2016) Degradation of the long-resistant pharmaceutical compounds carbamazepine and diatrizoate using mixed microbial culture. J Environ Sci Health Part A 51(6):467–471

Herzog B, Lemmer H, Horn H, Müller E (2013) Characterization of pure cultures isolated from sulfamethoxazole-acclimated activated sludge with respect to taxonomic identification and sulfamethoxazole biodegradation potential. BMC Microbiol 13(1):276

Höltge S, Kreuzig R (2007) Laboratory testing of sulfamethoxazole and its metabolite acetyl-sulfamethoxazole in soil. CLEAN–Soil Air Water 35(1):104–110

Huang M, Tian S, Chen D, Zhang W, Wu J, Chen L (2012) Removal of sulfamethazine antibiotics by aerobic sludge and an isolated Achromobacter sp. S-3. J Environ Sci 24(9):1594–1599

Illés E, Szabó E, Takács E, Wojnárovits L, Dombi A, Gajda-Schrantz K (2014) Ketoprofen removal by O3 and O3/UV processes: kinetics, transformation products, and ecotoxicity. Sci Total Environ 472:178–184

Jelic A, Gros M, Ginebreda A, Cespedes-Sánchez R, Ventura F, Petrovic M, Barcelo D (2011) Occurrence, partition, and removal of pharmaceuticals in sewage water and sludge during wastewater treatment. Water Res 45(3):1165–1176

Jiang J, Zhou Z (2013) Removal of pharmaceutical residues by ferrate (VI). PLoS ONE 8(2):e55729

Jiang B, Li A, Cui D, Cai R, Ma F, Wang Y (2014) Biodegradation and metabolic pathway of sulfamethoxazole by Pseudomonas psychrophila HA-4, a newly isolated cold-adapted sulfamethoxazole-degrading bacterium. Appl Microbiol Biotechnol 98(10):4671–4681

Kang SI, Kang SY, Hur HG (2008) Identification of fungal metabolites of anticonvulsant drug carbamazepine. Appl Microbiol Biotechnol 79(4):663

Khetan SK, Collins TJ (2007) Human pharmaceuticals in the aquatic environment: a challenge to green chemistry. Chem Rev 107(6):2319–2364

Kittelmann M, Lattmann R, Ghisalba O (1993) Preparation of 10, 11-epoxy-carbamazepine and 10, 11-dihydro-10-hydroxy-carbamazepine by microbial epoxidation and hydroxylation. Biosci Biotechnol Biochem 57(9):1589–1590

Kümmerer K (2009) The presence of pharmaceuticals in the environment due to human use–present knowledge and future challenges. J Environ Manage 90(8):2354–2366

Lertratanangkoon K, Horning MG (1982) Metabolismof carbamazepine. Drug Metab Dispos 10:1–10

Li A, Cai R, Cui D, Qiu T, Pang C, Yang J, Ren N (2013) Characterization and biodegradation kinetics of a new cold-adapted carbamazepine-degrading bacterium, Pseudomonas sp. CBZ-4. J Environ Sci 25(11):2281–2290

Li X, Xu J, de Toledo RA, Shim H (2016) Enhanced carbamazepine removal by immobilized Phanerochaete chrysosporium in a novel rotating suspension cartridge reactor under non-sterile condition. Int Biodeterior Biodegradation 115:102–109

Loos R, Locoro G, Comero S, Contini S, Schwesig D, Werres F, Balsaa P, Gans O, Weiss S, Blaha L, Bolchi M (2010) Pan-European survey on the occurrence of selected polar organic persistent pollutants in ground water. Water Res 44(14):4115–4126

Mao F, Liu X, Wu K, Zhou C, Si Y (2018) Biodegradation of sulfonamides by Shewanella oneidensis MR-1 and Shewanella sp. strain MR-4. Biodegradation 29(2):129–140

Marco-Urrea E, Pérez-Trujillo M, Vicent T, Caminal G (2009) Ability of white-rot fungi to remove selected pharmaceuticals and identification of degradation products of ibuprofen by Trametes versicolor. Chemosphere 74(6):765–772

Michael I, Rizzo L, McArdell CS, Manaia CM, Merlin C, Schwartz T, Dagot C, Fatta-Kassinos D (2013) Urban wastewater treatment plants as hotspots for the release of antibiotics in the environment: a review. Water Res 47(3):957–995

Monteiro SC, Boxall AB (2010) Occurrence and fate of human pharmaceuticals in the environment. Reviews of environmental contamination and toxicology. Springer, New York, NY, pp 53–154

Mulla SI, Sun Q, Hu A, Wang Y, Ashfaq M, Eqani SAMAS, Yu CP (2016) Evaluation of sulfadiazine degradation in three newly isolated pure bacterial cultures. PLoS ONE 11(10):e0165013

Muñoz I, Bueno MJM, Agüera A, Fernández-Alba AR (2010) Environmental and human health risk assessment of organic micro-pollutants occurring in a Spanish marine fish farm. Environ Pollut 158(5):1809–1816

Nguyen PY, Carvalho G, Reis AC, Nunes OC, Reis MAM, Oehmen A (2017) Impact of biogenic substrates on sulfamethoxazole biodegradation kinetics by Achromobacter denitrificans strain PR1. Biodegradation 28(2–3):205–217

Ottmar KJ, Colosi LM, Smith JA (2010) Development and application of a model to estimate wastewater treatment plant prescription pharmaceutical influent loadings and concentrations. Bull Environ Contam Toxicol 84(5):507–512

Ozdemir G, Aydin E, Topuz E, Yangin-Gomec C, Tas DO (2015) Acute and chronic responses of denitrifying culture to diclofenac. Biores Technol 176:112–120

Pan LJ, Li CX, Yu GW, Wang Y (2017) Biodegradation of sulfamethazine by an isolated thermophile–Geobacillus sp. S-07. World J Microbiol Biotechnol 33(5):85

Radke M, Ulrich H, Wurm C, Kunkel U (2010) Dynamics and attenuation of acidic pharmaceuticals along a river stretch. Environ Sci Technol 44(8):2968–2974

Reis PJ, Reis AC, Ricken B, Kolvenbach BA, Manaia CM, Corvini PF, Nunes OC (2014) Biodegradation of sulfamethoxazole and other sulfonamides by Achromobacter denitrificans PR1. J Hazard Mater 280:741–749

Ricken B, Corvini PF, Cichocka D, Parisi M, Lenz M, Wyss D, Kohler HPE (2013) Ipso-hydroxylation and subsequent fragmentation: a novel microbial strategy to eliminate sulfonamide antibiotics. Appl Environ Microbiol 79(18):5550–5558

Rodarte-Morales AI, Feijoo G, Moreira MT, Lema JM (2012) Biotransformation of three pharmaceutical active compounds by the fungus Phanerochaete chrysosporium in a fed batch stirred reactor under air and oxygen supply. Biodegradation 23(1):145–156

Shoaib Ahmad Shah S, Rivera G, Ashfaq M (2013) Recent advances in medicinal chemistry of sulfonamides. Rational design as anti-tumoral, anti-bacterial and anti-inflammatory agents. Mini Rev Med Chem 13(1):70–86

Sutton NB, Langenhoff AA, Lasso DH, van der Zaan B, van Gaans P, Maphosa F, Rijnaarts HH (2014) Recovery of microbial diversity and activity during bioremediation following chemical oxidation of diesel contaminated soils. Appl Microbiol Biotechnol 98(6):2751–2764

Tanna RN, Tetreault GR, Bennett CJ, Smith BM, Bragg LM, Oakes KD, McMaster ME, Servos MR (2013) Occurrence and degree of intersex (testis–ova) in darters (Etheostoma SPP.) across an urban gradient in the Grand River, Ontario, Canada. Environ Toxicol Chem 32(9):1981–1991
Tappe W, Hofmann D, Disko U, Koeppchen S, Kummer S, Vereecken H (2015) A novel isolated Terrabacter-like bacterium can mineralize 2-aminopyrimidine, the principal metabolite of microbial sulfadiazine degradation. Biodegradation 26(2):139–150
Tiwari B, Sellamuthu B, Ouarda Y, Drogui P, Tyagi RD, Buelna G (2017) Review on fate and mechanism of removal of pharmaceutical pollutants from wastewater using a biological approach. Biores Technol 224:1–12
Topp E, Chapman R, Devers-Lamrani M, Hartmann A, Marti R, Martin-Laurent F, Sumarah M (2013) Accelerated biodegradation of veterinary antibiotics in agricultural soil following long-term exposure, and isolation of a sulfamethazine-degrading Microbacterium sp. J Environ Qual 42(1):173–178
Wang S, Wang J (2018) Biodegradation and metabolic pathway of sulfamethoxazole by a novel strain Acinetobacter sp. Appl Microbiol Biotechnol 102(1):425–432
Yang CW, Hsiao WC, Chang BV (2016) Biodegradation of sulfonamide antibiotics in sludge. Chemosphere 150:559–565
Zhang W, Zhang M, Lin K, Sun W, Xiong B, Guo M, Cui X, Fu R (2012) Eco-toxicological effect of Carbamazepine on Scenedesmus obliquus and Chlorella pyrenoidosa. Environ Toxicol Pharmacol 33(2):344–352
Zimmerman MJ (2005) Occurrence of organic wastewater contaminants, pharmaceuticals, and personal care products in selected water supplies, Cape Cod, Massachusetts, June 2004 (No. USGS-OFR-2005–1206). GEOLOGICAL SURVEY RESTON VA
Zuccato E, Castiglioni S, Fanelli R (2005) Identification of the pharmaceuticals for human use contaminating the Italian aquatic environment. J Hazard Mater 122(3):205–209

Chapter 16
Oxidative Potential of Particulate Matter: A Prospective Measure to Assess PM Toxicity

Suman Yadav and Harish C. Phuleria

Abstract It is now broadly accepted that particulate matter exposure can lead to multiple adverse health effects. The capability of airborne particulate matter (PM) to generate reactive oxygen species (ROS), known as "oxidative potential" (OP) is suggested to be one of the most relevant indicators of PM toxicity. Redox active chemical species in PM, of both inorganic and organic nature, facilitate ROS generation, causing oxidative damages, which are harmful for cells, ultimately leading to different chronic diseases. Therefore, OP has been proposed as a new additional metric for PM toxicity which is better associated with biological responses to PM exposures, thus could be more informative than particulate mass alone. However, the mechanisms of toxicity and its relation with the physico-chemical properties of PM are still largely unknown and need further research. Several chemical assays exist to assess the oxidative potential of PM. They differ from each other in sensitivity to the ROS generating chemical constituents of PM. The consumption of dithiothreitol (DTT), which is primarily based on the capability of redox active compounds to transfer electrons from DTT to oxygen, is used as the most widely applicable acellular method to assess the OP of PM. The potential of PM to deplete antioxidants such as glutathione, ascorbic acid and uric acid are a few of the other methods used to measure OP with respect to time. Another method, electron spin resonance (ESR) with 5,5-dimethylpyrroline-N-oxide (DMPO) as a spin trap, measures the ability of PM to induce hydroxyl radicals in the presence of H_2O_2. The current understanding of oxidative potential of PM, its analysis methods, along with its spatial distribution across the globe are presented here. Effect of various particle sizes, chemical composition, and nature of origin in exhibiting oxidative potential and its impact on health are additionally discussed.

Keywords Oxidative potential · Particulate matter · Toxicity · Oxidative stress · Free radicals · Dithiothreitol assay

S. Yadav (✉) · H. C. Phuleria
IDP in Climate Studies, IIT Bombay, Mumbai, India
e-mail: suman.yadv@gmail.com

H. C. Phuleria
Centre for Environmental Science and Engineering, IIT Bombay, Mumbai, India

T. Gupta et al. (eds.), *Measurement, Analysis and Remediation of Environmental Pollutants*, Energy, Environment, and Sustainability,
https://doi.org/10.1007/978-981-15-0540-9_16

16.1 Introduction

Airborne particulate matter and its relation with adverse health effects particularly cardio respiratory end-points have been supported by several epidemiological studies (Bates et al. 2015; Abrams et al. 2017; Yang et al. 2016; Cesaroni et al. 2014; Wild et al. 2009; Magas et al. 2007; Delfino et al. 2010). PM_{10} and $PM_{2.5}$ (with aerodysnamic diameters smaller than 10 μm and 2.5 μm, respectively) are currently regulated on a mass basis, which are widely used in epidemiological studies to assess the health effects of PM. However, PM comprises of a broad range of chemical components with potentially deviating toxicity, such as redox-active metals (Yadav et al. 2016; Akhtar et al. 2010; Yang et al. 2014; Verma et al. 2014; Gasser et al. 2009), polycyclic aromatic hydrocarbons (PAHs) (Charrier and Anastasio 2012; Lundstedt et al. 2007), black carbon or elemental carbon (and linked species) (Choudhary et al. 2017; Chow et al. 2017; Kleinman et al. 2007) and other particular organic species (Verma et al. 2015; Kirillova et al. 2014; Nel et al. 2001). Other than chemical constituents, a few studies have suggested that size, sources, or surface area also have important implications on PM toxicity (Kelly and Fussell 2012; Nel 2005). Thus, PM toxicity is likely caused by synergistic effect of multiple components altogether. The studies utilising chemical components to assess whether it has higher impacts on health than mass concentration, does not account the interaction within the components and its synergistic effect.

Epidemiological studies have reported that exposure to PM having high oxidative potential leads to greater risk of cardiorespiratory health hazards (Bates et al. 2015; Abrams et al. 2017; Yang et al. 2016). Also, it has been reported in few studies that OP is more related to cardio respiratory problems than PM mass concentration (Weichenthal et al. 2016; Bates et al. 2015; Abrams et al. 2017). Moreover, OP measurement provides cumulative effect of different physico-chemical aspects of PM such as activity due to redox active metals, redox activity of organics, and oxidative stress due to PM size and surface thus giving integrated effect of PM toxicity. Thus, including oxidative potential of PM as an additional metric would be very useful for addressing the air quality and its adverse health effects.

The present chapter gives insight on the current knowledge of oxidative potential of PM, its mechanism for toxicity, measurement methods of OP, its relation with particle size, composition, emission sources and nature of origin along with its spatial distribution across the globe. Different measurement methods are mentioned in brief but the widely used acelluar assays methods are described in detail.

16.2 PM Oxidative Potential

Oxidative potential (OP) is defined as the capability of particles to generate ROS (viz., hydroxyl radicals ($^{\bullet}OH$), superoxide ($O_2^{\bullet -}$), alkoxyl radicals ($RO^{\bullet}$), peroxyl radicals ($RO_2^{\bullet}$) etc.) by oxidant generation or antioxidants consumption (Yang et al. 2014;

Ayres et al. 2008). A number of studies have linked OP with adverse health effects (Weichenthal et al. 2016; Bates et al. 2015; Øvrevik et al. 2015; Donaldson et al. 2005; Nel 2005; Shi et al. 2003). Weichenthal et al. (2016) studied ambient $PM_{2.5}$ oxidative potential (ability to deplete glutathione and ascorbic acid in artificial respiratory lining fluid) between the cities in Ontario province, Canada and reported that PM oxidative potential (OP) rather than PM mass concentration as a key determinant of emergency hospital visits due to asthma in children. OP of PM is linked to its chemical composition, which is dependent on point source (combustion, road side dust, soil) and climatic conditions (humidity, temperature, atmospheric salt content). PM rich in combustion derived polycyclic aromatic hydrocarbons (e.g., quinones) and metals (e.g., Fe, Cu, Mn, Cr, Cd, Ni) may exhibit higher OP (Schwarze et al. 2006; Mills et al. 2007).

Currently, OP of PM is measured by assessing its ability to oxidize glutathione or ascorbic acid, the major antioxidants in human respiratory lining fluid, by estimating the consumption of cellular reductant surrogate, direct measurement of radicals by EPR etc. "OP" measurement methods, generally utilize units of time rate-of-change per unit mass (nmol depletion min^{-1} μg^{-1} of PM) or per unit volume (nmol depletion min^{-1} m^{-3} of air) depending on the assay. Epidemiologic studies mostly use or relate the volume-based concentration rates of OP as they relate the adverse health effects with the volume of air inhaled whereas observational/toxicological studies make use of mass-based concentration rates of OP as the PM doses (in vitro or in vivo) are instilled on mass basis (Bates et al. 2019). Both approaches can provide evidence on the PM capacity and related sources to potentially initiate catalytic redox reactions in the body, which leads to health impacts associated with oxidative stress (Bates et al. 2019; Fang et al. 2016). Various OP measurement methods are discussed in detail in Sect. 16.5.

16.3 PM Composition and Oxidative Potential

16.3.1 Transition Metals

Transition metals are the group of elements which have partially filled subshell and variable valencies e.g., Cu, Fe, Co, Ni etc., and are considered possible mediators of PM induced toxicity by a number of studies (Yadav et al. 2016; Xiao et al. 2013; Park et al. 2013; Lu et al. 2011; Jung et al. 2006; Knaapen et al. 2002). Transition metals have been related with various kinds of health hazards including pulmonary and respiratory diseases (Verma et al. 2014; Alharbi et al. 2015), neurotoxicity and cancer (Chuang et al. 2012; Nagai and Toyokuni 2010; Monn and Becker 1999). Elevated amounts of these metals are observed in the surroundings, specifically close to industrial and high traffic areas. Besides their probable health effects, toxic heavy metals, in particular, are also of concern since they do not biodegrade, which means that after release into the atmosphere, they will continue to exist for re-introduction into the water, air and food chain. Select metals such as iron (Fe), copper (Cu), vanadium (V), and nickel (Ni) have been related as potential mediators of PM-induced respiratory inflammation and associated with mortality and increased hospital admissions (Kelly and Fussell 2012; Cesaroni et al. 2014). These metals have been shown to have association with OP. Especially, Fe and Cu are considered important in PM induced formation of ROS through the Fenton reaction (Schwarze et al. 2006; Vidrio et al. 2008). Soluble and insoluble transition metals with changing valencies have been correlated with measured oxidative potential but as only the soluble metal forms contribute in redox reactions so these correlations depend on metals solubility.

A few recent studies have shown significant correlations of OP measured by Ascorbic acid (AA) assay, (OP_{AA}) with both total and soluble metals, including lead, manganese, copper, iron and zinc (Fang et al. 2016; Godri et al. 2010; Janssen et al. 2014; Pant et al. 2015). It has been observed that OP determined by DTT method (OP_{DTT}) is not affected by iron likewise. Thus, the ROS generated by Fenton chemistry or synergistic effects may not be fully captured by OP_{DTT}, specifically with respects to $^{\bullet}OH$ generation (Xiong et al. 2017; Bates et al. 2019). Soluble manganese and copper also do not considerably influence OP_{DTT}(Bates et al. 2019). However, 80% of DTT loss in atmospheric PM samples collected in San Joaquin Valley, California is attributed to metals, particularly to Mn and Cu (Charrier and Anastasio 2012). Yadav et al. (2016) studied the impact of metals in generation of free radicals and related genotoxicity caused by fine PM in Pune, India. They suggest that Fe may be primarily accountable for DNA damage by radical production along with other metals like Cu and Zn. Thus, several studies provide the evidence that transition metals associated with PM play significant role in radical generation leading to oxidative stress and toxicity.

16.3.2 Organic PM

Organic carbon (OC) and elemental carbon (EC) constitute a large proportion of particle mass andhave been linked with hospital admissions and mortality (Kelly and Fussell 2012; Janssen et al. 2011). The OC fraction of PM includes compounds such as polycyclic aromatic hydrocarbons (PAHs) and quinones. Organic compounds such as quinones and semiquinones adsorbed onto the particle surface of PM can generate ROS by reducing oxygen (O_2^-, H_2O_2 and eventually $OH^\cdot$). Thus, especially quinones have the ability to catalyse the production of ROS (i.e. hydrogen peroxide) by undergoing redox recycling and reduce oxygen to produce superoxide radicals (Chung et al. 2006). In Banglore, India, Vreeland et al. (2016) examined redox activity of $PM_{2.5}$ related to near road waste burning. They assessed the organic and elemental carbon, brown carbon and related redox activity emissions from the burning of garbage near road. They found high OP_{DTT} in all the samples of garbage burning. Their overall results suggest that redox active PM exposure can be tremendously high in areas adjacent to trash-burning. Mondal et al. (2011) assessed genotoxicity of biomass burning emitted particles in buccal epithelial cells (BECs) of premenopausal women engaged in cooking with biomass (crop residues, dung, wood etc.). They found 51% increase in ROS generation in airway cells of women using biomass fuels.

Several studies globally (United States, China, India) depict a correlation amongst water soluble organic carbon (WSOC), organic carbon (OC) and OP_{DTT} of PM collected during various seasons (Verma et al. 2015; Hu et al. 2008; Velali et al. 2016; Verma et al. 2009, 2011; Vreeland et al. 2017; Patel and Rastogi 2018). The role of secondary organic aerosols (SOA), which are formed after photochemical oxidation of (volatile) organic precursors, is still unclear as presently it is difficult to distinguish between the toxicity of primary and secondary organic aerosols (Cesaroni et al. 2014). A few studies observe the effect of chemical aging in temporal and spatial variations of OP. Ultrafine PM samples collected in afternoon in Los Angeles showed higher mass-normalized OP_{DTT}. The afternoon PM is considered to have high contribution from secondary processes compared to freshly emitted PM in the morning, which suggest that photochemistry causes increased radical generation and redox active components (Saffari et al. 2016; Verma et al. 2009). A strong connection was observed between cellular based ROS assays with pure SOA in PM (Tuet et al. 2017; Saffari et al. 2014). Thus, these findings suggest that organic component of PM can also lead to ROS generation but further research is needed to study the role of secondary organic components on free radical generation.

16.4 Mechanism of PM Toxicity

The difficulty of studying the toxicity of PM partially lies in its inconsistency, complexity and its diversity in structure and composition which lead to distinct biological outcomes. The adverse effect caused due to exposure of PM to humans and animals can be assessed by an important metric, PM toxicity. Several toxicological studies have recognised that PM which is inhaled has the capability to cause oxidative stress within the lung, which could lead to the adverse cardio-respiratory health effects (Grevendonk et al. 2016; Hemmingsen et al. 2015; Xiao et al. 2013; Park et al. 2013; Nagai and Toyokuni 2010; Brook et al. 2010; Valavanidis et al. 2009). These toxicological studies play an important role in understanding the mechanism behind the toxicity of PM. To elucidate the adverse health effects of particulate pollutants, a number of mechanisms have been proposed viz., endotoxin effects, inflammation, autonomic nervous system activity, stimulation of capsaicin/irritant receptors, procoagulant effects, ROS generation and alteration of the components of cells (Nel et al. 2006; Li et al. 2008). Amongst these, the oxidative stress and ROS generation are the most considered mechanism (Rahal et al. 2014). ROS can destruct proteins, DNA and membrane lipids, which can lead to cellular death via either necrotic or apoptotic processes. Necrotic process of cell death is premature cell death due to external factors such as toxins, infection or trauma whereas apoptosis is type of cell death that is stimulated by regular processes in the body to eradicate dysfunctional cells. Elevated levels of oxidative stress lead to inflammatory response as various transcription factors are activated and production of cytokine is stimulated (Øvrevik et al. 2005; Billet et al. 2008). The respiratory disorders caused due to PM are mostly related to inflammation (Hetland et al. 2000, 2005; Dye et al. 2001).

16.4.1 Reactive Oxygen Species (ROS) and Oxidative Stress

ROS is chemically defined as 'the atomic species or group which has at least one or more unpaired electrons'. ROS comprise of hydroxyl radicals ($^{\bullet}OH$), superoxide ($O_2^{\bullet -}$), alkoxyl radicals ($RO^{\bullet}$) and peroxyl radicals ($RO_2^{\bullet}$). Moreover, some non-radicals which are oxidizing species or that can simply transform other radicals like O_3, peroxynitrite ($ONOO^{\bullet}$), hypochlorous acid (HOCl), hydrogen peroxide (H_2O_2) and singlet oxygen ($^1O_2$) are also considered as ROS. Various forms of ROS and reactive nitrogen species (RNS) are listed in Table 16.1. These species are either generated by components within the PM, or are present on the PM surface, or generated through the biological response endogenously (Li et al. 2003; Squadrito et al. 2001; Schins et al. 2004). The oxidative metabolism of mitochondria and the immune system cells of body in process of neutralising to foreign body can also produce free radicals. Environmental causes such as pollution, radiation and cigarette smoke can produce free radicals too. Biological systems are constantly interacting with the free radicals which are generated in body as a product to metabolism or from environmental factors. Free radicals can cause chain reaction when they react with a non-radical and can form a new radical giving rise to oxidative stress (Hemmingsen et al. 2015; Valavanidis et al. 2013; Xiao et al. 2013; Park et al. 2013; Mehta et al. 2008; Lu et al.

Table 16.1 Reactive oxygen and nitrogen species

Name	Formula	Characteristics
Hyperoxide/superoxide	$O_2^{-\bullet}$	Highly unstable, signalling function, synaptic plasticity
Hydrogen peroxide	H_2O_2	Cell toxicity, signalling function, generation of other ROS
Hydroxyl radical	$OH^{\cdot}$	Free radical, very reactive agent, highly unstable
Alkoxyl radical	$RO^{\bullet}$	Free radical, reaction product of lipids
Peroxyl radical	$ROO^{\bullet}$	Free radical, reaction product of lipids
Hypochlorite anion	OCl^{-}	Reactive chlorine species, reactive oxygen species, enzymatically produced by myeloperoxidase
Singlet oxygen	$^1O_2$	Induced/excited oxygen molecule, nonradical and radical form
Ozone	O_3	Environmental toxin
Peroxynitrite	$ONOO^{-}$	Highly reactive reaction intermediate of $^{\bullet}O_2$ and $^{\bullet}NO$
Nitrogen dioxide	$^{\bullet}NO_2$	Highly reactive radical, environmental toxin
Nitric oxide	$^{\bullet}NO$	Environmental toxin, endogenous signal molecule

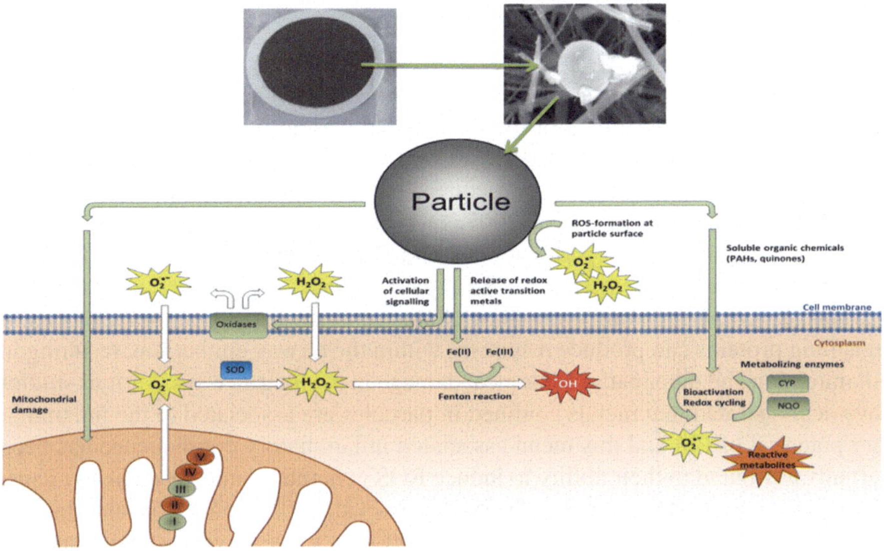

Fig. 16.1 Reactive oxygen species (ROS) formation in particle exposed cells (adapted from Øvrevik et al. 2015)

2006). An overview of ROS formation and oxidative stress in particle-exposed cells is presented in Fig. 16.1.

Oxidative stress occurs when the ROS are in excess in body and antioxidants are unable to counteract its harmful effects due to imbalance between ROS production and antioxidants. ROS mediated oxidative stress involves the process of aging and causes several metabolic disorders, gastrointestinal ulcer genesis, cystic fibrosis, hemorrhagic shock, rheumatoid, diabetes, arthritis, ischemia, neurodegenerative disease and even cancer (Bhattacharyya et al. 2014; Hensley and Floyd 2002; Lehucher-Michel et al. 2001; Das et al. 1997). Other diseases that can be caused by ROS are Parkinson's disease (Uttara et al. 2009; Guglielmotto et al. 2009; Wagner et al. 2013), Alzheimer's disease (Perry et al. 2002; Butterfield et al. 2007; Michel et al. 2012), Down's syndrome (Buscigilo et al. 1995; Michel et al. 2012; Rahman et al. 2012), and ischemic reperfusion injury in various organs like brain, liver, heart and kidney (Valko et al. 2007; Wagner et al. 2013). It has been established recently that ROS plays a major role in stress-induced inflammatory bowel diseases and gastric ulcer (Calder et al. 2009; Rahman et al. 2012). During normal physiological conditions, excessive reactive intermediates (free radicals) are scavenged or neutralized by antioxidants (Valavanidis et al. 2013).

16.4.2 Inflammation and Genotoxicity

Inflammation is resulted as a response of the organism to adverse stimuli. When it takes place with a high amount or for an extended duration, it can lead to inflammation of airways further leading to diseases such as COPD and asthma. Inflammation of airways is a critical short-term consequence of exposure to PM in humans (Jalava et al. 2007; Hetland et al. 2000, 2005; Dye et al. 2001). During the process of inflammation, pulmonary epithelial cells and alveolar macrophages (AM) release the pro-inflammatory intermediaries such as cytokines, chemokines, reactive oxygen species, etc. Depending upon intracellular generation of ROS, the cytokines or signalling proteins can produce responses within the airway epithelium, resulting in inflammation and other patho-biological damage in the lung airways. Various studies have demonstrated that metals confined in particles are associated to the inflammatory potential of PM and this metal associated inflammatory consequence has been partially attributed to their ability to induce ROS generation and consequently causing oxidative stress (Øvrevik et al. 2005; Becker et al. 2005; Donaldson and MacNee 2001; Hetland et al. 2000). During inflammation reactive oxygen species such as nitric oxide free radical ($NO^{\bullet}$), superoxide anion ($O_2^{-\bullet}$), hydrogen peroxide (H_2O_2), hydroxyl radical ($HO^{\bullet}$), peroxynitrite ($ONOO^{-}$) are generated critically for degradation of internalized pathogens (Rosanna and Salvatore 2012).

Genotoxicity describes the property of chemical agents that damages the genetic information within the cells which may lead to mutations and cancer. Diverse mechanisms have been described for DNA damage caused due to PM, which comprise of oxidative stress, direct interface of particles constituents with the DNA and organics metabolites formation. This DNA damage includes strand breakages (single and double), point mutations, chromosomal alterations (chromosomal and numerical), and cross linking. In addition to the formation of ROS, the P450 cytochromes (CYP1A1/1B1) causes metabolic activation of PAHs which form epoxides and reactive species that may form bulky DNA adducts (Billet et al. 2008). The modification of base pair and strand breakage due to free radical DNA damage alters the components of electron transport chain which results in the leakage of electrons and ultimately cell damage. The DNA damage due to oxidation also involves modification of DNA base and breakage of its strands which ultimately forms abnormal components of the electron transport chain. DNA strand breakage may be caused by a variety of reactions. The utmost important mode is direct scission of the DNA backbone by chemical or ROS attack. Within a definite period of time, if the cells are unable to restore the damage, further mechanisms may be activated to defend the organism; of which the utmost vital mechanisms are active cell death programs (Yang et al. 2004). Figure 16.2 depicts the mechanisms of PM-related DNA damage induction.

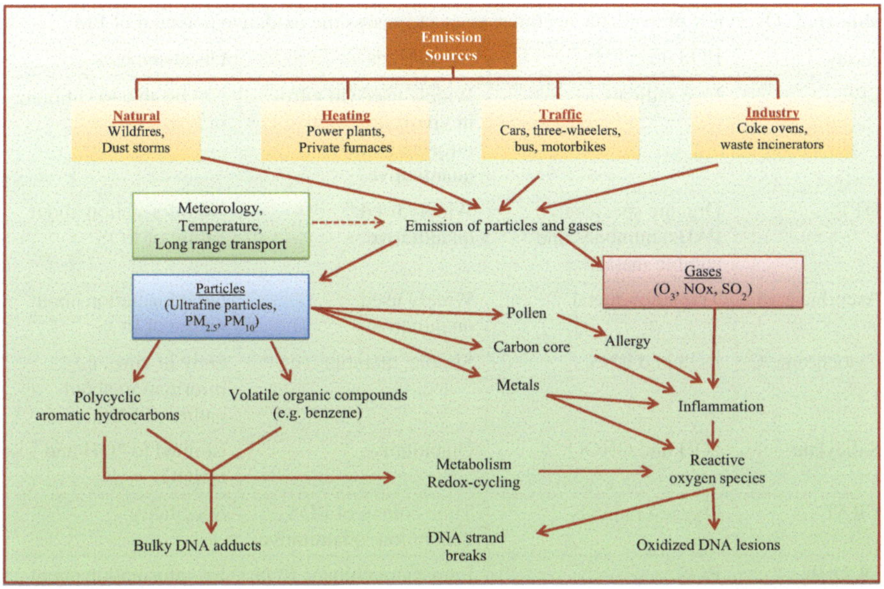

Fig. 16.2 Mechanisms of PM- related DNA damage induction (adapted from Møller et al. 2008)

16.5 Measurement Methods for Oxidative Potential

The oxidative potential of the PM can be estimated by both cellular (Kubátová et al. 2006; Antonini et al. 1998; Xia et al. 2006; Hiura et al. 1999; Bonvallot et al. 2001) and acellular (Frampton et al. 1999; Mudway et al. 2011; Jung et al. 2006; Venkatachari et al. 2005; Cho et al. 2005; Zomer et al. 2011) assays. Cellular assays measure the biochemical and molecular mechanisms related to PM cell toxicity. Numerous assays exist to assess PM-induced ROS production and oxidative stress. An indirect method involves using the DCFH-DA probe to monitor PM-induced ROS generation in rat alveolar macrophage cells (macrophage-ROS assay), and has been applied in several studies (Landreman et al.2008; Cheung et al. 2012; Daher et al. 2014; Saffari et al. 2013).

Acellular techniques require less controlled conditions and give quicker read-outs of PM oxidative potential. Table 16.2 depicts the overview of available acellular assays to assess the oxidative potential of PM. These assays usually include estimating the, (1) Consumption of cellular reductant surrogate, for example, the dithiothreitol (DTT) assay (Cho et al. 2005); (2) exhaustion of antioxidants, like ascorbic acid, reduced glutathione, urate and in a synthetic respiratory tract lining fluid model (RTLF) (Zielinski et al. 1999; Mudway et al. 2004; Yang et al. 2014); (3) $OH^\bullet$ radical formation and detection using electron spin/paramagnetic resonance (EPR/ESR) (Shi et al. 2003) or high-performance liquid chromatography (HPLC) (Jung et al. 2006); (4) concentration of ROS by means of chemiluminescent reductive acridinium triggering (CRAT) probe which is based on chemiluminescence emitted

Table 16.2 Overview of available acellular assays to assess the oxidative potential of PM

Assay	ROS detected	Advantages	Disadvantages
EPR	Free radicals	Widely used (in vitro, in vivo), structural information, quantitative	Not be able to calibrate in vivo
DTT	Organic species like PAHs, quinones and soluble metals	Widely used, quantitative	No information about nature of ROS
Ascorbic acid	Transition metal	Widely used, quantitative	No information about nature of ROS
Cytochrome C	ROS and RNS	Simple, quantitative	Only in vitro, no information about nature of ROS
Salicylate	$^{\bullet}OH$ and $ONOO^{-}$	Quantitative	Limited to $^{\bullet}OH$ and ONOO
CRAT	Oxygen radicals	Time course of ROS generation, quantitative	Specificity
DCDHF	ROS	Intra-extra cellular ROS formation, visualisation	No information about nature of ROS, Autocatalytic degradation

from acridinium compounds to measure the amount of H_2O_2 (Zomer et al. 2011), and fluorescent probes, such as dichlorofluorescin (DCFH) probe which is based on non-fluorescent DCFH which develop into fluorescent in the presence of ROS (Hung and Wang 2001; Venkatachari et al. 2005). Some of these methods viz., EPR, CRAT Salicylate assays can be used for cellular studies as well when used in the form of spin traps or probes (Hellack et al. 2017; Valgimigli et al. 2002). Figure 16.3 shows the overview of assays to assess the oxidative potential of PM based on its sensitivity to certain radicals. Most extensively used and comparatively low cost methods to study OP of PM (DTT and AA) are discussed in detail in the Sect. 16.5.1.

16.5.1 DTT and AA

Among the several existing methods for measuring aerosol OP, the dithiothreitol (DTT) and ascorbic acid (AA) assays are the most extensively used assays. Each method comprises incubating the aqueous filter extracts with DTT or AA at a biologically pertinent temperature (37 °C) and pH (7.4), and monitoring the depletion of DTT or AA over time. The quantification of DTT is attained by addition of 5,5′-dithiobis-(2-nitrobenzoic acid (DTNB) to form a colour species and assessing the light absorbance at 412 nm wavelength. The depletion of DTT happens when PM

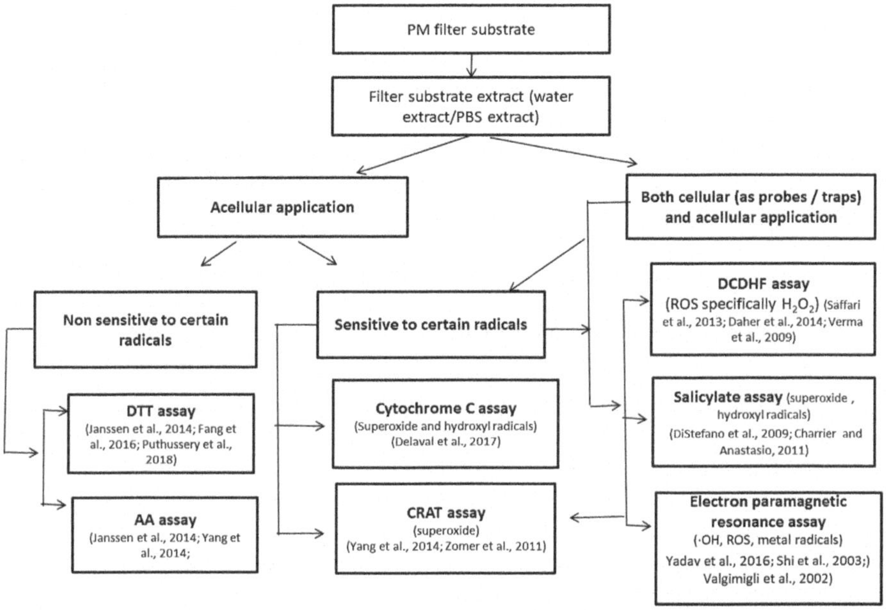

Fig. 16.3 Flow chart presenting different assays to assess the oxidative potential of PM

extracts catalytically transfer one electron from DTT to molecular oxygen and generate superoxide anion (Kumagai et al. 2002), mimicking the crucial initial step of producing ROS in vivo—the univalent reduction of molecular oxygen. AA itself absorbs light at 265 nm wavelength, thus absorbance of AA incubated with PM extracts is measured at 265 nm wavelength and depletion rate of ascorbic acid is calculated by linear regression of absorbance data against time. AA is among one of the physiological antioxidants in lung lining fluid that inhibits the oxidation of lung cells (Valko et al. 2005). Therefore, the in vitro measured OP depletion of DTT and AA by PM (OP_{DTT} and OP_{AA}) might represent the in vivo degree of ROS generation and direct antioxidant oxidation, respectively. This, in theory, offers a mechanistic association amongst aerosol OP_{DTT} or OP_{AA} and the capacity of particles to induce oxidative stress and lead to adverse health effects. As potential health indicators, it is important to understand the spatiotemporal variations, sources, major chemical players, and size-dependent characteristics of these OP. In several cases, when applying the AA or DTT assay, particles are extracted in water and then liquid filtration is done to remove insoluble or solid fractions, but studies indicate that OP_{DTT} can also be associated with insoluble constituents (Yang et al. 2014; McWhinney et al. 2013; Li et al. 2013; Verma et al. 2012; Daher et al. 2011). Therefore, it is also important to assess the contribution of insoluble species to aerosol OP_{DTT}.

Precise handling practices at every stage of the DTT assay reaction is needed for an accurate determination of the rate of DTT oxidation, which is time intensive and laborious analytical protocol. This limits its use as a routine protocol for large scale health studies. Some researchers have developed alternative approaches such as automation of DTT assay to address this concern. Fang et al. (2015) developed a semi-automated system for conducting DTT assay on several PM extracts. They used programmable syringe pumps with selector valves in this system. The system is based on a basic protocol in which DTT is oxidized by PM extract in a single vial. With this auto sampler the aqueous extracted particulate samples are analyzed in batches, roughly one hour per sample.

16.5.2 Semi Real Time Assessment of Oxidative Potential

A couple of studies have developed a microfluidic electrochemical sensor for online monitoring of OP_{DTT} (Sameenoi et al. 2012; Koehler et al. 2014) and a microfluidic paper-based analytical device using the DTT assay which skips the extraction procedure (Sameenoi et al. 2013). However, these systems have not been further tested in field studies. Eiguren-Fernandez et al. (2017) developed an online DTT system collecting particles directly into small volume of liquid using a Liquid Spot Sampler (LSS). This system was tested in the ambient environment only for 3 days. Puthussery et al. (2018) have developed an online instrument to measure OP of ambient PM using the DTT assay. The instrument collects the ambient PM in a mist chamber continuously and further estimates its DTT activity using an automated syringe pump system. The system has been evaluated in the field conditions for about 50 days.

16.6 Spatial Models for Assessment of Oxidative Potential

Modelling is normally used to assess the spatial distinctions in exposures which are challenging to be estimated from monitoring sites. In contrast to other estimated pollutant metrics (e.g. NO_X, NO_2, SO_2, $PM_{2.5}$, O_3), the modelling of OP is not established because of scarcity of data on source emissions. Instead, land use regression modelling (LUR)—utilizing multivariate regressions can model the spatial contrasts by building a relationship between the estimated concentrations of pollutants and geographical predictors (e.g. street traffic, land use, population distribution) (Hoek et al. 2008). Studies focusing on OP and its health impacts have utilized OP measurement data and modelling methods, source influenced regression and LUR, for OP estimation when and where measurements are not accessible (Bates et al. 2015; Fang et al. 2016; Yang et al. 2015).

LUR model uses statistical modeling and examines the relationship between the measured concentration at the monitoring site and geographic information systems (GIS) generated predictor variables, based on this it can explain the spatial differences (Bates et al. 2019). In several studies OP_{DTT} has been related to different acute cardio respiratory end points. In a study which was conducted for 14 years in the

Netherlands, Yang et al. (2016) related markers of respiratory health, comprising of asthma, forced expiratory volume and rhinitis, with OP_{DTT} rather than $PM_{2.5}$ as estimated by LUR. Another study utilized modeled OP_{DTT} 10-year data set in Atlanta, GA and established a relation between predicted OP_{DTT} exposure and emergency department (ED) visits due to wheeze, asthma and heart failure due to congestion (Bates et al. 2015). In Europe land-use regression (LUR) modeling has been used to predict OP in 10 different areas using several assays, including OP_{DTT} (Jedynska et al. 2017; Yang et al. 2015). However, LUR models established for one site may not be perfectly applicable to new or different locations.

16.7 Oxidative Potential Assessment Across the Globe

Although it is not possible to directly compare the numerical values for oxidative potential obtained by different assays, the overall trends for various aerosol types can be compared. Table 16.3 summarizes the OP activity of PM of different size fractions collected from different locations and sources. The oxidative potential for underground station was found to be highest followed by traffic emitted PM when compared to other sources in different parts of North America and Europe (Vreeland et al. 2017; Janssen et al. 2014; Gao et al. 2017). Oxidative potential of diesel particles are reported 0.042 nmol DTTmin^{-1} μg^{-1} (McWhinney et al. 2013) and 0.039 nmol DTTmin^{-1} μg^{-1} (Geller et al. 2006). The corresponding values at urban background locations were 0.09 nmol DTT min^{-1} μg^{-1} (Cho et al. 2005). Janssen et al. (2014) estimated the OP activity by both DTT and AA assay and found the OP at underground station was approximately 50 times higher than urban background in PM_{10} and nearly 15 times higher in $PM_{2.5}$ when estimated by DTT assay. Similar results were found with AA assay as well showing higher OP activity at underground station. In commute OP was studied by Vreeland et al. (2017) considering two different types of commute viz., sidestreet and highway commute, they reported slightly higher OP for highway commute. Gao et al. (2017) collected PM at roadside and urban site and compared the effect of filter substrates. Oxidative potential reported from Quartz extraction was higher at both sites compared to Teflon which suggests higher extraction efficiency for quartz filters.

Patel and Rastogi (2018) reported oxidative potential values ranging 1.3 to 7.2 nmol DTT min^{-1} m^{-3} with average of 4.2 nmol DTT min^{-1} m^{-3} at an urban site in India. The literature on oxidative potential of size segregated PM and their association with heavy metals of traffic and industrial origin—e.g. Fe, Zn, Cu, Sb, Cr, As, Pb—is very sparse. In particular, the association of oxidative potential with carbonaceous PM components has not been much investigated in low-and middle-income countries.

Table 16.3 Oxidative potential assessment across the globe

Sources	Location	OP activity nmol DTTmin^{-1} µg^{-1}/*nmol DTT min^{-1} m^{-3}		OP activity nmol AA min^{-1} µg^{-1}/*nmol AA min^{-1} m^{-3}		DCFH nmol H_2O_2 µg^{-1}
		PM_{10}	$PM_{2.5}$	PM_{10}	$PM_{2.5}$	
Urban site						
Urban[7]	Los Angeles		0.03#			
			0.55*			
Urban[8]	Atlanta	0.02				
Urban[6]	Georgia, Atlanta		0.13*			
Urban[9]	India (Patiala, Punjab)	4.25*				
Urban background						
Urban background[1]	Netherlands	1.7*	1.4*	41.9*	22.8*	
Urban background[10]	Los Angeles, California	0.09				
Underground station[1]	Netherlands	51.5*	18*	4793.0*	1840.3*	
Traffic emitted PM						
Traffic[1]	Netherlands					
Continous traffic:		2.6*	1.7*	39.6 *	22.8*	
Stop and go traffic:		3.7*	3.3*	172.5*	103.5*	
Roadside[6]	Georgia, Atlanta		0.18 *			
Commute (sample time 2 h*)*						
Sidestreet commute[5]	Atlanta		0.78*			
Highway commutes[5]	Atlanta		1.08*			
Source emitted PM						
Animal farm[1]	Netherlands	2.3*	2.7*	17.4*	19.0*	
Diesel exhaust[2]	Ottawa, Ontario, Canada	0.042				
Diesel exhaust[3]		0.039				
Dust[4]						
Brake dust	Rome	0.00015		0.0012		0.00005

(continued)

Table 16.3 (continued)

Sources	Location	OP activity nmol DTTmin^{-1} μg^{-1}/*nmol DTT min^{-1} m^{-3}		OP activity nmol AA min^{-1} μg^{-1}/*nmol AA min^{-1} m^{-3}		DCFH nmol H_2O_2 μg^{-1}
		PM_{10}	$PM_{2.5}$	PM_{10}	$PM_{2.5}$	
Coke	Rome	0.00018		0.00055		0.0001
Road dust	Rome					0.00018
Sahariana dust	Algeria (Sahara desert)	0.00003				0.000098
Soil	Rome (Rural)	0.00001				0.000025
Pellet ash	Refinery	0.00032		0.0026		0.000052

[1]Janssen et al. (2014); [2]McWhinney et al. (2013); [3]Geller et al. (2006); [4]Simonetti et al. (2018); [5]Vreeland et al. (2017); [6]Gao et al. (2017); [7]Saffari et al. (2016); [8]Verma et al. (2012); [9]Patel and Rastogi (2018);[10]Cho et al. (2005)
[#]Afternoon and night values are averaged
[*]indicates unit in nmol DTT min^{-1} m^{-3}

16.8 Research Gaps and Future Research Recommendations

The interest in acellular OP assays has increased drastically as a growing amount of health studies considering cardio-respiratory end-points have established an association with PM oxidative potential. OP of PM can be valuable health metric as it can provide the synergistic information on association of PM with which is not provided to that extent by PM mass alone. OP relates to the redox active species in PM which may be directly involved for certain adverse health effects, and thus should continue to serve as a better predictor of certain adverse health endpoints such as lung cancer, asthma and ischemic heart disease than PM mass concentration (Bates et al. 2015; Yang et al. 2016; Abrams et al. 2017; Weichenthal et al. 2016; Delfino et al. 2013). Moreover, OP gives perception of explicit routes that may facilitate PM caused health effects, comprising aging (solubilizing metals and quinones oxidation). It can also help deciding the regulatory activities like which are the better fuels options, and which PM emission sources need to be focused for mitigation. It would likewise be significant to examine the relationship of OP with other health end-points than just cardiorespiratory effects, for instance, birth outcomes, cancer etc. OP_{DTT} assay appears to be utmost sensitive to PM related organics; however more work is expected to examine synergistic or adversarial impacts of different PM constituents and sources.

Future work with monitoring and measurements in diverse study areas such as regional backgrounds, rural areas, particularly in indoor environments, could investigate comprehensive variations in OP and give insight into the association between PM composition and OP, as well as a better understanding of spatio—temporal variability in OP, and critical evidence for exposure studies. When different studies are compared it becomes very essential that the specific methods used to estimate OP are identified. Future work must evidently designate and utilize consistent OP assays so that the comparison of results between locations and times becomes easy. Eventually, evolving "standard" technique with enhanced conditions for various OP assays will help to give clear image of PM health impact. Thus OP may be a valuable indicator for elucidating the potential relationship between PM exposures and adverse health impacts related via an oxidative stress mechanism.

References

Abrams JY, Weber RJ, Klein M, Samat SE, Chang HH, Strickland MJ, Verma V, Fang T, Bates JT, Mulholland JA, Russell AG, Tolbert PE (2017) Associations between ambient fine particulate oxidative potential and cardiorespiratory emergency department visits. Environ Health Perspect 125(10):129001

Akhtar US, McWhinney RD, Rastogi N, Abbatt JPD, Evans GJ, Scott JA (2010) Cytotoxic and proinflammatory effects of ambient and source-related particulate matter (PM) in relation to the production of reactive oxygen species (ROS) and cytokine adsorption by particles. Inhalation Toxicol 22:37–47

Alharbi B, Shareef MM, Husain T (2015) Study of chemical characteristics of particulate matter concentrations in Riyadh, Saudi Arabia. Atmos Pollut Res 6:88–98

Antonini JM, Clarke RW, Krishna Murthy GG, Sreekanthan P, Jenkins N, Eagar TW, Brain JD (1998) Freshly generated stainless steel welding fume induces greater lung inflammation in rats as compared to aged fume. Toxicol Lett 98:77–86. https://doi.org/10.1016/S0378-4274(98)00103-9

Ayres JG, Borm P, Cassee FR, Castranova V, Donaldson K, Ghio A, Harrison RM, Hider R, Kelly F, Kooter IM (2008) Evaluating the toxicity of airborne particulate matter and nanoparticles by measuring oxidative stress potential-a workshop report and consensus statement. Inhalation Toxicol 20:75–99

Bates JT, Fang T, Verma V, Zeng L, Weber RJ, Tolbert PE, Russell AG (2019) Review of acellular assays of ambient particulate matter oxidative potential: methods and relationships with composition, sources, and health effects [review-article]. Environ Sci Technol 53(8):4003–4019. https://doi.org/10.1021/acs.est.8b03430

Bates JT, Weber RJ, Abrams J, Verma V, Fang T, Klein M, Strickland MJ, Sarnat SE, Chang HH, Mulholland JA, Tolbert PE, Russell AG (2015) Reactive oxygen species generation linked to sources of atmospheric particulate matter and cardiorespiratory effects. Environ Sci Technol 49:13605–13612. https://doi.org/10.1021/acs.est.5b02967

Becker S, Dailey LA, Soukup JM, Grambow SC, Devlin RB, Huang YCT (2005) Seasonal variations in air pollution particle-induced inflammatory mediator release and oxidative stress. Environ Health Persp 113:1032–1038

Bhattacharyya A, Chattopadhyay R, Mitra S, Crowe SE (2014) Oxidative stress: an essential factor in the pathogenesis of gastrointestinal mucosal diseases. Physiol Rev Published 94(2):329–354

Billet S, Abbas I, Le Goff J, Verdin A, Andrè V, Lafargue PE, Hachimi A, Cazier F, Sichel F, Shirali P, Garcon G (2008) Genotoxic potential of polycyclic aromatic hydrocarbons-coated onto airborne particulate matter (PM2.5) in human lung epithelial A549 cells. Cancer Lett 270:144–155

Bonvallot V, Baeza-Squiban A, Baulig A, Brulant S, Boland S, Muzeau F, Barouki R, Marano F (2001) Organic compounds from diesel exhaust particles elicit a proinflammatory response in human airway epithelial cells and induce cytochrome p450 1A1 expression. Am J Respir Cell Mol Biol 25:515–521

Brook RD, Rajagopalan S, Pope CA, Brook JR, Bhatnagar A, Diez-Roux AV, Holguin F, Hong Y, Luepker RV, Mittleman MA, Peters A, Siscovick D, Smith SC Jr, Whitsel L, Kaufman JD (2010)

Particulate matter air pollution and cardiovascular disease: an update to the scientific statement from the American Heart Association. Circulation 121:2331–2378

Buscigilo J, Lorenzo A, Yen J, Yanker B (1995) Amyloid fibrils induce the phosphorylation and loss of microtubule binding. Neuron 14:879–888

Butterfield DA, Reed TT, Perluigi M, De MC, Coccia R, Keller JN, Markesbery WR, Sultana R (2007) Elevated levels of 3-nitrotyrosine in brain from subjects with amnestic mild cognitive impairment: implications for the role of nitration in the progression of Alzheimer's disease. Brain Res 1148:243–248

Calder PC, Albers R, Antoine JM, Blum S, Bourdet-Sicard R, Ferns GA, Folkerts G, Friedmann PS, Frost GS, Guarner F, Løvik M, Macfarlane S, Meyer PD, M'Rabet L, Serafini M, van Eden W, van Loo J, Vas Dias W, Vidry S, Winklhofer-Roob BM, Zhao J (2009) Inflammatory disease processes and interactions with nutrition. Br J Nutr 101:S1

Cesaroni G, Forastiere F, Stafoggia M, Andersen ZJ, Badaloni C, Beelen R, Caracciolo B, de Faire U, Erbel R, Eriksen KT, Fratiglioni L, Galassi C, Hampel R, Heier M, Hennig F, Hilding A, Hoffmann B, Houthuijs D, Jockel KH, Korek M, Lanki T, Leander K, Magnusson PKE, Migliore E, Ostenson CG, Overvad K, Pedersen NL, Pekkanen JJ, Penell J, Pershagen G, Pyko A, Raaschou-Nielsen O, Ranzi A, Ricceri F, Sacerdote C, Salomaa V, Swart W, Turunen AW, Vineis P, Weinmayr G, Wolf K, de Hoogh K, Hoek G, Brunekreef B, Peters A (2014) Long term exposure to ambient air pollution and incidence of acute coronary events: prospective cohort study and meta-analysis in 11 European cohorts from the ESCAPE Project. Br Med J 348–412

Charrier JG, Anastasio C (2012) On dithiothreitol (DTT) as a measure of oxidative potential for ambient particles: evidence for the importance of soluble transition metals. Atmos Chem Phys 12(19):9321–9333

Choudhary V, Rajput P, Rajeev P, Gupta T (2017) Synergistic effect in absorption properties of brown carbon and elemental carbon over IGP during weak south-west monsoon. Aerosol Sci Eng 1(3):138–149

Chow JC, Wang X, Sumlin BJ, Choudhary V, Rajput P, Rajeev P, Gupta T (2017) Synergistic effect in absorption properties of brown carbon and elemental carbon over IGP during weak south-west monsoon. Aerosol Sci Eng 1(3):138–149

Cheung K, Shafer MM, Schauer JJ, Sioutas C (2012) Diurnal trends in oxidative potential of coarse particulate matter in the Los Angeles basin and their relation to sources and chemical composition. Environ Sci Technol 46:3779–3787. https://doi.org/10.1021/es204211v

Cho AK, Sioutas C, Miguel AH, Kumagai Y, Schmitz DA, Singh M, Eiguren-Fernandez A, Froines JR (2005) Redox activity of airborne particulate matter at different sites in the Los Angeles Basin. Environ Res 99:40–47. https://doi.org/10.1016/j.envres.2005.01.003

Chuang HC, Fan CW, Chen KY, Chang-Chien GP, Chan CC (2012) Vasoactive alteration and inflammation induced by polycyclic aromatic hydrocarbons and trace metals of vehicle exhaust particles. Toxicol Lett 214:131–136

Chung MY, Lazaro RA, Lim D, Jackson J, Lyon J, Rendulic D, Hasson AS (2006) Aerosol-borne quinones and reactive oxygen species generation by particulate matter extracts. Environ Sci Technol 40(16):4880–4886

Daher N, Ning Z, Cho AK, Shafer M, Schauer JJ, Sioutas C (2011) Comparison of the chemical and oxidative characteristics of particulate matter (PM) collected by different methods: filters, impactors, and biosamplers. Aerosol Sci Technol 45:1294–1304. https://doi.org/10.1080/02786826.2011.590554

Daher N, Saliba NA, Shihadeh AL, Jaafar M, Baalbaki R, Shafer MM, Schauer JJ, Sioutas C. (2014) Oxidative potential and chemical speciation of size-resolved particulate matter (PM) at near-freeway and urban background sites in the greater Beirut area. Sci Total Environ 470–471:417–426

Das D, Bandyopadhyay D, Bhattacharjee M, Banarjee RK (1997) Hydroxyl radical is the major causative factor in stress-induced gastric ulceration. Free Radic Biol Med 23:8–18

Delfino RJ, Staimer N, Tjoa T, Arhami M, Polidori A, Gillen DL, George SC, Shafer MM, Schauer JJ, Sioutas C (2010) Associations of primary and secondary organic aerosols with airway and systemic inflammation in an elderly panel cohort. Epidemiology 21:892–902

Delfino RJ, Staimer N, Tjoa T, Gillen DL, Schauer JJ, Shafer MM (2013) Airway inflammation and oxidative potential of air pollutant particles in a pediatric asthma panel. J Exposure Sci Environ Epidemiol 23(5):466–473

Donaldson K, MacNee W (2001) Potential mechanisms of adverse pulmonary and cardiovascular effects of particulate air pollution (PM10). Int J Hyg Environ Health 203:411–415

Donaldson K, Tran L, Jimenez L, Duffin R, Newby D, Mills N, MacNee W, Stone V (2005) Combustion-derived nanoparticles: a review of their toxicology following inhalation exposure. Part Fibre Toxicol 2:10. https://doi.org/10.1186/1743-8977-2-10

Dye JA, Lehmann JR, McGee JK, Winsett DW, Ledbetter AD, Everitt JI, Ghio AJ, Costa DL (2001) Acute pulmonary toxicity of particulate matter filter extracts in rats: coherence with epidemiological studies in Utah Valley residents. Environ Health Perspect Suppl 109(3):395–403

Eiguren-Fernandez A, Kreisberg N, Hering S (2017) An online monitor of the oxidative capacity of aerosols (o-MOCA). Atmos Measur Tech 10:633–644. https://doi.org/10.5194/amt-10-633-2017

Fang T, Verma V, Bates JT, Abrams J, Klein M, Strickland MJ, Sarnat SE, Chang HH, Mulholland JA, Tolbert PE, Russell AG, Weber RJ (2016) Oxidative potential of ambient water-soluble PM2.5 in the south eastern United States: contrasts in sources and health associations between ascorbic acid (AA) and dithiothreitol (DTT) assays. Atmos Chem Phys 16(6):3865–3879

Fang T, Verma V, Guo H, King LE, Edgerton ES, Weber RJ (2015) A semi-automated system for quantifying the oxidative potential of ambient particles in aqueous extracts using the dithiothreitol (DTT) assay: results from the Southeastern Center for Air Pollution and Epidemiology (SCAPE). Atmos Measur Tech 8:471–482. https://doi.org/10.5194/amt-8-471-2015

Frampton MW, Ghio AJ, Samet JM, Carson JL, Carter JD, Devlin RB (1999) Effects of aqueous extracts of PM10 filters from the Utah Valley on human airway epithelial cells. Am J Physiol Lung Cell Mol Physiol 277:L960–L967

Gasser M, Riediker M, Mueller L, Perrenoud A, Blank F, Gehr P, Rothen-Rutishauser B (2009) Toxic effects of brake wear particles on epithelial lung cells in vitro. Particle Fibre Toxicol 6:30. https://doi.org/10.1186/1743-8977-6-30

Gao D, Fang T, Verma V, Zeng L, Weber RJ (2017) A method for measuring total aerosol oxidative potential (OP) with the dithiothreitol (DTT) assay and comparisons between an urban and roadside site of water-soluble and total OP. Atmos Measur Tech 10(8):2821–2835. https://doi.org/10.5194/amt-10-2821-2017

Geller MD, Ntziachristos L, Mamakos A, Samaras Z, Schmitz DA, Froines JR, Sioutas C (2006) Physicochemical and redox characteristics of particulate matter (PM) emitted from gasoline and diesel passenger cars. Atmos Environ 40(36):6988–7004. https://doi.org/10.1016/j.atmosenv.2006.06.018

Godri KJ, Duggan ST, Fuller GW, Baker T, Green D, Kelly FJ, Mudway IS (2010) Particulate matter oxidative potential from waste transfer station activity. Environ Health Perspect 118(4):493–498

Grevendonk L, Janssen BG, Vanpoucke C, Lefebvre W, Hoxha M, Bollati V, Nawrot TS (2016) Mitochondrial oxidative DNA damage and exposure to particulate air pollution in mother-newborn pairs. Environ Health 15(10):1–8

Guglielmotto M, Tamagno E, Danni O (2009) Oxidative stress and hypoxia contribute to Alzheimer's disease pathogenesis: two sides of the same coin. Sci World J 9:781–791

Hemmingsen JG, Jantzen K, Møller P, Loft S (2015) No oxidative stress or DNA damage in peripheral blood mononuclear cells after exposure to particles from urban street air in overweight elderly. Mutagenesis 30:635–642

Hellack B, Nickel C, Albrecht C, Kuhlbusch TAJ, Boland S, Baeza-Squiban AW, Schins RPF (2017) Analytical methods to assess the oxidative potential of nanoparticles: a review. Environ Sci Nano 4:1920–1934. https://doi.org/10.1039/c7en00346c

Hensley K, Floyd RA (2002) Reactive oxygen species and protein oxidation in aging: a look back, a look ahead. Arch BiochemBiophys 397(2):377–383

Hetland RB, Cassee FR, Låg M, Dybing E, Schwarze PE (2005) Cytokine release from alveolar macrophages exposed to ambient particulate matter: heterogeneity in relation to size, city and season. Particle Fibre Toxicol 2:4

Hetland RB, Refsnes M, Myran T, Johansen BV, Uthus N, Schwartz PE (2000) Mineral and/or metal content as critical determinants of particle-induced release of IL-6 and IL-8 from A549 cells. J Toxicol Environ Health A 60:47–65

Hiura TS, Kaszubowski MP, Li N, Nel AE (1999) Chemicals in diesel exhaust particles generate reactive oxygen radicals and induce apoptosis in macrophages. J Immunol 163:5582–5591

Hoek G, Beelen R, de Hoogh K, Vienneau D, Gulliver J, Fischer P, Briggs D (2008) A review of land-use regression models to assess spatial variation of outdoor air pollution. Atmos Environ 42(33):7561–7578

Hu S, Polidori A, Arhami M, Shafer MM, Schauer JJ, Cho A, Sioutas C (2008) Redox activity and chemical speciation of size fractioned PM in the communities of the Los Angeles-Long Beach harbor. Atmos Chem Phys 8(21):6439–6451

Hung H-F, Wang C-S (2001) Experimental determination of reactive oxygen species in Taipei aerosols. J Aerosol Sci 32:1201–1211. https://doi.org/10.1016/S0021-8502(01)00051-9

Jalava PI, Salonen RO, Pennanen AS, Sillanpää M, Hälinen AI, Happo, M.S., Hillamo, R., Brunekreef, B, Katsouyanni K, Sunyer J, Hirvonen MR (2007) Heterogeneities in inflammatory and cytotoxic responses of RAW 264.7 macrophage cell line to urban air coarse, fine, and ultrafine particles from six European sampling campaigns. Inhalation Toxicol 19:213–225

Janssen NAH, Hoek G, Simic-Lawson M, Fischer P, Bree L, Brink H, Keuken M, Atkinson RW, Anderson HR, Brunekreef B, Cassee FR (2011) Black carbon as an additional indicator of the adverse health effects of airborne particles compared with PM10 and PM2.5. Environ Health Perspect 119:1691–1699

Janssen NAH, Yang AL, Strak M, Steenhof M, Hellack B, Gerlofs-Nijland ME, Kuhlbusch T, Kelly F, Harrison RM, Brunekreef B, Hoek G, Cassee F (2014) Oxidative potential of particulate matter collected at sites with different source characteristics. Sci Total Environ 472:572–581

Jedynska A, Hoek G, Wang M, Yang A, Eeftens M, Cyrys J, Keuken M, Ampe C, Beelen R, Cesaroni G, Forastiere F, Cirach M, de Hoogh K, De Nazelle A, Nystad W, Akhlaghi HM, Declercq C, Stempfelet M, Eriksen KT, Dimakopoulou K, Lanki T, Meliefste K, Nieuwenhuijsen M, Yli-Tuomi T, Raaschou-Nielsen O, Janssen NAH, Brunekreef B, Kooter IM (2017) Spatial variations and development of land use regression models of oxidative potential in ten European study areas. Atmos Environ 150:24–32

Jung H, Guo B, Anastasio C, Kennedy IM (2006) Quantitative measurements of the generation of hydroxyl radicals by soot particles in a surrogate lung fluid. Atmos Environ 40:1043–1052. https://doi.org/10.1016/j.atmosenv.2005.11.015

Kelly FJ, Fussell JC (2012) Size, source and chemical composition as determinants of toxicity attributable to ambient particulate matter. Atmos Environ 60:504–526. https://doi.org/10.1016/j.atmosenv.2012.06.039

Kirillova EN, Andersson A, Tiwari S, Srivastava AK, Bisht DS, Gustafsson Ö (2014) Water-soluble organic carbon aerosols during a full New Delhi winter: isotope-based source apportionment and optical properties. J Geophys Res Atmos 119(6):3476–3485

Kleinman MT, Sioutas C, Froines JR, Fanning E, Hamade A, Mendez L, Meacher D, Oldham M (2007) Inhalation of concentrated ambient particulate matter near a heavily trafficked road stimulates antigen-induced airway responses in mice. Inhalation Toxicol 19:117–126. https://doi.org/10.1080/08958370701495345

Knaapen AM, Shi T, Borm PJ, Schins RP (2002) Soluble metals as well as the insoluble particle fraction are involved in cellular DNA damage induced by particulate matter. Mol Cell Biochem 234/235(1–2):317–326

Koehler KA, Shapiro J, Sameenoi Y, Henry C, Volckens J (2014) Laboratory evaluation of a microfluidic electrochemical sensor for aerosol oxidative load. Aerosol Sci Technol 48(5):489–497

Kubátová A, Dronen LC, Picklo MJ, Hawthorne SB (2006) Midpolarity and nonpolar wood smoke particulate matter fractions deplete glutathione in RAW 264.7 macrophages. Chem Res Toxicol 19:255–261. https://doi.org/10.1021/tx050172f

Kumagai Y, Koide S, Taguchi K, Endo A, Nakai Y, Yoshikawa T, Shimojo N (2002) Oxidation of proximal protein sulfhydryls by phenanthraquinone, a component of diesel exhaust particles. Chem Res Toxicol 15:483–489. https://doi.org/10.1021/tx0100993

Landreman AP, Shafer MM, Hemming JC, Hannigan MP, Schauer JJ (2008) Macrophage-based method for the assessment of the Reactive Oxygen Species (ROS) activity of atmospheric Particulate Matter (PM) and application to routine (daily-24 h) aerosol monitoring studies. Aerosol Sci Technol 42:946

Lehucher-Michel MP, Lesgards JF, Delubac O, Stocker P, Durand P, Prost M (2001) Oxidative stress and human disease. Current knowledge and perspective for prevention. Presse Med 30(21):1076–81

Li N, Sioutas C, Cho A, Schmitz D, Misra C, Sempf J, Wang M, Oberly T, Froines J, Nel A (2003) Ultrafine particulate pollutants induce oxidative stress and mitochondrial damage. Environ Health Perspect 111:455–460

Li N, Xia T, Nel AE (2008) The role of oxidative stress in ambient particulate matter-induced lung diseases and its implications in the toxicity of engineered nanoparticles. Free Radic Biol Med 44:1689–1699

Li Q, Shang J, Zhu T (2013) Physicochemical characteristics and toxic effects of ozone-oxidized black carbon particles. Atmos Environ 81:68–75. https://doi.org/10.1016/j.atmosenv.2013.08.043

Lu S, Feng M, Yao Z, Jing A, Yufang Z, Wu M (2011) Physicochemical characterization and cytotoxicity of ambient coarse, fine and ultrafine particulate matters in Shanghai atmosphere. Atmos Environ 45:736–744

Lu S, Longyi S, Minghong W, Jones TP, Merolla L, Richard RJ (2006) Correlation between plasmid DNA damage induced by PM10 and trace metals in inhalable particulate matters in Beijing air. Sci in China Series D: Earth Sci 49(12):1323–1331

Lundstedt S, White PA, Lemieux CL, Lynes KD, Lambert IB, Öberg L, Haglund P, Tysklind M (2007) Sources, fate, and toxic hazards of oxygenated polycyclic aromatic hydrocarbons (PAHs) at PAH- contaminated sites. AMBIO J Hum Environ 36:475–485. https://doi.org/10.1579/0044-7447(2007)36%5b475:sfatho%5d2.0.co,2

Magas OK, Gunter JT, Regens JL (2007) Ambient air pollution and daily pediatric hospitalizations for asthma. Environ Sci Pollut Res 14:19–23

McWhinney RD, Badali K, Liggio J, Li S-M, Abbatt JPD (2013) Filterable redox cycling activity: a comparison between diesel exhaust particles and secondary organic aerosol constituents. Environ Sci Technol 47:3362–3369. https://doi.org/10.1021/es304676x

Mehta M, Chen L, Gordon T, Rom W, Tang M (2008) Particulate matter inhibits DNA repair and enhances mutagenesis. Mutat Res 657:116–121

Michel TM, Pülschen D, Thome J (2012) The role of oxidative stress in depressive disorders. Curr Pharm Des 18(36):5890–5899

Mills NL, Tornqvist H, Gonzalez MC, Vink E, Robinson SD, Soderberg S, Boon NA, Donaldson K, Sandstrom T, Blomberg A, Newby DE (2007) Ischemic and thrombotic effects of dilute diesel-exhaust inhalation in men with coronary heart disease. N Engl J Med 357:1075–1082

Møller P, Folkmann JK, Forchhammer L, Brauner EV, Danielsen PH, Risom L, Loft S (2008) Air pollution, oxidative damage to DNA, and carcinogenesis. Cancer Lett 266:84–97

Mondal NK, Bhattacharya P, Ray MR (2011) Assessment of DNA damage by comet assay and fast halo assay in buccal epithelial cells of Indian women chronically exposed to biomass smoke. Int J Hyg Environ Health 214(4):311–318

Monn C, Becker S (1999) Cytotoxicity and induction of proinflammatory cytokines from human monocytes exposed to fine (PM2.5) and coarse particles (PM10–2.5) in outdoor and indoor air. Toxicol Appl Pharmacol 155:245–252

Mudway IS, Fuller G, Green D, Dunster C, Kelly FJ (2011) Report: quantifying the London specific component of PM10 oxidative activity. University of London, Defra

Mudway IS, Stenfors N, Duggan ST, Roxborough H, Zielinski H, Marklund SL, Blomberg A, Frew AJ, Sandström T, Kelly FJ (2004) An in vitro and in vivo investigation of the effects of diesel exhaust on human airway lining fluid antioxidants. Arch Biochem Biophys 423:200–212. https://doi.org/10.1016/j.abb.2003.12.018

Nagai H, Toyokuni S (2010) Biopersistent fiber induced inflammation and carcinogenesis: Lessons learned from asbestos toward safety of fibrous nanomaterials. Arch Biochem Biophys 502:1–7

Nel AE, Diaz-Sanchez D, Li N (2001) The role of particulate pollutants in pulmonary inflammation and asthma: evidence for the involvement of organic chemicals and oxidative stress. Curr Opin Pulm Med 7:20–26. https://doi.org/10.1097/00063198-200101000-00004

Nel AE (2005) Air pollution-related illness: effects of particles. Science 308(804–806):2005. https://doi.org/10.1126/science.1108752

Nel A, Xia T, Madler L, Li N (2006) Toxic potential of materials at the nanolevel. Science 311:622–627

Øvrevik J, Myran T, Refsnes M, Lag M, Becher R, Hetland RB (2005) Mineral particles of varying composition induce differential chemokine release from epithelial lung cells: importance of physicochemical characteristics. Ann Occup Hyg 49:219–231

Øvrevik J, Refsnes M, Låg M, Holme JA, Schwarze PE (2015) Activation of proinflammatory responses in cells of the airway mucosa by particulate matter: oxidant- and non-oxidant-mediated triggering mechanisms. Biomolecules 5(3):1399–1440. https://doi.org/10.3390/biom5031399

Pant P, Baker SJ, Shukla A, Maikawa C, Pollitt KJG, Harrison RM (2015) The PM10 fraction of road dust in the UK and India: characterization, source profiles and oxidative potential. Sci Total Environ 530:445–452

Park YJ, Lim L, Song H (2013) Distinct oxidative damage of biomolecules by arrays of metals mobilized from different types of airborne particulate matters, SRM1648, Fine (PM2.5), and Coarse (PM10) fractions. Environ Eng Res 18(3):139–143

Patel A, Rastogi N (2018) Oxidative potential of ambient fine aerosol over a semi-urban site in the Indo-Gangetic Plain. Atmos Environ 175:127–134

Perry G, Cash AD, Smith MA (2002) Alzheimer's disease and oxidative stress. J Biomed Biotechnol. 2:120–123

Puthussery JV, Zhang C, Verma V (2018) Development and field testing of an online instrument for measuring the real-time oxidative potential of ambient particulate matter based on dithiothreitol assay. Atmos Measur Tech 11(10):5767–5780. https://doi.org/10.5194/amt-11-5767-2018

Rahal A, Kumar A, Singh V, Yadav B, Tiwari R, Chakraborty S, Dhama K (2014) Oxidative stress, prooxidants, and antioxidants: the interplay. Biomed Res Int 761264

Rahman T, Hosen I, Islam MMT, Shekhar HU (2012) Oxidative stress and human health. Adv Biosci Biotechnol 3:997–1019

Rosanna DP, Salvatore C (2012) Reactive oxygen species, inflammation, and lung disease. Curr Pharm Des 18:3889–3900

Saffari A, Daher N, Shafer MM, Schauer JJ, Sioutas C (2014) Seasonal and spatial variation in dithiothreitol (DTT) activity of quasi-ultrafine particles in the Los Angeles Basin and its association with chemical species. J Environ Sci Health, Part A: Toxic/Hazard Subst Environ Eng. 49(4):441–451

Saffari A, Hasheminassab S, Shafer MM, Schauer JJ, Chatila TA, Sioutas C (2016) Nighttime aqueous-phase secondary organic aerosols in Los Angeles and its implication for fine particulate matter composition and oxidative potential. Atmos Environ 133:112–122

Saffari A, Daher N, Samara C, Voutsa D, Kouras A, Manoli E, Karagkiozidou O, Vlachokostas C, Moussiopoulos N, Shafer MM, Schauer JJ, Sioutas C (2013) Increased biomass burning due to the economic crisis in Greece and its adverse impact on wintertime air quality in Thessaloniki. Environ Sci Technol 47:13313–13320

Sameenoi Y, Koehler K, Shapiro J, Boonsong K, Sun Y, Collett J, Volckens J, Henry CS (2012) Microfluidic electrochemical sensor for on-line monitoring of aerosol oxidative activity. J Am Chem Soc 134:10562–10568. https://doi.org/10.1021/ja3031104

Sameenoi Y, Panymeesamer P, Supalakorn N, Koehler K, Chailapakul O, Henry CS, Volckens J (2013) Microfluidic paper-based analytical device for aerosol oxidative activity. Environ Sci Technol 47(2):932–940

Schins RP, Lightbody JH, Borm PJ, Shi T, Donaldson K, Stone V (2004) Inflammatory effects of coarse and fine particulate matter in relation to chemical and biological constituents. Toxicol Appl Pharmacol 195:1–11

Schwarze PE, Øvrevik J, Låg M, Refsnes M, Nafstad P, Hetland RB, Dybing E (2006) Particulate matter properties and health effects: consistency of epidemiological and toxicological studies. Hum ExpToxicol 25:559–579

Shi T, Schins RP, Knaapen AM, Kuhlbusch T, Pitz M, Heinrich J, Borm PJ (2003) Hydroxyl radical generation by electron paramagnetic resonance as a new method to monitor ambient particulate matter composition. J Environ Monit JEM 5:550–556. https://doi.org/10.1039/B303928P

Simonetti G, Conte E, Massimi L, Frasca D, Perrino C, Canepari S (2018) Oxidative potential of particulate matter components generated by specific emission sources. J Aerosol Sci 126:99–109. https://doi.org/10.1016/j.jaerosci.2018.08.011

Squadrito GL, Cueto R, Dellinger B, Pryor WA (2001) Quinoid redox cycling as a mechanism for sustained free radical generation by inhaled airborne particulate matter. Free Radic Biol Med 31:1132–1138

Tuet WY, Chen YL, Xu L, Fok S, Gao D, Weber RJ, Ng NL (2017) Chemical oxidative potential of secondary organic aerosol (SOA) generated from the photooxidation of biogenic and anthropogenic volatile organic compounds. Atmos Chem Phys 17(2):839–853

Uttara B, Singh AV, Zamboni P, Mahajan RT (2009) Oxidative stress and neurodegenerative diseases: a review of upstream and downstream antioxidant therapeutic options. Curr Neuropharmacol 7:65–74

Valgimigli M, Valgimigli L, Trere D, Gaiani S, Pedulli GF, Gramantieri L, Bolondi L (2002) Oxidative stress EPR measurement in human liver by radical-probe technique. Correlation with etiology, histology and cell proliferation. Free Radic Res 36:939–948

Valavanidis A, Vlachogianni T, Fiotakis K, Loridas S (2013) Pulmonary oxidative stress, inflammation and cancer: respirable particulate matter, fibrous dusts and ozone as major causes of lung carcinogenesis through reactive oxygen species mechanisms. Int J Environ Res Public Health 10:3583–3590

Valavanidis A, Vlachogianni T, Fiotakis K (2009) Tobacco smoke: involvement of reactive oxygen species and stable free radicals in mechanisms of oxidative damage, carcinogenesis and synergistic effects with other respirable particles. Int J Environ Res Public Health 6(2):445–462

Valko M, Morris H, Cronin MT (2005) Metals, toxicity and oxidative stress. Curr Med Chem 12:1161–1208. https://doi.org/10.2174/0929867053764635

Valko MD, Leibfritz J, Moncol MTD, Cronin M, Mazur M, Telser J (2007) Free radicals and antioxidants in normal physiological functions and human disease. Int J Biochem Cell Biol 39(1):44–84

Velali E, Papachristou E, Pantazaki A, Choli-Papadopoulou T, Planou S, Kouras A, Manoli E, Besis A, Voutsa D, Samara C (2016) Redox activity and in vitro bioactivity of the water-soluble fraction of urban particulate matter in relation to particle size and chemical composition. Environ Pollut 208:774–786

Venkatachari P, Hopke PK, Grover BD, Eatough DJ (2005) Measurement of particle-bound reactive oxygen species in rubidoux aerosols. J Atmos Chem 50:49–58. https://doi.org/10.1007/s10874-005-1662-z

Verma MK, Poojan S, Sultana S, Kumar S (2014) Mammalian cell-transforming potential of traffic-linked ultrafine particulate matter PM0.056 in urban roadside atmosphere. Mutagenesis 29(5):335–340

Verma V, Ning Z, Cho AK, Schauer JJ, Shafer MM, Sioutas C (2009) Redox activity of urban quasi-ultrafine particles from primary and secondary sources. Atmos Environ 43(40):6360–6368

Verma V, Pakbin P, Cheung KL, Cho AK, Schauer JJ, Shafer MM, Kleinman MT, Sioutas C (2011) Physicochemical and oxidative characteristics of semi-volatile components of quasi-ultrafine particles in an urban atmosphere. Atmos Environ 45(4):1025–1033

Verma V, Rico-Martinez R, Kotra N, King LE, Liu J, Snell TW, Weber RJ (2012) Contribution of water-soluble and insoluble components and their hydrophobic/hydrophilic subfractions to the reactive oxygen species-generating potential of fine ambient aerosols. Environ Sci Technol 46:11384–11392. https://doi.org/10.1021/es302484r

Verma V, Wang Y, El-Afifi R, Fang T, Rowland J, Russell AG, Weber RJ (2015) Fractionating ambient humic-like substances (HULIS) for their reactive oxygen species activity—assessing the importance of quinones and atmospheric aging. Atmos Environ 120:351–359

Vidrio E, Jung H, Anastasio C (2008) Generation of hydroxyl radicals from dissolved transition metals in surrogate lung fluid solutions. Atmos Environ 42(18):4369–4379

Vreeland H, Schauerb JJ, Russellc AG, Marshalld JD, Fushimib A, Jainf G, Sethuramanf K, Verma V, Tripathi SN, Bergina MH (2016) Chemical characterization and toxicity of particulate matter emissions from roadside trash combustion in urban India. Atmos Environ 147:22–30

Vreeland H, Weber R, Bergin M, Greenwald R, Golan R, Russell AG, Verma V, Sarnat JA (2017) Oxidative potential of PM2.5 during Atlanta rush hour: measurements of in-vehicle dithiothreitol (DTT) activity. Atmos Environ 165:169–178

Wagner AP, Mitran S, Sivanesan S, Chang E, Buga AM (2013) ROS and brain diseases: the good, the bad, and the ugly. Oxidative Med Cell Longevity. Article ID 963520, 14 pp. http://dx.doi.org/10.1155/2013/963520

Weichenthal S, Crouse DL, Pinault L, Godri-Pollitt K, Lavigne E, Evans G, van Donkelaar A, Martin RV, Burnett RT (2016) Oxidative burden of fine particulate air pollution and risk of cause-specific mortality in the Canadian Census Health and Environment Cohort (CanCHEC). Environ Res 146:92–99. https://doi.org/10.1016/j.envres.2015.12.013

Wild P, Bourgkard E, Paris C (2009) Lung cancer and exposure to metals: the epidemiological evidence. Method Mol Biol 472:139–167

Xia T, Kovochich M, Brant J, Hotze M, Sempf J, Oberley T, Sioutas C, Yeh JI, Wiesner MR, Nel AE (2006) Comparison of the abilities of ambient and manufactured nanoparticles to induce cellular toxicity according to an oxidative stress paradigm. Nano Lett 6:1794–1807. https://doi.org/10.1021/nl061025k

Xiao Z, Shao L, Zhang N, Wang J, Wang J (2013) Heavy metal compositions and bioreactivity of airborne PM10 in a valley-shaped city in North western China. Aerosol Air Qual Res 13:1116–1125

Xiong QS, Yu HR, Wang RR, Wei JL, Verma V (2017) Rethinking dithiothreitol-based particulate matter oxidative potential: measuring dithiothreitol consumption versus reactive oxygen species generation. Environ Sci Technol 51(11):6507–6514

Yadav S, Jan R, Roy R, Satsangi PG (2016) Role of metals in free radical generation and genotoxicity induced by airborne particulate matter (PM2.5) from Pune (India). Environ Sci Poll Res 23:23854–23866

Yang A, Janssen NAH, Brunekreef B, Cassee FR, Hoek G, Gehring U (2016) Children's respiratory health and oxidative potential of PM2.5: the PIAMA birth cohort study. Occup Environ Med 73(3):154–160

Yang A, Jedynska A, Hellack B, Kooter I, Hoek G, Brunekreef B, Kuhlbusch TAJ, Cassee FR, Janssen NAH (2014) Measurement of the oxidative potential of PM2.5 and its constituents: the effect of extraction solvent and filter type. Atmos Environ 83:35–42. http://dx.doi.org/10.1016/j.atmosenv.2013.10.049

Yang A, Wang M, Eeftens M, Beelen R, Dons E, Leseman D, Brunekreef B, Cassee FR, Janssen NAH, Hoek G (2015) Spatial variation and land use regression modeling of the oxidative potential of fine particles. Environ Health Perspect 123(11):1187–1192

Yang J, Xu ZP, Huang Y, Hamrick HE, Duerksen-Hughes PJ, Yu YN (2004) ATM and ATR: sensing DNA damage. World J Gastroenterol 10:155–160

Zielinski H, Mudway IS, Berube KA, Murphy S, Richards R, Kelly FJ (1999) Modeling the interactions of particulates with epithelial lining fluid antioxidants. Am J Physiol Lung Cell Mol Physiol 277:L719–L726

Zomer B, Collé L, Jedyńska A, Pasterkamp G, Kooter I, Bloemen H (2011) Chemiluminescent reductive acridinium triggering (CRAT)—mechanism and applications. Anal Bioanal Chem 401:2945–2954. https://doi.org/10.1007/s00216-011-5342-3

Chapter 17
Monitoring and Processing of Data for Effective Wasteload Allocation Modeling in India

Dipteek Parmar and A. K. Keshari

Abstract Many rivers, especially in developing countries are getting polluted because of increased waste load emanating from industrial and urban sectors. Although many pollution abatement efforts have been taken up but no comprehensive effluent standards based on waste load allocation modeling (WLA) have been developed and implemented to control the river pollution. This chapter presents the overview of this method and the related impediments such as data collection, processing and its use to facilitate WLA modeling in India. It begins with the concept and components of waste load allocation modeling. The data required to carry out WLA modeling using QUAL2E simulation tool is also discussed. Finally it concludes with an emphasis on the need for a coordinated effort involving the policy makers, scientists, engineers and the academia. It is opined that incorporating the requisite data collection and processing in routine functioning by government agencies will pave the way for developing robust WLA models and facilitate state of the art research in this field. This eventually would lead us towards development of realistic pollution control management plans for the ailing rivers.

Keywords Waste load allocation · Simulation · Optimization · Cost function · Assimilative capacity · Effluent

17.1 Background

With the increase in population and consequent urbanization, the rivers along urban centers, especially in the developing countries are getting polluted because of discharge of untreated and partially treated wastewater. Several water quality management plans have been developed and implemented in last three decades but they

D. Parmar (✉)
Department of Civil Engineering, Harcourt Butler Technical University, Kanpur, UP 208002, India
e-mail: d_parmar@rediffmail.com

A. K. Keshari
Department of Civil Engineering, Indian Institute of Technology Delhi, New Delhi 110016, India
e-mail: akeshari@hotmail.com

T. Gupta et al. (eds.), *Measurement, Analysis and Remediation of Environmental Pollutants*, Energy, Environment, and Sustainability,
https://doi.org/10.1007/978-981-15-0540-9_17

haven't achieved desired results. This is because these plans have largely been based on unscientific and arbitrary ethics and approaches Keshari and Parmar (2006). The Waste load allocation (WLA) is a method of pollution control in rivers based on comprehensive scientific approach in a cost efficient manner. Unfortunately it has rarely been implemented in India. Two major reasons for WLA modeling not gaining much popularity especially in India are: (a) quantum/non availability of data and (b) lack of technical knowhow available with the agencies dealing with pollution control. In fact, the main thrust has been given on water quality monitoring (Goldar and Banerjee 2004) and experimental investigations (Jain 2004). **This monitoring is too simple and confined to few parameters only. However, some** studies carried out in the last decade have used models for simulating the river water quality (Ghosh 1996; Dikshit et al. 2000; Gupta et al. 2004; Babu et al. 2005). During the same period, studies have used simulation models for assessing the impact of various pollution abatement techniques (Kazmi and Hansen 1997; Kazmi and Agrawal 2005; Dhage et al. 2006; Paliwal et al. 2007; Parmar and Keshari 2012, 2018). The last decade has seen very few studies on WLA modeling (Mujumdar and Subbharao 2004; Subharao et al. 2004; Yandamuri et al. 2006; Singh et al. 2007; Parmar and Keshari 2014). It implies that the complexities involved in development of wasteload allocation model continue to bother environmental engineers and planners. This limits extensive use of these models to real life field problems especially in India. What could be done in such a situation? Can we, the engineers, scientists and the academia do something to address this problem? If yes-then how? That's the premise of this chapter. It is aimed at presenting the overview of implementation of WLA modeling with focus on data collection, processing and steps involved in simulation of water quality and optimal solutions.

The chapter is organized as follows: It begins with the present section on background. This is followed by the concept and components of waste load allocation modeling. The data required for WLA modeling is also discussed. This includes: water quality at check points in space and time, assimilative capacity of river, dispersion, reaeration, rate constants, sources, sinks, incremental flow in reaches, hydraulic routing- velocity, depth, width and corresponding discharge, data required for Initial and boundary conditions, calibration and validation, discharge, BOD loadings of all the drains, stream flow and temperature conditions for low flow conditions from the summer period etc. Thereafter the data required for optimization model is discussed. Lastly, the chapter concludes with recommendations for coordinated efforts amongst different stakeholders and multiple agencies.

17.2 Waste Load Allocation Modeling

The waste load allocation (WLA), deals with the process of finding the optimal BOD removal by number of treatment plants located along a rivers by number of sewage treatment plants located along a river to attain satisfactory water quality in the river at minimum cost (Burn and Yulianti 2001; Karmakar and Mujumdar 2004).

The model is solved using some form of optimization model in which the decision variables are the treatment levels at each of the treatment plants expressed as a BOD removal. The minimization of cost function is the objective function while the water quality response obtained from a simulation model constitutes the constraints. The classical single objective WLA model implies that the total cost (including operation and maintenance) for treating the wastewater before discharging into the river is minimized subject to compliance of the water quality standards in various reaches of the river as laid down by the Pollution Control agency (PCA). Figure 17.1 presents the methodological Framework of a Wasteload Allocation Model.

17.2.1 Components

17.2.1.1 Simulation Model

A simulation model attempts to represent the physical functioning and consequent effects of causative factors on the prototype system by a computerized algorithm. In a WLA model, a simulation model provides the quantitative formulation of the cause-and-effect relationship, where effluent quality is the cause and receiving water degradation is the effect. In river water quality management problems, this cause-effect relationship is usually considered in the form of a transfer coefficient. The finite volume transport model provides such a formulation. The finite volume model relates the steady-state transfer function, φ_{ij}, to the dissolved oxygen response in any volume to the BOD input in any other volume (Liebman 1972). Mathematically, it can be defined as the effect of a unit change in waste treatment at a particular discharge point on the water quality at another point in downstream reaches (Liebman 1972; Burn 1989). The outcome of this will be a transfer coefficient matrix: $A = \left| i\ x\ j \right|$, where i is the number of dischargers (i = 1, 2, …, N_j) and j = no. of locations (j = 1, 2, …, m) in the river stretch.

The transfer coefficients are used as coefficients of the decision variable in the linear constraints of the WLA model. The constraint of optimization involving the transfer coefficient, φ_{ij}, the increase in treatment r_i at discharge point i (i = 1, 2, 3,…,N_j), and the desired improvement (difference between DO standard and existing DO without any treatment) of water quality in terms of dissolved oxygen at the check points, b_j, can be written as:

$$\varphi_{1j}r_1 + \varphi_{2j}r_2 + \cdots + \varphi_{N_j j}r_{N_j} \geq b_j \tag{17.1}$$

Equations of the form (17.1) are written for (j − 1) reaches. The formulation assumes a steady-state condition, linearity and additivity of downstream water-quality improvement with respect to upstream waste reductions by wastewater treatment plants.

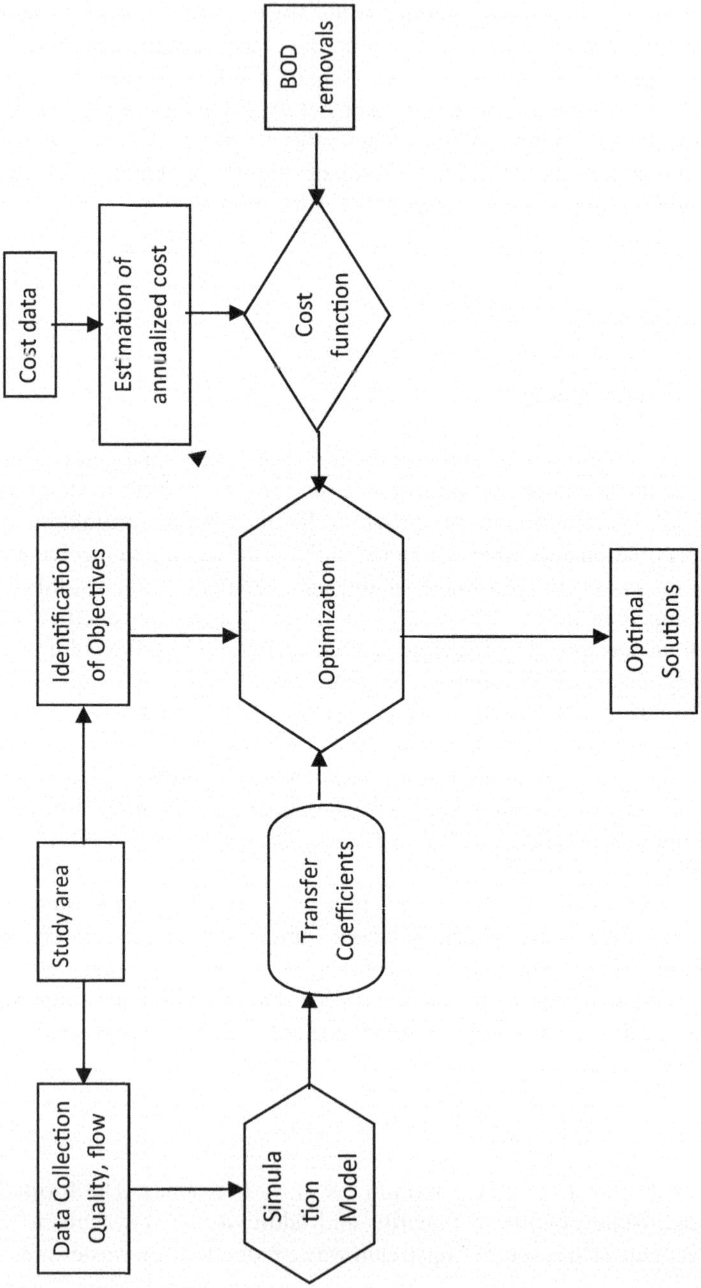

Fig. 17.1 Methodological framework of wasteload allocation model

17.2.1.2 Optimization Model

Majority of the optimization models dealing with river water quality management are multiobjective in nature because they include the water system (quantity and quality), an economic system (cost of wastewater treatment), and a decision making system (equity) (Cohon 1978; Lee and Wen 1996a). This implies that the goals of various bodies involved in decision making on water quality management plans are often non-commensurable and conflicting (Haimes and Hall 1975; Parmar 2006). However, planners have traditionally used a single objective of maximization of economic efficiency i.e. minimization of cost of wastewater treatment for solving waste load allocation problems. The methodology for development of the cost function is discussed below.

Cost function for wastewater treatment

Cost functions are important component of a WLA model as it forms the objective function of the optimization model (Smeers and Tyteca 1984). In a conventional cost curve, the cost is a function of the design flow rate only (Joeres et al. 1974). But the cost function normally used in WLA modeling is considers cost to be a function of BOD removal (Liebman 1972; Anderson and Day 1968; Smith and Morris 1969; Hass 1970; Ecker 1975; McNamara 1976; Burn and McBean 1985; Fujiwara et al. 1988; Burn 1989; Lee and Wen 1996a, b, 1997; Wen and Lee 1998; Burn and Yulianti 2001; Lee and Chang 2005; Yandamuri et al. 2006). This is because the water quality response obtained by a simulation model is in terms of BOD removal or DO improvement. It implies that the cost to be used for developing the function is the wastewater treatment cost. It is pertinent to note that due to inflation, costs are not constant over time. Thus, annualized costs are used to develop the cost functions. This cost comprises of amortized construction/capital cost and annual operation and maintenance cost. A typical cost curve for secondary treatment is of power form.

17.3 Data Requirement

The data requirements for the QUAL2E, water quality simulation model developed by USEPA has been detailed here for use in WLA modeling. Since a WLA model comprises of cost function, simulation model and optimization, the data sets as required for each of them is presented in Table 17.1.

17.4 Data Processing for Simulation Model

As discussed in Sect. 17.2.1.1, simulation models are used to generate the water quality response in space and time. Several water quality simulation models such as QUAL2E (USEPA 1995), WASP (Ambrose et al. 1993) and QUAL2 K (Pelletier et al.

Table 17.1 Details of data required for WLA modeling

Model	Broad data type	Specific/detailed data sets	Source of secondary data
Cost function	Cost data	Capital cost and year of construction of wastewater treatment plants	Respective state government department dealing with water supply and sewerage, National River Conservation Directorate (NRCD), National Mission for Clean Ganga (NMCG), CPCB
	Depreciation rate	Depreciation rate for plants	Economic Survey, GoI
	Operation and maintenance cost	Annual Operation and maintenance cost of all the plants	Respective state government department dealing with water supply and sewerage, NRCD, NMCG
	Annualized cost	Annualized cost for all plants	To be estimated by the user
	Annualized cost and corresponding BOD removal	Annualized cost and corresponding BOD removal of all plants	Central Pollution Control Board (CPCB); Respective state government department dealing with water supply and sewerage, NRCD, NMCG
Simulation model	Reach configuration	Chainage, length, size of computational element etc.	CPCB, literature or User/researcher defined
	Data for hydraulic routing	Velocity, depth, width of river and its corresponding discharge (sufficient data sets for all reaches so as to enable development of empirical equations for use in hydraulic routing)	Literature (Central Water Commission (CWC), CPCB or User/researcher defined

(continued)

Table 17.1 (continued)

Model	Broad data type	Specific/detailed data sets	Source of secondary data
	Initial Conditions for all reaches (during calibration, validation)	Flow, depth, BOD, DO, Temperature, mean monthly flow	CPCB, Literature or User/researcher defined
	Boundary conditions (during calibration and validation)	Flow, depth, BOD, DO, Temperature, mean monthly flow	CPCB, Literature or User defined
	Rate constants	K_1, K_2, K_3, K_4 for all river reaches	CPCB, Literature or User/researcher defined/estimated
	Geometric and hydraulic data	length, width, depth, discharge and velocity of all reaches	CPCB, CWC or Literature or User/researcher defined
	Point load (during calibration and validation) in form of dischargers/treatment plants located along river	BOD load of drains/discharges, discharge, BOD, DO, Temperature, percentage BOD removal by treatment plants located along the river	CPCB/State Pollution Control Board (SPCB), Literature or User/researcher defined
	Point withdrawal from river system	BOD load of drains/discharges, discharge, BOD, DO, Temperature, percentage BOD removal by treatment plants located along the river	CPCB/SPCB, Literature or User/researcher defined
Optimization model	Water quality standard	Water quality standard for all reaches (as per designated best use criteria) for the study stretch	Use defined
	Bounds of decision variable	Upper and lower bounds of BOD removal to be decided depending upon the treatment technology available	User defined

2006) are available these days. In this chapter, the methodology and data required for QUAL2E model is summarized under various steps and presented in the following sub-sections (USEPA 1995).

17.4.1 Conceptual Representation of a River System

The conceptual representation of a stream used in QUAL2E involves dividing the river system into reaches having uniform hydraulic characteristics, source/sink, and biochemical reaction. The length of various reaches has to be chosen and chainage of drains decided/adjusted in such a manner that it fits the QUAL2E requirement of equal size of computational elements for all the reaches. These computational elements are categorized into seven types. Details of these elements are presented elsewhere (USEPA 1995; Parmar 2006). The incremental flow can either be quantified for all reaches and fed in the software or it can be assumed zero. Thereafter the mass balance equations are solved for a steady state condition by using the implicit backward difference method.

17.4.2 Hydraulic Routing of River Flow

Hydraulic routing using QUAL2E requires developing equation of power form relating the discharge with the velocity, depth and width for each of the reaches. These equations relating velocity (v), depth (h), and width (w) with flow (Q) are written as follows:

$$V = aQ^b \tag{17.2a}$$

$$h = cQ^d \tag{17.2b}$$

$$w = eQ^f \tag{17.2c}$$

$$a \cdot c \cdot e = 1 \tag{17.2d}$$

$$b + d + f = 1 \tag{17.2e}$$

where,

$a,\ c,\ e$ velocity, depth, and width coefficients, respectively.
$b,\ d,\ f$ velocity, depth, and width exponents, respectively.

Table 17.2 Power equations relating mean velocity and depth to the flow

Reach No.	Velocity-discharge relation	Depth-discharge relation
1	$V = 0.0396\,Q^{0.5138}$	$h = 0.4411\,Q^{0.3374}$

Data related to velocity, discharge, depth, width and corresponding discharge need to be collected and empirical relationships of forms (17.2a) and (17.2b) developed for all the reaches. A sample of these equations is given in Table 17.2.

17.4.3 Governing Equations and Solution Technique

The basic equation solved by QUAL2E is the one-dimensional advection-dispersion mass transport equation. For any constituent, C, this equation can be written as (Brown and Barnwell 1987; Barnwell et al. 2004):

$$\frac{\partial C}{\partial t} = \frac{\partial\left(A_x D_l \frac{\partial C}{\partial x}\right)}{A_x \partial x} - \frac{\partial(A_x\, V\, C)}{A_x \partial x} + \frac{dC}{dt} + \frac{s}{V} \tag{17.3}$$

where

- M mass;
- x distance;
- t time;
- C concentration;
- A_x cross sectional area
- D_l dispersion coefficient;
- V mean velocity,
- s source/sink.

The first, second, third, and fourth *terms* on the right hand side of Eq. (17.3) represent dispersion, advection, constituent changes, and external sources/sinks and dilution, respectively. Under steady-state conditions, the local derivative $\frac{\partial C}{\partial t} = 0$; in other words, changes that occur to individual constituents are defined by the term, $\frac{dC}{dt}$, which is the individual constituent's change. These changes include the physical, chemical, and biological reactions and interactions which occur in a stream. Examples of these changes are reaeration, algal respiration, and photosynthesis. For each constituent, Eq. (17.7) is written i times, once for each of the i computational elements in the given stretch. The finite difference method utilizing the implicit backward difference scheme is used for solution of the governing equation.

17.4.4 Initial and Boundary Conditions

There are four types of data required for water quality modeling. These are: (a) initial conditions (b) boundary conditions (c) data for calibration and (d) data for validation. The initial conditions are the data pertaining to water quality conditions at the beginning of the simulation period (McCutcheon 1989). The flow, depth and all the water quality parameters to be simulated are to be specified for all the reaches. Normally for WLA modeling, data pertaining to BOD, DO and temperature is needed. However, other conservative minerals such as TDS, Chlorides, toxicants, and particulate form of nutrients can be also considered. The flow data during the study period is also given as the initial condition.

The flow and BOD loads entering the reaches (in form of drains) and other point/nonpoint sources that enter at diffused locations along the study stretch being simulated (McCutcheon 1989) forms the boundary conditions. In most real life problems, transport is unidirectional in nature, i.e. there is no significant transport upstream. Therefore, the concentration at some point just upstream from the beginning or end of the stream reach of interest can be used as the boundary condition.

17.4.5 Rate Constants

The health of a water body is governed by the presence of dissolved oxygen. *There are various sources and sinks of DO in a water body. The water quality modeling requires information about the rates of these sources and sinks. These rates are represented in the form of constants known as reaction rate constants. These are:* K_1 *(deoxygenation constant),* K_2 *(reaeration constant),* K_3 *(BOD settling rate) and* K_4 *(Sediment oxygen demand rate).*

The deoxygenation constant K_1 *(BOD decay rate) is the rate of biochemical decomposition of organic matter and can be obtained using* Eq. (17.8) (Ortolano 1984) *given below.*

$$K_1 = \frac{1}{x} \ln \frac{L_0}{L} \cdot V \tag{17.4}$$

where

L_o BOD at beginning of a reach
L BOD at end of the reach
V average velocity of flow in the reach
x distance from beginning to end of the reach.

Alternatively, the deoxygenation coefficient can be found by drawing a plot between BOD (on a semilog scale) and travel time (linear scale) and finding its slope. For municipal waste discharges the BOD rate, K_1, sometimes called a "bottle rate", would be expected to depend on the degree of treatment, that is, as the degree of treatment increases the residual material is increasingly more resistant to treatment; and, hence, the rate decreases. It should be noted that the "bottle rate", K_1 in general not equal to the deoxygenation rate for the BOD that occurs in natural water. Although for deep bodies of water, K_1 is a useful first approximation to the deoxygenation rate (Thomann and Muellar 1987).

Another rate constant is the reaeration rate coefficient (K_2) which is important for waste load allocation studies. It is the rate at which oxygen enters the water from the atmosphere (Jha et al. 2000). O'Connor and Dobbins (1958) formulation for the reaeration coefficient is

$$K_2 = \frac{K_L}{H} = \left(\frac{D_L U}{H^3}\right)^{1/2} \tag{17.5}$$

where D_L is the oxygen diffusivity at 20 °C, U is the average stream velocity, and H is the average depth. For a particular stretch of a river, the depth, H, is taken as the ratio of the volume to surface area. If U is in fps and H in feet; then for the reaeration coefficient in day^{-1} is given by,

$$K_2 = \frac{12.9U^{1/2}}{H^{3/2}} \tag{17.6}$$

British workers measured reaeration rates in several streams in England under controlled conditions where a sulfite dosing of the stream produced a transient well defined drop in DO. The time rate of recovery is a measure of reaeration rate. Owens et al. (1964) combined the British data with the data collected on the Tennessee Valley derived the following equation,

$$K_a = \frac{21.6U^{0.67}}{H^{1.85}} \tag{17.7}$$

Owens suggests this formula for the velocity range of 0.01–5.0 fps and depth range of 0.4–11.0 ft. The values of K_a computed from this formulation do not differ significantly from the O'Connor-Dobbins equation. In practice, the O'Connor-Dobbins formulation has formed a reasonable basis for estimating the reaeration coefficients although for small streams it appears to underestimate the coefficients. For small streams (i.e. flows less than about 10 cfs and depths less than about 1–3 ft), the work of Tsivolgou and Neal (1976) appears to compare well to observed reaeration rates. Note, however, for some of the equations K_2 approaches zero as the depth increases. This apparently implies that reaeration does not occur or is very low for deep bodies of water.

For DO modelling analyses, two courses of action are open: (a) calculation of the sensitivity of DO response to varying K_a values computed from different formulas or, (b) if the problem warrants, direct measurement of the reaeration using various field techniques.

Some recent studies (Kazmi and Agrawal 2005; Paliwal et al. 2007)*on the Delhi stretch have used the* O'Connor and Dobbins (1958)*equation for* K_2*estimation. However,* Melching and Flores (1999)*realized that available equations for* K_2 *estimation generally yield poor estimates when applied to stream conditions different from those for which they were derived.* Thus, on the basis of data obtained for Delhi stretch of river Yamuna, Parmar and Keshari (2012) developed the following predictive equation for calculating K_2.

$$K_2 = 4.27\frac{(V)^{0.47}}{(H)^{2.09}} \tag{17.8}$$

17.4.6 Calibration and Validation of QUAL2E

Calibration is one of the most important steps of modeling. A model has to be calibrated before it is put into use for simulation. It is accomplished by finding the model parameters during models runs, until the goodness of fit between predicted values and observed data is achieved. It is important to mention that QUAL2E is basically an indeterminate model and may not simulate water quality in rivers other than the ones for which it was developed. Indeterminate model means a model, which yields similar results under various combinations of model parameters. For example, if the simulated BOD compares well with the observed BOD under given set of value of model parameters, such as K_1 and K_3, the same simulation results can be obtained under different combinations of K_1 and K_3. When such indeterminately calibrated models are applied to the wastewater treatment scenarios, the model results become very sensitive to the indeterminately determined "calibrated" parameters. This is because of the fact that such models use a particular equation for finding the model parameters. These equations may not yield reliable results when applied to rivers different from the ones for which they were developed. For example, QUAL2E uses the O'Connor and Dobbins (1958) equation to estimate the value of K_2. This equation may not necessarily be the best to use for every river. Thus, the best way to deal with such indeterminate models is to measure as many parameters as possible in the field.

It may be noted that QUAL2E simulates ultimate BOD assuming first order kinetics and decay rate that can be specified by the user for each reach. Temperature correction factors are used in QUAL2E, since most processes are modeled using model coefficients that are temperature dependent. These factors are commonly accepted values from the scientific literature (USEPA 1995). Thereafter, using the velocity, flow, width, and depth obtained from the survey data, the exponent equations of the form of Eqs. (17.2a)–(17.2c) are obtained. The details of flow and water quality of the various drains discharging into the river are given as input during calibration. Once it is ensured that the model output (in terms of DO, BOD and temperature) matches the observed data, the model is considered to be calibrated. After calibration, the model has to be validated using datasets of different period. Only observed inputs are changed for the validation run. The initial conditions are also needed for validation. Once the model is calibrated and validated, and its performance is judged to be acceptable, the model can be used to carry out simulation.

17.5 Data Processing for Optimization Model

17.5.1 Objective Functions for the Proposed WLA

The first step in the development of a WLA model is deciding the objective functions to be used in the model. The most common objective function in WLA studies is the minimization of wastewater treatment cost. The cost minimization objective function implies that the total cost (including operation and maintenance) for treating the wastewater at STPs before discharging into the river is minimized. The cost is assumed to be a function of BOD removal. The function comprises of regression relation between annualized cost and BOD removal of the STP. Such relationships are required for all the proposed treatment plants. Thus, the overall cost function comprises of the summation of cost functions for all the plants located along the river stretch under consideration.

17.5.2 Formulation of WLA Model

In general, there is a Pollution Control agency (PCA) whose role is to look after the treatment of wastewater before discharging the effluent into the river in a manner such that water quality of the river is within permissible limits as per norms (Haimes et al. 1975). WLA The model considers determination of the optimal BOD removal at each treatment plant in order to minimize the total wastewater treatment cost subject to satisfying the water quality standard in all the river reaches. The decision variables, the removal fractions at each treatment plants are obtained from the feasible range. The model can be expressed as follows in Eqs. (17.9)–(17.11).

$$Minimize\, F_1 = \sum_{i=1}^{NS} C_i(r_i) \tag{17.9}$$

Subject to

(a) Water quality improvement constraints

$$\sum_{i=1}^{N_j} \varphi_{ij} r_i \geq b_j \; \forall \, i = 1, 2, \ldots, N_j \qquad \forall \, j = 1, 2, \ldots, m \tag{17.10}$$

(b) Inequality constraints (BOD removal constraints or bounds)

$$r_i^{(L)} \leq r_i \leq r_i^{(U)}, \quad \forall \, i = 1, 2, \ldots, NS \tag{17.11}$$

where,

F_1	Objective function for cost of wastewater treatment in the entire study stretch
C_i	Cost of wastewater treatment by plant i
r_i	BOD removal by the ith plant
NS	Total number of sources/treatment plants/dischargers
i	source number
j	reach number
m	total number of reaches
N_j	Number of dischargers upstream of j (highest i upstream of j).
b_j	Improvement in dissolved oxygen at point j due to removal r_i of BOD at all sources upstream of j.
φ_{ij}	transfer coefficient (It estimates the improvement in DO at j per unit increase in treatment or reduction in BOD load at i)
$r_i^{(L)}$	minimum allowable treatment level/lower bound.
$r_i^{(U)}$	maximum allowable treatment level/upper bound.

Equation (17.9) defines the objective representing total cost of wastewater treatment. As discussed earlier in Sect. 2.1.2. Equation (17.11) represents the feasible ranges (bounds) for the decision variables. The basic question answered by classical WLA model is: what optimal combination of waste effluent modifications along the river should be made to secure desired DO goals at least overall cost for the region? The optimum solution consists of those values of r_i that minimize the objective functions, i.e. Eq. (17.9) and simultaneously satisfy the constraint Eqs. (17.9)–(17.11).

17.6 Summary and Conclusions

The chapter presented the status of water quality modeling in India in general and wasteload allocation in particular. The concept and components of WLA have been discussed. An insight into the data required for WLA modeling, its processing and use has been illustrated using the QUAL2E water quality simulation model. All steps for using the simulation model to generate the water quality response in form of transfer coefficient has been elaborated. Since the secondary data required for WLA modeling has to be obtained from different agencies, it is emphasized that there has to be a coordinated effort involving the policy makers, scientists, engineers and the academia. It is strongly suggested that the academia (from various disciplines of engineering and sciences) should provide the knowledge and tools to carry out collection of comprehensive data required for WLA modeling for effective control of pollution. This, of course would require capacity building of technical manpower. It is opined that incorporating the data collection and processing in routine functioning by the government agencies will pave the way towards developing robust WLA models and carrying out state of the art research in the subject domain. This eventually would lead us towards development of realistic pollution control management plans of the ailing rivers.

References

Ambrose B, Wool TA, Martin JL (1993) The water quality analysis simulation program: WASP 5, Part A, Model documentation. USEPA, Center for Exposure Assessment and Modeling, Athens, GA

Anderson MV, Day HJ (1968) Regional management of water quality—a systems approach. J Water Pollut Control Feder 40(10):1679–1687

Babu MT, Das VK, Vethamony P (2005) BOD-DO modeling and water quality analysis of a wastewater outfall off Kochi, west coast of India. Environ Int 32(2):145–153

Barnwell TO, Brown LC, Whittemore RC (2004) Importance of field data in stream water quality modeling using QUAL2E-UNCAS. J Environ Eng 130(6):643–647

Brown LC, Barnwell TO (1987) The enhanced stream water quality models QUAL2E and QUAL2E-UNCAS: Documentation and User Manual, Report No. EPA/600/3-87/007, Environment Research Laboratory, USEPA, Athens, GA

Burn DH (1989) Water quality management through combined simulation optimization approach. J Environ Eng 15(5):1008–1023

Burn DH, McBean ED (1985) Optimzation modeling of water quality in uncertain environment. Water Resour Res 21(7):934–940

Burn DH, Yulianti JS (2001) Wasteload allocation using genetic algorithms. J Water Resour Plan Manag 127(2):121–129

Cohon JL (1978) Multiobjective programming and planning. Academic Press, New York

Dhage SS, Chandorkar AA, Kumar R, Srivastava A, Gupta I (2006) Marine water quality assessment at Mumbai West Coast. Environ Int 32:149–158

Dikshit AK, Anand K, Alam JB (2000) A water quality simulation model for river systems. In: Mehrotra R, Soni B, Bhatia KKS (eds) Proceedings of international conference on integrated water resources management for sustainable development. National Institute of Hydrology, New Delhi, pp 383–392

Ecker JG (1975) A geometric programming model for optimal allocation of stream dissolved oxygen. Manage Sci 21(6):658–668

Fujiwara O, Puangmaha W, Hanaki K (1988) River basin water quality management in stochastic environment. J Environ Eng 114(4):864–877

Ghosh NC (1996) Application of QUAL2E for water quality modeling of Kali River. Indian J Environ Health 38(3):160–170

Goldar B, Banerjee N (2004) Impact of informal regulation of pollution on water quality in rivers in India. J Environ Manage 73:117–130

Gupta I, Dhage S, Chandorkar AA, Srivastava A (2004) Numerical modeling of Thane Creek. Environ Model Softw 19:571–579

Haimes YY, Hall WA (1975) Analysis of multiple objectives in water quality. J Hydraulics Div ASCE 101(HY4):387–400
Haimes YY, Hall WA, Freedman HT (1975) Multiobjective optimization in water resources systems: The surrogate worth trade method. Elsevier Scientific Publishing Company, Amsterdam
Hass JE (1970) Optimal taxing for the abatement of water pollution. Water Resour Res 6(2):353–365
Jha R, Ojha CSP, Bhatia KKK (2000) Development of deoxygenation and reaeration rate coefficients for a small tributary of river Hindon, UP, India. In: Mehrotra R, Soni B, Bhatia KKS (eds) Proceedings of the international conference on integrated water resources management for sustainable development, New Delhi, pp 464–474
Jain CK (2004) Metal fractionation study on bed sediments of river Yamuna, India. Water Res 38:569–578
Joeres EF, Dressler J, Cho CC, Falkner CH (1974) Planning methodology for the design of regional waste water treatment systems. Water Resour Res 10(4):643–649
Karmakar S, Mujumdar PP (2004) Uncertainty in fuzzy membership functions for a river water quality management problem. In: Liong P, Babovic S (eds) International conference on hydroinformatics. World Scientific Publishing Company, pp 1–8
Kazmi AA, Agrawal L (2005) Strategies for water quality management of Yamuna River, India. In: Proceedings of third international symposium on South East Asian Water Environment, Bangkok, pp 70–80
Kazmi AA, Hansen IS (1997) Numerical models in water quality management: a case study for the Yamuna River (India). Water Sci Technol 36(5):193–200
Keshari AK, Parmar D (2006) Discussion on "Pollution management in the twentieth century". J Environ Eng 131(11):1543–1546
Lee CS, Chang SP (2005) Interactive fuzzy optimization for an economic and environmental balance in a river system. Water Res 39:221–231
Lee CS, Wen CG (1997) Fuzzy goal programming approach for water quality management in a river basin. Fuzzy Sets Syst 89:181–192
Lee CS, Wen CG (1996a) River assimilative capacity analysis via fuzzy linear programming. Fuzzy Sets Syst 79:191–201
Lee CS, Wen CG (1996b) Application of multiobjective programming to water quality management in a river basin. J Environ Manage 47:11–26
Liebman JC (1972) A simple management model for water quality. In: Keinath TM, Wanielista MP (eds) Proceedings of the mathematical modeling in environmental engineering, 8th Annual Workshop, Nassau, Bahamas, 18–22 Dec 1972, pp 469–476
McCutcheon SC (1989) Water quality modeling, vol. I, Transport and surface exchange in rivers. CRC Press, Boca Raton, FL
McNamara JR (1976) An optimization model for regional water quality management. Water Resour Res 12(2):125–134
Melching CS, Flores HE (1999) Reaeration equations derived from U.S. Geological Survey database. J Environ Eng 125(5):407–414
Mujumdar PP, Subbarao VRS (2004) Fuzzy waste load allocation model: simulation-optimization approach. J Comput Civil Eng 18(2):120–131
O'Connor DJ, Dobbins WE (1958) Mechanisms of reaeration in natural streams. Trans Am Soc Civil Eng 123:641–684
Ortolano L (1984) Environmental planning and decision making. Wiley, New York
Owens M, Edwards RW, Gibbs JM (1964) Some reaeration studies in streams. Int J Air Water Pollut 8:469–486
Paliwal R, Sharma P, Kansal A (2007) Water quality modeling of the river Yamuna (India) using QUAL2E-UNCAS. J Environ Manage 83(2):131–144
Parmar D (2006) Simulation and multiobjective optimization for river water quality management. PhD Thesis, Indian Institute of Technology, Delhi
Parmar D, Keshari AK (2012) Sensitivity analysis of water quality for Delhi stretch of river Yamuna, India. Environ Monitor Assess, Springer 184(3):1487–1508

Parmar D, Keshari AK (2014) Wasteload allocation using wastewater treatment and flow augmentation. Environ Model Assess, Springer 19(1):35–44

Parmar D, Keshari AK (2018) Simulating Strategic measures for managing water quality in the Delhi stretch of the River Yamuna, India. Sustain Water Resour Manage, Springer 4(4):1123–1133

Pelletier GJ, Chapra SC, Tao H (2006) QUAL2Kw—a framework for modeling water quality in streams and rivers using genetic algorithm for calibration. Environ Model Softw 21(3):419–425

Singh AP, Ghosh SK, Sharma P (2007) Water quality management of a stretch of river Yamuna: an interactive fuzzy multi-objective approach. Water Resour Manage 21:515–532

Smeers Y, Tyteca D (1984) A geometric programming model for the optimal design of wastewater treatment plants. Oper Res 32(2):314–342

Smith ET, Morris AR (1969) Systems analysis for optimal water quality management. J Water Pollut Control Feder 41(9):1635–1646

Subbarao VRS, Mujumdar PP, Ghosh S (2004) Risk evaluation in water quality management. J Water Resour Plan Manage 130(1):40–48

Thomann RV, Mueller JA (1987) Principles of surface water quality modeling and control. Harper & Row Publishers, New York

Tsivoglou EC, Neal LA (1976) Tracer measurement of reaeration: III. Predicting the reaeration capacity of inland streams. J Water Pollut Control Feder 48(12):2669–2689

USEPA (1995) QUAL2E Windows interface user's guide, United States Environmental Protection Agency, EPA/823/B/95/003

Wen CG, Lee CS (1998) A neural network approach to multiobjective optimization for water quality management in a river basin. Water Resour Res 34(3):427–436

Yandamuri SRM, Srinivasan K, Bhallamudi SM (2006) Multiobjective optimal wasteload allocation models for rivers using non-dominated sorting genetic algorithm-II. J Water Resour Plan Manage 132(3):133–143

Chapter 18
Methane Emission from Municipal Solid Waste Landfills—Estimation and Control

S. Rajesh, S. Roy and V. Khan

Abstract Methane gas can be generated from the Municipal solid waste (MSW) landfill due to the presence of organic fraction and bacterial activity occurring over a period. Methane production in landfills and the resulting emissions to the atmosphere, representing the second largest anthropogenic methane source; hence, there is a need to estimate methane generation rate accurately followed by devising various technique to mitigate the emission. As the methane generation rate is governed by various factors like waste temperature, waste composition and density, pH within landfill, concentration of substrate, moisture content and toxins, several numerical and mathematical tools have been developed considering one or more of the outlined factors to estimate the landfill gas generation. Moreover, the estimation of methane gas generation and emission is also essential for predicting the settlement, in specific, differential settlement that could occur in the engineered landfills. The differential settlement could lead to the initiation and propagation of cracks on the soil barrier. This in turn, might be able to set a free drainage path for the escape of methane gas. In this chapter, a brief assessment on various existing models to predict the estimation of methane generation rate is discussed. The accuracy of the predicted values obtained from various models is estimated from the experimentally observed dataset. This chapter also highlights the need for the evaluation of methane gas generation rate in modeling the settlement of MSW landfills. In addition, various techniques used to mitigate the emission of the methane from the landfill are also discussed.

Keywords Methane emission · Methane generation rate · Municipal solid waste · Engineered landfill · Cap barriers · Gas permeability

---

S. Rajesh (✉) · S. Roy · V. Khan
Department of Civil Engineering, Indian Institute of Technology Kanpur, Kanpur, India
e-mail: hsrajesh@iitk.ac.in

S. Roy
e-mail: sumanroy@iitk.ac.in

V. Khan
e-mail: vishwjit@iitk.ac.in

T. Gupta et al. (eds.), *Measurement, Analysis and Remediation of Environmental Pollutants*, Energy, Environment, and Sustainability,
https://doi.org/10.1007/978-981-15-0540-9_18

18.1 Introduction

Municipal solid waste (MSW) is a major concern for developing countries like India. Rapid population, economic and industrial growth has led to drastic changes in lifestyle of urban residents which has accelerated the generation of MSW in Indian cities. Daily waste generation in most of the megacities in India range from around 2,000 to 11,000 metric tons per day (TPD) which has increased by a huge amount since the last sixteen years (Joshi and Ahmed 2016; CPCB 2017). Management and disposal of this enormous quantity of waste therefore has become more challenging due to the complexities involved in waste management systems. Typical problems associated with the management of solid waste in developing countries mainly lies with the inefficiency in implanting funds effectively and lack of technical expertise for disposing waste efficiently (Narayana 2009; Sharma and Jain 2018). In addition, a successful waste management plan also requires proper knowledge about the constituents of waste, depending on which their transportation, treatment, storage and disposal needs to be organized. Extensive population growth rate has caused rapidly varying waste characterization and generation in India which is quite different from the waste generated in western countries (Joshi and Ahmed 2016). MSW composition in India presently consist of approximately 40–50% biodegradable materials, 20–30% inert waste, 5–10% paper, and almost 10% plastic/rubber materials which is steadily showing an increase with respect to time (Joshi and Ahmed 2016; Sharma and Jain 2018). Disposal of this kind of waste is mostly done through open dumping, which is highly unsatisfactory and can lead to surface and ground water contamination (Narayana 2009; Ramachandra et al. 2018). To avoid these environmental issues associated with open dumping; land filling has been opted extensively throughout India. Annual reports by CPCB based on state wise generation and collection of MSW reveals that almost 35% of the waste is disposed of in landfills (CPCB 2017).

Compared to other developed countries, India has a high amount of organic content in its waste (almost 50%) (Joshi and Ahmed 2016; Sharma and Jain 2018). In time, these organic wastes gradually decompose leading to generation of harmful gases and leachate. Decomposition in organic waste can be initiated by biological (aerobic or anaerobic biodegradation) or chemical processes (dissolution, oxidation/reduction and thermal degradation) (Kamalan et al. 2011; Jafari et al. 2017; Majdinasab et al. 2017). In open dumped sites, these extensively pollute the atmosphere and groundwater, especially during monsoon seasons. Even in engineered landfills, similar degradation process takes place, which gradually leads to the generation of landfill gases (LFG) and leachate within the landfill. However, unlike open dumped sites, engineered landfills are designed for top covers, bottom covers and side liners (Gourc et al. 2010a, b; Rajesh et al. 2015, 2017; Rajesh 2018a, b). These prevent the LFG from escaping into the atmospheres as well as groundwater contamination by restricting the flow of leachate (Rajesh et al. 2014, 2016; Roy and Rajesh 2018; Rajesh and Khan 2018). A top cover or capping system generally consists of a hydraulic barrier, gas-ventilating layer, water drainage layer, gas collection system, cover soil, and vegetation (Rajesh and Viswanadham 2012; Pawlowska 2014; Rajesh

2018b). The bottom cover or the base lining system on the other hand also consists of a hydraulic barrier along with a drainage blanket and leachate collection system to prevent the flow of contaminants into groundwater (Pawlowska 2014; Rajesh 2018b).

The hydraulic barrier in both top and bottom covers consists of a very low permeability material like compacted clays, compacted sand-bentonite, polymer amended compacted clays or sand—bentonite, geomembranes, geosynthetic clay liner, or combination of these materials (Rajesh and Khan 2018; Rajesh 2018a, b). These low permeability materials reduce the emission of LFG, mostly methane and carbon dioxide into the atmosphere, thus maintaining the air quality. Even then, inside the landfill continuous biodegradation can lead to elevated temperatures which can initiate a sudden fiery explosion, mostly due to the presence of methane. The gas collection system at the top eliminates the possibility of such accidents. Hence, this can be effectively used to safely monitor, collect and process landfill gases for environmental protection and energy generation.

Degradation of organic content within a landfill is also associated with volume reduction of MSW. Collapse of void space due to reorientation of particles, combustion, and oxidation, dissolution of solids into leachate and biological decomposition of organic materials often lead to deformation in the landfill liners. Moreover, the self-weight of MSW, creep in MSW skeleton and external loads on landfill can also create significant subsidence in the cover material. Excessive settlements can cause the cover material to fracture causing damage in the leachate or gas collection system (Sivakumar Babu et al. 2009; Rajesh et al. 2015). Often, they can lead to initiation and propagation of cracks on the soil barrier, which in turn, might set a free drainage path for the escape of LFG gas (Rajesh et al. 2014). These can lead to a catastrophic failure of the landfill site if proper precautions are not taken.

Maintenance of a landfill site therefore requires timely observations on biodegradation occurring within the landfill as they can lead to significant LFG emission and excessive settlements as shown in Fig. 18.1. Typical behavior of an MSW landfill over a long term can therefore be assessed using various mathematical models available in literature to avoid detrimental effects. These models can provide precise estimation for LFG emission and landfill settlement over the years depending on the rate of biodegradation for further maintenance of MSW landfills. Prediction of LFG emission from biodegradable wastes using mathematical models can provide crucial information on global warming, designing methane control systems and energy projects. Likewise, prediction of subsidence in landfill covers can provide an early warning for fractures in cover system. These can be especially taken care of during the landfill maintenance. It can provide a useful tool for proper care and maintenance of the landfill over long terms.

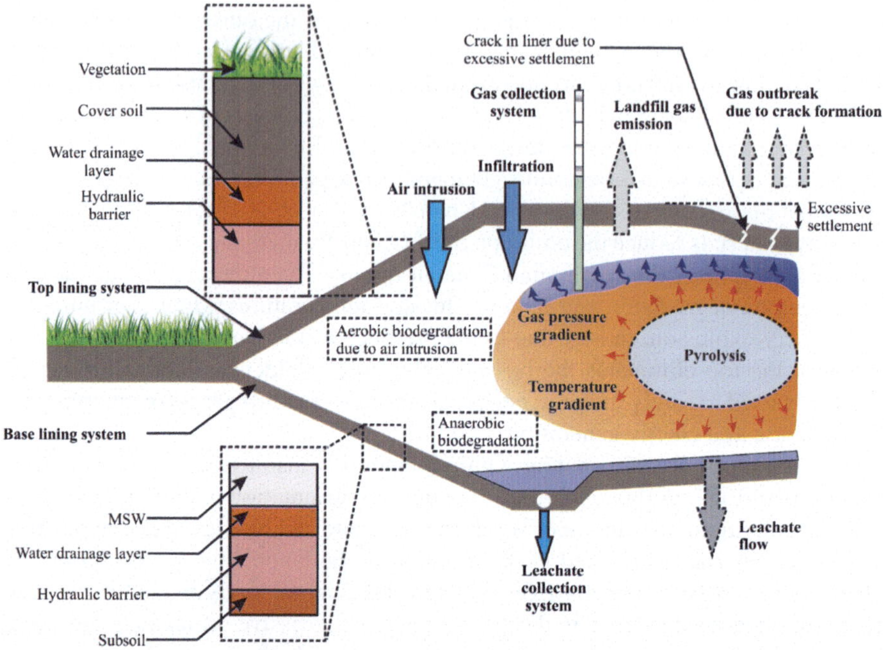

Fig. 18.1 Biological and chemical processes within a landfill creating LFG generation and excessive settlements. Modified after Jafari et al. (2017)

18.2 Estimation of LFG Generation and Emission in MSW Landfills

Gas emission in MSW landfills mainly occurs due to degradation of organic matter by bacterial activity over a prolonged period. Depending on the nature of waste, MSW dumped in a landfill at first undergoes aerobic decomposition stage with insignificant amount of methane generation. During this process, mostly carbon dioxide and water vapor is generated. With the end of aerobic decomposition stage, anaerobic conditions are established within the landfill. In anaerobic degradation stage, the bacteria continue to degrade the organic MSW with the generation of methane (methanogenesis). In addition, the chemical process like evaporation of volatile compounds and oxidation/reduction of the biodegradable materials also leads to LFG emission. Thermal degradation involving smoldering combustion and pyrolysis is also relevant in some MSW landfills due to elevated temperatures (Jafari et al. 2017). Degradation involving pyrolysis can continue for extended periods of time even in limited oxygen supply. This is typically influenced by the temperature and hot spot within the biodegradable mass.

The above discussion therefore suggests that generation and emission rate of LFG; mostly methane is affected by several factors like composition of waste, amount of biodegradable waste, moisture content, temperature and time (Majdinasab et al.

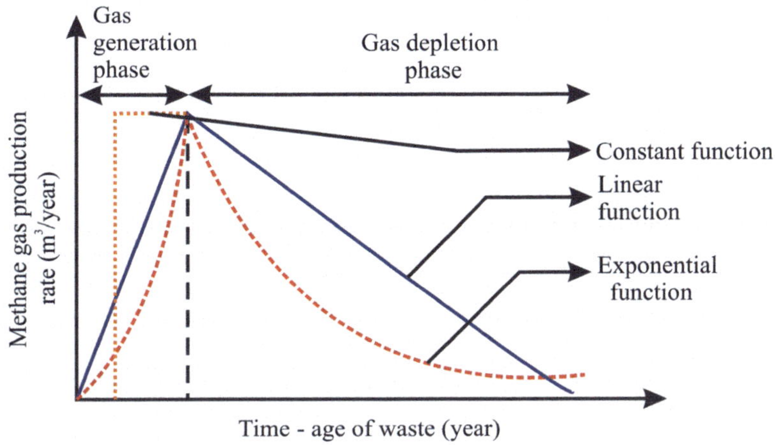

Fig. 18.2 Typical variation of methane gas production with time considering various functions accounting methane generation and depletion phases

2017). These parameters typically influence the kinetics of biodegradation within a landfill, thus affecting the emission rate of methane with respect to time. Modeling of methane emission in a landfill site therefore can be approached by representing the gas generation or emission in terms of a single mathematical function or combination of functions considering an overall kinetic parameter for biodegradation (Kamalan et al. 2011). For this type of typical representations, the overall kinetic parameter can be considered by observing the trend in methane production rates. As mentioned previously, during aerobic biodegradation very insignificant amount of methane is produced. This can be defined as the lag time or start up period. Following the lag period, active methane generation takes place, after which the methane generation gradually depletes with time. Different mathematical functions can be assumed to fit the trend in methane production during the active generation phase and depletion phase as shown in Fig. 18.2 (Natsev 1998). The formulation is based on Monod's equation (Kamalan et al. 2011; Majdinasab et al. 2017), a generalized kinetic expression for biodegradation process, which can be written as

$$\frac{dC}{dt} = \frac{KxC}{K_c + C} \tag{1}$$

where C is the mass of degradable organic waste per volume of waste at time t, x is the biomass or mass of microorganisms per volume of waste, K_c is the is the waste concentration at which the rate is one-half of the maximum rate of degradable organic waste utilization, and K is the maximum rate of degradable organic waste utilization per unit biomass. The above equation can be solved for two extreme cases, namely, zero order reaction kinetics and first order reaction kinetics, details of which are discussed below.

18.2.1 Zero Order Models

Zero order models primarily assume a constant value of biomass per volume of waste (x). For a large mass of degradable organic waste (C), the rate of degradation (dC/dt) can be assumed as constant. For zero order kinetics, therefore, concentration of landfill gas like methane may be independent of various factors affecting its production. According to Ham and Barlaz (1989), landfills may follow zero order kinetics, especially, during the period of active methane generation. In these cases, landfills produce almost same amount of the gas on a yearly basis. However, with the advent of time in different parts of the landfill, the collected gas may deplete gradually.

Methane generation rate (Q) from a typical zero order model, namely, Solid Waste Association of North America zero order model (SWANA Zero order) considering different kinds of waste can be given as

$$Q = \frac{ML_o}{(t_o - t_f)} \tag{2}$$

where M is the mass of the waste disposed at landfill, L_o is the methane generation potential per mass of the waste, t_o is the lag time and t_f is the final time till end of methane generation.

A similar zero order model for estimation of methane emission (Q) is the IPCC model based on degradable materials within the waste. Equation for IPCC model can be represented as

$$Q = [MSW_T \times MSW_F \times MCF \times DOC \times DOC_F \times F \times (16/12 - R) \times (1 - OX)] \tag{3}$$

where MSW_T is the total amount of MSW generated, MSW_F is the MSW fraction of disposed at landfill sites, MCF is the methane gas correction factor (default 0.6 by IPCC), DOC is the degradable organic biomass (default 15% w/w by IPCC), DOC_F is the fraction of decomposed DOC (default 77% w/w by IPCC), F is the fraction of methane in landfill gas (default 0.5 v/v by IPCC), R is the recovered methane and OX is the oxidation factor (default 0 by IPCC). The fraction 16/12 is the molecular weight ratio between methane and carbon. The MCF and DOC can vary for different type of site and waste even though certain default values are mentioned by IPCC (Kamalan et al. 2011).

18.2.2 First Order Models

The first order models are most commonly used by researchers world-wide to predict methane generation in a landfill. Like zero order models, the first order models also assume a constant volume of biomass per volume of waste (x). However, these models generally assume a small mass of degradable organic waste (C) compared to zero order models. Instead of constant degradation rate, a varying rate of degradation (dC/dt) is therefore obtained with respect to mass of degradable organic waste. First order models also consider various factors affecting the methane generation rate in landfills.

SWANA first order is a commonly used first order model for calculation of methane emission(Q) from landfills. The expression for the model can be given as

$$Q = kML_o e^{-k(t-t_o)} \tag{4}$$

where M is the mass of the waste disposed at landfill, k is the rate constant for first order degradation or methane generation rate, L_o is the methane generation potential per mass of the waste and t_o is the lag time. For this model, there is an exponential decrease in methane generation rate with time with deposited wastes in landfill.

IPCC model based on first order kinetics (Cakir et al. 2016; Datta and Zekkos 2019) for estimation of methane emission (Q) from biodegradable carbon for a particular waste category (j) can be written as

$$Q = \sum_j (Q_{j,T} - R_T) \times (1 - OX_T) \tag{5}$$

where R_T is the recovered methane in year T and OX_T is the oxidation factor in year T that reflects the amount of methane oxidized in the soil or into other material enclosing the waste. $Q_{j,T}$ is the amount of methane generated for a particular type of waste (j) in year T which can be estimated using the following equations

$$\begin{aligned} Q_{j,T} &= DDOC_{mdecomp_T} \times F \times (16/12) \\ DDOC_{mdecomp_T} &= DDOC_{ma_{T-1}} \times (1 - e^{-k}) \\ DDOC_{ma_T} &= DDOC_{md_T} \times (DDOC_{ma_{T-1}} \times e^{-k}) \end{aligned} \tag{6}$$

where $DDOC_{m\,decomp_T}$ is the mass of decomposed organic carbon in year T, $DDOC_{ma_T}$ and $DDOC_{ma_{T-1}}$ are the mass of decomposable organic carbon accumulated in the landfill at the end of year T and $T - 1$, $DDOC_{md_T}$ is the mass of decomposable organic carbon deposited in the landfill in year T, F is the fraction of methane in landfill gas, fraction 16/12 is the molecular weight ratio between methane and carbon and k is the rate constant for first order degradation given as

$$k = \frac{\ln(2)}{t_{1/2}} \tag{7}$$

where $t_{1/2}$ is the half life time of the waste (j).

The mass of decomposable organic carbon ($DDOC_{\mathrm{m}}$) can be expressed as per IPCC model as

$$DDOC_m = W \times DOC \times DOC_F \times MCF \tag{8}$$

where W is the mass of waste (j) deposited in the landfill, DOC is the degradable organic biomass/carbon, DOC_F is the fraction of decomposed DOC and MCF is the methane correction factor for aerobic decomposition in the year of waste (j) deposition.

LandGEM (Datta and Zekkos 2019; Fallahizadeh et al. 2019) is another first order model for estimating methane emissions from MSW landfills. In recent years, LandGEM has gained high attention as it can provide reliable estimation of methane generation. The program developed by USEPA, enables the user to calculate emissions over specific times based on the following inputs.

(i) Design capacity of landfill
(ii) Amount of waste in landfill (annual acceptance rate)
(iii) Rate constant for first order degradation (methane generation rate k) and methane generation potential per mass of waste (L_o)
(iv) Concentration of total and individual non-methane organic compounds ($NMOC$)
(v) Years of dumping waste into the specified landfill
(vi) Acceptance of hazardous waste in the specified landfill.

The values of k and L_o can be obtained specifically from field test measurements. The equation for methane emission as per LandGEM model can be given as,

$$Q = \sum_{i=1}^{n} k L_o M_i e^{-k(t-t_o)} \tag{9}$$

where M_i is the mass of the waste disposed at landfill for the year i, n is the number of years and t_o is the lag time. The mass of non-degradable solid waste can be subtracted from the total solid waste mass while calculating the mass of waste disposed at the landfill. However, as recommended by LandGEM this should be done only when proper documentation approved by a regulatory authority is provided. The LandGEM is also capable of predicting the estimated emissions for other pollutants such as non-methane organic compounds, carbon dioxide etc.

LFGGEN model, also known as landfill gas generation model, developed by University of Central Florida is a typical first order model where the methane generation phases are preceded by a lag phase (t_o). In this model, during methanogenesis, the methane emission linearly increases to a specific peak rate (Q_s) at the end of the year (t_p), after which it decreases exponentially to nearly zero rate at the end of biodegradation (t_{99}). The specific peak rate of methane emission can be represented according to this model as

$$Q_{sp} = L_o \frac{2k}{k(t_p - t_o) + 2} \tag{10}$$

where k is the rate constant for first order degradation or methane generation rate and L_o is the methane generation potential per mass of the waste. The exponential decreasing phase can be modeled by modifying the biodegradation rate constant as

$$k = -\frac{\ln 0.01}{(t_{99} - t_p)} = \frac{4.6052}{(t_{99} - t_p)} \tag{11}$$

where t_{99} is the time for the methane generation rate to reach one percent of specific peak rate (Q_{sp}). Annual methane emission rate (Q_{sj}) for a specific component of MSW (j) can be therefore expressed as

$$\begin{aligned} Q_{sj} &= 0 && 0 < t \le t_{oj} \\ Q_{sj} &= \frac{Q_{spj}}{2}\left[\frac{t-t_{oj}}{t_{pj}-t_{oj}} + \frac{(t-1)-t_{oj}}{t_{pj}-t_{oj}}\right] && t_{oj} < t < t_{pj} \\ Q_{sj} &= \frac{Q_{spj}}{2}\left[e^{-k_j(t-t_{pj})} + e^{-k_j(t-1-t_{pj})}\right] && t_{pj} < t < t_{99} \end{aligned} \tag{12}$$

Total methane emitted by the biodegradable wastes within the landfill can be written as

$$Q = \sum Q_{sj} \times M_j \tag{13}$$

where M_j is the mass of a specific MSW component (j) generating methane and Q_{sj} is the methane emission of the specific MSW component.

The first order exponential models can also be modified to incorporate the effect of varying rate of biodegradation for various organic wastes in various phases. The formulation of methane emission for this type of multiphase model can be written as

$$Q = ML_o[F_r k_r e^{-k_r(t-t_o)} + F_s k_s e^{-k_s(t-t_o)}] \tag{14}$$

where M is the mass of the waste disposed at landfill, L_o is the methane generation potential per mass of the waste, k_r and k_s are the rate constant for first order degradation of rapidly and slowly degradable waste (methane generation rate for rapidly and slowly degradable waste), while F_r and F_s are the fraction of rapidly and slowly degradable waste respectively and t_o is the lag time.

18.2.3 Second Order Models

The complex reactions of biodegradation within a landfill can also be modeled using second order models. In this case, a large number of first order models with varying rates are utilized to describe the biodegradation within a landfill. However, these types of systems are more complicated compared to ordinary first order models and are therefore generally avoided for estimating methane emission (Kamalan et al. 2011; Rajaram et al. 2011).

18.3 Estimation of Settlement in MSW Landfills

Apart from LFG gas emission, settlement is also an important parameter for governing the safety and health monitoring related to MSW landfills. Settlement in MSW can occur due to a variety of causes like mechanical loading, raveling, physiochemical changes, or anaerobic/aerobic decomposition of the waste materials within the landfill. Often, a mutual combination of the above factors can also induce huge depressions in the cover system of the landfill. Depending on the factors affecting its behavior, settlement in MSW landfills can be categorized in three different parts.

(i) Initial compression occurs under the effect of own weight or external mechanical loading (like soil cover, waste layers, final cover, roads or buildings). The effect is almost instantaneous (similar to settlement in sand) and mainly happens due to compaction of void spaces within the landfill.
(ii) Primary compression occurs due to consolidation of the MSW landfills after final load placement. However, since MSW remains mostly unsaturated dissipation of pore water pressure may not occur in this process. Rather, gradual expulsion of air from the voids occurs which induce the settlement in the landfill.
(iii) Secondary compression can be considered as the major factor affecting settlement behavior in MSW landfills. Secondary compression mostly occurs due to mechanical creep of the MSW, reorientation of the particles, physiochemical changes within the landfill, and degradation of the waste materials.

A conceptual model of time-settlement behavior in MSW landfills is shown in Fig. 18.3. Settlement behavior in landfills can be predicted using various mathematical and empirical models developed by researchers. The formulation for initial compression is generally not addressed in these typical models. Total settlement is considered to be the summation of only primary compression and secondary compression of the MSW over the years after the initial compression is over. Settlement models for MSW, therefore, can be classified in four broad categories following El-Fadel and Khoury (2000).

18.3.1 Models Based on Saturated Soil Mechanics

Simple models based on primary consolidation and secondary compression of saturated fine-grained soils (based on Terzaghi's consolidation theory) is utilized in this case to estimate settlement in MSW landfills. Settlement (ΔH) for MSW landfill

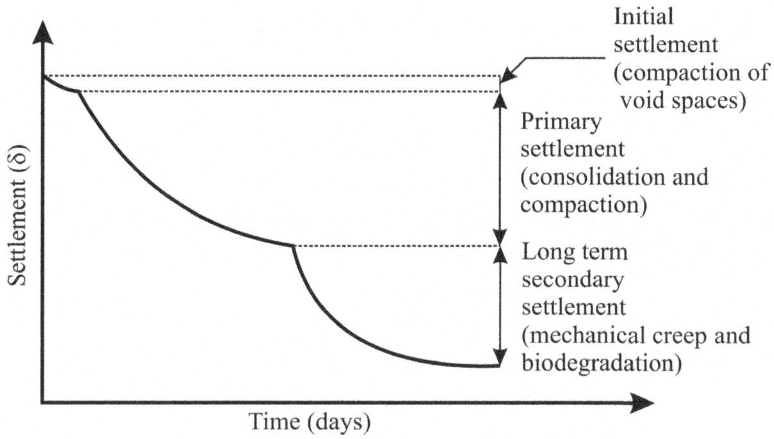

Fig. 18.3 Typical settlement response at different stages for MSW landfill

according to this model can be represented as

$$\Delta H = H_i C_{CR} \log\left(\frac{\sigma_o + \Delta\sigma}{\sigma_o}\right) + H_i C_\alpha \log\left(\frac{t_{final}}{t_{pr}}\right) \tag{15}$$

where H_i is the thickness of the waste layer after initial compression, C_{CR} is the primary compression ratio varying from 0.163 to 0.205 for MSW as reported by Sivakumar Babu et al. (2009), σ_o is the overburden pressure at the middle of the layer, $\Delta\sigma$ is the increment in overburden pressure at the middle of the layer due to construction of additional layer on top of it, C_α is the secondary compression index varying from 0.015 to 0.350 for MSW as reported by Sivakumar Babu et al. (2009), t_{pr} is the time for primary compression (can be assumed as 1–25 days) and t_{final} is the time during ending phase of long term settlement.

A modification of the above model in the secondary compression phase was introduced by Bjarngard and Edgers (1990) by subdividing into intermediate and long-term phase. Settlement (ΔH) for MSW landfill using this extended soil model can be expressed as,

$$\Delta H = H_i C_{CR} \log\left(\frac{\sigma_o + \Delta\sigma}{\sigma_o}\right) + H_i C_{\alpha 1} \log\left(\frac{t_1}{t_{pr}}\right) + H_i C_{\alpha 2} \log\left(\frac{t_{final}}{t_1}\right) \tag{16}$$

where H_i is the thickness of the waste layer after initial compression, C_{CR} is the primary compression ratio varying from 0.163 to 0.205 for MSW as reported by Sivakumar Babu et al. (2009), σ_o is the overburden pressure at the middle of the layer, $\Delta\sigma$ is the increment in overburden pressure at the middle of the layer due to construction of additional layer on top of it, $C_{\alpha 1}$ is the intermediate secondary compression index taken as 0.035 for MSW as reported by Sivakumar Babu et al. (2009), t_{pr} is the time for primary compression (can be assumed as 1–25 days), t_1

is the time for intermediate secondary compression (can be assumed as 200 days), $C_{\alpha 2}$ is the long term/final secondary compression index taken as 0.215 for MSW as reported by Sivakumar Babu et al. (2009), and t_{final} is the time during ending phase of long term settlement.

Settlement prediction using classic consolidation theory of saturated soils is generally avoided as most of the MSW is unsaturated in nature. Also, for utilization of this type of model precise determination of the parameters (like void ratio of MSW) present in the model are required.

18.3.2 Rheological Models

Rheological models consisting of springs, dashpots and sliders are also used to model settlements in MSW landfills. According to theory, a number of nonlinear or linear springs and dashpot arranged in continuous spectrum can perfectly fit a secondary compression curve of soils. Often, models consisting of only linear springs can be used which are simple in nature but deviate from actual field conditions. Therefore, to avoid complexities regarding multiple numbers of nonlinear spring and dashpot constants, only a single dashpot element is made nonlinear in rheological models for obtaining sensible results. Gibson and Lo (1961) model, a widely used rheological model to calculate settlements (ΔH) in MSW landfill, can be given as

$$\Delta H = H_i \Delta\sigma[a + b(1 - e^{-(\lambda/b)t})] \tag{17}$$

where H_i is the thickness of the waste layer after initial compression, $\Delta\sigma$ is the overburden pressure or compressive stress depending on waste height, a and b are the primary and secondary compressibility parameter, λ/b is the rate of secondary compression and t is the time of load application. For load applied in kPa, the parameter a has a value within the range of 2.5E−05 to 7E−05/kPa, while b has a value within the range of 1.6E−02 to 2E−03/kPa as per El-Fadel et al. (1999). The problem of incorporating rheological models in predicting MSW settlement lies in the fact that the constants themselves are stress dependent and hence may act differently than real field situations under various loading conditions.

18.3.3 Empirical Models

Empirical models use mathematical functions like logarithmic, power or hyperbolic functions to model the behavior of settlement in MSW landfills. These models consist of empirical parameters which rarely has a physical significance. These parameters are highly dependent on the site conditions where the settlement needs to be evaluated.

Settlement (ΔH) for MSW using the logarithmic function (Yen and Scanlon 1975) for empirical models can be written as

$$\Delta H = H_i\left[\alpha + \beta \log\left(t - \frac{t_C}{2}\right)\right] \tag{18}$$

where H_i is the thickness of the landfill after initial compression, α and β are the empirical fitting parameters, t_c is the construction period of landfill and t is the time since beginning of filling.

Settlement (ΔH) for MSW using the power function (Edil et al. 1990) for empirical models can be similarly represented as

$$\Delta H = H_i \Delta\sigma M\left(\frac{t}{t_r}\right)^N \tag{19}$$

where H_i is the thickness of the landfill after initial compression, M is the empirical parameter dependent on site termed as reference compressibility, N is a parameter dependent on age and placement conditions of waste referred to as rate of compression, t_r is reference time (taken as 1 day) and t is the time since beginning of filling.

Settlement (ΔH) for MSW at a particular time using the hyperbolic function (Ling et al. 1998) for empirical models can be obtained if ultimate settlement (S_{ult}) of MSW landfill is known beforehand. The expression can be given as

$$\Delta H = \frac{t}{(1/\rho_o) + (1/S_{ult})} \tag{20}$$

where ρ_o is the initial rate of settlement and t is the time since beginning of filling.

18.3.4 Biodegradation Models

Biodegradation models form an important part of long-term settlement in MSW landfills. The settlement in biodegradation models is directly proportional to the rate of degradation of the organic wastes following first order kinetics of Monod's model. These models can therefore be coupled with methane emission models following first order kinetics to evaluate the methane emission and settlement simultaneously in MSW landfills from organic wastes.

Park and Lee (1997) proposed a time dependent biodegradation model to evaluate the settlement in MSW landfills. Settlement (ΔH_{dec}) due to biodegradation using this model can be expressed as

$$\Delta H_{dec} = H_i \varepsilon_{tot\_dec}\left(1 - e^{-kt_{bio}}\right) \tag{21}$$

where H_i is the thickness of the waste layer after initial compression, ε_{total_dec} is the amount of compression that will occur due to decomposition of biodegradable wastes, k is the first order decomposition rate constant and t_{bio} is the time lapse from starting point (t_o) to which the settlement due to decomposition starts given as ($t - t_o$).

A similar settlement model dependent on biodegradation proposed by Hettiarachchi et al. (2009), considering four ($n = 4$) different groups of solid of unit cross sectional area in the MSW namely, rapidly degradable, moderately degradable, slowly degradable and non-degradable solids can be expressed as

$$\Delta H_{dec} = \left[\frac{M_{si}}{\rho_w} \sum_{i=1}^{n} \frac{f_{sj}}{G_{sj}} (1 - e^{-k_j t_{bio}}) \right] \quad (22)$$

where H_i is the thickness of the waste layer after initial compression, ρ_w is the density of water, M_{si} is the mass of solids for a particular group of MSW (j), f_{sj} is the solids fraction for each MSW group (j), G_{sj} is the specific gravity for each MSW group (j), k_j is the first order decomposition rate constant for each MSW group (j) and t_{bio} is the time lapse from starting point (t_o) to which the settlement due to decomposition starts given as ($t - t_o$).

18.4 Evaluation of Methane Emission and Settlement of an MSW Landfill—A Case Study (Gourc et al. 2010b)

In this section, a case study on MSW landfill after large scale experiments conducted by Gourc et al. (2010b) is presented. For this experiment, 'ELIA' large scale cell (C1) of 22 m^3 volume from Veolia Environment was utilized. The buried waste had a mass (M) of around 6.710–6.8 t DM containing fraction (f_{sj}) of 15.9% textile, 12.6% putrescibles, 13.8% paper, 6.8% cardboard, 4.2% textiles and 0.5% wood within the mass. Negligible vertical stress was applied on the cell and the mass within the cell was allowed to decompose gradually for 68 months. During this period, the amount of biogas (mostly methane) emitted and settlement of the cell was constantly measured. A plot of the daily biogas emission (Q/M) and settlement strain ($\Delta H/H$) data as measured by the authors for 68 months is shown in Fig. 18.4. The experimental data for biogas emission shows a similar trend to active generation and depletion phase of methane productions as shown in Fig. 18.2. The experimental data for both biogas emission and settlement strain is predicted with two different models namely the first order SWANA model and LFGGEN model. The IPCC model is purposely avoided as it requires comparatively a larger number of input parameters like DOC, DOC_F, and MCF which were not reported by the authors for the present case study.

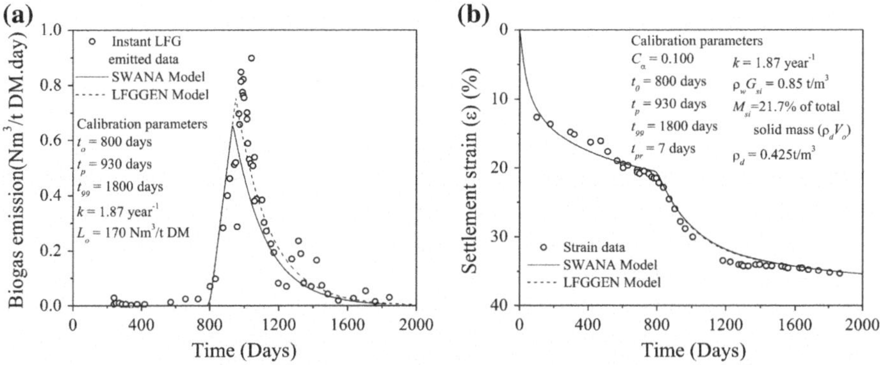

Fig. 18.4 **a** Daily biogas emission measured and predicted; **b** settlement strain measured and predicted for cell C1 using SWANA and LFGGEN model

18.4.1 Estimation of Biogas Emission

The biogas emission through SWANA model can be estimated from the value of methane generation rate (k) and methane generation potential per unit mass (L_o) provided by the authors as listed in calibration parameters of Fig. 18.4. The lag period (t_o) for biogas emission is 800 days, while the peak methane generation potential reached at 930 days (t_p). An important thing to mention here is that, as pointed out by Gourc et al. (2010b), since Eq. (4) of first order SWANA model is an exponentially decreasing function and predicts the depletion phase, instead of taking the actual value of t_o, it is better to take $t_o = t_p$ in the equation. This will help in better prediction of the experimental data as seen in Fig. 18.4a. The peak biogas emission predicted from SWANA model is about 12.5% lower than the biogas emission observed experimentally.

For LFGGEN model, the value of k as provided by the authors is taken for active generation phase, while a modified value of k is taken from t_p to t_{99} according to Eq. (11). The biogas emission for the entire period is predicted using Eq. (12). It is observed that, LFGGEN models predict a comparatively lower value of peak biogas emission (about 25.5% lower) compared to the emission observed experimentally. For this study related to daily biogas emission, the SWANA model with slight modifications performs comparatively better when compared to the LFGGEN model.

18.4.2 Estimation of Settlement Strain

Similar to biogas emission, for estimation of biodegraded settlement strain also the two first order models SWANA and LFGGEN was utilized. However, as mentioned in Sect. 18.3, the settlement in MSW is a combination of immediate, primary and secondary settlement. All these three must be estimated, in order to evaluate the total settlement of MSW.

The immediate and primary settlement of the MSW in this case will be minimal. This is because of the fact that negligible vertical stress or overburden pressure was applied at the top of the cell as mentioned earlier. The secondary settlement on the other hand will be induced due to the mechanical creep of MSW and degradation of the waste materials. To evaluate the total settlement strain a combination of Eqs. (15) and (22) will be utilized. Equation (15) is chosen because it is the simplest equation incorporating secondary compression of MSW. All other rheological models and empirical models require additional parameters which are site specific.

Evaluating settlement strain using Eq. (15) will only require the second term as the first term containing primary compression is negligible. The secondary compression index C_α is taken equivalent to the value provided by authors and lies within the range provided by Sivakumar Babu et al. (2009) for MSW. A time of 7 days (t_{pr}) was taken for secondary compression to initiate while after almost 800 days, the biodegradation settlement initiated (t_o). For evaluating the settlement due to biodegradation, the calibration parameters are listed in Fig. 18.4b. Equation (22) can then be utilized with the listed calibration parameters, to evaluate the strain due to biodegradation. For SWANA model, only one k is utilized throughout the entire duration, while for LFGGEN, two different values of k as suggested earlier for the active generation and depletion phase is utilized. Summation of settlement strains from Eqs. (15) and (22) will give the predicted curve as shown in Fig. 18.4b. It is observed from the results, that both the models predict the settlement close to the experimental data for the present case study.

18.5 Mitigation of LFG Emission from Landfills

LFG emission from the landfills, if not controlled properly is a major cause of environmental pollution. In order to avoid such emissions, the low permeable soil barrier at the top cover reduces the flow of generated gases into the atmosphere. Similarly, timely care and maintenance of the gas collection system at the top cover of the landfill is also important. This prevents high gas pressure build up within the landfill and reduces the progression of gases through soil barrier (Gourc et al. 2010a, b). A detailed mechanism for mitigating landfill gases using waste containment system is given by Rajesh (2018a). If the hydraulic barrier tends to crack either due to desiccation or differential settlement as shown in Fig. 18.1, the performance of waste containment system, in specific, cover liner in limiting the gas emission is highly questionable (Viswanadham and Rajesh 2009; Rajesh and Viswanadham 2012; Viswanadham et al. 2011, 2012). Moreover, gas transfer depends on degree of saturation of the hydraulic barrier. At low degree of saturation, advection predominates while at higher degree of saturation diffusion predominates (Rajesh et al. 2014, 2016; Rajesh and Khan 2018).

Apart from these conventional approaches, other methods are also available to reduce LFG emission from landfills. An anaerobic bioreactor landfill as shown in Fig. 18.5a can be used in place of a conventional landfill in order to mitigate LFG emission particularly methane into the atmosphere. The anaerobic bioreactor leads to a faster stabilization of the fresh waste by increasing the anaerobic decomposition rate. This is done by re-circulating the outgoing leachate into the interior of the landfill in a controlled manner so that the minimum average moisture content of the waste mass reaches 40 percent by weight (USEPA Clean air act, 40CFR 63, 1990, National Emissions Standards for Hazardous Air Pollutants). In case, the generated leachate

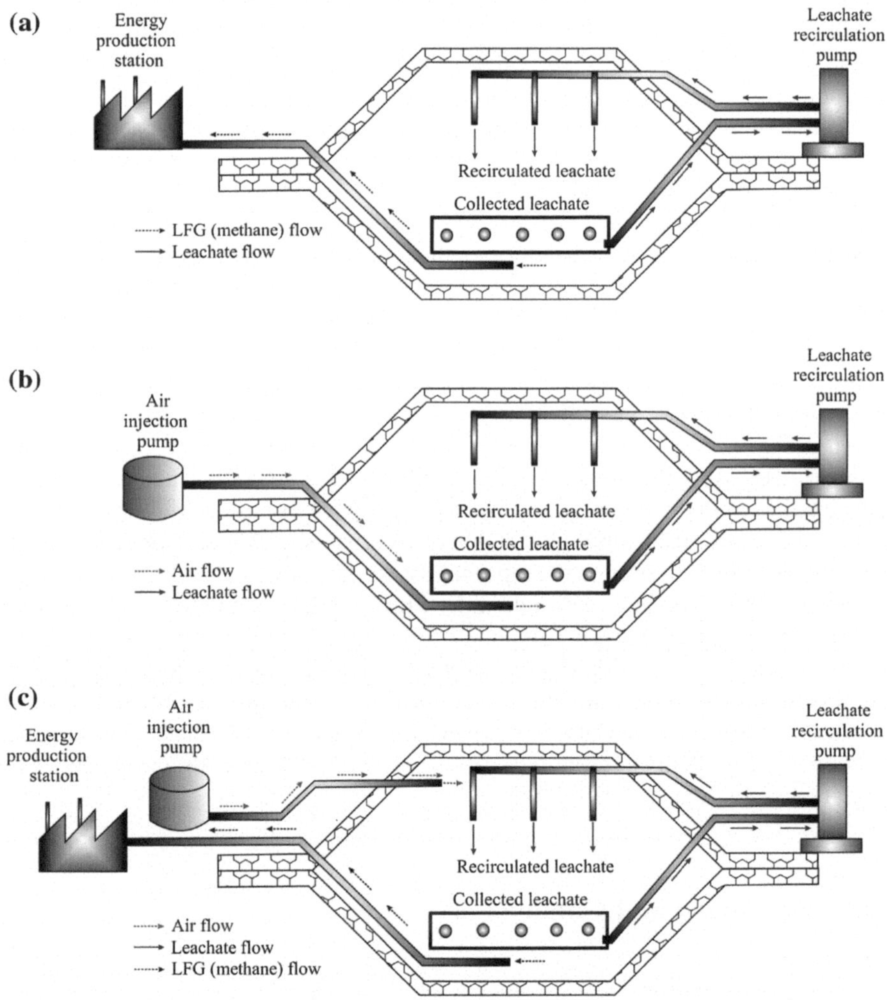

Fig. 18.5 **a** Anaerobic bioreactor landfills; **b** aerobic bioreactor landfills; **c** hybrid bioreactor landfills

is too small to achieve the required moisture content, then an additional water source (like municipal sewage sludge or industrial wastewater) can be used. The faster anaerobic decomposition leads to production of LFG in a controlled manner, which can then be used as an alternative energy source. Therefore, only carbon dioxide is released into the atmosphere, which has a significantly lower global warming potential compared to methane. In addition, the waste is stabilized within a much lower time span compared to conventional landfill.

Similar to anaerobic bioreactor landfill, an aerobic bioreactor landfill can also be used instead of a conventional landfill as shown in Fig. 18.5b. The aerobic bioreactor landfill is supplied with an air injecting pipe that helps to create aerobic conditions within the landfill. Also, leachate is collected from the bottom and re-circulated within the deposited waste. Aerobic conditions are faster than anaerobic processes which help to stabilize the waste rapidly. They reduce the leachate pollution and emission of methane from biodegradable masses within the landfill.

Alternative to aerobic/anaerobic bioreactor landfill, a hybrid bioreactor landfill can also be used. It uses a sequential aerobic anaerobic treatment for reduction of landfill gas emission from landfills. In this case, at first the aerobic phase in landfill is extended using air supply by horizontal wells or through vacuum induction. Products from the aerobic phase are utilized for methanogenesis in the lower chambers of the landfill where the anaerobic degradation takes place. During complete anaerobic phase within the landfill, a similar condition like anaerobic bioreactor landfill is established within a hybrid bioreactor. An advantage of using this type of landfill is that, methane generation starts at a much earlier phase compared to convention or anaerobic bioreactor landfills (Pawlowska 2014). A hybrid bioreactor landfill is shown in Fig. 18.5c.

Apart from using modified bioreactor landfills, a low-cost sustainable method of reducing the emission of LFG gases into the atmosphere is through biological oxidation of LFG components through suitable microorganisms. Methanotropic microorganisms in landfill cover or on the top soil possess the ability to break down methane in presence of oxygen into carbon dioxide, which has less global warming potential compared to the harmful parent compound. Suitable environment for growth for these types of organisms is provided by properly selecting the filling material, soil texture, temperature or increasing the amount of substrate within the landfill (Pawlowska 2014). Alternatively, often biofilters or biowindows inoculated with micro organisms (especially methanotropic population) are provided as daily covers to serve similar purpose (Scheutz et al. 2011).

18.6 Summary and Conclusions

In recent years, landfills have become an important element for disposal of municipal solid wastes, especially in developing countries like India. Compared to open trenches, they reduce a significant amount of air and water pollution by encasing the MSW within a low permeability soil barrier material. The landfill gases and leachate generated from the gradual decomposition of the organic wastes can also be contained within the gas collection system of the top cover and leachate collection system of the bottom cover. Even then, without proper monitoring the landfill is susceptible to detrimental failures which can lead to highly catastrophic events. Emitted LFG gas through soil cover, methane, pollutes the atmosphere to a large extent. Within the landfill, if not contained properly, it creates a high risk of explosion due to elevated temperatures within a landfill. In addition, the biodegradation of organic waste also creates uneven subsidence of the landfill which can lead to formation of cracks in the top cover or damages in the gas collection system due to differential settlement. An early warning system is possible, if the rate of biodegradation and methane emission at different times can be predicted beforehand. A brief literature review of several models in literature used for estimation of methane gas using zero and first order kinetics is discussed. The capability of various models under specific conditions is also emphasized in detail in this study. In addition, estimation of settlement using various available models in literature is also discussed. The biodegradation settlement models utilize first order kinetics similar to methane emission models. The coupled effect for both the models to estimate methane emission and total settlement due to biodegradation of organic waste was shown in a case study. Finally, remediation of methane emission from landfill is discussed in brief to provide an idea about timely management, monitoring and development of existing landfills for mitigation of pollutants and utilize it as an effective energy source in future.

References

Bjarngard A, Edgers L (1990) Settlement of municipal solid waste landfills. In: Proceedings of the 13th annual Madison waste conference, University of Wisconsin, Madison, Wisconsin, pp 192–205

Cakir AK, Gunerhan H, Hepbasli A (2016) A comparative study on estimating the landfill gas potential: modeling and analysis. Energy Sources Part A: Recovery Util Environ Eff 38:2478–2486

CPCB (2017) Consolidated annual review report on implementation of solid wastes management rules. Central Pollution Control Board, New Delhi

Datta S, Zekkos D (2019) Dependency of landfill gas generation parameters on waste composition based on large-size laboratory degradation experiments. In: Proceedings of the 8th international congress on environmental geotechnics, vol 2. Springer, Singapore, pp 186–193

Edil TB, Ranguette VJ, Wuellner WW (1990) Settlement of municipal refuse. Geotechnics of waste fills: theory and practice. ASTM, West Conshohocken, pp 225–239

El-Fadel M, Shazbak S, Saliby E, Leckie J (1999) Comparative assessment of settlement models for municipal solid waste landfill applications. Waste Manag Res 17(5):347–368

El-Fadel M, Khoury R (2000) Modeling settlement in MSW landfills: a critical review. Crit Rev Environ Sci Technol 30(3):327–361

Fallahizadeh S, Rahmatinia M, Mohammadi Z, Vaezzadeh M, Tajamiri A, Soleimani H (2019) Estimation of methane gas by LandGEM model from Yasuj municipal solid waste landfill, Iran. Methods X(6):391–398

Gibson RE, Lo KY (1961) A theory of soils exhibiting secondary compression. Acta Polytech Scand (C) 10:1–15

Gourc JP, Camp S, Viswanadham BVS, Rajesh S (2010a) Deformation behaviour of clay cap barriers of hazardous waste containment systems: full-scale and centrifuge tests. Geotext Geomembranes 28(3):281–291

Gourc JP, Staub MJ, Conte M (2010b) Decoupling MSW settlement into mechanical and biochemical processes—modeling and validation on large-scale setups. Waste Manag 30(8–9):1556–1568

Ham RK, Barlaz MA (1989) Chapter 3.1: Measurement and prediction of landfill gas quality and quantity. In: Sanitary landfilling: process, technology and environmental impact. Academic Press, pp 155–166

Hettiarachchi CH, Meegoda JN, Hettiarachchi P (2009) Effect of gas and moisture on modeling of bioreactor landfill settlement. Waste Manag 29:1018–1025

Jafari NH, Stark TD, Thalhamer T (2017) Spatial and temporal characteristics of elevated temperatures in municipal solid waste landfills. Waste Manag 59:286–301

Joshi R, Ahmed S (2016) Status and challenges of municipal solid waste management in India: a review. Cogent Environ Sci 2(1):1–18

Kamalan H, Sabour M, Shariatmadari N (2011) A review on available landfill gas models. J Environ Sci Technol 4:72–92

Ling HI, Leschchnsky D, Mohri I, Kawabata T (1998) Estimation of municipal solid waste landfill settlement. J Geotech Geoenviron Eng 124(1):21–28

Majdinasab A, Zhang Z, Yuan Q (2017) Modeling of landfill gas generation: a review. Rev Environ Sci Biotechnol 16:361–380

Narayana T (2009) Municipal solid waste management in India: from waste disposal to recovery of resources? Waste Manag 29:1163–1166

Natsev M (1998) Modeling landfill gas generation and migration in sanitary landfills and geological formations. Ph.D. thesis, LAVAL University

Park HI, Lee SR (1997) Long-term settlement behavior of landfills with refuse decomposition. J Resour Manag Technol 24(4):159–165

Pawlowska M (2014) Mitigation of landfill gas emissions. CRC Press, London, UK

Rajaram V, Siddiqui FZ, Khan ME (2011) From landfill gas to energy – technologies and challenges. CRC Press, Boca Raton, Florida

Rajesh S (2018a) Chapter 16: Waste containment system to limit landfill gas emission—mechanism, measurement, and performance assessment. In: Environmental contaminants. Energy, environment, and sustainability. Springer, Singapore, pp 371–389

Rajesh S (2018b) Comprehensive characteristics of fresh and processed MSW generated in Kanpur city. Geotechnics for natural and engineered sustainable technologies: GeoNEst (developments in geotechnical engineering). Springer, Singapore, pp 291–301

Rajesh S, Babel K, Mishra SK (2017) Reliability-based assessment of municipal solid waste landfill slope. J Hazard Toxic Radioact Waste 21(2):04016016

Rajesh S, Gourc JP, Viswanadham BVS (2014) Evaluation of gas permeability and mechanical behaviour of soil barriers of landfill cap covers through laboratory tests. Appl Clay Sci 97–98:200–214

Rajesh S, Khan V (2018) Characterisation of water sorption and retention behaviour of partially saturated GCLs using vapor equilibrium and filter paper methods. Appl Clay Sci 157:177–188

Rajesh S, Naik AA, Khan V (2016) Performance of compacted soil barriers under advective gas flow condition. ASCE Geotechnical Sp. Pub. No. 271, Geo-Chicago 2016, pp 212–221

Rajesh S, Rao BH, Sreedeep S, Arnepalli DN (2015) Environmental geotechnology: an Indian perspective. Environ Geotech 2(6):336–348

Rajesh S, Viswanadham BVS (2012) Centrifuge and numerical study on the behaviour of soil barriers under differential settlements. J Hazard Toxic Radioact Waste 16(4):284–279

Ramachandra T, Bharath HA, Kulkarni G, Han SS (2018) Municipal solid waste: generation, composition and GHG emissions in Bangalore, India. Renew Sustain Energy Rev 82:1122–1136

Roy S, Rajesh S (2018) Confining pressure dependency of air permeability of unsaturated soil barrier. ASCE Geotechnical Sp. Pub. No. 301, Pan America Unsat 2017, pp 114–123

Scheutz C, Fredenslund AM, Chanton J, Pedersen GB, Kjeldsen P (2011) Mitigation of methane emission from Fakse landfill using a biowindow system. Waste Manag 31:1018–1028

Sharma KD, Jain S (2018) Overview of municipal solid waste generation, composition, and management in India. J Environ Eng 145(3):04018143

Sivakumar Babu GL, Reddy KR, Chouskey SK, Kulkarni HS (2009) Prediction of long-term municipal solid waste landfill settlement using constitutive model. Pract Period Hazard Toxic Radioact Waste Manag 14(2):139–150

US EPA (2016) Inventory of US greenhouse gas emissions and sinks: 1990–2014. EPA 430-R-16-002, Washington

Viswanadham BVS, Rajesh S (2009) Centrifuge model test on clay based engineered barriers subjected to differential settlement. Appl Clay Sci 42(3–4):460–472

Viswanadham BVS, Rajesh S, Bouazza A (2012) Effect of differential settlements on the sealing efficiency of GCL compared to CCLs: centrifuge study. Geotech Eng J SEAGS AGSSEA 43(3):55–61

Viswanadham BVS, Rajesh S, Divya PV, Gourc JP (2011) Influence of randomly distributed geofibers on the integrity of clay-based landfill covers. Geosynth Int 18(5):255–271

Yen BC, Scanlon BS (1975) Sanitary landfill settlement rates. J Geotech Eng 101(5):475–487

Chapter 19
Low-Cost Adsorptive Removal Techniques for Pharmaceuticals and Personal Care Products

Dina Zaman, Manoj Kumar Tiwari and Swati Mishra

Abstract The production and consumption of pharmaceuticals and personal care products (PPCPs) have grown ominously over the last 3–4 decades. PPCPs, often considered as emerging contaminants, are being perceived as a serious risk to receiving environments, especially water bodies, due to their ecotoxicological effects. Further, many of the PPCPs are generally persistent, leading to their environmental accumulation, which is evident from the several PPCPs detected in rivers, lakes, groundwater, and soils at variable concentration levels. Although high-end and energy intensive systems like membrane processes are fairly effective in the removal of PPCPs from water or wastewater, conventional treatment technologies often fail to remove PPCPs, and hence treated effluents from various sewage treatment plants have been reported to contain PPCPs from parts per million (ppm) to parts per trillion (ppt) levels. This chapter will discuss the cost effective technologies, especially adsorptive removal methods, being developed for the remediation, recovery, and treatment of PPCPs. A series of low-cost natural and synthetic adsorbents are being investigated, and have shown variable effectiveness and potential for the removal of PPCPs. The chapter will include a state-of-art literature summary on various low-cost adsorbents tested for the removal of selective PPCPs.

Keywords PPCPs · Ecotoxicology · Remediation · Adsorption · Low-cost adsorbents

19.1 Introduction

Pharmaceuticals and personal care products (PPCPs) refer to a significantly large consortia of synthetic chemicals, primarily utilised as medicinal drugs for treating diseases and as constituents of personal health and hygiene products, as well as in cosmetic products (Boxall et al. 2012). PPCPs, especially those used in medical

---

D. Zaman · M. K. Tiwari (✉) · S. Mishra
School of Water Resources, Indian Institute of Technology
Kharagpur, Kharagpur 721302, India
e-mail: mktiwari@swr.iitkgp.ac.in

T. Gupta et al. (eds.), *Measurement, Analysis and Remediation of Environmental Pollutants*, Energy, Environment, and Sustainability,
https://doi.org/10.1007/978-981-15-0540-9_19

treatment have been a boon to human health and well-being and it is virtually impossible to entirely avoid them in day-to-day life. By and large, PPCPs are chemically persistent, biologically active, either water-soluble or lipid-soluble and non-volatile at a normal temperature (Barceló and Petrovic 2007). Most PPCPs are also polar organic complex molecules, varying widely in terms of molecular mass, degree of ionisation, photo-activity and response to pH (Rivera-Utrilla et al. 2013). Based on therapeutic use, pharmaceuticals are chiefly classified as analgesics, antibiotics, antimicrobials, estrogens and hormonal compounds, lipid regulators, stimulants and X-ray contrast media, which are further divided into several sub-classes (Bhandari et al. 2009). Similarly, personal care products are categorised as fragrances, photoinitiators, disinfectants, antiseptics, preservatives, anti-oxidants, surfactants, flame retardants, insect repellents and so on.

In pharmacology, several classification systems are in place to categorise PPCPs, based on their bio-chemical property, mode and mechanism of action in living bodies (Ku 2008). From the viewpoint of environmental technology, the Biopharmaceutics Classification System (BCS) is most appropriate, in which all pharmaceutical drugs are classified into four categories on the basis of solubility and permeability as—high solubility/high permeability (Class I), low solubility/high permeability (Class II), high solubility/low permeability (Class III) and low solubility/low permeability (Class IV) (Ku 2008). Infact, the solubility of PPCPs is the most crucial parameter, based on which the route of bioavailability of a particular compound in the living body is determined (Mannhold et al. 2009; Mantri and Sanghvi 2017). Likewise, the behaviour of PPCPs is highly dependent on pH, salinity, temperature and exposure to radiation in the environment (Ferguson et al. 2013; Oh et al. 2016). Owing to such variability in physico-chemical behaviour of the PPCPs, it has become a challenge to analyse and characterise these compounds in a common platform.

Although the first reported study, confirming the presence of PPCPs in the aquatic ecosystem was carried out way back in 1977 (Hignite and Azarnoff 1977), its complete metabolic cycle and long-term hazardous impacts on humans and other organisms are still not clearly understood. In environmental technology, PPCPs are categorised within a newly formulated group of trace impurities, known as 'emerging contaminants'. Other trace chemicals recognised as emerging contaminants are plasticizers, pesticides, nanomaterials, heavy metals and so on (Rivera-Utrilla et al. 2013). Owing to their reasonably high solubility and polar characteristics, most PPCPs remain in aqueous matrices for a significantly long time and act as a potential hazard, if non-target living organisms are continually exposed to unintended PPCPs.

The pharmaceutical industry is flourishing in developing countries like India, China, South Africa and Brazil at an unprecedented rate, due to competitive global markets (Osuji and Umahi 2012). Alongside, due to lack of proper regulations on the permissible limits of PPCPs in industrial effluent, pharmaceutical companies usually treat their effluents to reduce usual organic loads, and are not liable to remove PPCPs before discharging it into the natural water bodies. This has led to the entrance and subsequent accumulation of PPCP residues and their transformation products (metabolites and conjugates) in the geo-aquatic environment through various pathways, which act as an impending environmental risk to humans and overall environmental ecology (Saggioro et al. 2018). Primary and secondary processes in the conventional wastewater treatment plants (WWTPs) are rather inefficient in removing all PPCPs from industrial as well as municipal wastewater (Grassi et al. 2012). For the little removal that may take place in the WWTPs, its removal efficiency is unpredictable and widely varying from as low as 10% to as high as 90% for different contaminants and different treatment processes (Rivera-Utrilla et al. 2013). Specialised advanced or tertiary treatment units are now being investigated upon, for removing PPCPs as target contaminants (Hernández et al. 2019). These units may employ diverse techniques like advanced oxidation, membrane filtration, separation techniques and so on, which vary widely in terms of their operational and energy cost, removal efficiency as well as contaminant specificity (Rivera-Utrilla et al. 2013). Out of these, adsorption can act as an effective separation technique which is of easy operability, lower energy intensity, and cost efficient than other comparable methods.

19.2 Pharmaceuticals in the Environment

Presently, PPCPs have an unprecedented demand for use in humans, livestock, poultry, aquaculture, agricultural and cash crops. With time, PPCP residues have incessantly percolated into the geo-aquatic ecosystem from several points of usage through complicated pathways. Instances of occurrence of pharmaceutical residues have been reported in surface water bodies, seawater, groundwater and even in drinking water in several parts of the world (Kümmerer 2009a, b). The spatial presence of PPCP residues is so extensive that it has found its way even to the most remote locations on the earth like the Arctic and Antarctic environment (Hernández et al. 2019; Kallenborn et al. 2018).

After administering on human and animals, excess PPCP residues are excreted from the body in its original form or as metabolites and conjugates through bile or urine (Rivera-Utrilla et al. 2013). Overuse and misuse of PPCPs has also aggravated the residual excretion of PPCPs and its transformation products (Nielsen and Barratt 2009; Porter and Grills 2015). These compounds reach the influent of WWTPs, as an assortment of chemicals and may be partially removed in the primary and secondary treatment processes. In general, hydrophobic PPCPs may get accumulated in the sludge produced in the WWTPs, whereas hydrophilic PPCPs usually remain in soluble form in the effluent (Gomes et al. 2017). Thus, PPCPs find their way into the geo-aquatic ecosystem with the wastewater effluent being released by industries and municipalities to natural water bodies like rivers, ponds, lakes and oceans, without adequate or no treatment at all. Pharmaceuticals administered on aquatic organisms are directly introduced into the surface water bodies, from where it permeates to the soil and then percolates to the underlying groundwater table (Miller et al. 2018). Similarly, pharmaceuticals used for crops are directly applied to the soil, from where it can percolate to reach the groundwater or washed away by rain to reach surface water bodies (Panthi et al. 2019).

PPCPs may inadvertently assimilate in the soil and subsequently pass into the groundwater zone through land application of wastewater effluent as irrigation water (Kibuye et al. 2019). Accumulation of PPCPs in the soil and successive transport to the groundwater may also occur when contaminant laden bio-solids or sludge from WWTPs and manure of animals treated with veterinary pharmaceuticals are applied as organic fertilizer to agricultural soil (Wu et al. 2015; Kaczala and Blum 2016). The problem of PPCP contamination may be aggravated by the recent impetus on alternate use of wastewater effluent and sludge as irrigation water and bio-fertilizer, respectively in agriculture. Countries like Spain and Italy use a huge volume of recycled wastewater and nearly 50% of total irrigation water used in Israel is from reclaimed or recycled wastewater (Paltiel et al. 2016). Although the soil matrix may act as an efficient filter for containing salinity and pathogens from the wastewater effluent, trace level contaminants like heavy metals, PPCPs and fertilizer residues are only partially absorbed in the soil and the rest may percolate the geo-obstructions to reach the groundwater table (Ramirez-Fuentes et al. 2002; Candela et al. 2007). For instance, compounds like Bisphenol A and Triclosan were retained more efficiently in the soil matrix, whereas Ibuprofen easily percolated to the underlying groundwater zone (Xu et al. 2009). As an obvious presumption, PPCPs have been detected in treated drinking water, as raw water is sourced from surface or groundwater without examining it for the presence of PPCPs and conventional water treatment plants (WTPs) not ensuring their complete removal (Reis et al. 2019). Practically, all types of PPCPs have been detected in drinking water supplied through taps from several places in different continents, which proves the definitive consumption of these chemicals by human beings unsuspectingly (Davoli et al. 2018). Additionally, high concentration of PPCPs also occur in landfills from irresponsible dumping of unused and expired PPCPs, later appearing in the produced landfill leachate and is then transported to the soil media and underlying groundwater table (Tong et al. 2011; Vellinga et al. 2014; Yi et al. 2017). The major representative routes of fate and transport of PPCPs in the environment are depicted in Fig. 19.1.

India is one of the leading manufacturers of pharmaceutical products, in terms of volume and economic value, and the largest exporter of generic drugs (Carlsson et al. 2009). Manufacturing units are located in all major cities of India like Indore, Nashik, Aurangabad, Hyderabad, Chennai, Mumbai and so on. PPCPs and other such contaminants are on the rise in all water matrices in India, specifically in the vicinity of the pharmaceutical manufacturers. Detection of PPCPs in India was carried out by Larsson et al. (2007), in the pharmaceutical wastewater effluent from a common industrial wastewater treatment plant, set up for treating the effluent of 90 registered drug manufaturers. Out of the PPCPs screened, 11 of them were detected at a concentration higher than 100 μg/L. This study had also reported the highest observed concentration of Ciprofloxacin in any water matrix in the world, i.e. 31,000 μg/L. Mutiyar and Mittal (Mutiyar and Mittal 2014) summarised the literature available on the occurrence of PPCP in various water matrices in India,

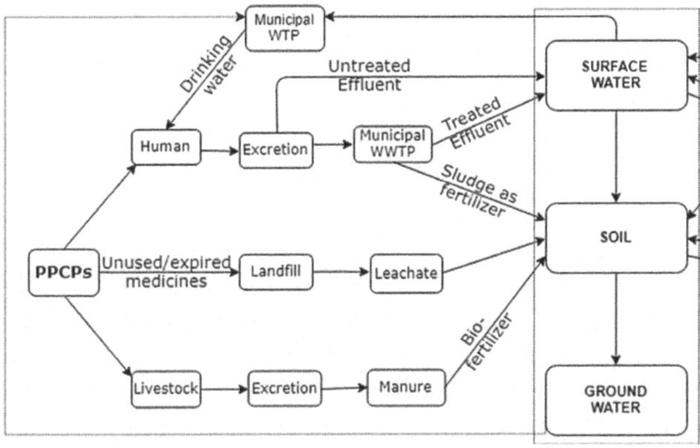

Fig. 19.1 Representative routes of fate and transport of PPCPs in the environment

from industrial effluents, lakes, rivers, hospital effluents, treated sewage and groundwater and utilised the secondary data for risk assessment. It was concluded in their study that industrial wastewater from pharmaceutical companies posed the greatest threat and Ciprofloxacin residues showed highest hazard quotient for all water bodies. PPCP and their metabolites were investigated in several sewage treatment plants in India by Subedi et al. (2015, 2017) and some of the drugs like Mefenamic acid, Paraxanthine, Atorvastatin were reported for the first time in India, in these studies. A recent study on the occurrence and distribution of 15 PPCPs in surface and groundwater along the Ganges river basin was carried out by Sharma et al. (2019). The concentration of the selected PPCP ranged from 54.7 to 826 ng/L in the Ganges river and from 34 to 293 ng/L in the groundwater. Caffeine was the highest detected PPCP in the Ganges, which is potentially associated to feminization of fishes (Larsson et al. 2007). Overall, studies on the occurrence and distribution of PPCPs were mostly concentrated to the rivers in Southern India (Shanmugam et al. 2014; Balakrishna et al. 2017). Investigative studies in groundwater were less frequently reported (Sharma et al. 2019; Jindal et al. 2015).

19.3 Ecotoxicological Effects of PPCPs

Although there are ample reported studies indicating the potentially hazardous consequences of PPCPs in the ecosystem, their complete chemical metabolism in the living body, long-term impacts of low-concentration exposure on different organisms and synergistic effects in presence of other compounds are still largely unknown (Kyzas et al. 2015). Owing to the non-volatility and polarity of most PPCP compounds, they are rather stable in the aqueous media, without degrading into their innocuous forms and persisting in the environment long after their release (Kyzas et al. 2015). At times, the transformation products—metabolites and conjugates formed from the PPCPs are potentially more harmful than the original compound (Michael et al. 2014). In other cases, PPCPs may lead to the formation of more toxic compounds, on reacting with other water-borne chemicals in the WTPs or WWTPs. For instance, PPCPs present in drinking water sources may react with chlorine and other disinfection by-products in the WTP to produce a plethora of organo-chlorinated sub-species, some of which are highly toxic (Leusch et al. 2019).

Owing to the persistent nature, PPCPs tend to assimilate and magnify in the environment and if consumed, they slowly bio-accumulate or bio-concentrate in the body leading to physiological changes in the living tissues and interfering with the functioning of the endocrine system (Zenker et al. 2014). These compounds may also be taken up by plants and crops and metabolised in their bodies, thus entering the food chain inconspicuously (Klampfl 2018). However, the greatest impact of environmental PPCPs is perhaps on the vast microbial population, both in water and soil matrices (Caracciolo et al. 2015). Due to overuse of these pharmaceuticals, antibiotic resistance in organisms has also posed as a significant health hazard (Laxminarayan and Chaudhury 2016). Instances of interference in the microbial community as well as the production of antibiotic resistant genes and bacteria is the outcome of exposure to incessant levels of PPCPs in the geo-aquatic environment.

The potential toxicity and long term impacts of PPCPs is dependent on the type of compound and its concentration in soluble form, duration of exposure, as well as the type of organism. A very low concentration of a compound may be less toxic for higher-order organisms like human beings, but may have adverse effects on lower-order organisms like fishes and birds. For instance, Carbamazepine compound has not been reported to register any health risk in humans, but it has been identified to cause severe organ damage in fishes at high concentrations (Triebskorn et al. 2007). Compounds like Paraben, Triclosan, Carbamazepine have been reported as hazardous to aquatic life in several instances (Ramaswamy et al. 2011). Till now, there has been no conclusive evidence of human health being severely impacted by PPCP contamination, except during childhood and pregnancy. Vertebrates like tadpoles have been observed to suffer from impaired growth due to PPCPs (Carlsson et al. 2006; Fick et al. 2009). The catastrophic decline of vulture population in Rajasthan (India) was attributed to indirect ingestion of Diclofenac compound provided to cattle (Prakash 1999; Oaks et al. 2004). Fick et al. (2009) predicted prominent antibiotic resistance in their study area near Hyderabad (India), owing to high broad-spectrum antibiotic levels in the water bodies. It was also anticipated in this study that the recycling of activated sludge in the treatment processes could induce the formation of more antibiotic-resistant bacteria. The continual rise in such genes due to constant exposure to a heavy dose of PPCPs, is an imminent health threat as it perpetuates a vicious cycle of producing drug resistant microbes which compel scientists to keep generating newer medicines, which again triggers the advent of new drug-resistant microbes (Chang et al. 2019). In 2015, there were about 480,000 new cases of multidrug resistant tuberculosis and another 100,000 people with rifampicin-resistant tuberculosis (RR-TB) who were also newly eligible for MDR-TB treatment (World Health Organization 2016).

The risk assessment of PPCPs and other substances of chemical origin are quantified in terms of common indicators—Risk Quotient (RQ), Hazard Quotient (HQ) and Hazard Index (HI) (EPA 2017). Different aquatic matrices like hospital effluent, industrial effluent, rivers, lakes and treated sewage in Indian states were inspected and Ciprofloxacin was obtained to have the highest HQ in all the matrices (Ashfaq et al. 2017). Similar risk assessment studies have been carried out for aquatic ecosystems in Lahore (Pakistan) for 11 common PPCPs and highest RQ values, far greater than one were obtained for several aquatic organisms like Daphnia, fishes, *Oncorhynchus mykiss*, *Oryziaslapites*, Pseudomonas putida, and *Microcystis aeruginosa* (Santos et al. 2010).

19.3.1 Analytical Techniques for the Detection of PPCPs

The development of new analytical techniques with high detection accuracy has greatly facilitated the analysis and monitoring of PPCPs, which typically occur in trace concentrations in the environment, ranging from ng/l to μg/L (Rivera-Utrilla et al. 2013). Commonly used separation techniques for PPCPs are solid phase extraction (SPE), solid phase microextraction (SPME), liquid liquid extraction (LLE), hollow fibre supported liquid phase microextraction (HF-LPME) (Zhang et al. 2018). Most frequently used identification techniques developed in the late 20th century for PPCP analysis are Gas-chromatography with mass spectroscopy (GC-MS), Gas chromatography with tandem mass spectrometry (GC-MS/MS), Gas chromatography-negative chemical ionization-mass spectrometry (GC-NCl-MS), Liquid-chromatography with fluorescence detection (LC-FD), Liquid-chromatography with mass-spectroscopy (LC-MS), Liquid-chromatography with tandem mass spectrometry (LC-MS/MS), High performance liquid chromatography with mass spectrometry (HPLC-MS), High performance liquid chromatography with tandem mass spectrometry (HPLC-MS/MS), Direct injection liquid chromatography with tandem mass spectrometry (DI-LC-MS/MS), Liquid chromatography quadrupole linear ion trap-mass spectrometry (LC-QqLIT-MS) and so on (Zhang et al. 2018). The limit of detection (LOD) or the limit of quantification of (LOQ) facilitated by these analytical techniques may lie in the range of 0.05 to 200 ng/L, depending on various contaminant or procedure-related factors. In recent developments, electrogenerated Chemiluminescence (ECL) biosensors like immunoassays, immunosensors, aptamer-based biosensors, enzyme-based biosensors, chiral sensors and molecularly imprinted polymers (MIP)-based sensors are also being used for clinical testing and pharmaceutical analysis (Clara et al. 2005).

19.4 Removal and Remediation Techniques for PPCPs

With the ongoing emphasis on reuse and reclamation of wastewater effluent for various purposes, it has become imperative to ensure that all toxic contaminant species be removed before its secondary application. Conventional municipal water and wastewater treatment plants (WTPs and WWTPs) do not directly facilitate the removal of PPCPs, although some of these contaminants are either unintentionally removed or transformed or degraded to other by-products during the treatment processes (Rivera-Utrilla et al. 2013). Even for the compounds removed in a WWTP, the removal efficiency is widely variant for different PPCP compounds as well as for the treatment processes employed. For instance, Diclofenac was removed by less than 15%, whereas Ibuprofen was removed by 98% in the same membrane bioreactor in a WWTP in Austria (Petrović et al. 2005). On the other hand, the same compound Diclofenac was removed by 86% in another WWTP (Zhang et al. 2008). Similarly, Carbamazepine is one of the most frequently detected PPCP in Indian surface water bodies and is only negligibly removed in the WWTP by about 10% (Ramaswamy et al. 2011; Zhang et al. 2008). Removal of PPCP compounds in WWTPs may occur due to adsorption or absorption during primary treatment or precipitation along with the bio-sludge during secondary treatment. The residual PPCPs and their metabolites remain soluble in the effluent. Further, chemicals used in the disinfection processes in a WTP may instead lead to the formation of more toxic products, on reaction with PPCPs (Leusch et al. 2019). Hence, it seems obvious that effluent from pharmaceutical industries and hospitals, etc. should be treated at its source of production itself, where the probable PPCP concentration is the highest, using specialised treatment processes, before being diluted with municipal wastewater (Lee et al. 2014).

19.4.1 Common Techniques for PPCPs Removal

Removal of micro-contaminants is enabled with the inclusion of advanced techniques, particularly for PPCPs as target contaminants, after applying conventional primary and secondary treatment processes to the wastewater effluent. Based on the inherent chemical nature of the PPCP compounds, its removal efficiency and remediation techniques will be different. The specific techniques commonly applied for PPCP removal are physico-chemical processes, advanced oxidation processes, membrane-based separation and biological processes or a synergistic combination of some of these processes, to enhance the performance (Rivera-Utrilla et al. 2013; Gomes et al. 2017; Kyzas et al. 2015).

Advanced oxidation processes (AOPs) enable efficient oxidation of the organic contaminants like PPCPs by employing highly reactive free radical species. Advanced oxidation can be carried out using ozone, UV radiation, gamma radiation, anodic oxidation or electro-oxidation, out of which ozonation-based techniques are most frequently used (Rivera-Utrilla et al. 2013). Oxidation of hospital effluent from an MBR containing 25 contaminants was carried out with ozone and ozone + H_2O_2, under pH controlled conditions, with satisfactory removal efficiencies (Lee et al. 2014). Several enhanced ozonation techniques have been proposed and investigated, to make the process more effective in the removal of contaminants. Some of these common hybrid techniques are catalytic ozonation using materials derived from carbon, alumina, manganese, etc. as heterogeneous catalysts. The ozonation process can be carried in combination with presence of $H_2O_2(O_3/H_2O_2)$, in presence of light—photo-aided ozonation (O_3/UV, O_3/UV/TiO_2), in presence of solar power—solar photo-aided ozonation (O_3/TiO_2/light, O_3/Fe(III)/light, O_3/Fe(III)/H_2O_2/light) and in presence of ultrasonic waves—untrasonication-aided ozonation (Gomes et al. 2017). Ozonation can also be combined with conventional physico-chemical and biological processes to augment the removal efficiencies of the combined systems. However, ozonation may lead to the generation of undesirable intermediates which may be more harmful than the parent compound itself (Kyzas et al. 2015).

Conventional biological treatment units like activated sludge process (ASP) is capable of partial removal of some PPCPs to a certain extent, owing to adsorption and biodegradation mechanism (Tiwari et al. 2017). Compounds like Ibuprofen, Naproxen and Estrone have high removal rates whereas, certain others like Diclofenac are scantily removed in the ASP. Removal efficiencies also tend to fluctuate seasonally which is difficult to explain conclusively. Membrane bioreactors (MBR) have been more efficient in the removal and biodegradation of PPCPs, in comparison to ASP (Mutiyar and Mittal 2014). The removal efficiency of MBR can be further enhanced by using an attached biofilm growth medium to facilitate the growth of assorted microbes. MBR can also be coupled with the optimal dosage of biologically active carbon (BAC) to enhance the filtration performance, by reducing fouling and enabling better retaining of microbes. Biodegradation of several classes of PPCPs can also be carried out by the application of three strains of white-rot fungi (Rodarte-Morales et al. 2011). Most PPCPs were effectively degraded by all the tested strains, with only a few exceptions. Use of bio-filters has also garnered attention in recent times, using materials like sand/anthracite (SA) or biologically activated carbon (BAC) as the filtering medium. A combined bio-adsorption process consisting of BAC was proposed for the removal of PPCPs from drinking water, which could successfully remove around 50% of the total PPCP mass (Fu et al. 2019).

19.4.2 Hybrid Techniques for PPCP Removal

It is evident from previous studies that more than one mechanism is responsible for the removal of PPCPs in any treatment process. These mechanisms work as complementary to each other and can enhance the performance efficiency of the removal systems. A combined nano-filtration and ozone-based advanced oxidation process was proposed by Liu et al. (2014), wherein the effluent from secondary treatment was passed through a nano-filtration unit and the subsequent concentrate was treated using three advanced oxidation processes—UV_{254} photolysis, ozonation, and UV/O_3. Out of these, UV/O_3 was most effective in degrading the PPCPs from the concentrate up to 87% and eliminating acute toxicity up to 58%. This method depicted the potential for successful expulsion of zero contaminant discharge, with further research. In another study, ozonation was combined with biological treatment in a three-step process as 'bio-ozone-bio' that is performance-enhancing as well as cost-effective (de Wilt et al. 2018). Contaminants with different biodegradability were effectively removed by the symbiotic arrangement of both processes, and removal efficiency increased with the increase in ozone dose and hydraulic retention time. Biological processes applied before ozonation resulted in less ozone requirement and the last stage biological treatment was likely to degrade the by-products of ozonation.

Both low-strength and high-strength real pharmaceutical effluent were efficiently treated using a hybrid approach, consisting of Fenton's treatment as a pre-treatment process, followed by aerobic biological treatment (Changotra et al. 2019a). Out of the different Fenton's processes applied, solar-driven Photo-Fenton combined with biological treatment was observed to obtain an average of 83% of COD removal and complete detoxification of the wastewater was observed for cytotoxicity assessment with different microbes using zone of inhibition test. Another such amalgamated approach for the treatment of real pharmaceutical effluent was proposed by the same authors, combining hybrid coagulation, gamma irradiation, and biological treatment to achieve degradation and detoxification of PPCPs (Changotra et al. 2019b). Gamma irradiation was employed as both pre- and post-treatment to biological process and a high TOC and COD removal as well as overall efficiency of more than 90% was obtained.

19.4.3 Full-Scale Studies for PPCP Removal

Full-scale studies for the removal of PPCPs from real wastewater are few and far between. An innovative full-scale system implementing the principle of natural biological purification was employed for PPCP removal in Spain (Hijosa-Valsero et al. 2010). The hybrid system consisted of a series of ponds, surface flow constructed wetland and horizontal subsurface flow constructed wetland and removal efficiency of 70% was obtained, which was comparable to the performance of a WWTP in the same region. Presence of multiple metabolic pathways and better photo-degradation was evident in these natural systems than in a WWTP. A full-scale membrane bioreactor for the monitoring and removal of 18 PPCPs in the leachate generated in a municipal solid-waste landfill in China was conducted by Sui et al. (2017). The removal efficiency was observed to be higher than that of municipal and hospital effluent, presumably due to the presence of lesser biodegradable substrates in the former and a much longer retention time of 9 days. However, the removal efficiency varied largely for the monitored PPCPs, with negligible removal for Carbamazepine and Trimethoprim, but high removal for all other compounds.

A full-scale hybrid constructed wetland system was proposed for the removal of PPCP, along with endocrine-disrupting chemicals and antibiotic-resistance genes from leachate generated in a solid-waste landfill in Singapore (Yi et al. 2017). The proposed system consisted of several units like equalisation tanks, aerobic lagoons, reed beds, and polishing ponds, out of which aerated lagoons and reed beds were the most effective in PPCP treatment. The combined system effectively removed 90% of the contaminants including antibiotic resistance genes. Another wetland in Singapore (Lorong Holus) constructed for the treatment of landfill leachate was investigated for the phytoremediation potential and bioaccumulation characteristics on eight plant species (Wang et al. 2019). Even recalcitrant compound like Bisphenol A was observed to be removed efficiently in this system. Combined effects of passive diffusion, phytoremediation, adsorption, photolysis, and microbial degradation appear to contribute to PPCP remediation.

19.5 Adsorptive Removal of PPCPs

Adsorption mechanism is one of the most extensively researched methods for the removal of trace aqueous impurities like heavy metals, dyes and paints, and other emerging contaminants (Burakov et al. 2018; Quesada et al. 2019; Zhou et al. 2019a). A wide range of materials has been utilised as adsorbents for the uptake of PPCPs from synthetic as well as real wastewater, out of which carbon-based or carbonaceous materials are the most popular (Xiang et al. 2019). Variants of commonly available carbon-based adsorbents are activated carbon (AC)—granular activated carbon (GAC) and powdered activated carbon (PAC), charcoal, graphene, and carbon nanotubes (Smith and Rodrigues 2015). Commercial activated carbon (CAC) is the most widely used carbonaceous material for the removal of several PPCPs and a high adsorption capacity has been observed in numerous previous studies, specifically for antibiotics (Ahmed 2017). Use of a new generation of novel adsorbents like nanomaterials, polymers, resins, and other innovative adsorbents have been a recent development in adsorption studies, which are comparatively costly but have high removal efficiency, consistent adsorbing properties and better regeneration (Ali 2012; Basheer 2018; Crini et al. 2019). Based on the origin of the material, adsorbents can be broadly classified as natural, modified and synthetic. However, this is not an exhaustive classification, as properties of adsorbent materials are too diverse to be included within a few classes. Adsorbents can also be sub-classified based on their chemical nature as organic, inorganic, organo-metallic adsorbents, based on comparative cost as costly and low-cost adsorbents, based on pore size as macroporous, mesoporous and microporous and so on.

Apart from the chosen adsorbate-adsorbent combination, the mode of contact system between the two also plays a crucial role in the effectiveness of the adsorption process. Most commonly used contact systems are batch mode and fixed-bed mode, while other available contact systems include pulsed bed, moving mat filters and fluidized beds (Ahmed et al. 2015). Competitive adsorption of Caffeine in presence of natural organic matter on three adsorbents—commercial activated carbon (F400), activated carbon synthesised from peach stones and commercial clay (sepiolite) was carried out in laboratory-scale (Álvarez-Torrellas et al. 2016). All adsorbents could remove Caffeine, although the presence of natural organic matter created competition for the target contaminant. Comparative adsorption using GAC and PAC was carried out for real municipal wastewater from three Swedish plants for treating 22 PPCPs (Xiang et al. 2019). Performance of PAC was slightly better than GAC, with most contaminants being removed by more than 80%. A full-scale granular activated-based adsorption system for the removal of PPCPs from sewage treatment plants in the UK was installed and monitored for the removal of 11 PPCPs (Smith and Rodrigues 2015). Removal of different compounds was between 17% to 98% in the GAC effluent, but it was likely to reduce with time as the GAC bed gets saturated. Mono- and multi-contaminant adsorption for Bisphenol A, Carbamazepine, and Chloroxylenol was carried out on a novel polymeric material (β-cyclodextrin) and highly satisfactory removal efficiencies were obtained, even for competitive adsorption, presence of fulvic acid, change in pH and ionic strength (Zhou et al. 2019b).

19.5.1 Low-Cost Adsorbent Materials

The primary expenditure of the adsorption process originates from the production and conveyance of the adsorbent material as well as the cost of its regeneration for reuse (Quesada et al. 2019). Using low-cost adsorbents is supposed to have a significant impact on cost optimisation of the treatment process. Owing to the potential environmental benefits, low-cost adsorbents are also called as ‘green adsorbents’ (Burakov et al. 2018). Low-cost adsorbents may be sourced from various different origins that include abundantly available geo-materials and minerals (like natural clay, bentonite and montmorillonite), freely available waste materials (like paper mill sludge, rice husk, straw), waste-derived materials (like sludge biochar, rice straw biochar, bone char, carbon black) and industrial by-products (like fly ash, red mud) (de Andrade et al. 2018; Burakov et al. 2018). Low-cost variants of adsorbents can also be produced by preparing activated carbon (AC) from waste materials, rather than using commercial activated carbon. ACs produced from waste materials like cork, olive stones, peach stones, plastic waste, husk, leaves pine chip and others have been extensively used for adsorption studies of PPCPs (Hijosa-Valsero et al. 2010).

Natural geo-sourced materials are perhaps the cheapest kind of adsorbents available. For instance, montmorillonite is around 20 times cheaper than AC (de Andrade et al. 2018). An extensive investigation of adsorptive removal of PPCPs using clay minerals was carried out by Chang et al. (2019). Several clay types with different mineral compositions were investigated for its adsorption potential and it was observed that clay was an excellent adsorbent for most common PPCPs, especially at low pH values and high organic content. Cation exchange was perceived to be the most dominant mechanism for removal of PPCPs using clays. PPCP compounds with fewer functional groups like macrolides and sulfonamides were poorly adsorbed on clays. Although these adsorbents are economic and easily sourced, their adsorptive capacity greatly varies for different target PPCPs, as it is difficult to obtain them with standard adsorbing characteristics. The adsorbing properties of natural materials can be further enhanced by modifying the original adsorbents through acid activation, thermal treatment, chemical treatment, pyrolysis and so on or by preparing composites of two or more materials with complementary adsorbing properties (de Andrade et al. 2018; Swarcewicz et al. 2013). Clays are easily modified by substitution of its interlayer cationic sites with organo-cations (HDTMA) to obtain *organo-clays*, or with polymeric metal species (PILCs) to obtain *metal oxide pillared clays* (Jiang and Ashekuzzaman 2012). Such modifications may bring about an increase in its cation exchange capacity and increase its affinity to organic compounds. It also increases the specificity of the adsorbent towards the removal of a particular species of contaminant, compared to its natural form (Cabrera-Lafaurie et al. 2014, 2015). Similarly, inorganic alumina-silicate minerals like natural zeolites and its modified counterparts also work as efficient low-cost adsorbents. Although natural zeolites are not very efficient in adsorption of organic contaminants like PPCPs, synthetically produced zeolites doped with surfactant and transition metal, hybrid zeolites with nanoparticle coatings have been specifically prepared and used for PPCP removal with satisfactory results (Cabrera-Lafaurie et al. 2014; Attia et al. 2013; Sun et al. 2017). Clay and its variants used for the removal of PPCPs are briefly discussed in Table 19.1.

Table 19.1 Natural adsorbents used for removal of PPCPs for aqueous media

Adsorbent	Target PPCPs	References	Applicability
Commercial sepiolite Sepiolite clay	Caffeine	(Álvarez-Torrellas et al. 2016; Sotelo et al. 2013)	***Advantages*** • Lowest cost • High removal efficiency • Readily available • Easily disposable • No secondary pollution ***Limitations*** • High specificity towards contaminant • Modifications required to enhance performance • Low reproducibility of results • Inconsistent properties • Low regeneration and reuse potential
Natural clay (smectite-kaolinite)	Ibuprofen, Naproxen, Carbamazepine	(Khazri et al. 2017)	
Bentonite Organobentonite	Ciprofloxacin, Amoxicillin	(Genç and Dogan 2015)	
Birnessite	Ciprofloxacin	(Jiang et al. 2013)	
Rectorite, Kaolinite	Ciprofloxacin	(Wang et al. 2011)	
Illite	Ciprofloxacin	(Wang et al. 2011)	
Montmorillonite	Ciprofloxacin, Trimethoprim	(de Andrade et al. 2018)	
Zeolite	Ciprofloxacin	(Genç and Dogan 2015)	
Pumice	Ciprofloxacin	(Genç and Dogan 2015)	

Waste materials like agricultural and industrial by-products may be utilised as adsorbent materials directly or to prepare waste-derived adsorbents after chemical modifications (Grassi et al. 2012). Likewise, agro-industrial wastes can also be employed in the production of biochar and low-cost activated carbon (Silva et al. 2018). Adsorbents can also be produced from industrial and municipal sewage (Nielsen and Bandosz 2016). Components in agri- and plant-based materials that facilitate adsorbing properties are cellulose, hemicellulose, lignin, lipids, proteins, sugars, hydrocarbon, starch, etc., whereas alumina and silica components in industrial by-products enable adsorption. Moreover, dead microbial biomass can also be used as an adsorbent to remove PPCPs, which is specifically known as biosorption (Daneshvar et al. 2018). Microalgae and seaweeds and microbial by-product from industrial fermentation processes act as potentially good biosorbents. The literature database on using waste-derived adsorbents is extensively large, as waste derivatives of all classes have been used for adsorption studies, depending on the waste materials produced in different regions having varied agricultural practices and industrial affluence (Silva et al. 2018). In the category of waste-derived adsorbents, biochar is perhaps the most promising and popular. Biochar is, to some extent structurally similar to activated carbon, but way inexpensive, as the former is prepared at a lower temperature (using slow pyrolysis) without any activation process (Tan et al. 2015). The average cost of biochar is around six times lesser than that of CAC. The adsorbing property of biochar is heavily dependent on the raw source of biomass used and the pyrolysis process employed, which predominantly determines its elemental chemical composition. Recent works on waste-derived materials used for the adsorption of PPCPs have been discussed in Table 19.2.

Apart from the natural and derived adsorbents, synthetically prepared inorganic solids with multi-layered structure and high functional agents, called layered double hydroxides have garnered attention as a low-cost and non-toxic adsorbent for PPCP removal (Elhalil et al. 2017; Gupta et al. 2019). A relatively new, low-cost polymeric adsorbent is metal organic frameworks (MOFs), which can be used either in its original or modified form (Drout et al. 2019). Their structural and chemical properties like size, porosity, functionality can be controlled during synthesis (de Andrade et al. 2018). Hence, they are fairly consistent in their adsorbing properties than the natural, unmodified adsorbents and have better reuse and regeneration potential. The most popular adsorbent in this category is perhaps MIL-101 and its subsequent functional modifications, some of which perform way better than AC (Hasan et al. 2012). Some of the recent MOF-based adsorption studies are mentioned in Table 19.3.

It is often observed that the adsorptive removal of natural adsorbents is lower than their commercial counterparts. Such materials can be modified through different functionalization methods to increase their removal capacity. Natural clays are chemically modified to attach functional cationic species, to increase their contaminant affinity and specificity (de Andrade et al. 2018). High-charge swelling mica and organo-functionlized mica were prepared by functionalizing natural clay and utilized for contaminant removal (Martín et al. 2018). Such organo-functionalization of adsorbent surfaces allow for interstitial adsorption of non-ionic pollutants. In

Table 19.2 Waste-derived materials used as adsorbents for removal of PPCPs

Source	Adsorbent	PPCP	References	Applicability
Silkworm production	Sericin	Ibuprofen	(Verma and Subbiah 2017)	***Advantages*** • Low cost • High efficiency • Easy disposal • Promotes sustainability • Facilitates circular economy • Carbon negative or carbon neutral • Waste-to-use process ***Limitations*** • Modifications required to increase efficiency • High specificity towards contaminant • Secondary pollution • Low regeneration and reuse
Wine production	Gape stalk, Isabel grape bagasse	Diclofenac Paracetamol	(Antunes et al. 2012) (Villaescusa et al. 2011)	
Coffee production	Coffee waste	Tetracycline	(Oladipo et al. 2016)	
	Fish and sewage sludge composite	Carbamazepine Sulfamethoxazole Trimethoprim	(Nielsen and Bandosz 2016)	
Coal power plant	Flyash	Ciprofloxacin	(Zhang et al. 2011)	
Aluminium industry	Red mud	Ciprofloxacin	(Balarak et al. 2017)	
Paper industry	Mill sludge	Citalopram	(Calisto et al. 2014)	
Any carbon-rich biomass	Biochar	Most PPCPs	(Inyang et al. 2016)	

Table 19.3 Metal organic frameworks used as adsorbents for removal of PPCPs

Adsorbent	Target PPCPs	References	Applicability
MIL-101 Modified MIL-101 – *MIL-101-OH* – *MIL-101-(OH)*$_3$	p-chloro-m-xylenol Triclosan Ketoprofen Naproxen	(Song and Jhung 2017)	***Advantages*** • High porosity & surface area • Large pore size • Easy desorption • High regeneration • Easy syntheses • Easy functionalisation • Easy modification ***Limitations*** • Large-scale production • Water stability • Reusability • Secondary pollution from metal leaching
UiO-66-COOH	Triclosan Oxybenzone	(Song et al. 2016)	
MIL-101 Modified MIL-101 – *MIL-101-OH* – *MIL-101-(OH)*$_2$ – *MIL-101-NH*$_2$	Naproxen Oxybenzone Ibuprofen	(Seo et al. 2016)	
MIL-101 Modified MIL-101 – *ED-MIL-101* – *AMSA-MIL-101* – *MIL-101-Fe*	Naproxen Clofibric acid	(Hasan et al. 2013) (Hasan et al. 2012)	

MIL Material of Institute Lavosier; *UiO* University of Oslo; *AMSA* Aminomethanesulfonic; *ED* Ethylenediamine addition to the production as well as modification cost of adsorbent, the other predominant factor dictating the overall adsorption cost is the regeneration potential of the spent adsorbent which is accomplished either by desorption or decomposition of the contaminant (Crini et al. 2019). Based on the suitability, regeneration of saturated adsorbents may be carried out by several methods like thermal regeneration, solvent regeneration using inorganic acids and bases (using HCl, H_2SO_4, NaOH) or organic solvents (using C_6H_6, C_2H_5OH, CH_3OH) or any supercritical fluid like CO_2, and microwave heating (Dai et al. 2019). Only a few adsorption-based studies reported the regeneration potential of the used adsorbents (Silva et al. 2018). The desorption potential of used adsorbents is depicted by its desorption constant, which is experimentally determined. Desorption studies available for low-cost adsorbents are even lesser. These adsorbents may also be judiciously disposed, if the cost of regeneration is more than the cost of production (Tan et al. 2015). Adsorbents like biochar used for the uptake of non-toxic contaminants may be used as soil fertilizer and those with toxic contaminant can be disposed off by incineration or in landfill, after appropriate treatment. The economic feasibility of management of spent adsorbent is yet to be properly investigated and needs further attention in research. Moreover, only a few life cycle assessment studies in case of adsorption process with low-cost adsorbents are available (Alhashimi and Aktas 2017). Comparative performance-based analysis of low-cost adsorbents with their standard and commercial counterparts, for better cost-benefit analysis are also scarce (Huggins et al. 2016). Inspite of enormous investigation on the adsorptive removal of pollutants, these studies are mostly restricted to the generation of a new adsorbent and implementing it only in batch scale for a single contaminant. Very few studies have explored multi-component adsorption studies as well as continuous column-based experiments, which are more useful towards carving a path for the real-scale application of the adsorption process.

19.6 Discussion and Future Perspectives

Unlike conventional water-borne impurities, the emerging contaminants are potentially more hazardous due to some of their intrinsic properties. Owing to the persistent nature of PPCPs, these compounds tend to bioaccumulate and at times biomagnify by forming various transformation products on reacting with other co-existing compounds (Rehman et al. 2015). In several instances, the accumulated PPCPs in the living body have interfered with the metabolic activities by altering the normal functioning of the vital organ systems and impairing the immune system (Zenker et al. 2014). The prime challenge of removal of PPCPs from different environmental matrices is the characteristic specificity of these compounds along with their metabolic behaviour, that is unpredictable and not well-established. This problem is further magnified when a blend of different PPCPs and their metabolites occur simultaneously in an aqueous medium, leading to the synergistic production of more toxic by-products.

Adsorption is popularly used as an effective means of containing and removing PPCPs from water matrices. Kinetic, equilibrium and thermodynamic analysis in batch-flow mode have been performed for a variety of PPCP compounds, utilising a wide range of adsorbent materials. However, investigation in continuous-flow mode and pilot-scale studies are still in nascent stages (Cabrera-Lafaurie et al. 2014). Full-scale analysis of adsorptive removal of real PPCP-laden effluent is also scarcely available (Grover et al. 2011). Cost of the adsorbent material dominates the cost of the entire process of adsorptive removal. To economise the adsorption process for feasible commercial use, low-cost adsorbents has been initiated. The list of different adsorbent materials used for PPCP removal is difficult to be summarised in a single study. Out of these, 'low-cost adsorbents' is an umbrella terminology used to denote all adsorbents cheaper than commercial adsorbents (Crini et al. 2019). These adsorbents are also termed as 'green adsorbents' as they do not cause any negative environmental impact (Crini et al. 2019). Using low-cost adsorbents also work in synergy with waste management goals. Waste-derived adsorbents beautifully align with the principles of environmental sustainability and successfully attaches an economic value to the waste. These adsorbents transform a burden into a commodity, thereby promoting circular economy with negligible additional investment (Silva et al. 2018).

Although significant investigation into the performance characteristics of different low-cost adsorbents has been initiated, there is a need for further in-depth research before using these alternative, non-conventional materials in pilot-based and full-scale studies, to ensure their practical use in commercial scale. It is also necessary to comparatively analyse the cost of production to the adsorbing efficiency of the prepared adsorbent material, using lifecycle assessment (Arena et al. 2016). Further, the issue of regeneration, disposal, and scaling-up for commercial application of adsorption process may be relatively more critical in case of low-cost adsorbents, which needs to be analysed (Dai et al. 2019). No matter how efficient an adsorbent is, it is of no practical value if it cannot be regenerated and reused with ease and convenience. Life cycle assessment studies of low-cost adsorbent utilization as well as their comparative economic performance concerning to commercial adsorbents which are scantily available at present, needs to be further researched upon (Alhashimi and Aktas 2017). In spite of such exhaustive literature on adsorbent-based studies, only a few of these focus on the cost-benefit analysis for the feasible real-time use of proposed adsorbent material. Multi-contaminant adsorption studies are scarce and adsorption of real wastewater effluent having a confluence of contaminants is not readily available (Xiang et al. 2019; Cabrera-Lafaurie et al. 2015). Pilot-scale, as well as full-scale studies for the most promising adsorbents, need to be taken up, using real wastewater effluent with multi-contaminants and actual environmental conditions, to reap the benefits of extensive research and innovation in the field of adsorption.

19.7 Conclusion

Pharmaceutical and personal care products (PPCPs) are some of the most essential synthetic chemicals produced for myriad anthropogenic uses, but chiefly for maintaining physiological health and treating illnesses. Over time, incessant use of PPCPs has led to the permeation of their residues and conjugates and their subsequent accumulation in the geo-aquatic environment through various ambiguous pathways. Subsequently, these compounds have also entered the food chain, leading to metabolic alterations in consistently exposed lower-order organisms like microbes, fishes and birds. The harmful effects of PPCPs present in the environment have been seriously recognised, although their long-term implications have not been well established through research. Removal of PPCP residues from the aquatic environment has garnered serious attention in recent times. Several removal techniques based on different physico-chemical, activated oxidation, membrane-based, and biological processes like constructed wetlands have been initiated, which are rather costly and energy intensive.

Adsorptive removal of PPCPs has demonstrated satisfactory removal efficiency when using synthetic as well as real pharmaceutical effluent. Use of commercial as well as low-cost non-conventional adsorbents has been extensively analysed for removal of various PPCPs. Commercial adsorbents are prepared to be highly porous with large surface area, thus requiring a low dosage for removal. They are mostly contaminant-specific and depict high removal efficiencies as well as a high rate of adsorption, under suitable conditions. However, manufacturing and regeneration cost of commercial adsorbents averts its rampant use in real systems. Simultaneously, the development of sophisticated adsorbents, especially nano-materials, is evident from recent research communications. Although highly effective in terms of rate and efficiency of adsorption, these materials are costly, pose an indirect risk of nanotoxic contamination and need to be further scrutinized for feasible practical applications.

A simultaneous exploration of peculiar materials of diverse origins is being carried out for their use as suitable adsorbents. These are mostly low-cost, waste-derived materials with reasonable adsorptive capacity, comparatively less contaminant specificity and selectivity, but significantly economical. Such adsorbent materials are abundantly available natural materials or waste products with little utility or no profitable applicability. Use of such materials for adsorption ensures the benefit of reducing the overall cost of the adsorption process as well as resourceful utilisation of a waste product. Performance of low-cost adsorbents can be enhanced through chemical modifications and preparation of composite adsorbents. Production cost, as well as cost of handling, regeneration and ultimate disposal of exhausted adsorbents, is comparatively low for such adsorbents. In the event of the complexity of the regeneration process, these adsorbents can be effectively disposed off, instead of reusing it. Thus, the use of low-cost adsorbents economises the entire operational cycle of contaminant removal and appear to be one of the most effective means, concerning large scale application of adsorptive removal of contaminants like PPCPs.

References

Ahmed MJ (2017) Adsorption of non-steroidal anti-inflammatory drugs from aqueous solution using activated carbons. J Environ Manage 190:274–282

Ahmed MB, Zhou JL, Ngo HH, Guo W (2015) Adsorptive removal of antibiotics from water and wastewater: progress and challenges. Sci Total Environ 532:112–126

Alhashimi HA, Aktas CB (2017) Life cycle environmental and economic performance of biochar compared with activated carbon: a meta-analysis. Resour Conserv Recycl 118:13–26

Ali I (2012) New generation adsorbents for water treatment. Chem Rev 112(10):5073–5091

Álvarez-Torrellas S, Rodríguez A, Ovejero G, Gómez JM, García J (2016) Removal of caffeine from pharmaceutical wastewater by adsorption: influence of NOM, textural and chemical properties of the adsorbent. Environ Technol 37(13):1618–1630

Antunes M, Esteves VI, Guégan R, Crespo JS, Fernandes AN, Giovanela M (2012) Removal of diclofenac sodium from aqueous solution by Isabel grape bagasse. Chem Eng J 192:114–121

Arena N, Lee J, Clift R (2016) Life cycle assessment of activated carbon production from coconut shells. J Clean Prod 125:68–77

Ashfaq M, Khan KN, Rehman MSU, Mustafa G, Nazar MF, Sun Q, Iqbal J, Mulla SI, Yu CP (2017) Ecological risk assessment of pharmaceuticals in the receiving environment of pharmaceutical wastewater in Pakistan. Ecotoxicol Environ Saf 136:31–39

Attia TMS, Hu XL, Yin DQ (2013) Synthesized magnetic nanoparticles coated zeolite for the adsorption of pharmaceutical compounds from aqueous solution using batch and column studies. Chemosphere 93(9):2076–2085

Balakrishna K, Rath A, Praveenkumarreddy Y, Guruge KS, Subedi B (2017) A review of the occurrence of pharmaceuticals and personal care products in Indian water bodies. Ecotoxicol Environ Saf 137:113–120

Balarak D, Mostafapour FK, Joghataei A (2017) Kinetics and mechanism of red mud in adsorption of ciprofloxacin in aqueous solution. Biosci Biotechnol Res Commun 10:241–248

Barceló D, Petrovic M (2007) Pharmaceuticals and personal care products (PPCPs) in the environment. Anal Bioanal Chem 387:1141–1142

Basheer AA (2018) New generation nano-adsorbents for the removal of emerging contaminants in water. J Mol Liq 261:583–593

Bhandari A, Surampalli RY, Adams CD, Champagne P, Ong SK, Tyagi RD, Zhang T (eds) (2009) Contaminants of emerging environmental concern. American Society of Civil Engineers

Boxall AB, Rudd MA, Brooks BW, Caldwell DJ, Choi K, Hickmann S, Innes E, Ostapyk K, Staveley JP, Verslycke T, Ankley GT (2012) Pharmaceuticals and personal care products in the environment: what are the big questions? Environ Health Perspect 120(9):1221–1229

Burakov AE, Galunin EV, Burakova IV, Kucherova AE, Agarwal S, Tkachev AG, Gupta VK (2018) Adsorption of heavy metals on conventional and nanostructured materials for wastewater treatment purposes: a review. Ecotoxicol Environ Saf 148:702–712

Cabrera-Lafaurie WA, Román FR, Hernández-Maldonado AJ (2014) Removal of salicylic acid and carbamazepine from aqueous solution with Y-zeolites modified with extraframework transition metal and surfactant cations: equilibrium and fixed-bed adsorption. J Env Chem Eng 2(2):899–906

Cabrera-Lafaurie WA, Román FR, Hernández-Maldonado AJ (2015) Single and multi-component adsorption of salicylic acid, Clofibric acid, carbamazepine and caffeine from water onto transition metal modified and partially calcined inorganic–organic pillared clay fixed beds. J Hazard Mater 282:174–182

Calisto V, Ferreira CIA, Santos SM, Gil MV, Otero M, Esteves VI (2014) Production of adsorbents by pyrolysis of paper mill sludge and application on the removal of citalopram from water. Bioresour Technol 166:335–344

Candela L, Fabregat S, Josa A, Suriol J, Vigués N, Mas J (2007) Assessment of soil and groundwater impacts by treated urban wastewater reuse. A case study: application in a golf course (Girona, Spain). Sci Total Env 374(1):26–35

Caracciolo AB, Topp E, Grenni P (2015) Pharmaceuticals in the environment: biodegradation and effects on natural microbial communities. A review. J Pharm Biomed Anal 106:25–36

Carlsson C, Johansson AK, Alvan G, Bergman K, Kühler T (2006) Are pharmaceutical potent environmental pollutants. Part I: environmental risk assessments of selected active pharmaceutical ingredients. Sci Total Environ 364(13):67–87

Carlsson G, Örn S, Larsson DJ (2009) Effluent from bulk drug production is toxic to aquatic vertebrates. Environ Toxicol Chem 28(12):2656–2662

Chang PH, Li Z, Jiang WT, Sarkar B (2019) Clay minerals for pharmaceutical wastewater treatment. In: Modified clay and zeolite nanocomposite materials. Elsevier, pp 167–196

Changotra R, Rajput H, Dhir A (2019a) Treatment of real pharmaceutical wastewater using combined approach of Fenton applications and aerobic biological treatment. J Photochem Photobiol A 376:175–184

Changotra R, Rajput H, Guin JP, Varshney L, Dhir A (2019b) Hybrid coagulation, gamma irradiation and biological treatment of real pharmaceutical wastewater. Chem Eng J 370:595–605

Clara M, Strenn B, Gans O, Martinez E, Kreuzinger N, Kroiss H (2005) Removal of selected pharmaceuticals, fragrances and endocrine disrupting compounds in a membrane bioreactor and conventional wastewater treatment plants. Water Res 39(19):4797–4807

Crini G, Lichtfouse E, Wilson LD, Morin-Crini N (2019) Conventional and non-conventional adsorbents for wastewater treatment. Environ Chem Lett 17(1):195–213

Dai Y, Zhang N, Xing C, Cui Q, Sun Q (2019) The adsorption, regeneration and engineering applications of biochar for removal organic pollutants: a review. Chemosphere 223:12–27

Daneshvar E, Zarrinmehr MJ, Hashtjin AM, Farhadian O, Bhatnagar A (2018) Versatile applications of freshwater and marine water microalgae in dairy wastewater treatment, lipid extraction and tetracycline biosorption. Biores Technol 268:523–530

Davoli E, Zuccato E, Castiglioni S (2018) Illicit drugs in drinking water. Curr Opin Env Sci Health 7:92–97

de Andrade JR, Oliveira MF, da Silva MG, Vieira MG (2018) Adsorption of pharmaceuticals from water and wastewater using nonconventional low-cost materials: a review. Ind Eng Chem Res 57(9):3103–3127

de Wilt A, van Gijn K, Verhoek T, Vergnes A, Hoek M, Rijnaarts H, Langenhoff A (2018) Enhanced pharmaceutical removal from water in a three step bio-ozone-bio process. Water Res 138:97–105

Drout RJ, Robison L, Chen Z, Islamoglu T, Farha OK (2019) Zirconium metal–organic frameworks for organic pollutant adsorption. Trends Chem 1:304–317

Elhalil A, Elmoubarki R, Machrouhi A, Sadiq M, Abdennouri M, Qourzal S, Barka N (2017) Photocatalytic degradation of caffeine by ZnO–$ZnAl_2O_4$ nanoparticles derived from LDH structure. J Env Chem Eng 5(4):3719–3726

EPA (2017) Technical overview of ecological risk assessment: risk characterization. https://www.epa.gov/pesticide-science-and-assessing-pesticide-risks/technical-overview-ecological-risk-assessment-risk. Assessed on 30 May 2019

Ferguson PJ, Bernot MJ, Doll JC, Lauer TE (2013) Detection of pharmaceuticals and personal care products (PPCPs) in near-shore habitats of southern Lake Michigan. Sci Total Environ 458:187–196

Fick J, Soderstrom H, Lindberg R, Phan C, Tysklind M, Larsson D (2009) Contamination of surface, ground and drinking water from pharmaceutical production. Environ Toxicol Chem 28:2522

Fu J, Lee WN, Coleman C, Nowack K, Carter J, Huang CH (2019) Removal of pharmaceuticals and personal care products by two-stage bio-filtration for drinking water treatment. Sci Total Environ 664:240–248

Genç N, Dogan EC (2015) Adsorption kinetics of the antibiotic ciprofloxacin on bentonite, activated carbon, zeolite, and pumice. Desalin Water Treat 53(3):785–793

Gomes J, Costa R, Quinta-Ferreira RM, Martins RC (2017) Application of ozonation for pharmaceuticals and personal care products removal from water. Sci Total Environ 586:265–283

Grassi M, Kaykioglu G, Belgiorno V, Lofrano G (2012) Removal of emerging contaminants from water and wastewater by adsorption process. Emerging compounds removal from wastewater. Springer, Dordrecht, pp 15–37
Grover DP, Zhou JL, Frickers PE, Readman JW (2011) Improved removal of estrogenic and pharmaceutical compounds in sewage effluent by full scale granular activated carbon: impact on receiving river water. J Hazard Mater 185(2–3):1005–1011
Gupta K, Huo JB, Yang JCE, Fu ML, Yuan B, Chen Z (2019) $(MoS_4)^{2-}$ intercalated $CAMoS_4$· LDH material for the efficient and facile sequestration of antibiotics from aqueous solution. Chem Eng J 355:637–649
Hasan Z, Jeon J, Jhung SH (2012) Adsorptive removal of naproxen and clofibric acid from water using metal-organic frameworks. J Hazard Mater 209:151–157
Hasan Z, Choi EJ, Jhung SH (2013) Adsorption of naproxen and clofibric acid over a metal–organic framework MIL-101 functionalized with acidic and basic groups. Chem Eng J 219:537–544
Hernández F, Calısto-Ulloa N, Gómez-Fuentes C, Gómez M, Ferrer J, González-Rocha G, Bello-Toledo H, Botero-Coy AM, Boıx C, Ibáñez M, Montory M (2019) Occurrence of antibiotics and bacterial resistance in wastewater and sea water from the Antarctic. J Hazard Mater 363:447–456
Hignite C, Azarnoff DL (1977) Drugs and drug metabolites as environmental contaminants: chlorophenoxyisobutyrate and salicylic acid in sewage water effluent. Life Sci 20(2):337–341
Hijosa-Valsero M, Matamoros V, Martín-Villacorta J, Bécares E, Bayona JM (2010) Assessment of full-scale natural systems for the removal of PPCPs from wastewater in small communities. Water Res 44(5):1429–1439
Huggins TM, Haeger A, Biffinger JC, Ren ZJ (2016) Granular biochar compared with activated carbon for wastewater treatment and resource recovery. Water Res 94:225–232
Inyang MI, Gao B, Yao Y, Xue Y, Zimmerman A, Mosa A, Pullammanappallil P, Ok YS, Cao X (2016) A review of biochar as a low-cost adsorbent for aqueous heavy metal removal. Crit Rev Environ Sci Technol 46:406–433
Jiang JQ, Ashekuzzaman SM (2012) Development of novel inorganic adsorbent for water treatment. Curr Opin Chem Eng 1(2):191–199
Jiang WT, Chang PH, Wang YS, Tsai Y, Jean JS, Li Z, Krukowski K (2013) Removal of ciprofloxacin from water by birnessite. J Hazard Mater 250:362–369
Jindal K, Narayanam M, Singh S (2015) A systematic strategy for the identification and determination of pharmaceuticals in environment using advanced LC–MS tools: application to groundwater samples. J Pharm Biomed Anal 108:86–96
Kaczala F, Blum SE (2016) The occurrence of veterinary pharmaceuticals in the environment: a review. Curr Anal Chem 12(3):169–182
Kallenborn R, Brorström-Lundén, E Reiersen L-O, Wilson S (2018) Pharmaceuticals and personal care products (PPCPs) in Arctic environments: indicator contaminants for assessing local and remote anthropogenic sources in a pristine ecosystem in change. Env Sci Pollut Res 1–13
Khazri H, Ghorbel-Abid I, Kalfat R, Trabelsi-Ayadi M (2017) Removal of ibuprofen, naproxen and carbamazepine in aqueous solution onto natural clay: equilibrium, kinetics, and thermodynamic study. Appl Water Sci 7(6):3031–3040
Kibuye FA, Gall HE, Elkin KR, Ayers B, Veith TL, Miller M, Jacob S, Hayden KR, Watson JE, Elliott HA (2019) Fate of pharmaceuticals in a spray-irrigation system: From wastewater to groundwater. Sci Total Environ 654:197–208
Klampfl CW (2018) Metabolization of pharmaceuticals by plants after uptake from water and soil: a review. TrAC Trends Anal Chem 111:13–26
Ku MS (2008) Use of the biopharmaceutical classification system in early drug development. AAPS J 10(1):208–212
Kümmerer K (2009a) Antibiotics in the aquatic environment—a review–part I. Chemosphere 75(4):417–434
Kümmerer K (2009b) Antibiotics in the aquatic environment–a review–part II. Chemosphere 75(4):435–441

Kyzas GZ, Fu J, Lazaridis NK, Bikiaris DN, Matis KA (2015) New approaches on the removal of pharmaceuticals from wastewaters with adsorbent materials. J Mol Liq 209:87–93
Larsson DJ, de Pedro C, Paxeus N (2007) Effluent from drug manufactures contains extremely high levels of pharmaceuticals. J Hazard Mater 148(3):751–755
Laxminarayan R, Chaudhury RR (2016) Antibiotic resistance in India: drivers and opportunities for action. PLoS Med 13(3):e1001974
Lee Y, Kovalova L, McArdell CS, von Gunten U (2014) Prediction of micropollutant elimination during ozonation of a hospital wastewater effluent. Water Res 64:134–148
Leusch FD, Neale PA, Busetti F, Card M, Humpage A, Orbell JD, Ridgway HF, Stewart MB, van de Merwe JP, Escher BI (2019) Transformation of endocrine disrupting chemicals, pharmaceutical and personal care products during drinking water disinfection. Sci Total Environ 657:1480–1490
Liu P, Zhang H, Feng Y, Yang F, Zhang J (2014) Removal of trace antibiotics from wastewater: a systematic study of nano-filtration combined with ozone-based advanced oxidation processes. Chem Eng J 240:211–220
Mannhold R, Kubinyi H, Folkers G (2009) Drug bioavailability: estimation of solubility, permeability, absorption and bioavailability, vol 40. Wiley
Mantri RV, Sanghvi R (2017) Solubility of pharmaceutical solids. In: Developing solid oral dosage forms. Academic Press, pp 3–22
Martín J, del Mar Orta M, Medina-Carrasco S, Santos JL, Aparicio I, Alonso E (2018) Removal of priority and emerging pollutants from aqueous media by adsorption onto synthetic organo-funtionalized high-charge swelling micas. Environ Res 164:488–494
Michael I, Vasquez MI, Hapeshi E, Haddad T, Baginska E, Kümmerer K, Fatta-Kassinos D (2014) Metabolites and transformation products of pharmaceuticals in the aquatic environment as contaminants of emerging concern. In: Advanced mass spectrometry-based techniques for the identification and structure elucidation of transformation products of emerging contaminants: Wiley, pp 413–459
Miller TH, Bury NR, Owen SF, MacRae JI, Barron LP (2018) A review of the pharmaceutical exposome in aquatic fauna. Environ Pollut 239:129–146
Mutiyar PK, Mittal AK (2014) Risk assessment of antibiotic residues in different water matrices in India: key issues and challenges. Environ Sci Pollut Res 21(12):7723–7736
Nielsen L, Bandosz TJ (2016) Analysis of the competitive adsorption of pharmaceuticals on waste derived materials. Chem Eng J 287:139–147
Nielsen S, Barratt MJ (2009) Prescription drug misuse: is technology friend or foe? Drug Alcohol Rev 28(1):81–86
Oaks JL, Gilbert M, Virani MZ, Watson RT, Meteyer CU, Rideout B (2004) Diclofenac residues as a cause of population decline of White-backed Vultures in Pakistan. Nature 2004(427):630–633
Oh S, Shin WS, Kim HT (2016) Effects of pH, dissolved organic matter, and salinity on ibuprofen sorption on sediment. Environ Sci Pollut Res 23(22):882–889
Oladipo AA, Abureesh MA, Gazi M (2016) Bifunctional composite from spent "Cyprus coffee" for tetracycline removal and phenol degradation: Solar-Fenton process and artificial neural network. Int J Biol Macromol 90:89–99
Osuji OK, Umahi OT (2012) Pharmaceutical companies and access to medicines—social integration and ethical CSR resolution of a global public choice problem. J Glob Ethics 8(2–3):139–167
Paltiel O, Fedorova G, Tadmor G, Kleinstern G, Maor Y, Chefetz B (2016) Human exposure to wastewater-derived pharmaceuticals in fresh produce: a randomized controlled trial focusing on carbamazepine. Environ Sci Technol 50(8):4476–4482
Panthi S, Sapkota AR, Raspanti G, Allard SM, Bui A, Craddock HA, Murray R, Zhu L, East C, Handy E, Callahan MT (2019) Pharmaceuticals, herbicides, and disinfectants in agricultural water sources. Environ Res 174:1–8
Petrović M, Hernando MD, Díaz-Cruz MS, Barceló D (2005) Liquid chromatography–tandem mass spectrometry for the analysis of pharmaceutical residues in environmental samples: a review. J Chromatogr A 1067(1–2):1–14

Porter G, Grills N (2015) Medication misuse in India: a major public health issue in India. J Pub Health 38(2):150–157
Prakash V (1999) Status of vultures in Keoladeo National Park, Bharatpur, Rajasthan with special reference to population crash in Gyps species. J Bombay Nat Hist Soc 96(4):365–378
Quesada HB, Baptista ATA, Cusioli LF, Seibert D, de Oliveira Bezerra C, Bergamasco R (2019) Surface water pollution by pharmaceuticals and an alternative of removal by low-cost adsorbents: a review. Chemosphere 222:766–780
Ramaswamy BR, Shanmugam G, Velu G, Rengarajan B, Larsson DJ (2011) GC–MS analysis and ecotoxicological risk assessment of Triclosan, carbamazepine and parabens in Indian rivers. J Hazard Mater 186(2–3):1586–1593
Ramirez-Fuentes E, Lucho-Constantino C, Escamilla-Silva E, Dendooven L (2002) Characteristics, and carbon and nitrogen dynamics in soil irrigated with wastewater for different lengths of time. Biores Technol 85(2):179–187
Rehman MSU, Rashid N, Ashfaq M, Saif A, Ahmad N, Han JI (2015) Global risk of pharmaceutical contamination from highly populated developing countries. Chemosphere 138:1045–1055
Reis EO, Foureaux AFS, Rodrigues JS, Moreira VR, Lebron YA, Santos LV, Amaral MC, Lange LC (2019) Occurrence, removal and seasonal variation of pharmaceuticals in Brasilian drinking water treatment plants. Environ Pollut 250:773–781
Rivera-Utrilla J, Sánchez-Polo M, Ferro-García MÁ, Prados-Joya G, Ocampo-Pérez R (2013) Pharmaceuticals as emerging contaminants and their removal from water. a review. Chemosphere 93(7):1268–1287
Rodarte-Morales AI, Feijoo G, Moreira MT, Lema JM (2011) Degradation of selected pharmaceutical and personal care products (PPCPs) by white-rot fungi. World J Microbiol Biotechnol 27(8):1839–1846
Saggioro EM, Bila DM, Satyro S (2018) Ecotoxicology of pharmaceutical and personal care products (PPCPs). Ecotoxicology 79–110
Santos LH, Araújo AN, Fachini A, Pena A, Delerue-Matos C, Montenegro MCBSM (2010) Ecotoxicological aspects related to the presence of pharmaceuticals in the aquatic environment. J Hazard Mater 175(1–3):45–95
Seo PW, Bhadra BN, Ahmed I, Khan NA, Jhung SH (2016) Adsorptive removal of pharmaceuticals and personal care products from water with functionalized metal-organic frameworks: remarkable adsorbents with hydrogen-bonding abilities. Sci Rep 6:34462
Shanmugam G, Sampath S, Selvaraj KK, Larsson DJ, Ramaswamy BR (2014) Non-steroidal anti-inflammatory drugs in Indian rivers. Environ Sci Pollut Res 21(2):921–931
Sharma BM, Bečanová J, Scheringer M, Sharma A, Bharat GK, Whitehead PG, Klánová J, Nizzetto L (2019) Health and ecological risk assessment of emerging contaminants (pharmaceuticals, personal care products, and artificial sweeteners) in surface and groundwater (drinking water) in the Ganges River Basin, India. Sci Total Env 646:1459–1467
Silva CP, Jaria G, Otero M, Esteves VI, Calisto V (2018) Waste-based alternative adsorbents for the remediation of pharmaceutical contaminated waters: Has a step forward already been taken? Biores Technol 250:888–901
Smith SC, Rodrigues DF (2015) Carbon-based nanomaterials for removal of chemical and biological contaminants from water: a review of mechanisms and applications. Carbon 91:122–143
Song JY, Jhung SH (2017) Adsorption of pharmaceuticals and personal care products over metal-organic frameworks functionalized with hydroxyl groups: quantitative analyses of H-bonding in adsorption. Chem Eng J 322:366–374
Song JY, Ahmed I, Seo PW, Jhung SH (2016) UiO-66-type metal–organic framework with free carboxylic acid: versatile adsorbents via H-bond for both aqueous and nonaqueous phases. ACS Appl Mater Interfaces 8(40):27394–27402
Sotelo JL, Ovejero G, Rodríguez A, Álvarez S, García J (2013) Study of natural clay adsorbent sepiolite for the removal of caffeine from aqueous solutions: batch and fixed-bed column operation. Water Air Soil Pollut 224(3):1466

Subedi B, Balakrishna K, Sinha RK, Yamashita N, Balasubramanian VG, Kannan K (2015) Mass loading and removal of pharmaceuticals and personal care products, including psychoactive and illicit drugs and artificial sweeteners, in five sewage treatment plants in India. J Environ Chem Eng 3(4):2882–2891

Subedi B, Balakrishna K, Joshua DI, Kannan K (2017) Mass loading and removal of pharmaceuticals and personal care products including psychoactives, antihypertensives, and antibiotics in two sewage treatment plants in southern India. Chemosphere 167:429–437

Sui Q, Zhao W, Cao X, Lu S, Qiu Z, Gu X, Yu G (2017) Pharmaceuticals and personal care products in the leachates from a typical landfill reservoir of municipal solid waste in Shanghai, China: occurrence and removal by a full-scale membrane bioreactor. J Hazard Mater 323:99–108

Sun K, Shi Y, Wang X, Li Z (2017) Sorption and retention of diclofenac on zeolite in the presence of cationic surfactant. J Hazard Mater 323:584–592

Swarcewicz MK, Sobczak J, Paździoch W (2013) Removal of carbamazepine from aqueous solution by adsorption on fly ash-amended soil. Water Sci Technol 67:1396–1402

Tan X, Liu Y, Zeng G, Wang X, Hu X, Gu Y, Yang Z (2015) Application of biochar for the removal of pollutants from aqueous solutions. Chemosphere 125:70–85

Tiwari B, Sellamuthu B, Ouarda Y, Drogui P, Tyagi RD, Buelna G (2017) Review on fate and mechanism of removal of pharmaceutical pollutants from wastewater using biological approach. Biores Technol 224:1–12

Tong AY, Peake BM, Braund R (2011) Disposal practices for unused medications around the world. Environ Int 37(1):292–298

Triebskorn R, Casper H, Scheil V, Schwaiger J (2007) Ultrastructural effects of pharmaceuticals (carbamazepine, clofibric acid, metoprolol, diclofenac) in rainbow trout (Oncorhynchus mykiss) and common carp (Cyprinus carpio). Anal Bioanal Chem 387(4):1405–1416

Vellinga A, Cormican S, Driscoll J, Furey M, O'Sullivan M, Cormican M (2014) Public practice regarding disposal of unused medicines in Ireland. Sci Total Environ 478:98–102

Verma VK, Subbiah S (2017) Prospects of silk sericin as an adsorbent for removal of ibuprofen from aqueous solution. Ind Eng Chem Res 56:10142–10154

Villaescusa I, Fiol N, Poch J, Bianchi A, Bazzicalupi C (2011) Mechanism of paracetamol removal by vegetable wastes: the contribution of $\pi-\pi$ interactions, hydrogen bonding and hydrophobic effect. Desalination 270:135–142

Wang CJ, Li Z, Jiang WT (2011) Adsorption of ciprofloxacin on 2:1 dioctahedral clay minerals. Appl Clay Sci 53(4):723–728

Wang Y, Yin T, Kelly BC, Gin KYH (2019) Bioaccumulation behaviour of pharmaceuticals and personal care products in a constructed wetland. Chemosphere 222:275–285

World Health Organization (2016) WHO treatment guidelines for drug-resistant tuberculosis. World Health Organization

Wu X, Dodgen LK, Conkle JL, Gan J (2015) Plant uptake of pharmaceutical and personal care products from recycled water and bio-solids: a review. Sci Total Environ 536:655–666

Xiang Y, Xu Z, Wei Y, Zhou Y, Yang X, Yang Y, Yang J, Zhang J, Luo L, Zhou Z (2019) Carbon-based materials as adsorbent for antibiotics removal: mechanisms and influencing factors. J Environ Manage 237:128–138

Xu J, Wu L, Chang AC (2009) Degradation and adsorption of selected pharmaceuticals and personal care products (PPCPs) in agricultural soils. Chemosphere 77(10):1299–1305

Yi X, Tran NH, Yin T, He Y, Gin KYH (2017) Removal of selected PPCPs, EDCs, and antibiotic resistance genes in landfill leachate by a full-scale constructed wetlands system. Water Res 121:46–60

Zenker A, Cicero MR, Prestinaci F, Bottoni P, Carere M (2014) Bioaccumulation and biomagnification potential of pharmaceuticals with a focus to the aquatic environment. J Environ Manage 133:378–387

Zhang Y, Geissen SU, Gal C (2008) Carbamazepine and diclofenac: removal in wastewater treatment plants and occurrence in water bodies. Chemosphere 73(8):1151–1161

Zhang CL, Qiao GL, Zhao F, Wang Y (2011) Thermodynamic and kinetic parameters of ciprofloxacin adsorption onto modified coal fly ash from aqueous solution. J Mol Liq 163(1):53–56

Zhang Y, Zhang R, Yang X, Qi H, Zhang C (2018) Recent advances in electrogenerated chemiluminescence biosensing methods for pharmaceuticals. J Pharm Anal 9(1):9–19

Zhou Y, Cheng G, Chen K, Lu J, Lei J, Pu S (2019a) Adsorptive removal of bisphenol A, chloroxylenol, and carbamazepine from water using a novel β-cyclodextrin polymer. Ecotoxicol Environ Saf 170:278–285

Zhou Y, Lu J, Zhou Y, Liu Y (2019b) Recent advances for dyes removal using novel adsorbents: a review. Environ Pollut 252:352–365

Chapter 20
Measurement, Analysis, and Remediation of Bisphenol-A from Environmental Matrices

Sukanya Krishnan, Ansaf V. Karim, Swatantra Pratap Singh and Amritanshu Shriwastav

Abstract Bisphenol-A (BPA) is one of the important emerging contaminants, which has been widely used as a raw material for the preparation of epoxy and polycarbonate. They are mainly present in our daily use products such as the lining of water containers, canned food and beverages, infant bottles, and medical devices due to its heat resistance and elasticity property. BPA is an alkyl phenol, find its primary route to environmental matrices by leaching out from the final consumer product containers or during the manufacturing process. The various factors responsible for leaching of BPA include temperature, the presence of acids, and storage time. It is an endocrine disruptor and causes an adverse impact on humans as well as aquatic organisms. Therefore, many countries have banned their usage, especially in infant bottles and other food containers. Conventional wastewater treatment technologies have been inefficient for degrading these type of persistent compounds. Advanced treatment techniques are sustainable approaches for the removal of persistent compounds from water. Further, the measurement and analysis of BPA and their conjugates requires sophisticated analytical instrumentation. This chapter provides a detailed review of available measurement and analysis methods for determination of bisphenol-A. Further, the chapter also reviews the various treatment technologies for the removal of BPA from environmental matrices.

Keywords Bisphenol-A · Toxicity · Measurement · Remediation

20.1 Introduction

Bisphenol-A (BPA) [2,2-bis (4-hydroxyphenyl) propane] is an alkyl phenol which has been widely used as a raw material for the manufacture of consumer goods (Sharma et al. 2015). It's a white solid and is low volatile in nature at ambient temperature (Lazim et al. 2015; Dehghani et al. 2016). The chemical formula of

S. Krishnan · A. V. Karim · S. P. Singh · A. Shriwastav (✉)
Environmental Science and Engineering Department, Indian Institute of Technology Bombay, Mumbai 400076, India
e-mail: amritan@iitb.ac.in

T. Gupta et al. (eds.), *Measurement, Analysis and Remediation of Environmental Pollutants*, Energy, Environment, and Sustainability,
https://doi.org/10.1007/978-981-15-0540-9_20

BPA is $(CH_3)_2C(C_6H_4OH)_2$, consisting of 2 methyl functional groups connected by a bridge with two unsaturated phenolic rings, as shown in Fig. 20.1. BPA has received enhanced attention as a monomer in the polymer industry owing to its good mechanical properties and thermal stability (Aristiawan et al. 2015; Wirasnita et al. 2014). They are extensively used in the production of polycarbonates, epoxy resins, and other goods (Bautista-Toledo et al. 2005; Kamaraj et al. 2014). The physical properties of BPA are presented in Table 20.1. The annual production of BPA is estimated at 8 million metric tons in 2016 and is expected to reach 10.6 million metric tons by 2022 globally (Noszczyńska and Piotrowska-Seget 2018). They are mainly present in our daily use products such as the lining of water containers, canned food and beverages, compact disks, thermal paper, infant bottles, and medical devices due to its heat resistance and elasticity property.

The primary route of BPA in environmental matrices is via discharges of industrial wastewater, sewage treatment plants, or by leaching out from the products made by BPA resins (Wirasnita et al. 2014). Another way is from landfills or sewage sludge from treatment plants (Moriyoshi et al. 2008). Monteagudo et al. (2018) reported that the temperature and pH changes of BPA containing materials could result in the hydrolysis of ester bonds of BPA ending up in leaching of BPA. As a result of the far greater solubility of BPA than $EC_{50,}$ it reaches the natural water sources from the discharge of untreated wastewater which causes endocrine disruption to the aquatic ecosystem (Dehghani et al. 2016). The trace level of BPA is frequently found in different sources such as wastewater, municipal sewage sludge, sediments, etc. Seyhi et al. (2012) reported that BPA concentration was detected in the range of 0.16– 0.36 $\mu g\ L^{-1}$ in wastewater treatment plants, while in municipal sludge it varied from 0.033–36.7 $\mu g\ g^{-1}$total solids (TS) in Canada and 30–330 $\mu g\ g^{-1}$ TS in

Fig. 20.1 Structure of BPA

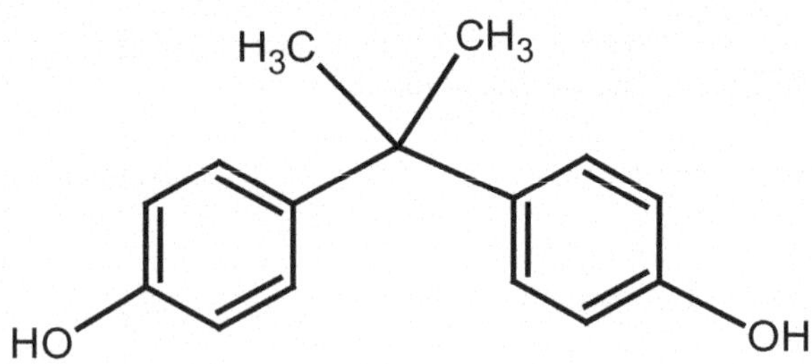

Table 20.1 Physical properties of BPA (Escalona et al. 2014)

Molecular weight (g mol^{-1})	228.28
Specific gravity at 25 1C (g cm^{-3})	1.06
Log K_{ow}	3.32
pKa	9.6–10.2
Water solubility (mg L^{-1})	120–300
Melting Point (°C)	150–155

Germany. Also, as reported by Escalona et al. (2014), a maximum concentration of 17.2 mg L^{-1} was detected in hazardous landfill leachates while 27.3 mg L^{-1} in natural surface water. Gattullo et al. (2012) reported that BPA concentration was detected between 1 and 150 $\mu g\,L^{-1}$ in industrial effluent and between 1 and 21 $\mu g\,L^{-1}$ in rivers. The data obtained from biomonitoring methods can provide reliable information about BPA exposure without identifying its sources (Vandenberg et al. 2007).

Researchers have reported that the low biodegradability and higher resistance to chemical decomposition, which makes it one of the important emerging contaminants as an endocrine-disrupting chemical (EDC) (Acosta et al. 2018). Because of the wide range use of BPA, its exposure is a serious health concern to both humans and animals and has attracted considerable attention (Kuo et al. 2010). Numerous studies have reported that the endocrine disrupting nature causes abnormalities in the functions of endocrine systems, demonstrated reproductive toxicities in humans and animals even at low concentrations (Soni and Padmaja 2014; Dong et al. 2010). Unlike for other contaminants, the regulatory standards for micropollutants such as BPA in wastewater discharge streams are not established. Thus, reducing its toxicity level before discharging in aqueous environments is a priority goal (Petrie et al. 2019). The European Commission Scientific Committee on Food (EC SCF) and US Environmental Protection Agency (EPA) established a maximum acceptable dose of 0.05 $mgkg^{-1}$ body weight/day for BPA in the context of its use in food contact plastics (Yüksel et al. 2013).

This chapter mainly focuses on the measurement and analysis techniques of BPA from environmental matrices. Also, the chapter covers the toxicity and exposure of BPA on living organisms. Finally, the chapter reviews the different remediation techniques for BPA such as physicochemical, biological and advanced oxidation processes.

20.2 BPA Toxicity and Exposure

Environmental distribution and daily human exposure are of great concern since BPA is classified as an endocrine disruptor that alters natural hormonal functions and is considered to be toxic (Zühlke et al. 2016). Studies have reported that the BPA might cause potential reproductive toxicity because of its estrogenic activity (Li et al. 2012). Continuous exposure to BPA can interfere with endocrine system and cause changes in developmental processes, cardiovascular diseases, immune effects in human and wildlife (Omoike et al. 2013; Dasssi et al. 2016). In addition, it affects the nervous system and immune system and also results in feminisation in both humans and wildlife (Heran et al. 2013; Senthilkumar et al. 2017). The daily human intake of BPA is found to be less than 1 μgkg^{-1} body weight/day, and the primary route of human exposure of BPA is mainly through the intake of contaminated food and water (Vandenberg et al. 2007; Li et al. 2010). For instance, the study by Li et al. (2010) reported the presence of BPA in drinking water, tap water, and baby bottles. It was observed that BPA got leached into the water from baby bottles at room temperature, 40 and 100 °C. Also, they have estimated the daily intake of BPA for adults and infants with drinking tap water as 148 ng $day^{-1}$and 1340 ng day^{-1} respectively. This shows that its daily intake for infants is approximately ten times that of adults. BPA exposure to humans is primarily through the contamination from food and beverage cans coated with epoxy resins, polycarbonate bottles, etc. Koduru et al. (2016) reported that hormone-dependent cancer caused by BPA decreases the sperm quality in human beings and hence considered it as an environmental phenolic estrogen pollutant. Very low doses of BPA can cause proliferation of prostate cancer cells, heart diseases, abnormalities in liver enzymes, etc. in humans (Dehghani et al. 2016).

Liao and Kannan (2011) have estimated the average daily intake for the common population and occupationally exposed individuals of about 0.219 and 16.3 ng kg^{-1} BW/day respectively. As a step further, detailed studies can be conducted to examine the transfer of BPA from different paper products into air and food, and thereby in humans. Additionally, the risk of exposure and transport of BPA from mothers to infants has been investigated by the researchers in Japan (Kuruto-Niwa et al. 2007). They have collected the initial breast milk (colostrum) from 101 healthy mothers immediately after delivery and detected an average concentration of BPA about 3.41 $\pm$ 0.13 ng L^{-1}. A recent study by Zhang et al. (2019) showed a comparison of the effects of various types of bisphenols on catalase molecules and human blood cells. The different analogues of bisphenol-A were used in the study, namely bisphenol S (BPS), bisphenol F(BPF), bisphenol B(BPB). They have concluded that BPB may be used as the BPA substitute in the manufacture, and the concentration of BPB should be controlled within one μM. The structural analogues of BPA such as BPS, bisphenol AF (BPAF), and tetrabromobisphenol A (TBBPA) are increasingly being used in consumer products. However, these analogues may exert similar adverse effects on the reproductive system, and their toxicological data are still limited (Siracusa et al. 2018). Apart from humans, marine lives are adversely affected by plastic contamination to oceans. In the study of Li et al. (2009), it is reported 8.65 $\pm$ 0.26 mg $L^{-1}$of BPA concentration at 96-hr EC_{50} exposure resulted in decreased chlorophyll-a content of marine algae *Stephanodiscus hantzschii* at concentrations higher than 3.00 mg L^{-1}.

The effect of BPA exposure to other living organisms has been extensively investigated by the researchers. BPA alters the normal hormonal system of animals and causes a variety of effects on reproduction and development in animals even at trace concentrations. These endocrine disrupting chemicals can lead to sexual dysfunction, genital malformations, prostate weight and cancer, low body weight, protein induction in the uterus, reduction of fertility etc. (Gattullo et al. 2012; Dekant and Völkel 2008). These compound can also lead to the disturbances in the immune system and can stimulate a cellular response by a variety of pathways (Lazim et al. 2015). The effects of BPA on sympathetic nerve fibers in the uterine wall of the domestic pig were studied by Liliana et al. (2019) and they observed that even at low concentrations of BPA it changes the uterine sympathetic nerve characteristics. They concluded that fluctuations of the neurochemical characterization of the uterine intramural nerves might be the first subclinical signs of harmful exposure to BPA. Similarly, the chronic effect of BPA exposure to animal *Caenorhabditis elegans* was studied by Zhou et al. (2016) and found that long-term exposure has significant adverse effects on *C. elegans* at the physiological, molecular and population level. However, based on different studies, researchers have concluded that apart from food sources, there are several other means for BPA exposure (Liao and Kannan 2011; Geens et al. 2011). Due to the low vapour pressure of BPA, inhalation exposure has a minor contribution along with dust ingestion to the overall exposure compared to food (Dekant and Völkel 2008; Geens et al. 2009). It is also reported that dermal absorption also contributes to the overall exposure of BPA on human beings (Geens et al. 2011). The contribution of BPA from additives such as polycarbonate, thermal paper and epoxy resins are also released and potential for dermal exposure. Another indirect way of exposure to BPA on human beings is through the consumption of contaminated fish and mammals (Lazim et al. 2015).

20.3 Extraction of BPA

To quantify BPA from environmental samples containing complex materials, the samples have to be extracted before chromatographic analysis. Different methods were established for the extraction of BPA from water samples as well as from biological fluids including plasma, urine etc. (Coughlin et al. 2011). Based on the matrix where BPA is present, the method of extraction, its accuracy and precision varies. Extraction of BPA from different sources could enhance the effectiveness of preceding chromatography analysis. Mostly widely used extraction process for separation and clean up for BPA in various samples are Solid phase extraction (SPE) and Liquid-liquid Extraction (LLE). Solid phase extraction is the predominant isolation method due to its high recovery, high pre-concentration factors, low consumption of organic solvents, simplicity, easy automation, and operation (Coughlin et al. 2011). Apart from this, micro-extraction techniques owing to their advantages of low sample consumption and ease of automation, Solid phase micro extraction (SPME), and Liquid Phase Micro Extraction (LPME) are also suitable for extraction (Sun et al. 2016).

Omoike et al. (2013) used solid phase extraction (SPE) of pre-concentrated BPA on a C-18 column dried with the use of 3 mL of mobile phase containing 5 mL of methanol and 20 mL of ultrapure water. Prior to sample extraction, BPA standards were extracted with the same mobile phase and observed 99% extraction yield. In another study, SPE was used for the extraction of BPA from a leachate sample passing through glass cartridges packed with 2 g Amberlite XAD-2 and eluted with 10 mL of 5 mL of methanol/acetone in 30:70 ratio (He et al. 2009). Solid-liquid extraction (SLE) was used for extraction of BPA from dust samples by addition of 3 mL n-hexane-acetone (3:1) followed by the sonication and vortexing of the sample. Further, the upper layer of the samples was evaporated in a nitrogen atmosphere to 0.5 mL. The final extracted samples were cleaned up before injecting in a Liquid Chromatography with tandem mass spectrometry (LC-MS/MS) for BPA analysis. Hu et al. (2018)

extracted BPA from sediment samples using SLE before Liquid Chromatography-Mass Spectrometry (LC-MS) analysis. The residual BPA in sediments were extracted by 10 mL solvent mixture of methanol-acetone (1:1, v/v) prior to centrifuging at 4000 rpm (10 min), and was further sonicated for 20 min, evaporated and redissolved the organic phases in 2 mL of acetonitrile for analysis. To extract BPA from biological samples, different processes are followed. Zühlke et al. (2016) centrifuged the culture supernatant with ethyl acetate at pH 7 followed by acidification of the aqueous residue to pH 2. The residues were dissolved in methanol for further analysis. A simple method of mixing 1.5 g of plant part mixed with 10 mL of 24 h resulted in extracting BPA from a different plant during the phytoremediation (Saiyood et al. 2010).

20.4 Measurement and Analysis Techniques for BPA

The determination and separation of BPA in various environmental matrices requires highly sensitive analytical methods such as High Performance Liquid Chromatography (HPLC) (Ohko et al. 2001; Rivas et al. 2009; Da Oh et al. 2018), HPLC-UV (He et al. 2009), High Performance Liquid Chromatography-Electrospray Tandem Mass Spectrometry (HPLC-ESI-MS) (Yu 2018), Gas Chromatography-Mass Spectrometry(GC-MS) (Li et al. 2010), LC-MS (Alnaimat et al. 2019), and Ultra Performance Liquid Chromatography-Tandem Mass Spectrometer (UPLC-MS/MS) (Petrie et al. 2019).

The presence of BPA and their detection methods have been reported in a variety of environmental samples such as river water, tap water, wastewater effluents, tea bag samples, thermal receipt papers, etc. as presented in Table 20.2. HPLC analysis is one of the widely used simple and rapid methods for the determination of BPA equipped with various UV and fluorescent detectors and the commonly used mobile phase as methanol/acetonitrile/water. For instance, Da Oh et al. (2018) used HPLC analysis for the measurement of BPA in the water sample after photocatalysis treatment by using Hypersil Gold C18 reverse phase column with mobile phase as aqueous solute ion of methanol and ultrapure water (60:40 v/v) at $\lambda_{max} = 220$ nm, and flow-rate of 0.6 mL min^{-1}. In another photocatalysis treatment study, acetonitrile and water (60:50 v/v) were used as mobile phase for BPA determination at $\lambda_{max} = 275$ nm with Inertsil ODS-3 column followed by LC-MS analysis for intermediate identification (Ohko et al. 2001). Also, UV-vis spectrophotometer is a simple and rapid method for the determination of BPA at $\lambda_{max} = 277$ nm (Neogy et al. 2019). But the detection becomes more sensitive and complex when it comes to real samples with extremely low BPA concentration in a variety of environmental matrices (Sun et al. 2016). In such cases, the higher selective and sensitive techniques such as LC-MS and GC-MS analysis is often applied to detect BPA.

Mass spectrometric methods have higher specificity and are ideal for analyzing the intensity of fragments during chromatographic separation which includes simple dilution of aqueous samples using polar organic solvents, extraction of bisphenol A, followed by chromatography (Dekant and Völkel 2008). For GC-MS analysis of

Table 20.2 Methods used for the determination of BPA from various environmental matrices

Sample	Detected BPA concentration	Method used	Location	References
Tidal water of futian mangrove nature reserve	0.34–4.01 $\mu g\ L^{-1}$	Gas chromatography-flame ionization detector (GC-FID)	China	Li et al. (2009)
Bottled water and tap water	17.6–324 $ng\ L^{-1}$ in bottled water, 317 $ng\ L^{-1}$ (max) in Tap water	Derivatization followed by GC-MS analysis	Guangzhou	Li et al. (2010)
Tea bag samples	0.072–0.24 $\mu g\ L^{-1}$	Liquid Chromatography Coupled with Mass Spectrometry	Portugal, Spain, Jordan, India and Azerbaijan	Alnaimat et al. (2019)
Influent and effluent wastewater, river water and digested sludge	Influent: 100 $\mu g\ L^{-1}$ Effluent: 62–892 $ng\ L^{-1}$ Digested sludge: 4.6–38.7 $\mu g\ g^{-1}$	UPLC-MS/MS analysis	South West UK	Petrie et al. (2019)
Yong River water	15–1415 $ng\ L^{-1}$	High-performance liquid chromatography system coupled to a tandem mass spectrometer with an electrospray ionization interface (HPLC-ESI-MS)	China	Yu (2018)
Thermal receipt papers	1 ng/g to 13.9 mg/g	High-performance liquid chromatography (HPLC)	USA, Japan, Korea, and Vietnam	Liao and Kannan (2011)

BPA, the derivatization process is required as a sample pretreatment after extraction to enhance the volatilization of the analytes and thereby increase the peak intensity and resolution (Sun et al. 2016). Li et al. (2009) used the derivatization agent N, O-bis (trimethylsilyl) trifluoroacetamide (BSTFA) after solid phase microextraction of BPA from tidal water samples. Similarly, in another study by Li et al. (2012), the biodegradation metabolites of BPA were identified using gas chromatography coupled with a mass selective detector (GC-MSD) after derivatization with BSTFA.

LC coupled with UV, Fluorescence and MS detectors is also another commonly used method for the quantification of BPA because of its higher selectivity of detector compared to HPLC at low concentrations in ng levels (Sun et al. 2016). For instance, in the study of Deborde et al. (2008) used LC-MS and MS/MS for the identification of intermediates during ozonation treatment of BPA. Further, the combination of LC-MS and GC-MS provides an effective and powerful method for the identification of intermediates and quantification (Kusvuran and Yildirim 2013).

20.5 Remediation Techniques for BPA

Due to the recalcitrant nature and endocrine disrupting potential of BPA, its reduction to non-toxic levels is particular importance. The removal of BPA has been studied in several perspectives using different technologies such as physico-chemical, biological, advanced oxidation processes, etc.

20.5.1 Adsorption

Adsorption is a promising approach for the removal of BPA from water and wastewater. The physical and chemical properties of activated carbon make it the most effective and widely used adsorbent displaying a superior adsorption capacity (Bautista-Toledo et al. 2005; Zhou et al. 2011). The mechanism of adsorption of BPA on adsorbent may be assumed to occur as follows: the transport of BPA molecules towards adsorbent surface followed by diffusion through the pores of adsorbent and attachment to active sites (Han et al. 2012). The potential of the activated carbon prepared from different materials such as empty fruit bunch of oil palm wastes, palm shell, chitosan, etc. has advantages over the other methods including relatively lower operating cost, simplicity of design and ease of operation (Dehghani et al. 2016; Wirasnita et al. 2014; Acosta et al. 2018; Soni and Padmaja 2014). Two commercial activated carbons and one prepared from almond shells were used by (Bautista-Toledo et al. 2005) for the removal of BPA from the water.

Surfactant modified Zeolite synthesized from coal fly ash was examined for the adsorption of BPA from the water and showed high retention capacity for BPA. The adsorption process primarily depended on the pH where the adsorption of the nonionic organic solutes got enhanced at alkaline pH due to its ionizable character in aqueous solutions (Dong et al. 2010). In a similar kind of study, Zhou et al. (2011) modified the fibric peat with hexadecyltrimethylammonium bromide and observed maximum sorption of 31.40 mg g^{-1} at an initial concentration of 45 mg L^{-1}. The chemical modification improved the hydrophobic interaction between the biosorbent and BPA molecules which enhanced the sorption capacity. Lignin isolated from black liquor was able to remove 237.07 mg g^{-1} BPA within 5 h where pH above 7.5 inhibited BPA adsorption due to the electrostatic repulsion between bisphenolate anion and lignin surface (Han et al. 2012). In another study, a composite material goethite iron oxide particles impregnated on activated carbon (GPAC) was prepared to remove and observed a predominant multilayer chemical adsorption on the heterogeneous surfaces with an enhanced removal of BPA in the presence of natural organic matter (NOM) (Koduru et al. 2016). Adsorption of BPA on novel magnesium ascorbyl phosphate graphene-based monolith was contributed by the combined electrostatic interactions, hydrophobic effects, hydrogen bonding, and π-π interactions with a saturated adsorption capacity of 324 mg g^{-1} of BPA.

The following Table 20.3 show some of the studies conducted with different adsorbents for BPA removal

Table 20.3 BPA removal by different Adsorbents and major findings

Adsorbent	Major findings	References
Palm shell activated carbon (PAC)	The interaction of BPA on PAC was observed to be driven by hydrogen bonding and π- π interactions	Soni and Padmaja (2014)
KOH-activated tyre pyrolysis char	The maximum adsorption capacity of 123 mg g^{-1} was observed, and lower temperature favoured more adsorption	Acosta et al. (2018)
Cetylpyridinium bromide modified natural zeolites	pH played a significant role in controlling the monolayer and bilayer adsorption of BPA on the surfactant-modified natural zeolite adsorbent	Li et al. (2014)
Metal-organic frameworks	The adsorption kinetics and capacity of BPA on the adsorbent depend on average pore size and specific surface area	Qin et al. (2015)
Reduced graphene oxide (rGO)	rGO prepared by thermal exfoliation method exhibited at last 2.5dd times larger adsorption capacity than rGO prepared by a chemical reduction method using hydrazine towards BPA	Kwon and Lee (2015)

20.5.2 Biological Processes

Biological treatment based on utilizing the metabolic potential of microorganism or mechanisms by intact higher plants is an effective physicochemical process to remove BPA (Omoike et al. 2013). The bacterial consortia posses species of different properties and had good resistance to substrate inhibition resulting degradation of BPA (Eio et al. 2014). Kang and Kondo (2002) isolated two strains of bacteria (*Pseudomonas* sp. and *Pseudomonas putida*x) from three river water samples and used them under aerobic and anaerobic conditions for the degradation of BPA. Rapid degradation occurred under aerobic conditions and the half-life for BPA degradation ranged between 2–3 days, while only <10% degradation was observed under anaerobic conditions. In a similar study, bacterial strain *Sphingomonas* sp. isolated from the soil of a vegetable-growing field was able to degrade 115 $\mu g\ mL^{-1}$ BPA in 6 h in aqueous medium and the metabolites remaining in the culture were below the HPLC detection level after 8 h.

A facultative anaerobic bacterial strain, *Bacillus species* was used for the decontamination of BPA under the optimized aerobic condition and observed complete degradation of compounds with 51% mineralization with seven metabolites identified (Li et al. 2012). An aquatic fungus, *Heliscus lugdunensis* was evaluated in live and heat-inactivated fungal cultures by Omoike et al. (2013) for its ability to utilize BPA as a carbon source. They have observed around 70% removal of BPA in 12 days by the live fungus cultures, while heat inactivation showed no significant growth. Dasssi et al. (2016) studied the potential of biocatalysts fungal laccases for the degradation of BPA and observed rapid oxidation of the compound by *Coriolopsis gallica* laccase among the different fungal laccases tested. *Monoraphidium braunii*, a green alga was tested for its ability to tolerate and remove the endocrine disruptor bisphenol A either in natural organic matter (NOM)-free or NOM-containing media at different concentrations (Gattullo et al. 2012). After 4 days of growth, good removal of BPA was observed at lower concentrations, whereas it was strongly inhibited by the highest concentration with a consistent reduction of algal content with no significant influence of NOM on the removal of BPA at any concentration (Gattullo et al. 2012). However, the Total Organic Carbon mineralisation potential of biodegradation is limited.

Utilizing the metabolic potentials of microbial communities is a safer, economical, and widely applicable alternative method for biodegradation of BPA (Wojnowska-Baryła et al. 2014; Ferro Orozco et al. 2016). The microorganisms in the wastewater treatment systems can transform the compound through a range of metabolic pathways and thus reducing the possible risk of BPA (Yang et al. 2014). Microbes use the BPA as the sole source of carbon and energy and grow on it while the microbes that are unable to biodegrade BPA can help in enhancing the action of BPA degraders through co-metabolism (Yang et al. 2014; Cydzik-Kwiatkowska et al. 2017). An activated sludge rich in high nitrifiers was used by Yoo et al. (Yoo et al. 2007) for the removal of BPA to verify if the ammonium or nitrite oxidizing activity plays a role in the degradation of BPA. It was observed that the removal of BPA was mostly mediated by biological activity along with the acclimation of heterotrophs to specific compounds. Recently, Seyhi et al. (2012) investigated the removal of BPA using an immersed membrane activated sludge process and observed an effective removal of BPA due to adsorption and biodegradation processes with 99% Chemical Oxygen Demand removal irrespective of the hydraulic retention time. Further, the effect of acclimation strategy and sludge age on the acclimation of activated sludge in the presence of biogenic substrates for degradation of BPA was studied by Ferro Orozco et al. (2013). Regardless of the acclimation strategy, the degradation activity begins after a lag period and varying the initial BPA concentration decreased the biodegradation efficiency. However, the biological treatment methods play a crucial role in the degradation of BPA; it is difficult to eliminate BPA through biological processes leading to the occurrence of low concentration aqueous sources (Cydzik-Kwiatkowska et al. 2017).

20.5.3 Advanced Oxidation Processes (AOPs)

20.5.3.1 Photocatalysis

Among the various AOPs available for the treatment recalcitrant compounds like BPA, photocatalysis is being actively investigated by the researchers because it completely mineralizes the organic pollutant into CO_2 and H_2O without the help of hazardous oxidants with less production of residual waste (Deng and Zhao 2015). Several studies are available for the TiO_2 based photocatalysis for the degradation of BPA with and without various surface modification techniques. For instance, in a study carried out by Ohko et al. (2001) the complete mineralization of BPA, as well as reduction in estrogenic activity, was achieved by photocatalysis with pure TiO_2 under UV light irradiation for 20 h. Similarly, Monteagudo et al. (2018) showed the photocatalytic degradation of BPA with TiO_2 under various irradiation sources such as UV-LED, solar, and UV-Black light (BL), etc. and observed that LED-driven photocatalysis gave the highest reaction rates, followed by TiO_2/solar, and TiO_2/UV-BL systems.

Platinum metal deposited TiO_2 was used in the study of Chiang et al. (2004) for the degradation of BPA under two extreme pH conditions at 3 and 10. The performance of Pt/TiO_2 was 3–6 times higher than pure TiO_2 at optimum operational condition of 0.2–1 wt% Pt/TiO_2 under 120 min UV irradiation with initial BPA concentration of 20 ppm at pH 3. Whereas, the degradation was about 20–30% when operated under pH 10, keeping all other parameters the same as described. This is due to high stable intermediates formed during pH 10 as compared with pH 3. At pH 3, the BPA molecules were more adsorbed to the catalyst surface in unionized form whereas, at pH 10, the negatively charged BPA molecules repelled from the TiO2 surface resulting in lower degradation. The complete mineralization of BPA was achieved by Zr-doped TiO_2 (TiZr0.2) at an optimum concentration of 15 mg L^{-1} of BPA at pH 9 under UV irradiation with the intensity of 5.3 mW cm^{-2}

for 120 min (Gao et al. 2010). As the Zr content increases, there was a shift in the conduction band, thereby giving enhanced photocatalytic activity. In another study, N, Co-codoped TiO_2 was developed for the mineralization of Bisphenol-A (BPA) under visible light irradiation and found that 97% TOC removal at the end of 140 min (Neogy et al. 2019).

Apart from doping with metals and non-metals, heterogeneous composite, the graphene modified hybrid nanomaterials have gained more attention among the researchers since it provides large surface area and suppresses the electron-hole pair recombination (Lai et al. 2016). In the study by (Lai et al. 2016), graphene/TiO_2 was modified by molecularly imprinted technique with BPA as a template molecule and observed enhanced adsorption and photocatalytic activity. The BPA degradation was maximum at pH 5 to 6; initial concentration is 4 mg L^{-1} and catalyst concentration of 1.0 g L^{-1} under visible light irradiation of 180 min. Ag metal deposition along with S-doped g-C_3N_4 composites was developed for the degradation of bisphenol-A under solar irradiation and found that the photocatalytic performance was three times than that of the pure g-C_3N_4 due to the ability of Ag to act as electron storage and minimize recombination rate (Da Oh et al. 2018). Also, heterogeneous composite using $ZnFe_2O_4$ and TiO_2 showed almost 20.8–21.4 times higher degradation of BPA than that of commercial TiO_2 photocatalysts under visible light irradiation (Nguyen et al. 2019).

The wide industrial application of the photocatalytic system on a large scale is restricted due to difficulty in the separation of the catalyst by filtration after treatment. For instance, Wang et al. (Wang et al. 2009) developed a reactor setup where TiO_2 was immobilized on polyurethane foam in a horizontal circulating bed photocatalytic reactor (HCBPR) for the degradation of BPA. They observed BPA removal of 97% at optimum operational conditions with pH 12.3 and initial BPA concentration of 10 ppm after 6 h. of UV irradiation. This is a promising technique where particles are immobilized on various support materials and used in the photocatalytic studies. Therefore, there is significant progress in the field of various surface modified photocatalysis as described above. But there exist ample of scope in the surface modifications such as dye sensitization, quantum dot sensitization, and plasmon-based photocatalysts for the degradation of BPA which is not being investigated in detail by the researchers till yet (Reddy et al. 2018).

20.5.3.2 Ozonation

The ozonation process is one of the faster and cheaper process, and does not require any extra chemicals like the process of photocatalysis, wet-air oxidation, and Fenton process (Kusuma and Mahfud 2017). Many researchers have demonstrated the effective removal of BPA by ozonation (Rivas et al. 2009; Deborde et al. 2008; Kusuma and Mahfud 2017; Garoma and Matsumoto 2009). The basic mechanism of ozonation is the reaction of molecular ozone and hydroxyl radicals with the high molecular weight compounds, thereby forming low molecular weight compounds (Umar et al.

2013). It was observed that the rate of BPA removal from the aqueous solution is linearly proportional to ozone dosage (Garoma and Matsumoto 2009).

For instance, Kusvuran and Yildirim (2013) showed complete degradation of BPA with an initial concentration of 0.509 mM within 25 min of ozonation treatment at pH = 3. Also they have identified the various intermediates were formed during ozonation namely oxalic acid, malonic acid, 1,2-Dihydroxy benzene, 2-(p-Hydroxyphenyl)-2-propanol, 4-Hydroxy-benzoic acid, 1,4-Dihydroxy benzene, Trihydroxybenzene, 4-(p-Hydroxy phenyl), 4,4-dimethyl,1-buten-3-on, 4-(2-(4-Hydroxy phenyl)propan-2yl)cyclohexa-3,5-diene-1,2-dione, (2E,4Z)-3-(2-(4-Hydroxyphenyl)propan-2-yl) hexa-2,4-dienedioic acid, etc. Out of which oxalic acid and malonic acids were formed in the first 5 min of ozonation treatment and represents the mineralization processs. Therefore, a wide range of by-products and intermediates are formed by the ozonation of BPA in water matrices. But it is very difficult to identify the ozonation by-products of BPA in the presence of other organic matter as in real wastewater (Umar et al. 2013).

The secondary effluent of municipal wastewater treatment plant containing BPA of in the range of 0.14–0.43 mg L^{-1} was significantly removed by ozonation (Bertanza et al. 2010). It was observed that 70% removal in the BPA could be achieved from an initial concentration of 1 mg L^{-1} with a significant reduction in the estrogenic activity due to synergistic effects of BPA and other similar compounds in the effluent. Additionally, the combination of ozonation and catalyst in the system is used to improve the mineralization of BPA. Rivas et al. (2009) have investigated the degree of mineralization achieved by ozonation with and without UV light and activated carbon modified TiO_2. It was observed that the combination of O_3 and UV-C, significantly increases the mineralization of BPA. Also, the rate of mineralization got improved by the addition of Activated carbon modified TiO_2 to the system. Similarly, Keykavoos et al. (2013) reported up to 90% degradation of BPA with catalytic-ozonation with alumina, whereas the treatment with only ozonation gave 35% removal within 4 min. Further detailed studies need to be conducted to establish the effect of co-existing organic matter on the removal efficiency and intermediate formation of BPA by ozonation (Umar et al. 2013).

20.5.3.3 Fenton Processes

Fenton process is considered as one of the promising AOPs for the treatment of recalcitrant and toxic organic compounds. BPA presence has been detected also in leachates from landfills apart from water and wastewater streams. The study was done by He et al. (2009) and was mainly focused on the detection of BPA from leachates and its remediation by Fenton oxidation process. The trace amount of BPA concentration causes the Fenton's process to be more challenging and observed after adding the spiked concentration of BPA(1.5 mg L^{-1}) to the system; the removal efficiencies were enhanced from 62 to 88%. Also, BPA has been widely present in soil and sediments because of the low water solubility. BPA from soil sediments was removed by developing a heterogeneous Fenton system with H_2O_2 and hydroxylamine with high pH value (Yu et al. 2019). It was observed that at high pH values, secondary Fe (II) precipitate was formed and which enhanced the decomposition of H_2O_2 to produce hydroxyl radicals, thereby maximum BPA removal was achieved.

One of the major limitations of the Fenton process is that the precipitation of inert iron at neutral pH (Yu et al. 2019). The study of Chen et al. (2017) was focused mainly on the Fenton oxidation at neutral pH with and without the presence of Cyclodextrins (CDs). It was observed the maximum of 57.2% BPA degradation by Fenton oxidation at neutral pH without the presence of Cyclodextrins (CDs) in the system. Whereas the Fenton process with CDs produces less toxic BPA intermediates, thereby making the process more feasible. In the recent study by Li et al. (2016), the traditional Fenton process was modified with the incorporation of Mn ion and microwave heating to the system and used for the removal of BPA. A maximum removal efficiency of 99.7% was observed with BPA at an initial concentration of 100 mg L^{-1} at optimal operational conditions (pH = 4, H_2O_2 concentration of 34.0 mg L^{-1}, Fe^{2+} ion concentration of 2.1 mg L^{-1}, Mn^{2+} ion concentration of 2.7 mg L^{-1}).

20.5.4 Membrane Processes

Apart from conventional treatment technologies and AOPs, other treatment technologies such as membrane processes are also effective in the degradation of BPA. Membrane processes can overcome the difficulties associated with the removal of low molecular weight organic micropollutants. Parameters such as molecular size, polarity, and compositions of the solution are important factors which affect the efficiency of a membrane process (Zhang et al. 2006). Bing-zhi et al. (2010) reported that the adsorption mechanism plays a significant role in the performance of low-pressure membrane such as ultrafiltration and microfiltration in removing organic BPA. The efficiency and mechanism involved in a nanofiltration membrane for BPA removal were studied by Zhang et al. (2006) and observed an overall 90% retention at the beginning of filtration and was reduced to 50% due to saturation of the membrane by the adsorbed BPA molecules. In another study, hollow fiber microfiltration (MF) membrane was used for the removal of BPA from drinking water observed significant amount on BPA removal due to adsorption mechanism and backwashing effectively recover BPA removal efficiency. The effectiveness of coupling of enzyme polymerization using Peroxidase from horseradish and laccase before nanofiltration to enhance BPA removal was studied by Escalona et al. (2014). Effect of pH and enzyme dose were studied and a quick BPA removal 95% was observed at optimized conditions.

20.5.5 Chemical Oxidation Processes

Chemical oxidation methods are also effective in the degradation of BPA. Han et al. (2015) studied the degradation of BPA using ferrate(VI) oxidizing process and observed rapid degradation of 97.5% within 10 min. The maximum degradation was observed at a pH 7 and ferrate(VI) to BPA molar ratio of 8.0 and have identified 12 intermediates. In another study, the oxidative capacity of MnO_2 was explored by Lin et al. (2009) for the removal of BPA, and the effect of different parameters was studied. Nearly complete degradation of BPA was observed in 6 min of reaction time at pH, presence of co existing metal ions such as Ca^+, Mn^+, $Fe^{3+,}$ etc. inhibited the removal capacity of MnO_2. Several other oxidants such as zero-valent iron(ZVI), zero valent aluminium, permanganate, etc. were also used for the degradation of BPA (Clark et al. 2012; Liu et al. 2011; Zhang et al. 2013). Clark et al. (2012) studied the potential of degradation of bisphenol-A (BPA) in water by ZVI, and a higher amount of BPA was degraded at lower iron concentration. In another study, ZVI supported on organo-montmorillonite nanoparticles resulted in nearly 100% degradation of BPA within 60 min under optimized conditions. The partial oxidation of ZVI resulted in the formation of ferrous ions resulted in mineralization of BPA.

20.5.6 Hybrid Processes

Hybrid advanced oxidation processes by combining sonolysis, photocatalysis, and Fenton process are one of the promising techniques for the degradation of BPA from various environmental matrices. The sono-photo-Fenton process was used for the degradation of BPA under visible light irradiation with an iron-containing perovskite catalyst ($LaFeO_3$) and showed 21.8% degradation of BPA and 11.2% COD removal (Dükkancı 2019). Hybrid process of sonolysis and visible light photocatalysis for BPA removal have been investigated by Sunasee et al.(2017) using graphitic carbon nitride (g-C_3N_4) and compared with the TiO_2 based sonophotocatalytic system. It was observed that the efficiency of BPA removal for the hybrid process is two times higher than TiO_2 based sonophotocatalytic process. The graphitic carbon nitride (g-C_3N_4) showed significantly higher visible light activity along with effective exfoliation of g-C_3N_4 by the application of ultrasound. Also, Sunasee et al. (2017) they have studied TiO_2 based sono-photocatalysis under UV light and found about 2.2 of synergy index for sono-photocatalysis at the optimum operational condition of 35 kHz, 50 W and 300 rpm stirring speed. A synergistic effect on removal was observed when TiO_2/wood charcoal composites were used during the adsorption and photocatalytic degradation of BPA (Luo et al. 2015). The composites had a strong ability for adsorption of BPA and good photocatalytic activity resulting in a maximum of 80% removal.

The study of Chakma and Moholkar (2014) was focused on the degradation of BPA using hybrid process including sonolysis, Fenton process along with UV irradiation also compared with the individual combination of processes. The maximum degradation was observed by the sono-Fenton UV assisted process due to the enhanced interaction of BPA with hydroxyl radicals by ultrasound cavitation process and generation of hydroxyl radicals by Fe(III)-hydroxy complexes under UV irradiation. A similar study by Torres et al. (2007) using combined ultrasound-UV-iron(II) process observed a significantly higher removal of BPA than the other individual processes alone. The BPA removal efficiency after 30 min of reaction was observed to be in the order of Ultrasound/UV/Fe^{2+} (95%) > Ultrasound/Fe^{2+} (89%) > Ultrasound/UV (82%) > Ultrasound (77%) > UV (25%). Torres-Palma et al.(2010) have reported about 93% degradation of BPA after 4 h. By hybrid AOP that includes sonolysis, Fe^{2+}(5.6 mg $L^{-1)}$, and TiO_2 (10 mg L^{-1}). The efficiency of the hybrid process was compared with the individual processes such as TiO_2 photocatalysis, ultrasound, and photo-Fenton and found the removal of only 5, 6, and 22% respectively. Hu et al. (2018) have developed a novel treatment technology for the removal of BPA from river sediments by Fenton process combined with bioremediation (*P. chrysosporium*). They observed 58.23% removal after 24 days of combined treatment.

20.6 Conclusion

Bisphenol A is an important synthetic monomer widely used in the manufacture of polycarbonate plastics, epoxy resins, etc. It is toxic in nature and possess endocrine disrupting effect. The main source of BPA in environmental matrices are from industrial wastewater, wastewater treatment plants, leaching from daily used products such as plastic water bottles, food cans, etc. Continuous exposure to BPA can cause severe health effects on developmental changes, nervous system, and adverse effects on the reproductive system etc. on human and other living species. Quantification of BPA from environmental samples containing complex materials require extraction before chromatographic analysis. The most commonly used methods of extraction of BPA from environmental matrices is solid phase extraction (SPE) due to its simplicity, high recovery and low consumption of organic solvents. Apart from this, liquid-liquid extraction, micro-extraction techniques (SPME, LPME) are also used. Highly sensitive analytical instruments such as HPLC, LC-MS, LC-MS/MS and GC-MS are used for the quantification and identification of degradation intermediates of BPA. But the detection becomes more complex for real samples at extremely low BPA concentrations. In such cases, the higher selective and sensitive techniques such as LC-MS and GC-MS analyses are often applied. Remediation of BPA from different sources is usually achieved by adsorption, biological processes and advanced oxidation processes. The selection of treatment option varies with the concentration of BPA and the cost of treatment technologies. The ability of the combined process to achieve complete degradation of BPA in environmental matrices have to be explored in future.

References

Acosta R, Nabarlatz D, Sánchez-Sánchez A, Jagiello J, Gadonneix P, Celzard A, Fierro V (2018) Adsorption of Bisphenol A on KOH-activated tyre pyrolysis char. J Environ Chem Eng 6:823–833. https://doi.org/10.1016/j.jece.2018.01.002

Alnaimat AS, Barciela-Alonso MC, Bermejo-Barrera P (2019) Determination of bisphenol A in tea samples by solid phase extraction and liquid chromatography coupled to mass spectrometry. Microchem J. https://doi.org/10.1016/j.microc.2019.03.026

Aristiawan Y, Aryana N, Putri D, Styarini D (2015) Analytical method development for bisphenol a in Tuna by using high performance liquid chromatography-UV. Procedia Chem 16:202–208. https://doi.org/10.1016/j.proche.2015.12.042

Bautista-Toledo I, Ferro-García MA, Rivera-Utrilla J, Moreno-Castilla C, Fernández FJV (2005) Bisphenol A removal from water by activated carbon. Effects of carbon characteristics and solution chemistry. Environ Sci Technol 39:6246–6250. https://doi.org/10.1021/es0481169

Bertanza G, Pedrazzani R, Papa M, Mazzoleni G, Steimberg N, Caimi L, Montani C, Dilorenzo D (2010) Removal of BPA and NPnEOs from secondary effluents of municipal WWTPs by means of ozonation. Ozone Sci Eng 32:204–208. https://doi.org/10.1080/01919511003795303

Bing-zhi D, Hua-qiang C, Lin W, Sheng-ji X, Nai-yun G (2010) The removal of bisphenol A by hollow fiber microfiltration membrane. Desalination 250:693–697. https://doi.org/10.1016/j.desal.2009.05.022

Chakma S, Moholkar VS (2014) Investigations in synergism of hybrid advanced oxidation processes with combinations of sonolysis + fenton process + UV for degradation of bisphenol A. Ind Eng Chem Res 53:6855–6865. https://doi.org/10.1021/ie500474f

Cheng JR, Wang K, Yu J, Yu, ZX, Yu ZB. Zhang Z (2018) Distribution and fate modeling of 4-nonylphenol, 4-t-octylphenol, and bisphenol A in the Yong River of China. Chemosphere 195:594–605. https://doi.org/10.1016/j.chemosphere.2017.12.085

Chen W, Zou C, Liu Y, Li X (2017) The experimental investigation of bisphenol A degradation by Fenton process with different types of cyclodextrins. J Ind Eng Chem 56:428–434. https://doi.org/10.1016/j.jiec.2017.07.042

Chiang K, Lim TM, Tsen L, Lee CC (2004) Photocatalytic degradation and mineralization of bisphenol A by TiO_2 and platinized TiO_2. Appl. Catal. A Gen. 261:225–237. https://doi.org/10.1016/j.apcata.2003.11.004

Clark CJ, Cooper AT, Martin CL, Pipkin L (2012) Evaluation of Potential Degradation of Bisphenol A by Zero-Valent Iron (ZVI). Environ Forensics 13:248–254. https://doi.org/10.1080/15275922.2012.702331

Coughlin JL, Winnik B, Buckley B (2011) Measurement of bisphenol A, bisphenol A -d-glucuronide, genistein, and genistein 4′—D-glucuronide via SPE and HPLC-MS/MS. Anal Bioanal Chem 401:995–1002. https://doi.org/10.1007/s00216-011-5151-8

Cydzik-Kwiatkowska A, Bernat K, Zielińska M, Bułkowska K, Wojnowska-Baryła I (2017) Aerobic granular sludge for bisphenol A (BPA) removal from wastewater. Int Biodeterior Biodegrad 122:1–11. https://doi.org/10.1016/j.ibiod.2017.04.008

Da Oh W, Lok LW, Veksha A, Giannis A, Lim TT (2018) Enhanced photocatalytic degradation of bisphenol A with Ag-decorated S-doped g-C3N4 under solar irradiation: performance and mechanistic studies. Chem Eng J 333:739–749. https://doi.org/10.1016/j.cej.2017.09.182

Dasssi D, Prieto A, Zouari-Mechichi H, Martínez MJ, Nasri M, Mechichi T (2016) Degradation of bisphenol A by different fungal laccases and identification of its degradation products. Int Biodeterior Biodegrad 110:181–188. https://doi.org/10.1016/j.ibiod.2016.03.017

Deborde M, Rabouan S, Mazellier P, Duguet JP, Legube B (2008) Oxidation of bisphenol A by ozone in aqueous solution. Water Res 42:4299–4308. https://doi.org/10.1016/j.watres.2008.07.015

Dehghani MH, Ghadermazi M, Bhatnagar A, Sadighara P, Jahed-Khaniki G, Heibati B, McKay G (2016) Adsorptive removal of endocrine disrupting bisphenol a from aqueous solution using chitosan. J Environ Chem Eng 4:2647–2655. https://doi.org/10.1016/j.jece.2016.05.011

Dekant W, Völkel W (2008) Human exposure to bisphenol A by biomonitoring: methods, results and assessment of environmental exposures. Toxicol Appl Pharmacol 228:114–134. https://doi.org/10.1016/j.taap.2007.12.008

Deng Y, Zhao R (2015) Advanced oxidation processes (AOPs) in wastewater treatment. Curr Pollut Reports 1:167–176. https://doi.org/10.1007/s40726-015-0015-z

Dong Y, Wu D, Chen X, Lin Y (2010) Adsorption of bisphenol A from water by surfactant-modified zeolite. J Colloid Interface Sci 348:585–590. https://doi.org/10.1016/j.jcis.2010.04.074

Dükkancı M (2019) Heterogeneous sonocatalytic degradation of Bisphenol-A and the influence of the reaction parameters and ultrasonic frequency. Water Sci Technol 79:386–397. https://doi.org/10.2166/wst.2019.065

Eio EJ, Kawai M, Tsuchiya K, Yamamoto S, Toda T (2014) Biodegradation of bisphenol A by bacterial consortia. Int Biodeterior Biodegrad 96:166–173. https://doi.org/10.1016/j.ibiod.2014.09.011

Escalona I, de Grooth J, Font J, Nijmeijer K (2014) Removal of BPA by enzyme polymerization using NF membranes. J Memb Sci 468:192–201. https://doi.org/10.1016/j.memsci.2014.06.011

Ferro Orozco AM, Lobo CC, Contreras EM, Zaritzky NE (2013) Biodegradation of bisphenol-A (BPA) in activated sludge batch reactors: Analysis of the acclimation process. Int Biodeterior Biodegrad 85:392–399. https://doi.org/10.1016/j.ibiod.2013.09.005

Ferro Orozco AM, Contreras EM, Zaritzky NE (2016) Biodegradation of bisphenol A and its metabolic intermediates by activated sludge: Stoichiometry and kinetics analysis, Int. Biodeterior. Biodegrad 106:1–9. https://doi.org/10.1016/j.ibiod.2015.10.003

Gao B, Lim TM, Subagio DP, Lim TT (2010) Zr-doped TiO_2 for enhanced photocatalytic degradation of bisphenol A. Appl Catal A Gen 375:107–115. https://doi.org/10.1016/j.apcata.2009.12.025

Garoma T, Matsumoto S (2009) Ozonation of aqueous solution containing bisphenol A: effect of operational parameters. J Hazard Mater 167:1185–1191. https://doi.org/10.1016/j.jhazmat.2009.01.133

Gattullo CE, Bährs H, Steinberg CEW, Loffredo E (2012) Removal of bisphenol A by the freshwater green alga Monoraphidium braunii and the role of natural organic matter. Sci Total Environ 416:501–506. https://doi.org/10.1016/j.scitotenv.2011.11.033

Geens T, Roosens L, Neels H, Covaci A (2009) Assessment of human exposure to Bisphenol-A, Triclosan and Tetrabromobisphenol-A through indoor dust intake in Belgium. Chemosphere 76:755–760. https://doi.org/10.1016/j.chemosphere.2009.05.024

Geens T, Goeyens L, Covaci A (2011) Are potential sources for human exposure to bisphenol-A overlooked? Int J Hyg Environ Health 214:339–347. https://doi.org/10.1016/j.ijheh.2011.04.005

Han W, Luo L, Zhang S (2012) Adsorption of bisphenol A on lignin: Effects of solution chemistry. Int J Environ Sci Technol 9:543–548. https://doi.org/10.1007/s13762-012-0067-1

Han Q, Wang H, Dong W, Liu T, Yin Y, Fan H (2015) Degradation of bisphenol A by ferrate(VI) oxidation: kinetics, products and toxicity assessment. Chem Eng J 262:34–40. https://doi.org/10.1016/j.cej.2014.09.071

He PJ, Zheng Z, Zhang H, Shao LM, Tang QY (2009) PAEs and BPA removal in landfill leachate with Fenton process and its relationship with leachate DOM composition. Sci Total Environ 407:4928–4933. https://doi.org/10.1016/j.scitotenv.2009.05.036

Heran M, Seyhi B, Azaïs A, Drogui P, Buelna G (2013) Contribution of a submerged membrane bioreactor in the treatment of synthetic effluent contaminated by Bisphenol-A: mechanism of BPA removal and membrane fouling. Environ Pollut 180:229–235. https://doi.org/10.1016/j.envpol.2013.05.028

Hu C, Huang D, Zeng G, Cheng M, Gong X, Wang R, Xue W, Hu Z, Liu Y (2018) The combination of Fenton process and Phanerochaete chrysosporium for the removal of bisphenol A in river sediments: mechanism related to extracellular enzyme, organic acid and iron. Chem Eng J 338:432–439. https://doi.org/10.1016/j.cej.2018.01.068

Kamaraj M, Ranjith KS, Sivaraj R, Rajendra Kumar RT, Abdul Salam H (2014) Photocatalytic degradation of endocrine disruptor Bisphenol-A in the presence of prepared $Ce_x Zn_{1-x}O$ nanocomposites under irradiation of sunlight. J Environ Sci 26:2362–2368. https://doi.org/10.1016/j.jes.2014.09.022

Kang JH, Kondo F (2002) Bisphenol A degradation by bacteria isolated from river water. Arch Environ Contam Toxicol 43:265–269. https://doi.org/10.1007/s00244-002-1209-0

Keykavoos R, Mankidy R, Ma H, Jones P, Soltan J (2013) Mineralization of bisphenol A by catalytic ozonation over alumina. Sep Purif Technol 107:310–317. https://doi.org/10.1016/j.seppur.2013.01.050

Koduru JR, Lingamdinne LP, Singh J, Choo KH (2016) Effective removal of bisphenol-A (BPA) from water using a goethite/activated carbon composite. Process Saf Environ Prot 103:87–96. https://doi.org/10.1016/j.psep.2016.06.038

Kuo CY, Wu CH, Lin HY (2010) Photocatalytic degradation of bisphenol A in a visible light/TiO2 system. Desalination 256:37–42. https://doi.org/10.1016/j.desal.2010.02.020

Kuruto-Niwa R, Tateoka Y, Usuki Y, Nozawa R (2007) Measurement of bisphenol A concentrations in human colostrum. Chemosphere 66:1160–1164. https://doi.org/10.1016/j.chemosphere.2006.06.073

Kusuma HS, Mahfud M (2017) GC-MS analysis of essential oil of Pogostemon cablin growing in Indonesia extracted by microwave-assisted hydrodistillation. Int Food Res J 24:1525–1528

Kusvuran E, Yildirim D (2013) Degradation of bisphenol A by ozonation and determination of degradation intermediates by gas chromatography-mass spectrometry and liquid chromatography-mass spectrometry. Chem Eng J 220:6–14. https://doi.org/10.1016/j.cej.2013.01.064

Kwon J, Lee B (2015) Bisphenol A adsorption using reduced graphene oxide prepared by physical and chemical reduction methods. Chem Eng Res Des 104:519–529. https://doi.org/10.1016/j.cherd.2015.09.007

Lai C, Wang MM, Zeng GM, Liu YG, Huang DL, Zhang C, Wang RZ, Xu P, Cheng M, Huang C, Wu HP, Qin L (2016) Synthesis of surface molecular imprinted TiO_2/graphene photocatalyst and its highly efficient photocatalytic degradation of target pollutant under visible light irradiation. Appl Surf Sci 390:368–376. https://doi.org/10.1016/j.apsusc.2016.08.119

Lazim ZM, Hadibarata T, Puteh MH, Yusop Z (2015) Adsorption characteristics of bisphenol a onto low-cost modified phyto-waste material in aqueous solution, Water. Air Soil Pollut 226. https://doi.org/10.1007/s11270-015-2318-5

Li R, Chen GZ, Tam NFY, Luan TG, Shin PKS, Cheung SG, Liu Y (2009) Toxicity of bisphenol A and its bioaccumulation and removal by a marine microalga Stephanodiscus hantzschii. Ecotoxicol Environ Saf 72:321–328. https://doi.org/10.1016/j.ecoenv.2008.05.012

Li X, Ying GG, Su HC, Yang XB, Wang L (2010) Simultaneous determination and assessment of 4-nonylphenol, bisphenol A and triclosan in tap water, bottled water and baby bottles. Environ Int 36:557–562. https://doi.org/10.1016/j.envint.2010.04.009

Li G, Zu L, Wong PK, Hui X, Lu Y, Xiong J, An T (2012) Biodegradation and detoxification of bisphenol A with one newly-isolated strain *Bacillus* sp. GZB: kinetics, mechanism and estrogenic transition, Bioresour Technol 114:224–230. https://doi.org/10.1016/j.biortech.2012.03.067

Li J, Zhan Y, Lin J, Jiang A, Xi W (2014) Removal of bisphenol A from aqueous solution using cetylpyridinium bromide (CPB)-modified natural zeolites as adsorbents. Environ Earth Sci 72:3969–3980. https://doi.org/10.1007/s12665-014-3286-6

Li S, Zhang G, Wang P, Zheng H, Zheng Y (2016) Microwave-enhanced Mn-Fenton process for the removal of BPA in water. Chem Eng J 294:371–379. https://doi.org/10.1016/j.cej.2016.03.006

Liao C, Kannan K (2011) Widespread occurrence of bisphenol A in paper and paper products: implications for human exposure. Environ Sci Technol 45:9372–9379. https://doi.org/10.1021/es202507f

Liliana R, Slawomir G, Tomasz J, Joanna W, Andrzej P (2019) The effects of Bisphenol A (BPA) on sympathetic nerve fibers in the uterine wall of the domestic pig. Reprod Toxicol 84:39–48. https://doi.org/10.1016/j.reprotox.2018.12.004

Lin K, Liu W, Gan J (2009) Oxidative removal of bisphenol A by manganese dioxide: Efficacy, products, and pathways. Environ Sci Technol 43:3860–3864. https://doi.org/10.1021/es900235f
Liu W, Zhang H, Cao B, Lin K, Gan J (2011) Oxidative removal of bisphenol A using zero valent aluminum-acid system. Water Res 45:1872–1878. https://doi.org/10.1016/j.watres.2010.12.004
Luo L, Yang Y, Xiao M, Bian L, Yuan B, Liu Y, Jiang F, Pan X (2015) A novel biotemplated synthesis of TiO_2/wood charcoal composites for synergistic removal of bisphenol A by adsorption and photocatalytic degradation. Chem Eng J 262:1275–1283. https://doi.org/10.1016/j.cej.2014.10.087
Monteagudo JM, Chatzisymeon E, Davididou K, Durán A, Nelson R, Expósito AJ (2018) Photocatalytic degradation of bisphenol-A under UV-LED, blacklight and solar irradiation. J Clean Prod 203:13–21. https://doi.org/10.1016/j.jclepro.2018.08.247
Moriyoshi K, Sakai K, Ohmoto T, Yamanaka H, Ohe T (2008) Efficient microbial degradation of bisphenol a in the presence of activated carbon. J Biosci Bioeng 105:157–160. https://doi.org/10.1263/jbb.105.157
Neogy A, Rani S, Garg A, Singhania T, Sangal VK, Yadav SR, Singh A, Sharma S, Garg N (2019) Photocatalytic degradation of bisphenol-A using N, Co codoped TiO_2 catalyst under solar light. Sci. Rep 9:1–13. https://doi.org/10.1038/s41598-018-38358-w
Nguyen TB, Huang CP, Doong R (2019) Photocatalytic degradation of bisphenol A over a $ZnFe_2O_4/TiO_2$ nanocomposite under visible light. Sci Total Environ 646:745–756. https://doi.org/10.1016/j.scitotenv.2018.07.352
Noszczyńska M, Piotrowska-Seget Z (2018) Bisphenols: application, occurrence, safety, and biodegradation mediated by bacterial communities in wastewater treatment plants and rivers. Chemosphere 201:214–223. https://doi.org/10.1016/j.chemosphere.2018.02.179
Ohko Y, Ando I, Niwa C, Tatsuma T, Yamamura T, Nakashima T, Kubota Y, Fujishima A (2001) Degradation of bisphenol A in water by TiO2 photocatalyst. Environ Sci Technol 35:2365–2368. https://doi.org/10.1021/es001757t
Omoike A, Wacker T, Navidonski M (2013) Biodegradation of bisphenol A by Heliscus lugdunensis, a naturally occurring hyphomycete in freshwater environments. Chemosphere 91:1643–1647. https://doi.org/10.1016/j.chemosphere.2012.12.045
Petrie B, Lopardo L, Proctor K, Youdan J, Barden R, Kasprzyk-Hordern B (2019) Assessment of bisphenol-A in the urban water cycle. Sci Total Environ 650:900–907. https://doi.org/10.1016/j.scitotenv.2018.09.011
Qin FX, Jia SY, Liu Y, Li HY, Wu SH (2015) Adsorptive removal of bisphenol A from aqueous solution using metal-organic frameworks. Desalin Water Treat 54:93–102. https://doi.org/10.1080/19443994.2014.883331
Reddy PVL, Kim KH, Kavitha B, Kumar V, Raza N, Kalagara S (2018) Photocatalytic degradation of bisphenol A in aqueous media: a review. J Environ Manage 213:189–205. https://doi.org/10.1016/j.jenvman.2018.02.059
Rivas FJ, Encinas Á, Acedo B, Beltrán FJ (2009) Mineralization of bisphenol A by advanced oxidation processes. J Chem Technol Biotechnol 84:589–594. https://doi.org/10.1002/jctb.2085
Saiyood S, Vangnai AS, Thiravetyan P, Inthorn D (2010) Bisphenol A removal by the Dracaena plant and the role of plant-associating bacteria. J Hazard Mater 178:777–785. https://doi.org/10.1016/j.jhazmat.2010.02.008
Senthilkumar P, Sivamani S, Vasantharaj S, Mophin-Kani K, Vijayalakshmi V, Sivarajasekar N (2017) Bio-degradation of Bisphenol A by Pseudomonas aeruginosa PAb1 isolated from effluent of thermal paper industry: kinetic modeling and process optimization. J Radiat Res Appl Sci 11:56–65. https://doi.org/10.1016/j.jrras.2017.08.003
Seyhi B, Drogui P, Buelna G, Blais JF (2012) Removal of bisphenol-A from spiked synthetic effluents using an immersed membrane activated sludge process. Sep Purif Technol 87:101–109. https://doi.org/10.1016/j.seppur.2011.11.029
Sharma J, Mishra IM, Dionysiou DD, Kumar V (2015) Oxidative removal of bisphenol a by uv-c/peroxymonosulfate (pms): Kinetics, influence of co-existing chemicals and degradation pathway. Chem Eng J 276:193–204. https://doi.org/10.1016/j.cej.2015.04.021

Siracusa JS, Yin L, Measel E, Liang S, Yu X (2018) Effects of bisphenol A and its analogs on reproductive health: a mini review. Reprod Toxicol 79:96–123. https://doi.org/10.1016/j.reprotox.2018.06.005

Soni H, Padmaja P (2014) Palm shell based activated carbon for removal of bisphenol A: an equilibrium, kinetic and thermodynamic study. J Porous Mater 21:275–284. https://doi.org/10.1007/s10934-013-9772-5

Sun F, Kang L, Xiang X, Li H, Luo X, Luo R, Lu C, Peng X (2016) Recent advances and progress in the detection of bisphenol A. Anal Bioanal Chem 408:6913–6927. https://doi.org/10.1007/s00216-016-9791-6

Sunasee S, Leong KH, Wong KT, Lee G, Pichiah S, Nah I, Jeon B-H, Yoon Y, Jang M (2017) Sonophotocatalytic degradation of bisphenol A and its intermediates with graphitic carbon nitride, Env. Sci Pollut Res. https://doi.org/10.1007/s11356-017-8729-7

Sunasee S, Wong KT, Lee G, Pichiah S, Ibrahim S, Park C, Kim NC, Yoon Y, Jang M (2017b) Titanium dioxide-based sonophotocatalytic mineralization of bisphenol A and its intermediates. Environ Sci Pollut Res 24:15488–15499. https://doi.org/10.1007/s11356-017-9124-0

Torres RA, Pétrier C, Combet E, Moulet F, Pulgarin C (2007) Bisphenol A mineralization by integrated ultrasound-UV-iron (II) treatment. Environ Sci Technol 41:297–302. https://doi.org/10.1021/es061440e

Torres-Palma RA, Nieto JI, Combet E, Petrier C, Pulgarin C (2010) An innovative ultrasound, Fe_2 + and TiO_2 photoassisted process for bisphenol a mineralization. Water Res 44:2245–2252. https://doi.org/10.1016/j.watres.2009.12.050

Umar M, Roddick F, Fan L, Aziz HA (2013) Application of ozone for the removal of bisphenol A from water and wastewater—A review. Chemosphere 90:2197–2207. https://doi.org/10.1016/j.chemosphere.2012.09.090

Vandenberg LN, Hauser R, Marcus M, Olea N, Welshons WV (2007) Human exposure to bisphenol A (BPA). Reprod Toxicol 24:139–177. https://doi.org/10.1016/j.reprotox.2007.07.010

Wang R, Ren D, Xia S, Zhang Y, Zhao J (2009) Photocatalytic degradation of Bisphenol A (BPA) using immobilized TiO_2 and UV illumination in a horizontal circulating bed photocatalytic reactor (HCBPR). J Hazard Mater 169:926–932. https://doi.org/10.1016/j.jhazmat.2009.04.036

Wirasnita R, Hadibarata T, Yusoff ARM, Yusop Z (2014) Removal of bisphenol a from aqueous solution by activated carbon derived from oil palm empty fruit bunch. Water Air Soil Pollut 225. https://doi.org/10.1007/s11270-014-2148-x

Wojnowska-Baryła I, Bernat K, Bułkowska K, Zielińska M, Cydzik-Kwiatkowska A (2014) Removal of bisphenol A (BPA) in a nitrifying system with immobilized biomass. Bioresour Technol 171:305–313. https://doi.org/10.1016/j.biortech.2014.08.087

Yang Y, Wang Z, Xie S (2014) Aerobic biodegradation of bisphenol A in river sediment and associated bacterial community change. Sci Total Environ 470–471:1184–1188. https://doi.org/10.1016/j.scitotenv.2013.10.102

Yoo I-K, Cha GC, Choe WS, Ryu K, Kim EJ, Kim JY (2007) Degradation of bisphenol A and nonylphenol by nitrifying activated sludge. Process Biochem 42:1470–1474. https://doi.org/10.1016/j.procbio.2007.06.010

Yu Q, Feng L, Chai X, Qiu X, Ouyang H, Deng G (2019) Enhanced surface fenton degradation of BPA in soil with a high pH. Chemosphere 335–343. https://doi.org/10.1016/j.chemosphere.2018.12.141

Yüksel S, Kabay N, Yüksel M (2013) Removal of bisphenol A (BPA) from water by various nanofiltration (NF) and reverse osmosis (RO) membranes. J Hazard Mater 263:307–310. https://doi.org/10.1016/j.jhazmat.2013.05.020

Zhang Y, Causserand C, Aimar P, Cravedi JP (2006) Removal of bisphenol A by a nanofiltration membrane in view of drinking water production. Water Res 40:3793–3799. https://doi.org/10.1016/j.watres.2006.09.011

Zhang J, Sun B, Guan X (2013) Oxidative removal of bisphenol A by permanganate: Kinetics, pathways and influences of co-existing chemicals. Sep Purif Technol 107:48–53. https://doi.org/10.1016/j.seppur.2013.01.023

Zhang X, Li C, Pan J, Liu R, Cao Z (2019) Searching for a bisphenol A substitute: Effects of bisphenols on catalase molecules and human red blood cells. Sci Total Environ 669:112–119. https://doi.org/10.1016/j.scitotenv.2019.03.129

Zhou Y, Chen L, Lu P, Tang X, Lu J (2011) Removal of bisphenol A from aqueous solution using modified fibric peat as a novel biosorbent. Sep Purif Technol 81:184–190. https://doi.org/10.1016/j.seppur.2011.07.026

Zhou D, Yang J, Li H, Cui C, Yu Y, Liu Y, Lin K (2016) The chronic toxicity of bisphenol A to Caenorhabditis elegans after long-term exposure at environmentally relevant concentrations. Chemosphere 154:546–551. https://doi.org/10.1016/j.chemosphere.2016.04.011

Zühlke MK, Schlüter R, Henning AK, Lipka M, Mikolasch A, Schumann P, Giersberg M, Kunze G, Schauer F (2016) A novel mechanism of conjugate formation of bisphenol A and its analogues by Bacillus amyloliquefaciens: detoxification and reduction of estrogenicity of bisphenols. Int Biodeterior Biodegrad 109:165–173. https://doi.org/10.1016/j.ibiod.2016.01.019

Chapter 21
Removal of Chromium Ions from Water Using Eco-friendly Based Adsorbents

Karthik Rathinam and Swatantra Pratap Singh

Abstract In recent years, individual access to clean water becomes one of the most critical challenges around the world owing to the presence of various toxic pollutants in water. Chromium is one of the important heavy metal pollutants and posing significant threat to human beings due to its severe adverse effects. Because of its potential adverse effects on humans and other species, remediation becomes an absolute necessity. Whilst many techniques have been reported for the removal of chromium ions from polluted water, adsorption-based technologies gained more consideration as being inexpensive, simple, easy handling, etc. Furthermore, bio-based adsorbents are being extensively studied over other adsorbents for the removal of chromium due to their excellent adsorption capability, non-toxicity, biodegradability and availability. Therefore, the foremost aim of this chapter is to review the various bio-based adsorbents used for the removal of chromium ions from polluted water.

Keywords Chromium · Biopolymers · Eco-friendly · Adsorption

21.1 Introduction

Water is one of the most important fluids required by human beings and other living species to survive. Nevertheless, a major factor that limits individual access to clean water is water pollution by various toxic pollutants (Rathinam et al. 2017, 2018, 2019; Karthik and Meenakshi 2014). Water pollution by the heavy metals receiving considerable attention as they cause various health diseases to humans (Jaishankar et al. 2014; Jarup 2003). Chromium (Cr) is widely used in chromium plating, leather

K. Rathinam (✉)
University of Duisburg-Essen, Chair of Mechanical Process Engineering/Water Technology, Lotharstr. 1, D-47057, Duisburg, Germany
e-mail: drkarthikrathinam@gmail.com

S. P. Singh (✉)
Department of Environmental Science and Engineering, Indian Institute of Technology Bombay, Powai, Mumbai 400076, India
e-mail: swatantra@iitb.ac.in

T. Gupta et al. (eds.), *Measurement, Analysis and Remediation of Environmental Pollutants*, Energy, Environment, and Sustainability,
https://doi.org/10.1007/978-981-15-0540-9_21

tanning, wood preserving, dyes, and pigment industries. An inappropriate discharge of the effluent from these industries to the environment significantly pollute the water systems with chromium (Barnhart 1997).

Chromium often exists as Cr(VI) and Cr(III) oxidation states in nature. Moreover, the former oxidation state is almost 1000 times more toxic than the Cr(III) oxidation state (Karthik and Meenakshi 2015). According to the world health organization, the maximum allowable limits of Cr(VI) in potable water and industrial wastewater is 0.05 and 0.25 mg/L, respectively (World Health Organization 2006). In view of its non-biodegradability in nature, carcinogenic and mutagenic effects to the humans and other living species, remediation of chromium becomes an absolute mandatory. Furthermore, the dissolved Cr(VI) often exist as oxyanions such as $Cr_2O_7^{2-}$, $HCrO_4^-$ and CrO_4^{2-} with respect to different pH conditions and concentration (Karthik and Meenakshi 2014). Several methods have been applied for the removal of Cr(VI) ions from water and wastewater. Among the various remediation technologies, adsorption-based technology received more recognition due to its inexpensive, simplicity, easy handling and availability of a broad range of adsorbents.

Chromium removal has been studied using plenty of materials as adsorbent, however, eco-friendly based adsorbents are preferred over other conventional adsorbents because of its non-toxic, more abundant and biodegradable in nature (Karthik and Meenakshi 2015, 2016). Therefore, the foremost aim of this chapter is to review the contribution of eco-friendly based adsorbents for the removal of chromium ions from water or wastewater using adsorption technique. Nevertheless, the focus of this chapter will be discussed after a brief introduction about the various analytical techniques used for the determination of Cr(VI) ions in water.

21.2 Chromium Analysis in Water

As mentioned before, chromium may be existing either as Cr(III) or Cr(VI) or in both forms under different ambient conditions. Sometimes either the reduction of Cr(VI) into Cr(III) or the oxidation of Cr(III) into Cr(VI) may also be possible in solution under given conditions (Karthik and Meenakshi 2014). Analytical techniques reported for the determination of chromium in water samples are given in Table 2.1.

21.3 Removal of Cr(VI) Ions Using Eco-Friendly Based Bio-Adsorbents

21.3.1 Adsorbents Derived from Chitin

Chitin is one of the most abundant polysaccharides in nature consisting of β-(1→4)-2-acetamido-2-deoxy-d-glucopyranose units. Chitin is extracted from the shell of the

Table 21.1 Analytical techniques used for chromium determination in water samples

S. No	Analytical method	Analyte	References
1	Atomic absorption spectrophotometry (AAS)	Cr(VI)	EPA (1983)
2	Inductive coupled plasma-mass spectrometry (ICP-MS)	Cr(III)	Wilbur et al. (2012)
		Cr(VI)	
		Total Cr	
3	Inductive coupled plasma-atomic emission spectrometry (ICP-AES)	Total Cr	EPA (1996)
4	UV-Vis spectrometry (1,5-diphenyl carbazide method)	Cr(VI)	(EPA 2011)

shrimps, prawns, and craps (Percot et al. 2003; Devi and Dhamodharan 2018). Chitin is less studied as adsorbent due to its low adsorption capacity and poor solubility limitations (Otuonye et al. 2014; Singh and Nagendran 2016). Therefore, to increase the metal uptake capacity, chitin was subjected to several chemical modifications and blended with many other materials. For instance, Kousalya et al. prepared protonated chitin, carboxylated chitin and grafted chitin and used for the removal of Cr(VI) ions (Kousalya et al. 2010). The modified forms of chitin showed higher adsorption capacity for Cr(VI) ions than the unmodified chitin. Furthermore, the pH of the solution has played a big role on the removal of Cr(VI) ions using modified chitins. Chitin obtained from Bargi fish was used as an adsorbent for the uptake of Cr(VI) ions and the adsorption capacity of the chitin was increased with the increase in agitation time and increase in pH from 2 to 6 (Otuonye et al. 2014). Moreover, chitin flakes were packed in a column reactor and used for the dynamic removal of Cr(VI) ions from water (Sağ and Aktay 2001). In another study, removal of Cr(III) ions from a simulated solution was investigated using chitin as an adsorbent and the results showed that 49.98% Cr(III) was removed within 20 min of contact time (Singh and Nagendran 2016).

Saravanan et al. investigated the removal of Cr(VI) ions using chitin/bentonite bio-composite as adsorbent (Saravanan et al. 2013) and found that the Cr(VI) removal was highly pH dependent and the maximum Cr(VI) removal was obtained at pH 4.0. Chitin was modified with conduction polymers such as polyaniline and polypyrrole and used for the Cr(VI) uptake studies (Karthik and Meenakshi 2014, 2015). It was observed that the crystalline structure of the chitin was retained even after subjection to the chemical modification with polyaniline and polypyrrole. Furthermore, the conducting polymer coated chitin showed enhanced adsorption capacity towards Cr(VI) with reasonable regeneration ability. Kousalya et al. prepared a chitin hybrid nano-hydroxyapatite composite (n-HApC) for the removal of Cr(VI) ions. The adsorption capacity of the n-HApC was influenced by the solution pH but not in the presence of other co-ions (Kousalya et al. 2010).

Furthermore, the biosorption of chromium from chromated copper arsenate (CCA)-treated wood was studied using chitin as an adsorbent and the results showed that 2.5 g of chitin removed 62% chromium after 10 days of time (Kartal and Imamura

2005). Chitin micro-particles (CMP) and chitin nanofibrils (CNF) were developed as adsorbents for the removal of Cr(III) ions. CNF exhibited adsorption capacity of 2.5 times higher than the CMP owing to higher surface area and widely distributed pores (Liu et al. 2013). Mohamed Abdel Salam physically mixed the chitin with magnetic nanoparticles and multiwalled carbon nanotubes (MWCNTs) to obtain a chitin/magnetic/MWCNTs nanocomposite (CMM) for the removal of Cr(VI) ions. The adsorption study results showed that the adsorption capacity of the chitin was significantly improved upon preparation of the CMM nanocomposite (Salam 2017). Santosa et al. immobilized humic acid (HA) on chitin and used as an adsorbent for the removal of Cr(III) ions in the effluent of tannery wastewater treatment. The adsorption study results exhibited that 1 g of chitin-HA could able to remove 10.82 mg of Cr(III) ions from the real wastewater (Santosa et al. 2008). Quaternized chitin was crosslinked with polyethylenimine using epichlorohydrin as a crosslinker and tested as biosorbent for the effective Cr(VI) ions remediation from water. Owing to the presence of highly porous structures, quaternary ammonium, and amino groups the modified chitin revealed excellent adsorption of Cr(VI) ions. The XPS study showed that after the adsorption of Cr(VI) ions onto biosorbent, some of the bounded Cr(VI) ions were reduced to Cr(III) by the presence of amines and hydroxyl groups. On the other hand, the adsorption capacity for Cr(VI) ions was 4.2% decreased even after 7th cycles which shows the excellent reuse-ability of the prepared adsorbent (Liang et al. 2018). The list of adsorbents derived from chitin for the removal of chromium ions are given in Table 2.2.

21.3.2 Adsorbents Derived from Chitosan

Chitosan is a cationic biopolymer obtained from the deacetylation of chitin (Younes and Rinaudo 2015). Like chitin, chitosan polymer is also consisting of β-(1→4)-2-acetamido-2-deoxy-d-glucopyranose units. Unlike chitin, chitosan is being widely used as an adsorbent for the removal of heavy metals due to excellent metal chelating properties, higher solubility and higher adsorption potential (Wang and Chen 2014; Kanmani et al. 2017). Furthermore, chitosan can be chemically modified due to the presence of $–CH_2$ OH and $–NH_2$ groups for the removal of various heavy metal ions (Karthik and Meenakshi 2015; Wang and Chen 2014; Sharma and Lalita Singh 2017).

The removal of Cr(III) and Cr(VI) ions from water was conducted by adopting the batch adsorption method using chitosan beads as adsorbent. The adsorption results exhibited that the maximum removal of Cr(III) and Cr(VI) was obtained between pH 5 and pH 3, respectively. The adsorption capacity was found to be changed with respect to contact time and initial concentration (Ngah et al. 2006). Preethi et al. synthesized chitosan assisted mixed oxyhydroxide material and used as an adsorbent for the effective removal of Cr(VI) ions. The maximum Cr(VI) removal was achieved between pH 4-5 with the adsorbent dosage of 2 g/L. Overall, the chitosan assisted mixed oxyhydroxide material showed higher adsorption capacity for Cr(VI) ions than

Table 21.2 List of chitin-based adsorbents used for the removal of Cr(VI) and Cr(III) ions

Adsorbent	Adsorbate	Analytical method	References
Chitin	Cr(III)	AAS	Singh and Nagendran (2016)
Chitin flakes	Cr(VI)	UV-Vis spectrometry	Sağ and Aktay (2001)
Chitin/bentonite biocomposite	Cr(VI)	AAS	Saravanan et al. (2013)
Polyaniline coated chitin	Cr(VI)	UV-Vis spectrometry	Karthik and Meenakshi (2015)
Polypyrrole functionalized chitin	Cr(VI)	UV-Vis spectrometry	Karthik and Meenakshi (2014)
Chitin micro-particles	Cr(III)	AAS	Liu et al. (2013)
Chitin nanofibrils	Cr(III)	AAS	Liu ct al. (2013)
Chitin/magnetite/ multiwalled carbonnanotubes magnetic nanocomposite	Cr(VI)	UV-Vis spectrometry	Salam (2017)
Chitin-humic acid hybrid	Cr(III)	AAS	Santosa et al. (2008)
Nano-hydroxyapatite/chitin composite	Cr(VI)	UV-Vis spectrometry	Kousalya et al. (2010)
Chitin from Bargi fish	Cr(VI)	UV-Vis spectrometry	Otuonye et al. (2014)
Protonated chitin	Cr(VI)	UV-Vis spectrometry	Kousalya et al. (2010)
Carboxylated chitin	Cr(VI)	UV-Vis spectrometry	Kousalya et al. (2010)
Grafted chitin	Cr(VI)	UV-Vis spectrometry	Kousalya et al. (2010)
Quaternized chitin/branched polyethylenimine biosorbent	Cr(VI)	UV-Vis spectrometry	Liang et al. (2018)

the pure mixed oxyhydroxide (Preethi et al. 2017). Polyamidoamine functionalized chitosan beads were prepared to enhance the chitosan adsorption capacity for Cr(VI) ions. The zirconium (Zr) loaded chitosan beads showed higher Cr(VI) removal efficiency than the other modified chitosan beads (Rajiv Gandhi and Meenakshi 2013). Chauhan et al. prepared thiocarbamoyl chitosan and evaluated its adsorption performance towards the removal of Cr(VI) ions from electroplating wastewater. The maximum Cr(VI) removal was obtained at pH 2.0 (Chauhan et al. 2012). Chitosan microparticles were blended with Henna and used as adsorbent towards the removal of Cr(VI) ions. The adsorption capacity did not significantly change from pH 2 to 9 (Davarnejad et al. 2018).

Moussout et al. synthesized bentonite/chitosan nanocomposite for the uptake of Cr(VI) ions from synthetic wastewater and tanning wastewater and the adsorption capacity was compared with local chitosan (Moussout et al. 2018). Similarly, quaternized chitosan coated bentonite was prepared for the removal of Cr(VI) ions. The adsorption process was found to be feasible, endothermic and spontaneous (Hu et al. 2016). Chitosan nanoclay composite was prepared using a solvent casting method and investigated as an adsorbent for Cr(VI) ions. The prepared composite exhibited good regeneration ability and might be a potential option for removing Cr(VI) ions from wastewater (Kahraman 2017). The adsorption behavior of Cr(VI) ions onto chitosan graft acrylonitrile copolymer was investigated and the results revealed that the adsorption of Cr(VI) ions was enhanced with the increase in contact time and Cr(VI) ion concentration (Shankar et al. 2014). Reticulated chitosan nano/microparticles were synthesized by ionic gelation of chitosan using tripolyphosphate and tested for the removal of Cr(VI) ions. The results revealed that the adsorbed toxic Cr(VI) on reticulated chitosan was then reduced to less toxic Cr(III) which is one of the advantages of the prepared adsorbent (Dima et al. 2015).

Likewise, diepoxyoctane crosslinked chitosan (Vakili et al. 2018), diethylenetriaminepentaacetic acid crosslinked chitosan (Bhatt et al. 2015), trimesic acid crosslinked chitosan (Bhatt et al. 2017), epichlorohydrin crosslinked chitosan resin (Wu et al. 2012), chitosan-coated fly ash composite (Wen et al. 2011), chitosan/alumina biocomposite (Rajiv Gandhi et al. 2010), chitosan-based hydrogel (Vilela et al. 2018), chitosan coated diatomaceous earth beads (Salih and Ghosh 2018), silk protein sericin-chitosan conjugate (Singh et al. 2018), chitosan-lysozyme biocomposite (Rathinam et al. 2018), chitosan coated onto ceramic alumina (Boddu et al. 2003), modified forms of chitosan beads (Kousalya et al. 2010), titanium crosslinked chitosan composite (Zhang et al. 2015), chitosan engraved iron and lanthanum mixed oxyhydroxide (Preethi et al. 2019), self-assembled chitosan-zirconium phosphate nanostructures (Bhatt et al. 2019), chitosan/polymethylmethacrylate composite (Li et al. 2016), chitosan/polycaprolactam nanofibrous (Li et al. 2016), phosphonium salt crosslinked chitosan (Sessarego et al. 2019) and nanosized chitosan fibers (Li et al. 2015) were the other chitosan contained adsorbents used for the removal of Cr(VI) ions from water.

Development of iron (Fe)-based chitosan composites for the removal of Cr(VI) ions received more attention due to easy separation, higher adsorption capacity, and relatively higher regeneration performances. Shen et al. (2013) studied the detoxification of Cr(VI) ions using chitosan-Fe(III) complex and they found that the –OH on C-6 of chitosan served as an electron donor for the reduction of Cr(VI) to Cr(III). Moreover, the mechanism of Cr(VI) removal by chitosan-Fe(III) complex was elucidated. Similarly, Fe-crosslinked chitosan was used as an adsorbent for the removal of Cr(VI) ions. The adsorption capacity of the Fe- crosslinked chitosan was decreased from 295 mg/g to 205 mg/g while increasing the temperature from 25 to 65 °C. The Cr(VI) removal was influenced by the presence of coexisting ions (Zimmermann et al. 2010). Also, fixed-bed column studies were performed for the uptake of Cr(VI) ions using crosslinked chitosan-Fe(III) complex and influence of several parameters including column bed height, pH, flow rate and Cr(VI) concentration on the adsorption Cr(VI) ions were also studied (Demarchi et al. 2015).

Similarly, plenty of chitosan-iron based adsorbents including nitrogen-doped chitosan-Fe(III) composite (Zhu et al. 2016), Fe^0 nanorod modified with chitosan (Sun et al. 2015), chemically modified magnetic chitosan resins (Elwakeel 2010), magnetic silica particles encapsulated chitosan microspheres (Sun et al. 2016), magnetic chitosan nanoparticles (Thinh et al. 2013), thiourea modified magnetic chitosan composite (Cai et al. 2019), chitosan/magnetic graphene oxide composite (Zhang et al. 2018), nano iron oxide impregnated chitosan bead (Lu et al. 2017), Zr(IV) crosslinked chitosan magnetic microspheres (Chen et al. 2017), magnetic chitosan modified with diethylenetriamine (Li et al. 2011), chitosan/MWCNTs/Fe_3O_4 nanofibers (Beheshti et al. 2016), crosslinked chitosan/MWCNT/Fe film (Neto et al. 2019), Fe_3O_4 nanoparticles functionalized polyvinyl alcohol/chitosan magnetic composite hydrogel (Yan et al. 2018), quaternary ammonium salt modified chitosan magnetic composite (Li et al. 2016), magnetic chitosan-iron(III) hydrogel (Yu et al. 2013), magnetic Fe_3O_4@SiO_2–chitosan (Jiang et al. 2019), magnetic chitosan composite (Cai et al. 2018), 1,6-hexanediamine modified magnetic chitosan microspheres (Yue et al. 2018), chitosan/montmorillonite–Fe_3O_4 hybrid (Chen et al. 2013), modified magnetic chitosan chelating resin (Abou El-Reash et al. 2011), Fe-loaded chitosan carbonized rice husk composite beads (Sugashini et al. 2015) and crosslinked magnetic chitosan (Huang et al. 2009) have been explored as adsorbents for the remediation of Cr(VI) ions from water.

In recent years, chitosan-carbon based composites were fabricated extensively and studied as adsorbents for the removal of heavy metal ions. For instance, chitosan/modified MWCNTs nanocomposite was prepared for fast and efficient removal of Cr(VI) ions. The authors found that the adsorption of Cr(VI) ions by the composite was mainly driven by electrostatic adsorption as well as chemical redox mechanisms (Huang et al. 2018). Yu et al. self-assembled sponge-like chitosan/reduced graphene oxide/montmorillonite composite hydrogels for effective Cr(VI) sorption and the results revealed that the Cr(VI) sorption was highly pH dependent and increased with the increase in temperature (Yu et al. 2017). Modified graphene oxide/ chitosan composite was explored as an adsorbent to remove Cr(VI) ions and the experiment results exhibited the excellent removal of Cr(VI) ions at pH 2. The further regeneration experiments indicated a 5% decrease in Cr(VI) removal efficiency after 7 reuse cycles (Zhang et al. 2016). Furthermore, triethylenetetramine modified graphene oxide/chitosan composite was successfully synthesized by microwave irradiation and the adsorption of Cr(VI) ion by the composite was investigated. The effects of various equilibrium parameters on the Cr(VI) adsorption was examined and the adsorption process was fast and 85% Cr(VI) was removed in the first 20 min (Ge and Ma 2015). Kim et al. synthesized new chitosan-based functional gel comprising of multiwall carbon nanotubes and used as adsorbent towards the removal of Cr(VI). The prepared gel showed a synergistic effect on the Cr(VI) removal. Also, the plausible mechanism behind the Cr(VI) uptake by the gel was explained (Kim et al. 2015). Chitosan-1,2-cyclohexylenedinitrilotetraacetic acid-graphene oxide nanocomposite was prepared to remove Cr(VI) ions and the results showed that the maximum Cr(VI) removal was obtained with the adsorbent dosage of 2 g/L at pH 3.5 and contact time of 60 min (Ali 2018).

It is worthwhile noting that the development of chitosan composite consisting of different conducting polymers gained substantial attention for the removal of Cr(VI) ions as these composites have various functional groups that actively interacts with

Cr(VI) during the Cr(VI) removal process. Karthik and Meenakshi conducted the polymerization of aniline in the presence of crosslinked and non-crosslinked chitosan and prepared crosslinked and non-crosslinked chitosan-grafted-polyaniline composites and evaluated its adsorption performance for the removal of Cr(VI) ions. The adsorption capacity of the crosslinked chitosan polyaniline composite was higher for Cr(VI) compared to chitosan polyaniline composite. Cr(VI) removal by these composites was mainly driven by electrostatic adsorption coupled reduction (Karthik and Meenakshi 2014). In another study by this group, pyrrole was polymerized in the presence of chitosan and prepared chitosan/polypyrrole composite for the detoxification of Cr(VI) ions. The adsorption study results revealed that the adsorption capacity of the prepared composite for Cr(VI) was varied with different pH conditions and the maximum Cr(VI) removal has occurred at lower pH. Furthermore, the Cr(VI) removal was increased with the increase in temperature which indicated the endothermic nature of the adsorption process (Karthik and Meenakshi 2015). Substituted polyaniline/chitosan composites were developed by Yavuz et al and used for the removal of Cr(VI) ions. The different substituted polyaniline composites exhibited superior adsorption capacity than the other reported adsorbents for the Cr(VI) removal (Yavuz et al. 2011). Like aforementioned adsorbents, polypyrrole/chitosan nanocomposite aerogel monolith (Ji et al. 2018), chitosan-grafted-polyaniline montmorillonite (Ahmad et al. 2017) and chitosan coated with poly-3-methyl thiophene (Hena 2010) are important chitosan conducting polymer-based adsorbents developed and used for the removal of Cr(VI) ions from water.

Chitosan-based adsorbents with high surface area, high porosity, high functionality, high mechanical strength, and high regeneration abilities are much needed for the effective detoxification of Cr(VI) and Cr(III) ions from water and wastewater. Silica is a better choice over other materials due to strong attractive physical and chemical properties therefore, adsorbents consisting of chitosan and silica are developed and used for Cr(VI) removal. For example, chitosan-silica composite was synthesized using a sol-gel method through the hydrolysis of the tetraethoxysilane in the presence of chitosan and tested as an adsorbent for the remediation of Cr(VI) ions. The pH of the adsorption medium played a major role during the adsorption process (Budnyak et al. 2015). Chitosan-g-poly(butyl acrylate)/silica gel nanocomposite was developed via the sol-gel method for toxic Cr(VI) removal and the adsorption results revealed that the prepared adsorbent showed better Cr(VI) removal compared to other reported adsorbents (Nithya et al. 2016). Shi et al. prepared crosslinked modified silicon materials and investigated its adsorption capability for Cr(VI) and the results showed that the removal of Cr(VI) by the prepared adsorbents were 6- fold and 4- fold higher than the zeolite and montmorillonite (Shi et al. 2017). Chitosan/silica gel and lanthanum loaded chitosan/silica gel composite were synthesized and the adsorption of Cr(VI) ions onto these composites were investigated under a batch method. The lanthanum loaded chitosan/silica gel composite showed higher removal of Cr(VI) ions than the chitosan/silica gel composite and the Cr(VI) removal occurred through electrostatic adsorption coupled reduction/ion exchange mechanism (Gandhi and Meenakshi 2012). The layer-by-layer silicate-chitosan composite was developed and tested as biosorbent for the removal of Cr(III) and Cr(VI) ions. The silicate-chitosan film ended with the silica layer showed better adsorption capacities for Cr(III) and Cr(VI) ions (Copello et al. 2008). The list of adsorbents derived from chitosan for the removal of chromium ions are listed in Table 2.3.

21.3.3 Adsorbents Derived from Sodium Alginate

Sodium alginate is a non-toxic linear biopolymer obtained from brown algae and seaweed (Fertah 2017) and widely studied as an adsorbent for the removal of Cr(VI) ions. Owing to the immense presence of –COOH groups, sodium alginate is negatively charged at above pH 4.5 which limits one of its applications as an efficient adsorbent for the removal of oxyanion such as Cr(VI). Therefore, it was imperative to modify the sodium alginate to enhance its adsorption capacity for Cr(VI). For example, a biocomposite composed of alginate and alumina was synthesized and tested as an adsorbent for the effective removal of Cr(VI) ions by a batch method. The alginate/alumina biocomposite showed higher removal efficiency for Cr(VI) ions than the calcium crosslinked sodium alginate. The Cr(VI) adsorption by the composite was mainly controlled by the electrostatic forces. The presence of Al^{3+} in the composite electrostatically adsorbed the $HCrO_4^-$ ions and thereby removed Cr(VI) ions (Gopalakannan et al. 2016). Sodium alginate/polyaniline nanofibers were prepared through in-situ polymerization of aniline in the presence of sodium alginate and evaluated its adsorption performance for Cr(VI). The adsorption capacity decreased with the increase in solution pH and the presence of co-ions compete with the Cr(VI) ions for the adsorption sites on the nanofibers (Karthik and Meenakshi 2015). Vu et al. designed encapsulated magnetite graphene oxide inside alginate beads and studied their enhanced removal performance for Cr(VI) ions. The removal of Cr(VI) ions by the prepared beads varied with the pH and it can be reused at least for 5 cycles (Vu et al. 2017). Fe^0 nanoparticles embedded into reduced graphene oxide alginate beads and demonstrated as an efficient adsorbent for the detoxification of Cr(VI) ions. The prepared beads showed robust performance in Cr(VI) removal and the performance of the beads was controlled by several critical factors such as ionic strength, temperature, pH and initial Cr(VI) concentration (Lv et al. 2017). The alginate beads were chemically modified using polyethylenimine and employed for the adsorption of Cr(VI) ions from water. The Cr(VI) removal experiments were conducted under both batch and column methods. The Cr(VI) removal efficiency was significantly affected by the pH of the working solution. Overall, the modified alginate beads showed higher adsorption capacity for Cr(VI) ions than the unmodified alginate beads (Yan et al. 2017).

The remediation of Cr(VI) from aqueous solution using magnetic alginate–layered double hydroxide composites was investigated by a batch method and the experimental results demonstrated the excellent removal of Cr(VI) ions by the composite. The Cr(VI) removal efficiency was not significantly affected in the solution pH range between 4.1 and 9.5, nevertheless, the calcination of the composite played a vital role on the Cr(VI) ions removal efficiency (Lee et al. 2013). Hydrothermal assisted

Table 21.3 List of chitosan-based adsorbents used for the removal of Cr(VI) and Cr(III) ions

Adsorbent	Adsorbate	Analytical method	References
Chitosan	Cr(VI)	UV-Vis spectrometry	Jung et al. (2013)
Chitosan/MWNT/poly (acrylic acid)/poly(4-amino diphenylamine)functional gel	Cr(VI)	UV-Vis spectrometry	Kim et al. (2015)
Chitosan-coated fly-ash composite	Cr(VI)	UV-Vis spectrometry	Wen et al. (2011)
Chitosan and alginate nanocomposite	Cr(VI)	AAS	Gokila et al. (2017)
Chemically modified magnetic chitosan microspheres	Cr(VI)	ICP-OES	Sun et al. (2016)
Alkyl-substituted polyaniline/chitosan composites	Cr(VI)	AAS	Yavuz et al. (2011)
Reticulated chitosan micro/nanoparticles	Cr(VI)	FAAS	Dima et al. (2015)
1,2:7,8-diepoxyoctanecrosslinked chitosan	Cr(VI)	UV-Vis spectrometry	Vakili et al. (2018)
Alumina/chitosan biocomposite	Cr(VI)	UV-Vis spectrometry	Rajiv Gandhi et al. (2010)
Modified magnetic chitosan chelating resin	Cr(VI)	ICP-OES	Abou El-Reash et al. (2011)
Cross-linked chitosan-g-acrylonitrile copolymer	Cr(VI)	AAS	Shankar et al. (2014)
Bentonite/chitosan nano biocomposite	Cr(VI)	UV-Vis spectrometry	Moussout et al. (2018)
Chitosan-grafted-polyaniline composite	Cr(VI)	UV-Vis spectrometry	Karthik and Meenakshi (2014)
Quaternized chitosan coated bentonite	Cr(VI)	UV-Vis spectrometry	Hu et al. (2016)
Chitosan-nanoclay composite	Cr(VI)	UV-Vis spectrometry	Kahraman (2017)
Henna blended chitosan microparticles	Cr(VI)	AAS	Davarnejad et al. (2018)
Chitosan engraved iron and lanthanum mixedoxyhydroxide	Cr(VI)	UV-Vis spectrometry	Preethi et al. (2019)
Thiocarbamoyl chitosan	Cr(VI)	UV-Vis spectrometry	Chauhan et al. (2012)

(continued)

Table 21.3 (continued)

Adsorbent	Adsorbate	Analytical method	References
Amino terminated polyamidoamine functionalized chitosan beads	Cr(VI)	UV-Vis spectrometry	Rajiv Gandhi and Meenakshi (2013)
Triethylenetetramine modified graphene oxide/chitosan composite	Cr(VI)	UV-Vis spectrometry	Ge and Ma (2015)
Ionic solid impregnated phosphated chitosan	Cr(VI)	UV-Vis spectrometry	Kahu et al. (2016)
Chitosan-lysozyme biocomposite	Cr(VI)	UV-Vis spectrometry	Rathinam et al. (2018)
Phosphonium crosslinked chitosan	Cr(VI)	ICP-AES	Sessarego et al. (2019)
Chitosan crosslinked modified silicon materials	Cr(VI)	UV-Vis spectrometry	Shi et al. (2017)
Chitosan-Fe(III) complex	Cr(VI)	UV-Vis spectrometry	Shen et al. (2013)
Cross-linked xanthated chitosan	Cr(VI)	UV-Vis spectrometry	Sankararamakrishnan et al. (2006)
Chitosan assisted mixed oxyhydroxide	Cr(VI)	UV-Vis spectrometry	Preethi et al. (2017)
Chitosan and crosslinked graft copolymers	Cr(VI)	UV-Vis spectrometry	Sharma and Lalita Singh (2017)
Chitosan	Cr(VI)	UV-Vis spectrometry	Jung et al. (2013)
Chitosan beads	Cr(III) and Cr(VI)	AAS	Ngah et al. (2006)
Chitosan	Cr(III) and Cr(VI)	UV-Vis spectrometry	Ba et al. (2018)
Chitosan-based hydrogel	Cr(VI)	FAAS	Vilela et al. (2018)
Chitosan-coated diatomaceous earth	Cr(VI)	ICP-MS	Salih and Ghosh (2018)
Silk protein sericin-chitosan conjugate	Cr(VI)	UV-Vis spectrometry	Singh et al. (2018)
2-hydroxyethyltrimethyl ammonium chloride grafted chitosan	Cr(VI)	FAAS	Dai et al. (2012)
1,6-hexanediamine modified magnetic chitosan microspheres	Cr(VI)	UV-Vis spectrometry	Yue et al. (2018)

(continued)

Table 21.3 (continued)

Adsorbent	Adsorbate	Analytical method	References
Chitosan/magnetite graphene oxide composites	Cr(VI)	UV-Vis spectrometry	(Zhang et al. 2018)
Chitosan cross-linked with diethylenetriaminepen-taacetic acid	Cr(VI)	UV-Vis spectrometry	Bhatt et al. (2015)
Trimesic acid crosslinked chitosan	Cr(VI)	UV-Vis spectrometry	Bhatt et al. (2017)
Crosslinked chitosan resin	Cr(VI)	UV-Vis spectrometry	Wu et al. (2012)
Chitosan-coated ceramic alumina	Cr(VI)	UV-Vis spectrometry	Boddu et al. (2003)
Modified forms of chitosan beads	Cr(VI)	UV-Vis spectrometry	Kousalya et al. (2010)
Titanium crosslinked chitosan composite	Cr(VI)	UV-Vis spectrometry	Zhang et al. (2015)
Chitosan engraved iron and lanthanum mixed oxyhydroxide	Cr(VI)	UV-Vis spectrometry	Preethi et al. (2019)
Self-assembled chitosan-zirconium phosphate nanostructures	Cr(VI)	UV-Vis spectrometry	Bhatt et al. (2019)
Chitosan/ polymethylmethacrylate composite	Cr(VI)	UV-Vis spectrometry	Li et al. (2016)
Chitosan/polycaprolactam nanofibrous	Cr(VI)	UV-Vis spectrometry	Li et al. (2016)
Phosphonium salt crosslinked chitosan	Cr(VI)	ICP-AES	Sessarego et al. (2019)
Nanosized chitosan fibers	Cr(VI)	UV-Vis spectrometry	Li et al. (2015)
Fe-crosslinked chitosan	Cr(VI)	UV-Vis spectrometry	Zimmermann et al. (2010)
Nitrogen-doped chitosan-Fe(III) composite	Cr(VI)	UV-Vis spectrometry	Zhu et al. (2016)
Fe^0 nanorod modified with chitosan	Cr(VI)	UV-Vis spectrometry	Sun et al. (2015)
Chemically modified magnetic chitosan resins	Cr(VI)	UV-Vis spectrometry	Elwakeel (2010)

(continued)

Table 21.3 (continued)

Adsorbent	Adsorbate	Analytical method	References
Chemically modified chitosan microspheres	Cr(VI)	ICP-OES	Sun et al. (2016)
Magnetic chitosan nanoparticles	Cr(VI)	UV-Vis spectrometry	Thinh et al. (2013)
Thiourea modified magnetic chitosan composite	Cr(VI)	UV-Vis spectrometry	Cai et al. (2019)
Chitosan/magnetic graphene oxide composite	Cr(VI)	UV-Vis spectrometry	Zhang et al. (2018)
Nano iron oxide impregnated chitosan bead	Cr(VI)	UV-Vis spectrometry	Lu et al. (2017)
Zr(IV) crosslinked chitosan magnetic microspheres	Cr(VI)	UV-Vis spectrometry	Chen et al. (2017)
Magnetic chitosan modified with diethylenetriamine	Cr(VI)	AAS	Li et al. (2011)
Amino-functionalized graphene oxide decorated with Fe3O4 nanoparticles	Cr(VI)	ICP-MS	Zhao et al. (2016)
Chitosan/MWCNTs/Fe_3O_4 nanofibers	Cr(VI)	ICP-AES	Beheshti et al. (2016)
Crosslinked chitosan/MWCNT/ Fe film	Cr(VI)	UV-Vis spectrometry	Neto et al. (2019)
Fe_3O_4 nanoparticles functionalized polyvinyl alcohol/chitosan magnetic composite hydrogel	Cr(VI)	UV-Vis spectrometry	Yan et al. (2018)
Quaternary ammonium salt modified chitosan magnetic composite	Cr(VI)	AAS	Li et al. (2016)
Magnetic chitosan-iron(III) hydrogel	Cr(VI)	UV-Vis spectrometry	Yu et al. (2013)
Magnetic Fe_3O_4@SiO_2–chitosan	Cr(VI)	UV-Vis spectrometry	Jiang et al. (2019)
Magnetic chitosan composite	Cr(VI)	UV-Vis spectrometry	Cai et al. (2018)

(continued)

Table 21.3 (continued)

Adsorbent	Adsorbate	Analytical method	References
Chitosan/montmorillonite-Fe_3O_4 hybrid	Cr(VI)	UV-Vis spectrometry	Chen et al. (2013)
Modified magnetic chitosan chelating resin	Cr(VI)	ICP-OES	Abou El-Reash et al. (2011)
Fe-loaded chitosan carbonized rice husk composite beads	–	–	Sugashini et al. (2015)
Crosslinked magnetic chitosan beads	Cr(VI)	UV-Vis spectrometry	(Huang et al. 2009)
Chitosan/modified MWCNTs nanocomposite	Cr(VI)	ICP-MS	Huang et al. (2018)
Self-assembled sponge-like chitosan/reduced graphene oxide/montmorillonite composite hydrogel	Cr(VI)	UV-Vis spectrometry	Yu et al. (2017)
Modified graphene oxide/chitosan composite	Cr(VI)	UV-Vis spectrometry	Zhang et al. (2016)
Chitosan-1,2-cyclohexylene dinitrilotetraacetic acid-graphene oxide nanocomposite	Cr(VI)	ICP-MS	Ali (2018)
Graphene oxide functionalized with magnetic cyclodextrin-chitosan	Cr(VI)	AAS	Li et al. (2013)
Chitosan/polypyrrole composite	Cr(VI)	UV-Vis spectrometry	Karthik and Meenakshi (2015)
Substituted polyaniline/chitosan composite	Cr(VI)	AAS	Yavuz et al. (2011)
Polypyrrole/chitosan nanocomposite aerogel monolith	Cr(VI)	UV-Vis spectrometry	Ji et al. (2018)
Chitosan grafted polyaniline montmorillonite	Cr(VI)		Ahmad et al. (2017)
Chitosan-coated with poly-3-methyl thiophene	Cr(VI)	UV-Vis spectrometry	Hena (2010)

(continued)

Table 21.3 (continued)

Adsorbent	Adsorbate	Analytical method	References
Chitosan-silica composite	Cr(VI)	UV-Vis spectrometry	Budnyak et al. (2015)
Chitosan-g-poly(butyl acrylate)/silica gel nanocomposite	–	–	Nithya et al. (2016)
Crosslinked chitosan modified silicon materials	Cr(VI)	UV-Vis spectrometry	Shi et al. (2017)
Chitosan/silica gel and lanthanum loaded chitosan/silica gel composite	Cr(VI)	UV-Vis spectrometry	Gandhi and Meenakshi (2012)
Silicate-chitosan composite	Cr(VI)	AAS	Copello et al. (2008)
Zirconium crosslinked chitosan composite	Cr(VI)	ICP-OES	Zhang et al. (2013)
Ethylenediamine modified chitosan microspheres	Cr(VI)	UV-Vis spectrometry	Chethan and Vishalakshi (2015)
Chitosan grafted graphene oxide nanocomposite	Cr(VI)	UV-Vis spectrometry	Samuel et al. (2019)
Graphene oxide/chitosan/ferrite nanocomposite	Cr(VI)	UV-Vis spectrometry	Samuel et al. (2018)
Strong base anion exchanger microspheres embedded into chitosan/poly(vinylamine) cryogels	Cr(VI)	AAS	Dragan et al. (2017)

magnetic nano-hydroxyapatite encapsulated alginate beads were developed for the efficient Cr(VI) uptake from water and the results demonstrated that the Cr(VI) removal by the beads was high at acidic pH conditions than the alkaline pH conditions (Periyasamy et al. 2018). Adsorptive removal of Cr(III) and Cr(VI) ions from binary solution was investigated using alginate-goethite beads and the adsorption of Cr(III) ions was increased in the pH range from 2 to 4 while for Cr(VI) ions moderately decreased. Higher removal of Cr(VI) and Cr(III) was observed with higher mixing rates, higher adsorbent dosage and smaller bead size (Lazaridis and Charalambous 2005). In the same way various alginate based adsorbents such as hydrocalumite/alginate composite (Periyasamy and Viswanathan 2018), sodium alginate-derived magnetic carbonaceous (Lei et al. 2015), metal ions (Ca^{2+}, Ce^{3+}, and Zr^{4+}) anchored alginate-gelatin binary biocomposite (Gopalakannan and Viswanathan 2016), metal ions anchored alginate-bentonite biocomposite (Gopalakannan et al. 2016), grape stalks entrapped alginate beads (Escudero et al. 2017), brewers draff and grape waste encapsulated calcium alginate beads (Šillerová et al. 2015), zirconium oxide-immobilized alginate beads (Kumar et al. 2018), adapted bacterial consortia immobilized alginate beads (Samuel et al. 2013), baker's yeast strain biomass immobilized alginate beads (Mahmoud and Mohamed 2017), magnetic alginate hybrid beads (Gopalakannan and Viswanathan 2015), Fe^0-Fe_3O_4 nanocomposites embedded polyvinyl alcohol/sodium alginate beads (Lv et al. 2013) and *Pseudomonas aeruginosa* immobilized polyvinyl alcohol/sodium alginate/MWCNTs (Pang et al. 2011) were prepared and demonstrated as further adsorbents for the effective removal of Cr(VI) ions from the aqueous solution. The list of sodium alginate-based adsorbents investigated for the removal of chromium ions are listed in Table 2.4.

21.3.4 Adsorbents Derived from Cellulose

Cellulose is a linear biopolymer consisting of β-D-glucopyranose units joined by β-1,4 glycosidic linkages. Cellulose is abundant in nature and is derived from plenty of sources such as woods, animals, cotton, etc. (Hokkanen et al. 2016). Cellulose is serving as an adsorbent for the removal of Cr(VI) and Cr(III) ions due to the existence plenty of -OH groups, however, the adsorption capacity is significantly affected by the physio-chemical modifications. It is worth mentioning that most of the modified cellulose-based adsorbents showed higher adsorption capacity for Cr(VI) and Cr(III) ions (Miretzky and Cirelli 2010; Wang and Lee 2011). For instance, cellulose was interacted with the surfactant- modified sodium montmorillonite and tested as an adsorbent for the detoxification of Cr(VI) ions. The prepared biocomposite showed excellent removal of Cr(VI) ions and the composite could be regenerated using sodium hydroxide as a regeneration medium. In addition, the advantage of this composite for the removal of Cr(VI) ions from a real chrome tannery wastewater was also demonstrated (Kumar et al. 2012). Hokkanen et al. synthesized calcium hydroxyapatite microfibrillated cellulose composite and evaluated its Cr(VI) uptake abilities from Cr(VI) contaminated water. The influence of various parameters such as pH, temperature, and Cr(VI) ion concentration on Cr(VI) removal was established under optimized conditions (Hokkanen et al. 2016). Cellulose was coated with polyaniline was prepared and demonstrated as an efficient adsorbent for Cr(VI) ions than the pure cellulose. With the increase in polyaniline loading, the Cr(VI) removal efficiency was also increased. Notably, the prepared material could remove both Cr(VI) and Cr(III) ions through adsorption and also the adsorbed Cr(VI) was then further reduced to Cr(III) through surface reactions (Qiu et al. 2014).

Adsorption of Cr(VI) ions onto 2-aminomethyl pyridine modified spherical cellulose-based adsorbent was conducted through batch and column methodologies and the results showed that the prepared adsorbent could selectively, simultaneously remove and detoxify Cr(VI) ions (Dong et al. 2016). Yu et al. developed an aminated cellulose microsphere through radiation- induced graft polymerization and used for efficient Cr(VI) uptake. The adsorption capacity was found to be increased with the increase in temperature which indicated the endothermic nature of the process. The ion-exchange was the main mechanism involved during the adsorption process (Yu

Table 21.4 List of sodium alginate-based adsorbents used for the removal of Cr(VI) and Cr(III) ions

Adsorbent	Adsorbate	Analytical method	References
Alginate/alumina biocomposite	Cr(VI)	UV-Vis spectrometry	Rajiv Gandhi et al. (2010)
Sodium alginate/polyaniline nanofibers	Cr(VI)	UV-Vis spectrometry	Karthik and Meenakshi (2015)
Magnetite graphene oxide encapsulated alginate beads	Cr(VI)	Inductively coupled plasma optical emission spectrometry (ICP-OES)	Vu et al. (2017)
Fe^0 nanoparticles embedded reduced graphene oxide alginate beads	Cr(VI)	UV-Vis spectrometry	Lv et al. (2017)
Flexible core-shell/bead-like alginate@PEI	Cr(VI)	UV-Vis spectrometry	Yan et al. (2017)
Magnetic alginate-layered double hydroxide composite	Cr(VI)	UV-Vis spectrometry	Lee et al. (2013)
Magnetic nano-hydroxyapatite encapsulated alginate beads	Cr(VI)	UV-Vis spectrometry	Periyasamy et al. (2018)
Alginate-goethite beads	Cr(III) and Cr(VI)	UV-Vis spectrometry	Lazaridis and Charalambous (2005)
Hydrocalumite/alginate composite	Cr(VI)	UV-Vis spectrometry	Periyasamy and Viswanathan (2018)
Sodium alginate-derived magnetic carbonaceous	Cr(VI)	UV-Vis spectrometry	Lei et al. (2015)
Metal ions (Ca^{2+}, Ce^{3+}, and Zr^{4+}) anchored alginate-gelatin binary biocomposite	Cr(VI)	UV-Vis spectrometry	Gopalakannan and Viswanathan (2016)
Metal ions (Ca^{2+}, Ce^{3+}, and Zr^{4+}) anchored alginate-bentonite biocomposite	Cr(VI)	UV-Vis spectrometry	Gopalakannan et al. (2016)

(continued)

Table 21.4 (continued)

Adsorbent	Adsorbate	Analytical method	References
Grape stalks entrapped alginate beads	Cr(VI) and Cr(III)	Flame Atomic Absorption Spectrometry (FAAS)	Escudero et al. (2017)
Brewer's draff and grape waste encapsulated calcium alginate beads	Cr(VI)	UV-Vis spectrometry	Šillerová et al. (2015)
Zirconium oxide-immobilized alginate beads	Cr(VI)	ICP-OES	Kumar et al. (2018)
Pseudomonas aeruginosa immobilized polyvinyl alcohol/sodium alginate/MWCNTs	Cr(VI)	UV-Vis spectrometry	Pang et al. (2011)
Magnetic alginate hybrid beads	Cr(VI)	UV-Vis spectrometry	(Gopalakannan and Viswanathan 2015)
Fe^0-Fe_3O_4nanocomposites embedded polyvinyl alcohol/sodium alginate beads	Cr(VI)	UV-Vis spectrometry	Lv et al. (2013)
Baker's yeast strain biomass immobilized alginate beads	Cr(VI)	FAAS	Mahmoud and Mohamed (2017)

et al. 2016). Guo et al. developed polyethyleneimine functionalized cellulose aerogel beads and studied the efficient dynamic removal of highly toxic Cr(VI) ions from water. The influence of several parameters on the adsorption capacity of the developed adsorbent for Cr(VI) ions was systematically investigated. The functionalized cellulose aerogel exhibited 12 times higher adsorption capacity than the unmodified cellulose aerogel which underlined the advantage of the functionalized cellulose (Guo et al. 2017). Succinylatedmercerized cellulose functionalized with quaternary ammonium groups was prepared and investigated its Cr(V) adsorption performance. The maximum sorption capacity was seen at pH 3.1 and decreased with the increase in pH (Gurgel et al. 2009). Adsorptive reduction and detoxification of Cr(VI) contaminated water using polypyrrole/cellulose fiber composite was conducted and the adsorption study results indicated that the prepared composite was highly effective in Cr(VI) ions removal. In addition, the XPS results revealed that the toxic Cr(VI) was reduced to less toxic Cr(III) in solution at lower pH conditions and then adsorbed onto the composite (Lei et al. 2012). Liu et al. adopted response surface methodology coupled with central composite design and demonstrated the removal of Cr(III) ions using highly porous glutamate-modified cellulose beads. The optimum conditions for the high Cr(III) removal performance were established using different models

(Liu et al. 2011). Amine functionalized cellulose acetate/silica composite nanofibrous membranes were fabricated for the removal of Cr(VI) ions under static and flow conditions. The prepared membranes exhibited high surface area and porosity with the maximum adsorption capacity of 19.46 mg/g for Cr(VI) (Taha et al. 2012).

Tetrabutylammonium bromide surfactant anchored cellulose was prepared through microwave irradiation and evaluated its adsorption potential for Cr(VI) ions removal and the interactions between the Cr(VI) ions and the adsorbent was observed using thermal, analytical and spectroscopy techniques. The adsorbent interacts with the Cr(VI) ions through electrostatic attraction and hydrogen bonding and thereby facilitated the Cr(VI) removal (Kalidhasan et al. 2012). Cellulose was combined with ionic liquids and explored as an adsorbent for the removal of Cr(VI) ions. Kalidhasan et al. developed a crystalline cellulose-methyltrioctylammonium chloride, an ionic liquid-based adsorbent by ultrasonication method and assessed its Cr(VI) removal ability. The maximum Cr(VI) removal has occurred in the pH range between 2 and 4. Moreover, the prepared adsorbent was packed in a column and the Cr(VI) removal was examined under flow conditions. The cellulose-ionic liquid blend adsorbent showed superior adsorption capacity than the other cellulose-based adsorbents (Kalidhasan et al. 2012). Similarly, ionic-liquid functionalized cellulose was prepared and tested its adsorption performance for the removal of Cr(VI) ions in both adsorption and column methods. The functionalized cellulose material exhibited excellent removal performance for Cr(VI) ions and the XPS results indicated the ion-exchange reduction mechanism behind the Cr(VI) ions removal (Dong and Zhao 2018).

Cellulose was chemically modified with different materials to have anion exchange property. Anirudhan et al. conducted the adsorption of Cr(VI) ions from water and wastewater using an adsorbent synthesized through ethylation of glycidyl methacrylate grafted aminated titanium dioxide densified cellulose. The Cr(VI) ions were removed through the ion-exchange mechanism and the Cr(VI) removal was significantly affected by the increase in the ionic strength and increase in the pH from 4.5 (Anirudhan et al. 2013). Likewise, cellulose powder was chemically modified with β-cyclodextrin and quaternary ammonium groups and used for the detoxification of Cr(VI) ions. The synthesized adsorbent exhibited excellent Cr(VI) adsorption performance and notably, the adsorbent could selectively remove Cr(VI) ions even in the presence of other competing ions (Zhou et al. 2011). Table 2.5 shows the list of cellulose derived adsorbents used for the removal of chromium ions.

Table 21.5 List of cellulose-based adsorbents used for the removal of Cr(VI) and Cr(III) ions

Adsorbent	Adsorbate	Analytical method	References
Cellulose-montmorillonite composite	Cr(VI)	UV-Vis spectrometry	Kumar et al. (2012)
Calcium hydroxyapatite microfibrillated cellulose composite	Cr(VI)	ICP-OES	Hokkanen et al. (2016)
Polyaniline coated ethyl cellulose	Cr(VI)	UV-Vis spectrometry	Qiu et al. (2014)
2-aminomethyl pyridine modified spherical cellulose	Cr(VI)	UV-Vis spectrometry	Dong et al. (2016)
Aminated cellulose microsphere	Cr(VI)	UV-Vis spectrometry	Yu et al. (2016)
Polyethyleneimine functionalized cellulose aerogel beads	Cr(VI)	UV-Vis spectrometry	Guo et al. (2017)
Succinylatedmercerized cellulose functionalized with quaternary ammonium groups	Cr(VI)	AAS	Gurgel et al. (2009)
Polypyrrole/cellulose fiber composite	Cr(VI)	UV-Vis spectrometry	Lei et al. (2012)
Glutamate-modified cellulose beads	Cr(III)	ICP-OES	Liu et al. (2011)
Amine functionalized cellulose acetate/silica composite	Cr(VI)	ICP-OES	Taha et al. (2012)
Surfactant anchored cellulose	Cr(VI)	UV-Vis spectrometry	Kalidhasan et al. (2012)
Crystalline cellulose-ionic liquid blend polymeric material	Cr(VI)	UV-Vis spectrometry	Kalidhasan et al. (2012)
Ionic-liquid functionalized cellulose	Cr(VI)	ICP-OES	Dong and Zhao (2018)
Glycidyl methacrylate-grafted-densified cellulose with quaternary ammonium groups	Cr(VI)	UV-Vis spectrometry	Anirudhan et al. (2013)
Cellulose powder modified with β-cyclodextrin and quaternary ammonium groups	Cr(VI)	UV-Vis spectrometry	Zhou et al. (2011)

21.4 Conclusions

This chapter comprehensively reviewed the remediation of Cr(VI) ions using eco-friendly adsorbents derived from chitin, chitosan, sodium alginate and cellulose. Besides, the analytical method used for the determination of chromium is also discussed. Notably, chitin, chitosan, sodium alginate and cellulose possessed very low adsorption capacity in its original form but upon subjection to several chemical modifications their adsorption capacity was much more pronounced for Cr(VI) ions. The presence of enormous amount of $-NH_2$, $-CH_2OH$ and –OH functional groups in the chitin and chitosan, –COOH and –OH functional groups in the sodium alginate and –OH functional groups in the cellulose interacts with the Cr(VI) ions through electrostatic forces, ion-exchange and chemical redox reactions. In most of the cases, the bounded toxic Cr(VI) was further reduced to less toxic Cr(III) and this is yet another advantage of using the eco-friendly adsorbents. Furthermore, these bio-adsorbents retained their original adsorption capacity even after several numbers of reuse cycles and thus showed high regeneration ability and cost-effective of the process. Overall, owing to all these notorious advantages, eco-friendly based adsorbents are well in a good position to replace the conventional adsorbents used for the removal of Cr(VI) ions from water and wastewater.

References

Abou El-Reash YG, Otto M, Kenawy IM, Ouf AM (2011) Adsorption of Cr(VI) and As(V) ions by modified magnetic chitosan chelating resin. Int J Biol Macromolecules 49(4):513–522. https://doi.org/10.1016/j.ijbiomac.2011.06.001

Ahmad R, Hasan I, Mittal A (2017) Adsorption of Cr (VI) and Cd (II) on chitosan grafted polyaniline-OMMT nanocomposite: isotherms, kinetics and thermodynamics studies. Desalin Water Treat 58:144–153

Ali MEA (2018) Synthesis and adsorption properties of chitosan-CDTA-GO nanocomposite for removal of hexavalent chromium from aqueous solutions. Arab J Chem 11(7):1107–1116. https://doi.org/10.1016/j.arabjc.2016.09.010

Anirudhan TS, Nima J, Divya PL (2013) Adsorption of chromium(VI) from aqueous solutions by glycidylmethacrylate-grafted-densified cellulose with quaternary ammonium groups. Appl Surf Sci 279:441–449. https://doi.org/10.1016/j.apsusc.2013.04.134

Ba S, Alagui A, Hajjaji M (2018) Retention and release of hexavalent and trivalent chromium by chitosan, olive stone activated carbon, and their blend. Environ Sci Pollut Res 25(20):19585–19604. https://doi.org/10.1007/s11356-018-2196-7

Barnhart J (1997) Occurrences, uses, and properties of Chromium. Regul Toxicol Pharmacol 26(1):S3–S7. https://doi.org/10.1006/rtph.1997.1132

Beheshti H, Irani M, Hosseini L, Rahimi A, Aliabadi M (2016) Removal of Cr (VI) from aqueous solutions using chitosan/MWCNT/Fe3O4 composite nanofibers-batch and column studies. Chem Eng J 284:557–564. https://doi.org/10.1016/j.cej.2015.08.158

Bhatt R, Sreedhar B, Padmaja P (2015) Adsorption of chromium from aqueous solutions using crosslinked chitosan–diethylenetriaminepentaacetic acid. Int J Biol Macromolecules 74:458–466. https://doi.org/10.1016/j.ijbiomac.2014.12.041

Bhatt R, Sreedhar B, Padmaja P (2017) Chitosan supramolecularly cross linked with trimesic acid—Facile synthesis, characterization and evaluation of adsorption potential for Chromium(VI). Int J Biol Macromolecules 104:1254–1266. https://doi.org/10.1016/j.ijbiomac.2017.06.067

Bhatt R, Ageetha V, Rathod SB, Padmaja P (2019) Self-assembled chitosan-zirconium phosphate nanostructures for adsorption of chromium and degradation of dyes. Carbohydr Polym 208:441–450. https://doi.org/10.1016/j.carbpol.2018.12.077

Boddu VM, Abburi K, Talbott JL, Smith ED (2003) Removal of hexavalent Chromium from wastewater using a new composite Chitosan biosorbent. Environ Sci Technol 37(19):4449–4456. https://doi.org/10.1021/es021013a

Budnyak TM, Pylypchuk IV, Tertykh VA, Yanovska ES, Kolodynska D (2015) Synthesis and adsorption properties of chitosan-silica nanocomposite prepared by sol-gel method. Nanoscale Res Lett 10:87. https://doi.org/10.1186/s11671-014-0722-1

Cai W, Xue W, Jiang Y (2018) Facile preparation of magnetic Chitosan coprecipitated by ethanol/NH_3 H_2O for highly efficient removal toward Cr(VI). ACS Omega 3(5):5725–5734. https://doi.org/10.1021/acsomega.8b00393

Cai W, Zhu F, Liang H, Jiang Y, Tu W, Cai Z, Wu J, Zhou J (2019) Preparation of thiourea-modified magnetic chitosan composite with efficient removal efficiency for Cr(VI). Chem Eng Res Des 144:150–158. https://doi.org/10.1016/j.cherd.2019.01.031

Chauhan D, Jaiswal M, Sankararamakrishnan N (2012) Removal of Cadmium and hexavalent Chromium from electroplating waste water using thiocarbamoyl chitosan. Carbohydr Polym 88(2):670–675. https://doi.org/10.1016/j.carbpol.2012.01.014

Chen D, Li W, Wu Y, Zhu Q, Lu Z, Du G (2013) Preparation and characterization of chitosan/montmorillonite magnetic microspheres and its application for the removal of Cr (VI). Chem Eng J 221:8–15. https://doi.org/10.1016/j.cej.2013.01.089

Chen X, Zhang W, Luo X, Zhao F, Li Y, Li R, Li Z (2017) Efficient removal and environmentally benign detoxification of Cr(VI) in aqueous solutions by Zr(IV) cross-linking chitosan magnetic microspheres. Chemosphere 185:991–1000. https://doi.org/10.1016/j.chemosphere.2017.07.113

Chethan PD, Vishalakshi B (2015) Synthesis of ethylenediamine modified chitosan microspheres for removal of divalent and hexavalent ions. Int J Biol Macromolecules 75:179–185. https://doi.org/10.1016/j.ijbiomac.2015.01.032

Copello GJ, Varela F, Vivot RM, Díaz LE (2008) Immobilized chitosan as biosorbent for the removal of Cd(II), Cr(III) and Cr(VI) from aqueous solutions. Bioresour Technol 99(14):6538–6544. https://doi.org/10.1016/j.biortech.2007.11.055

Dai J, Ren F, Tao C (2012) Adsorption of Cr(VI) and speciation of Cr(VI) and Cr(III) in aqueous solutions using chemically modified chitosan. Int J Environ Res Public Health 9(5):1757–1770. https://doi.org/10.3390/ijerph9051757

Davarnejad R, Karimi Dastnayi Z, Kennedy JF (2018) Cr(VI) adsorption on the blends of Henna with chitosan microparticles: Experimental and statistical analysis. Int J Biol Macromolecules 116:281–288. https://doi.org/10.1016/j.ijbiomac.2018.04.189

Demarchi CA, Debrassi A, Magro JD, Nedelko N, Ślawska-Waniewska A, Dłużewski P, Greneche J-M, Rodrigues CA (2015) Adsorption of Cr(VI) on crosslinked chitosan–Fe(III) complex in fixed-bed systems. J Water Process Eng 7:141–152. https://doi.org/10.1016/j.jwpe.2015.05.003

Devi R, Dhamodharan R (2018) Pretreatment in hot glycerol for facile and green separation of Chitin from prawn shell waste. ACS Sustain Chem Eng 6(1):846–853. https://doi.org/10.1021/acssuschemeng.7b03195

Dima JB, Sequeiros C, Zaritzky NE (2015) Hexavalent chromium removal in contaminated water using reticulated chitosan micro/nanoparticles from seafood processing wastes. Chemosphere 141:100–111. https://doi.org/10.1016/j.chemosphere.2015.06.030

Dong Z, Zhao L (2018) Covalently bonded ionic liquid onto cellulose for fast adsorption and efficient separation of Cr(VI): Batch, column and mechanism investigation. Carbohydr Polym 189:190–197. https://doi.org/10.1016/j.carbpol.2018.02.038

Dong Z, Zhao J, Du J, Li C, Zhao L (2016) Radiation synthesis of spherical cellulose-based adsorbent for efficient adsorption and detoxification of Cr(VI). Radiat Phys Chem 126:68–74. https://doi.org/10.1016/j.radphyschem.2016.05.013

Dragan ES, Humelnicu D, Dinu MV, Olariu RI (2017) Kinetics, equilibrium modeling, and thermodynamics on removal of Cr(VI) ions from aqueous solution using novel composites with strong base anion exchanger microspheres embedded into chitosan/poly(vinyl amine) cryogels. Chem Eng J 330:675–691. https://doi.org/10.1016/j.cej.2017.08.004

Elwakeel KZ (2010) Removal of Cr(VI) from alkaline aqueous solutions using chemically modified magnetic chitosan resins. Desalination 250(1):105–112. https://doi.org/10.1016/j.desal.2009.02.063

EPA (1996) Test methods for evaluating solid waste. 3rd ed. Washington, DC: U.S. Environmental Protection Agency, Office of Solid Waste and Emergency Response. Method 6010B: Inductively coupled plasma-atomic emission spectrometry. SW-846

EPA (2011) Method 218.7: Determination of hexavalent chromium in drinking water by ion chromatography with post-column derivatization and UV-visable spectroscopic detection. U.S. Environmental Protection Agency (EPA815R11005)

EPA. Method 218.4 (Atomic absorption, chelation-extraction). Methods for chemical analysis of water and wastes. Environmental Protection Agency, Cincinnati, OH: U.S. (1983b.Chromium. EPA600479020)

Escudero C, Fiol N, Villaescusa I, Bollinger J-C (2017) Effect of chromium speciation on its sorption mechanism onto grape stalks entrapped into alginate beads. Arab J Chem 10:S1293–S1302. https://doi.org/10.1016/j.arabjc.2013.03.011

Fertah M (2017) Chapter 2—isolation and characterization of alginate from seaweed. In: Venkatesan J, Anil S, Kim S-K (eds) Seaweed polysaccharides. Elsevier, pp 11–26

Gandhi MR, Meenakshi S (2012) Preparation and characterization of La(III) encapsulated silica gel/chitosan composite and its metal uptake studies. J Hazard Mater 203–204:29–37. https://doi.org/10.1016/j.jhazmat.2011.11.062

Ge H, Ma Z (2015) Microwave preparation of triethylenetetramine modified graphene oxide/chitosan composite for adsorption of Cr(VI). Carbohydr Polym 131:280–287. https://doi.org/10.1016/j.carbpol.2015.06.025

Gokila S, Gomathi T, Sudha PN, Anil S (2017) Removal of the heavy metal ion chromiuim(VI) using Chitosan and Alginate nanocomposites. Int J Biol Macromolecules 104:1459–1468. https://doi.org/10.1016/j.ijbiomac.2017.05.117

Gopalakannan V, Viswanathan N (2015) Synthesis of magnetic alginate hybrid beads for efficient chromium (VI) removal. Int J Biol Macromolecules 72:862–867. https://doi.org/10.1016/j.ijbiomac.2014.09.024

Gopalakannan V, Viswanathan N (2016) One pot synthesis of metal ion anchored alginate–gelatin binary biocomposite for efficient Cr(VI) removal. Int J Biol Macromolecules 83:450–459. https://doi.org/10.1016/j.ijbiomac.2015.10.010

Gopalakannan V, Periyasamy S, Viswanathan N (2016a) One pot eco-friendly synthesis of highly dispersed alumina supported alginate biocomposite for efficient chromium(VI) removal. J Water Process Eng 10:113–119. https://doi.org/10.1016/j.jwpe.2016.02.005

Gopalakannan V, Periyasamy S, Viswanathan N (2016b) Synthesis of assorted metal ions anchored alginate bentonite biocomposites for Cr(VI) sorption. Carbohydr Polym 151:1100–1109. https://doi.org/10.1016/j.carbpol.2016.06.030

Guo D-M, An Q-D, Xiao Z-Y, Zhai S-R, Shi Z (2017) Polyethylenimine-functionalized cellulose aerogel beads for efficient dynamic removal of chromium(vi) from aqueous solution. RSC Adv 7(85):54039–54052. https://doi.org/10.1039/C7RA09940A

Gurgel LVA, Perin de Melo JC, de Lena JC, Gil LF (2009) Adsorption of chromium (VI) ion from aqueous solution by succinylated mercerized cellulose functionalized with quaternary ammonium groups. Bioresour Technol 100(13):3214–3220. https://doi.org/10.1016/j.biortech.2009.01.068

Hena S (2010) Removal of chromium hexavalent ion from aqueous solutions using biopolymer chitosan coated with poly 3-methyl thiophene polymer. J Hazard Mater 181(1):474–479. https://doi.org/10.1016/j.jhazmat.2010.05.037

Hokkanen S, Bhatnagar A, Sillanpää M (2016a) A review on modification methods to cellulose-based adsorbents to improve adsorption capacity. Water Res 91:156–173. https://doi.org/10.1016/j.watres.2016.01.008

Hokkanen S, Bhatnagar A, Repo E, Lou S, Sillanpää M (2016b) Calcium hydroxyapatite microfibrillated cellulose composite as a potential adsorbent for the removal of Cr(VI) from aqueous solution. Chem Eng J 283:445–452. https://doi.org/10.1016/j.cej.2015.07.035

Hu P, Wang J, Huang R (2016) Simultaneous removal of Cr(VI) and Amido black 10B (AB10B) from aqueous solutions using quaternized chitosan coated bentonite. Int J Biol Macromolecules 92:694–701. https://doi.org/10.1016/j.ijbiomac.2016.07.085

Huang G, Zhang H, Shi JX, Langrish TAG (2009) Adsorption of Chromium(VI) from aqueous solutions using cross-linked magnetic chitosan beads. Ind Eng Chem Res 48(5):2646–2651. https://doi.org/10.1021/ie800814h

Huang Y, Lee X, Macazo FC, Grattieri M, Cai R, Minteer SD (2018) Fast and efficient removal of chromium (VI) anionic species by a reusable chitosan-modified multi-walled carbon nanotube composite. Chem Eng J 339:259–267. https://doi.org/10.1016/j.cej.2018.01.133

Jaishankar M, Tseten T, Anbalagan N, Mathew BB, Beeregowda KN (2014) Toxicity, mechanism and health effects of some heavy metals. Interdisc Toxicol 7(2):60–72. https://doi.org/10.2478/intox-2014-0009

Jarup L (2003) Hazards of heavy metal contamination. Br Med Bull 68:167–182

Ji J, Xiong H, Zhu Z, Li L, Huang Y, Yu X (2018) Fabrication of Polypyrrole/Chitosan nanocomposite aerogel monolith for removal of Cr(VI). J Polym Environ 26(5):1979–1985. https://doi.org/10.1007/s10924-017-1095-1

Jiang Y, Cai W, Tu W, Zhu M (2019) Facile cross-link method to synthesize magnetic Fe_3O_4@SiO_2–Chitosan with high adsorption capacity toward hexavalent chromium. J Chem Eng Data 64(1):226–233. https://doi.org/10.1021/acs.jced.8b00738

Jung C, Heo J, Han J, Her N, Lee S-J, Oh J, Ryu J, Yoon Y (2013) Hexavalent chromium removal by various adsorbents: Powdered activated carbon, chitosan, and single/multi-walled carbon nanotubes. Sep Purif Technol 106:63–71. https://doi.org/10.1016/j.seppur.2012.12.028

Kahraman HT (2017) Development of an adsorbent via chitosan nano-organoclay assembly to remove hexavalent chromium from wastewater. Int J Biol Macromolecules 94:202–209. https://doi.org/10.1016/j.ijbiomac.2016.09.111

Kahu SS, Shekhawat A, Saravanan D, Jugade RM (2016) Two fold modified chitosan for enhanced adsorption of hexavalent chromium from simulated wastewater and industrial effluents. Carbohydr Polym 146:264–273. https://doi.org/10.1016/j.carbpol.2016.03.041

Kalidhasan S, Gupta PA, Cholleti VR, Santhana Krishna Kumar A, Rajesh V, Rajesh N (2012) Microwave assisted solvent free green preparation and physicochemical characterization of surfactant-anchored cellulose and its relevance toward the effective adsorption of chromium. J Colloid Interface Sci 372(1):88–98. https://doi.org/10.1016/j.jcis.2012.01.013

Kalidhasan S, Santhana KrishnaKumar A, Rajesh V, Rajesh N (2012b) Ultrasound-assisted preparation and characterization of crystalline cellulose–ionic liquid blend polymeric material: a prelude to the study of its application toward the effective adsorption of chromium. J Colloid Interface Sci 367(1):398–408. https://doi.org/10.1016/j.jcis.2011.09.062

Kanmani P, Aravind J, Kamaraj M, Sureshbabu P, Karthikeyan S (2017) Environmental applications of chitosan and cellulosic biopolymers: a comprehensive outlook. Bioresour Technol 242:295–303. https://doi.org/10.1016/j.biortech.2017.03.119

Kartal SN, Imamura Y (2005) Removal of copper, chromium, and arsenic from CCA-treated wood onto chitin and chitosan. Bioresour Technol 96(3):389–392. https://doi.org/10.1016/j.biortech.2004.03.004

Karthik R, Meenakshi S (2014a) Synthesis, characterization and Cr(VI) uptake studies of polypyrrole functionalized Chitin. Synth Metals 198:181–187. https://doi.org/10.1016/j.synthmet.2014.10.012

Karthik R, Meenakshi S (2014b) Facile synthesis of cross linked-chitosan–grafted-polyaniline composite and its Cr(VI) uptake studies. Int J Biol Macromolecules 67:210–219. https://doi.org/10.1016/j.ijbiomac.2014.03.035

Karthik R, Meenakshi S (2015a) Adsorption study on removal of Cr(VI) ions by polyaniline composite. Desalin Water Treat 54(11):3083–3093. https://doi.org/10.1080/19443994.2014.909330

Karthik R, Meenakshi S (2015b) Removal of Pb(II) and Cd(II) ions from aqueous solution using polyaniline grafted chitosan. Chem Eng J 263:168–177. https://doi.org/10.1016/j.cej.2014.11.015

Karthik R, Meenakshi S (2015c) Synthesis, characterization and Cr(VI) uptake study of polyaniline coated Chitin. Int J Biol Macromolecules 72:235–242. https://doi.org/10.1016/j.ijbiomac.2014.08.022

Karthik R, Meenakshi S (2015d) Removal of hexavalent chromium ions from aqueous solution using chitosan/polypyrrole composite. Desalin Water Treat 56(6):1587–1600. https://doi.org/10.1080/19443994.2014.951964

Karthik R, Meenakshi S (2015e) Removal of Cr(VI) ions by adsorption onto sodium alginate-polyaniline nanofibers. Int J Biol Macromolecules 72:711–717. https://doi.org/10.1016/j.ijbiomac.2014.09.023

Karthik R, Meenakshi S (2016) Biosorption of Pb(II) and Cd(II) ions from aqueous solution using polyaniline/chitin composite. Sep Sci Technol 51(5):733–742. https://doi.org/10.1080/01496395.2015.1130060

Kim MK, Shanmuga Sundaram K, Anantha Iyengar G, Lee K-P (2015) A novel chitosan functional gel included with multiwall carbon nanotube and substituted polyaniline as adsorbent for efficient removal of chromium ion. Chem Eng J 267:51–64. https://doi.org/10.1016/j.cej.2014.12.091

Kousalya GN, Gandhi MR, Meenakshi S (2010a) Preparation of modified Chitin for the removal of Chromium(VI). Bioremediat J 14(4):208–218. https://doi.org/10.1080/10889868.2010.515136

Kousalya GN, Gandhi MR, Meenakshi S (2010b) Removal of toxic Cr(VI) ions from aqueous solution using nano-hydroxyapatite-based Chitin and Chitosan hybrid composites. Adsorpt Sci Technol 28(1):49–64. https://doi.org/10.1260/0263-6174.28.1.49

Kousalya GN, Rajiv Gandhi M, Meenakshi S (2010c) Sorption of chromium(VI) using modified forms of chitosan beads. Int J Biol Macromolecules 47(2):308–315. https://doi.org/10.1016/j.ijbiomac.2010.03.010

Kumar ASK, Kalidhasan S, Rajesh V, Rajesh N (2012) Application of cellulose-clay composite biosorbent toward the effective adsorption and removal of chromium from industrial wastewater. Ind Eng Chem Res 51(1):58–69. https://doi.org/10.1021/ie201349h

Kumar R, Kim S-J, Kim K-H, Lee S-H, Park H-S, Jeon B-H (2018) Removal of hazardous hexavalent chromium from aqueous phase using zirconium oxide-immobilized alginate beads. Appl Geochem 88:113–121. https://doi.org/10.1016/j.apgeochem.2017.04.002

Lazaridis NK, Charalambous C (2005) Sorptive removal of trivalent and hexavalent chromium from binary aqueous solutions by composite alginate–goethite beads. Water Res 39(18):4385–4396. https://doi.org/10.1016/j.watres.2005.09.013

Lee C-G, Park J-A, Lee I, Kang J-K, Yoon S-Y, Kim S-B (2013) Preparation of magnetic alginate–layered double hydroxide composite adsorbents and removal of Cr(VI) from aqueous solution. Water Supply 13(3):846–853. https://doi.org/10.2166/ws.2013.069

Lei Y, Qian X, Shen J, An X (2012) Integrated reductive/adsorptive detoxification of Cr(VI)-contaminated water by polypyrrole/cellulose fiber composite. Ind Eng Chem Res 51(31):10408–10415. https://doi.org/10.1021/ie301136g

Lei Z, Zhai S, Lv J, Fan Y, An Q, Xiao Z (2015) Sodium alginate-based magnetic carbonaceous biosorbents for highly efficient Cr(vi) removal from water. RSC Adv 5(95):77932–77941. https://doi.org/10.1039/C5RA13226F

Li H, Bi S, Liu L, Dong W, Wang X (2011) Separation and accumulation of Cu(II), Zn(II) and Cr(VI) from aqueous solution by magnetic chitosan modified with diethylenetriamine. Desalination 278(1):397–404. https://doi.org/10.1016/j.desal.2011.05.056

Li L, Fan L, Sun M, Qiu H, Li X, Duan H, Luo C (2013) Adsorbent for chromium removal based on graphene oxide functionalized with magnetic cyclodextrin–chitosan. Colloids Surf B Biointerfaces 107:76–83. https://doi.org/10.1016/j.colsurfb.2013.01.074

Li L, Li Y, Cao L, Yang C (2015) Enhanced Chromium (VI) adsorption using nanosized chitosan fibers tailored by electrospinning. Carbohydr Polym 125:206–213. https://doi.org/10.1016/j.carbpol.2015.02.037

Li Z, Li T, An L, Fu P, Gao C, Zhang Z (2016a) Highly efficient Chromium(VI) adsorption with nanofibrous filter paper prepared through electrospinning chitosan/polymethylmethacrylate composite. Carbohydr Polym 137:119–126. https://doi.org/10.1016/j.carbpol.2015.10.059

Li Z, Li T, An L, Liu H, Gu L, Zhang Z (2016b) Preparation of chitosan/polycaprolactam nanofibrous filter paper and its greatly enhanced Chromium(VI) adsorption. Colloids Surf A Physicochem Eng Aspects 494:65–73. https://doi.org/10.1016/j.colsurfa.2016.01.021

Li K, Li P, Cai J, Xiao S, Yang H, Li A (2016c) Efficient adsorption of both methyl orange and chromium from their aqueous mixtures using a quaternary ammonium salt modified chitosan magnetic composite adsorbent. Chemosphere 154:310–318. https://doi.org/10.1016/j.chemosphere.2016.03.100

Liang X, Fan X, Li R, Li S, Shen S, Hu D (2018) Efficient removal of Cr(VI) from water by quaternized chitin/branched polyethylenimine biosorbent with hierarchical pore structure. Bioresour Technol 250:178–184. https://doi.org/10.1016/j.biortech.2017.10.071

Liu J, Yan M, Zhang Y-K, Du K-F (2011) Study of glutamate-modified cellulose beads for Cr(III) adsorption by response surface methodology. Ind Eng Chem Res 50(18):10784–10791. https://doi.org/10.1021/ie200857n

Liu D, Zhu Y, Li Z, Tian D, Chen L, Chen P (2013) Chitin nanofibrils for rapid and efficient removal of metal ions from water system. Carbohydr Polym 98(1):483–489. https://doi.org/10.1016/j.carbpol.2013.06.015

Lu J, Xu K, Yang J, Hao Y, Cheng F (2017) Nano iron oxide impregnated in chitosan bead as a highly efficient sorbent for Cr(VI) removal from water. Carbohydr Polym 173:28–36. https://doi.org/10.1016/j.carbpol.2017.05.070

Lv X, Jiang G, Xue X, Wu D, Sheng T, Sun C, Xu X (2013) Fe^0-Fe_3O_4 nanocomposites embedded polyvinyl alcohol/sodium alginate beads for chromium (VI) removal. J Hazard Mater 262:748–758. https://doi.org/10.1016/j.jhazmat.2013.09.036

Lv X, Zhang Y, Fu W, Cao J, Zhang J, Ma H, Jiang G (2017) Zero-valent iron nanoparticles embedded into reduced graphene oxide-alginate beads for efficient chromium (VI) removal. J Colloid Interface Sci 506:633–643. https://doi.org/10.1016/j.jcis.2017.07.024

Mahmoud MS, Mohamed SA (2017) Calcium alginate as an eco-friendly supporting material for Baker's yeast strain in chromium bioremediation. HBRC J 13(3):245–254. https://doi.org/10.1016/j.hbrcj.2015.06.003

Miretzky P, Cirelli AF (2010) Cr(VI) and Cr(III) removal from aqueous solution by raw and modified lignocellulosic materials: a review. J Hazard Mater 180(1):1–19. https://doi.org/10.1016/j.jhazmat.2010.04.060

Moussout H, Ahlafi H, Aazza M, El Akili C (2018) Performances of local chitosan and its nanocomposite 5%Bentonite/Chitosan in the removal of chromium ions (Cr(VI)) from wastewater. Int J Biol Macromolecules 108:1063–1073. https://doi.org/10.1016/j.ijbiomac.2017.11.018

Neto JDOM, Bellato CR, de Castro Silva D (2019) Iron oxide/carbon nanotubes/chitosan magnetic composite film for chromium species removal. Chemosphere 218:391–401. https://doi.org/10.1016/j.chemosphere.2018.11.080

Ngah WSW, Kamari A, Fatinathan S, Ng PW (2006) Adsorption of chromium from aqueous solution using chitosan beads. Adsorption 12(4):249–257. https://doi.org/10.1007/s10450-006-0501-0

Nithya R, Gomathi T, Sudha PN, Venkatesan J, Anil S, Kim S-K (2016) Removal of Cr(VI) from aqueous solution using chitosan-g-poly(butyl acrylate)/silica gel nanocomposite. Int J Biol Macromolecules 87:545–554. https://doi.org/10.1016/j.ijbiomac.2016.02.076

Otuonye UC, Barminas JT, Magomya AM, Kamba EA, Andrew C (2014) Removal of Chromium (VI) as a heavy metal from aqueous solution using Chitin obtained from Bargi fish (Heterotis Miloticus) scale. Sci-Afric J Sci Issues 2(3):128–131

Pang Y, Zeng G-M, Tang L, Zhang Y, Liu Y-Y, Lei X-X, Wu M-S, Li Z, Liu C (2011) Cr(VI) reduction by Pseudomonas aeruginosa immobilized in a polyvinyl alcohol/sodium alginate matrix containing multi-walled carbon nanotubes. Bioresour Technol 102(22):10733–10736. https://doi.org/10.1016/j.biortech.2011.08.078

Percot A, Viton C, Domard A (2003) Optimization of Chitin extraction from shrimp shells. Biomacromol 4(1):12–18. https://doi.org/10.1021/bm025602k

Periyasamy S, Viswanathan N (2018) Hydrothermal synthesis of hydrocalumite assisted biopolymeric hybrid composites for efficient Cr(vi) removal from water. New J Chem 42(5):3371–3382. https://doi.org/10.1039/C7NJ04524G

Periyasamy S, Gopalakannan V, Viswanathan N (2018) Hydrothermal assisted magnetic nano-hydroxyapatite encapsulated alginate beads for efficient Cr(VI) uptake from water. J Environ Chem Eng 6(1):1443–1454. https://doi.org/10.1016/j.jece.2018.01.007

Preethi J, Prabhu SM, Meenakshi S (2017) Effective adsorption of hexavalent chromium using biopolymer assisted oxyhydroxide materials from aqueous solution. React Funct Polym 117:16–24. https://doi.org/10.1016/j.reactfunctpolym.2017.05.006

Preethi J, Vigneshwaran S, Meenakshi S (2019) Performance of Chitosan engraved iron and lanthanum mixed oxyhydroxide for the detoxification of hexavalent chromium. Int J Biol Macromolecules. https://doi.org/10.1016/j.ijbiomac.2019.02.101

Qiu B, Xu C, Sun D, Yi H, Guo J, Zhang X, Qu H, Guerrero M, Wang X, Noel N, Luo Z, Guo Z, Wei S (2014) Polyaniline coated ethyl cellulose with improved hexavalent chromium removal. ACS Sustain Chem Eng 2(8):2070–2080. https://doi.org/10.1021/sc5003209

Rajiv Gandhi M, Meenakshi S (2013) Preparation of amino terminated polyamidoamine functionalized chitosan beads and its Cr(VI) uptake studies. Carbohydr Polym 91(2):631–637. https://doi.org/10.1016/j.carbpol.2012.08.028

Rajiv Gandhi M, Viswanathan N, Meenakshi S (2010) Preparation and application of alumina/chitosan biocomposite. Int J Biol Macromolecules 47(2):146–154. https://doi.org/10.1016/j.ijbiomac.2010.05.008

Rathinam K, Singh SP, Li Y, Kasher R, Tour JM, Arnusch CJ (2017) Polyimide derived laser-induced graphene as adsorbent for cationic and anionic dyes. Carbon 124:515–524. https://doi.org/10.1016/j.carbon.2017.08.079

Rathinam K, Singh SP, Arnusch CJ, Kasher R (2018) An environmentally-friendly chitosan-lysozyme biocomposite for the effective removal of dyes and heavy metals from aqueous solutions. Carbohydr Polym 199:506–515. https://doi.org/10.1016/j.carbpol.2018.07.055

Rathinam K, Jayaram P, Sankaran M (2019) Synthesis and characterization of magnetic Chitin composite and its application towards the uptake of Pb(II) and Cd(II) ions from aqueous solution. Environ Prog Sustain Energy 38(s1):S288–S297. https://doi.org/10.1002/ep.13013

Sağ Y, Aktay Y (2001) Application of equilibrium and mass transfer models to dynamic removal of Cr(VI) ions by Chitin in packed column reactor. Process Biochem 36(12):1187–1197. https://doi.org/10.1016/S0032-9592(01)00150-9

Salam MA (2017) Preparation and characterization of chitin/magnetite/multiwalled carbon nanotubes magnetic nanocomposite for toxic hexavalent chromium removal from solution. J Mol Liquids 233:197–202. https://doi.org/10.1016/j.molliq.2017.03.023

Salih SS, Ghosh TK (2018) Preparation and characterization of Chitosan-coated diatomaceous earth for hexavalent Chromium removal. Environ Process 5(1):23–39. https://doi.org/10.1007/s40710-017-0280-5

Samuel J, Pulimi M, Paul ML, Maurya A, Chandrasekaran N, Mukherjee A (2013) Batch and continuous flow studies of adsorptive removal of Cr(VI) by adapted bacterial consortia immobilized in alginate beads. Bioresour Technol 128:423–430. https://doi.org/10.1016/j.biortech.2012.10.116

Samuel MS, Shah SS, Subramaniyan V, Qureshi T, Bhattacharya J, Pradeep Singh ND (2018) Preparation of graphene oxide/chitosan/ferrite nanocomposite for Chromium(VI) removal from aqueous solution. Int J Biol Macromolecules 119:540–547. https://doi.org/10.1016/j.ijbiomac.2018.07.052

Samuel MS, Bhattacharya J, Raj S, Santhanam N, Singh H, Pradeep Singh ND (2019) Efficient removal of Chromium(VI) from aqueous solution using chitosan grafted graphene oxide (CS-GO) nanocomposite. Int J Biol Macromolecules 121:285–292. https://doi.org/10.1016/j.ijbiomac.2018.09.170

Sankararamakrishnan N, Dixit A, Iyengar L, Sanghi R (2006) Removal of hexavalent chromium using a novel cross linked xanthated chitosan. Bioresour Technol 97(18):2377–2382. https://doi.org/10.1016/j.biortech.2005.10.024

Santosa SJ, Siswanta D, Sudiono S, Utarianingrum R (2008) Chitin–humic acid hybrid as adsorbent for Cr(III) in effluent of tannery wastewater treatment. Appl Surf Sci 254(23):7846–7850. https://doi.org/10.1016/j.apsusc.2008.02.102

Saravanan D, Gomathi T, Sudha PN (2013) Sorption studies on heavy metal removal using chitin/bentonite biocomposite. Int J Biol Macromolecules 53:67–71. https://doi.org/10.1016/j.ijbiomac.2012.11.005

Sessarego S, Rodrigues SCG, Xiao Y, Lu Q, Hill JM (2019) Phosphonium-enhanced chitosan for Cr(VI) adsorption in wastewater treatment. Carbohydr Polym 211:249–256. https://doi.org/10.1016/j.carbpol.2019.02.003

Shankar P, Gomathi T, Vijayalakshmi K, Sudha PN (2014) Comparative studies on the removal of heavy metals ions onto cross linked chitosan-g-acrylonitrile copolymer. Int J Biol Macromolecules 67:180–188. https://doi.org/10.1016/j.ijbiomac.2014.03.010

Sharma RK, Lalita Singh AP (2017) Sorption of Pb(II), Cu(II), Fe(II) and Cr(VI) metal ions onto cross-linked graft copolymers of chitosan with binary vinyl monomer mixtures. React Funct Polym 121:32–44 (2017). https://doi.org/10.1016/j.reactfunctpolym.2017.10.015

Shen C, Chen H, Wu S, Wen Y, Li L, Jiang Z, Li M, Liu W (2013) Highly efficient detoxification of Cr(VI) by chitosan–Fe(III) complex: process and mechanism studies. J Hazard Mater 244–245:689–697. https://doi.org/10.1016/j.jhazmat.2012.10.061

Shi T, Yang D, Yang H, Ye J, Cheng Q (2017) Preparation of chitosan crosslinked modified silicon material and its adsorption capability for chromium(VI). Appl Clay Sci 142:100–108. https://doi.org/10.1016/j.clay.2016.11.023

Šillerová H, Komárek M, Liu C, Poch J, Villaescusa I (2015) Biosorbent encapsulation in calcium alginate: effects of process variables on Cr(VI) removal from solutions. Int J Biol Macromolecules 80:260–270. https://doi.org/10.1016/j.ijbiomac.2015.06.032

Singh P, Nagendran R (2016) A comparative study of sorption of chromium (III) onto chitin and chitosan. Appl Water Sci 6(2):199–204. https://doi.org/10.1007/s13201-014-0218-2

Singh SP, Rathinam K, Kasher R, Arnusch CJ (2018) Hexavalent chromium ion and methyl orange dye uptake via a silk protein sericin–chitosan conjugate. RSC Adv 8(48):27027–27036. https://doi.org/10.1039/C8RA03907K

Sugashini S, Begum KMMS, Ramalingam A (2015) Removal of Cr(VI) ions using Fe-loaded chitosan carbonized rice husk composite beads (Fe-CCRCB): experiment and quantum chemical calculations. J Mol Liquids 208:380–387. https://doi.org/10.1016/j.molliq.2015.04.048

Sun L, Yuan Z, Gong W, Zhang L, Xu Z, Su G, Han D (2015) The mechanism study of trace Cr(VI) removal from water using Fe^0 nanorods modified with chitosan in porous anodic alumina. Appl Surf Sci 328:606–613. https://doi.org/10.1016/j.apsusc.2014.12.094

Sun X, Li Q, Yang L, Liu H (2016) Chemically modified magnetic chitosan microspheres for Cr(VI) removal from acidic aqueous solution. Particuology 26:79–86. https://doi.org/10.1016/j.partic.2015.11.003

Taha AA, Wu Y-N, Wang H, Li F (2012) Preparation and application of functionalized cellulose acetate/silica composite nanofibrous membrane via electrospinning for Cr(VI) ion removal from aqueous solution. J Environ Manage 112:10–16. https://doi.org/10.1016/j.jenvman.2012.05.031
Thinh NN, Hanh PTB, Ha LTT, Anh LN, Hoang TV, Hoang VD, Dang LH, Khoi NV, Lam TD (2013) Magnetic chitosan nanoparticles for removal of Cr(VI) from aqueous solution. Mater Sci Eng, C 33(3):1214–1218. https://doi.org/10.1016/j.msec.2012.12.013
Vakili M, Deng S, Li T, Wang W, Wang W, Yu G (2018) Novel crosslinked chitosan for enhanced adsorption of hexavalent chromium in acidic solution. Chem Eng J 347:782–790. https://doi.org/10.1016/j.cej.2018.04.181
Vilela PB, Dalalibera A, Duminelli EC, Becegato VA, Paulino AT (2018) Adsorption and removal of chromium (VI) contained in aqueous solutions using a chitosan-based hydrogel. Environ Sci Pollut Res. https://doi.org/10.1007/s11356-018-3208-3
Vu HC, Dwivedi AD, Le TT, Seo S-H, Kim E-J, Chang Y-S (2017) Magnetite graphene oxide encapsulated in alginate beads for enhanced adsorption of Cr(VI) and As(V) from aqueous solutions: role of crosslinking metal cations in pH control. Chem Eng J 307:220–229. https://doi.org/10.1016/j.cej.2016.08.058
Wang J, Chen C (2014) Chitosan-based biosorbents: modification and application for biosorption of heavy metals and radionuclides. Bioresour Technol 160:129–141. https://doi.org/10.1016/j.biortech.2013.12.110
Wang S-L, Lee J-F (2011) Reaction mechanism of hexavalent chromium with cellulose. Chem Eng J 174(1):289–295. https://doi.org/10.1016/j.cej.2011.09.031
Wen Y, Tang Z, Chen Y, Gu Y (2011) Adsorption of Cr(VI) from aqueous solutions using chitosan-coated fly ash composite as biosorbent. Chem Eng J 175:110–116. https://doi.org/10.1016/j.cej.2011.09.066
Wilbur S, Abadin H, Fay M et al (2012) Toxicological profile for Chromium. Atlanta (GA): agency for toxic substances and disease registry (US). (2012 Sep.7, ANALYTICAL METHODS)
World Health Organization (2006) Guidelines for drinking-water quality, 3rd edn. Geneva
Wu Z, Li S, Wan J, Wang Y (2012) Cr(VI) adsorption on an improved synthesised cross-linked chitosan resin. J Mol Liquids 170:25–29. https://doi.org/10.1016/j.molliq.2012.03.016
Yan Y, An Q, Xiao Z, Zheng W, Zhai S (2017) Flexible core-shell/bead-like alginate@PEI with exceptional adsorption capacity, recycling performance toward batch and column sorption of Cr(VI). Chem Eng J 313:475–486. https://doi.org/10.1016/j.cej.2016.12.099
Yan E, Cao M, Ren X, Jiang J, An Q, Zhang Z, Gao J, Yang X, Zhang D (2018) Synthesis of Fe_3O_4 nanoparticles functionalized polyvinyl alcohol/chitosan magnetic composite hydrogel as an efficient adsorbent for chromium (VI) removal. J Phys Chem Solids 121:102–109. https://doi.org/10.1016/j.jpcs.2018.05.028
Yavuz AG, Dincturk-Atalay E, Uygun A, Gode F, Aslan E (2011) A comparison study of adsorption of Cr(VI) from aqueous solutions onto alkyl-substituted polyaniline/chitosan composites. Desalination 279(1):325–331. https://doi.org/10.1016/j.desal.2011.06.034
Younes I, Rinaudo M (2015) Chitin and chitosan preparation from marine sources. Structure, properties and applications. Mar Drugs 13(3):1133–1174. https://doi.org/10.3390/md13031133
Yu Z, Zhang X, Huang Y (2013) Magnetic Chitosan–Iron(III) Hydrogel as a Fast and Reusable Adsorbent for Chromium(VI) Removal. Ind Eng Chem Res 52(34):11956–11966. https://doi.org/10.1021/ie400781n
Yu T, Liu S, Xu M, Peng J, Li J, Zhai M (2016) Synthesis of novel aminated cellulose microsphere adsorbent for efficient Cr(VI) removal. Radiat Phys Chem 125:94–101. https://doi.org/10.1016/j.radphyschem.2016.03.019
Yu P, Wang H-Q, Bao R-Y, Liu Z, Yang W, Xie B-H, Yang M-B (2017) Self-assembled sponge-like chitosan/reduced graphene oxide/montmorillonite composite hydrogels without cross-linking of chitosan for effective Cr(VI) sorption. ACS Sustain Chem Eng 5(2):1557–1566. https://doi.org/10.1021/acssuschemeng.6b02254

Yue R, Chen Q, Li S, Zhang X, Huang Y, Feng P (2018) One-step synthesis of 1,6-hexanediamine modified magnetic chitosan microspheres for fast and efficient removal of toxic hexavalent chromium. Sci Rep 8(1):11024. https://doi.org/10.1038/s41598-018-29499-z

Zhang L, Xia W, Teng B, Liu X, Zhang W (2013) Zirconium cross-linked chitosan composite: preparation, characterization and application in adsorption of Cr(VI). Chem Eng J 229:1–8. https://doi.org/10.1016/j.cej.2013.05.102

Zhang L, Xia W, Liu X, Zhang W (2015) Synthesis of titanium cross-linked chitosan composite for efficient adsorption and detoxification of hexavalent chromium from water. J Mater Chem A 3(1):331–340. https://doi.org/10.1039/C4TA05194G

Zhang L, Luo H, Liu P, Fang W, Geng J (2016) A novel modified graphene oxide/chitosan composite used as an adsorbent for Cr(VI) in aqueous solutions. Int J Biol Macromolecules 87:586–596. https://doi.org/10.1016/j.ijbiomac.2016.03.027

Zhang B, Hu R, Sun D, Wu T, Li Y (2018) Fabrication of chitosan/magnetite-graphene oxide composites as a novel bioadsorbent for adsorption and detoxification of Cr(VI) from aqueous solution. Sci Rep 8(1):15397. https://doi.org/10.1038/s41598-018-33925-7

Zhao D, Gao X, Wu C, Xie R, Feng S, Chen C (2016) Facile preparation of amino functionalized graphene oxide decorated with Fe_3O_4 nanoparticles for the adsorption of Cr(VI). Appl Surf Sci 384:1–9. https://doi.org/10.1016/j.apsusc.2016.05.022

Zhou Y, Jin Q, Zhu T, Akama Y (2011) Adsorption of chromium (VI) from aqueous solutions by cellulose modified with β-CD and quaternary ammonium groups. J Hazard Mater 187(1):303–310. https://doi.org/10.1016/j.jhazmat.2011.01.025

Zhu C, Liu F, Zhang Y, Wei M, Zhang X, Ling C, Li A (2016) Nitrogen-doped chitosan-Fe(III) composite as a dual-functional material for synergistically enhanced co-removal of Cu(II) and Cr(VI) based on adsorption and redox. Chem Eng J 306:579–587. https://doi.org/10.1016/j.cej.2016.07.096

Zimmermann AC, Mecabô A, Fagundes T, Rodrigues CA (2010) Adsorption of Cr(VI) using Fe-crosslinked chitosan complex (Ch-Fe). J Hazard Mater 179(1):192–196. https://doi.org/10.1016/j.jhazmat.2010.02.078

The manufacturer's authorised representative in the EU is Springer
Nature Customer Service Centre GmbH, Europaplatz 3, 69115 Heidelberg,
Germany. If you have any concerns regarding our products, please
contact ProductSafety@springernature.com

Printed and bound by CPI Group (UK) Ltd, Croydon, CR0 4YY
25/09/2026
02222886-0007